Viral Diarrheas of Man and Animals

Editors

Linda J. Saif, Ph.D.
Associate Professor

Kenneth W. Theil, Ph.D.
Associate Professor

Food Animal Health Research Program
Ohio Agricultural Research and Development Center
Wooster, Ohio
and
Department of Veterinary Preventive Medicine
The Ohio State University
Columbus, Ohio

CRC Press, Inc.
Boca Raton, Florida

Library of Congress Cataloging-in-Publication Date

Viral diarrheas of man and animals / editors, Linda J. Saif, Kenneth W.
Theil.
p. cm.
Includes bibliographies and index.
ISBN 0-8493-6640-2
1. Viral diarrhea--Microbiology. 2. Diarrhea in animals-
-Microbiology. 3. Enteroviruses--Pathogenicity. 4. Viral diarrhea-
-Preventive inoculation--Research. I. Saif, Linda J. II. Theil,
Kenneth W.
[DNLM: 1. Diarrhea--immunology. 2. Diarrhea--microbiology.
3. Enteroviruses--immunology. 4. Enteroviruses--pathogenicity. WI
407 V813]
QR201.V53V57 1989
616,3'427--dc19
DNLM/DLC
for Library of Congress

89-805
CIP

This book represents information obtained from authentic and highly regarded sources. Reprinted material is quoted with permission, and sources are indicated. A wide variety of references are listed. Every reasonable effort has been made to give reliable data and information, but the author and the publisher cannot assume responsibility for the validity of all materials or for the consequences of their use.

Direct all inquiries to CRC Press, Inc., 2000 Corporate Blvd., N.W., Boca Raton, Florida, 33431.

International Standard Book Number 0-8493-6640-2

Library of Congress Card Number 89-805
Printed in the United States

PREFACE

The intestinal mucosa represents an extraordinarily vast and complex internal interface situated between the animal and its environment. The epithelium comprising this mucosa is constantly undergoing rapid renewal as new cells emerge from discrete proliferative zones to differentiate into fully functional cells which are eventually extruded into the lumen. Viruses that infect these cells, disrupting their functions, and the orderly replacement necessary for normal digestion and absorption, are the subjects of this book. Most of these viruses were recognized only within the last two decades, making them relative newcomers within the realm of virology. Consequently, while great strides have been made in our understanding concerning some of these viral infections, our knowledge regarding the others is fragmentary and much remains to be elucidated.

The discovery of viruses that induce diarrhea in man and animals had to await the application of new techniques and procedures to the examination of clinical specimens. As the widespread availability of cell culture techniques for virus isolation led to the detection of many previously unrecognized enteric viruses during the 1950s, the use of negative stain electron microscopy techniques in the last 20 years to examine diarrheic stools has revealed a multitude of hitherto unrecognized, fastidious viruses associated with enteric illness. Medical and veterinary research leading to a more comprehensive understanding of these viral infections has often been intertwined, so inclusion of chapters dealing with viruses of both medical and veterinary importance is appropriate for a more unified appreciation of our current knowledge regarding them. In each chapter, the authors, each of whom have made significant contributions to our understanding of these viral infections, emphasize the current state of knowledge regarding a group of viruses now recognized as major etiologic agents of diarrhea in man or animals. Additional chapters cover more specialized, but important, facets of intestinal infections including the comparative aspects of viral pathogenesis, mixed infections, the intestinal immune system and its response to infection, and the development of efficacious vaccines against viral infections of the intestinal tract.

We hope that this book will serve as a valuable source of information for those interested in obtaining a greater appreciation of enteric viral infections of medical and veterinary importance. If the reader recognizes the many significant gaps in our understanding of these infections and is stimulated to undertake additional research, especially that involving the newer, more sophisticated approaches, then we will consider this book a success.

THE EDITORS

Linda J. Saif, Ph.D., is an Associate Professor in the Food Animal Health Research Program of the Department of Veterinary Preventive Medicine, The Ohio State University, at the Ohio Agricultural Research and Development Center, in Wooster.

Dr. Saif received her B.A. degree from the College of Wooster, Wooster, Ohio in 1969. She obtained her M.S. and Ph.D. degrees in 1971 and 1976, respectively, from the Department of Microbiology, The Ohio State University, Columbus. After doing her post-doctoral research in the Veterinary Science Department, at the Ohio Agricultural Research and Development Center in Wooster, she was appointed an Assistant Professor of Immunology and Virology in the Veterinary Preventive Medicine Department, The Ohio State University, Wooster in 1979. She became an Associate Professor in the same department in 1985.

Dr. Saif is a member of the American Society for Microbiology, the American Society for Virology, the Society for Mucosal Immunology, the American Association of Veterinary Immuologists, the Conference of Research Workers in Animal Diseases, the National Honorary Society Phi Kappa Phi, and the Veterinary Medicine National Honor Society, Phi Zeta.

Dr. Saif was the recipient of Graduate Research Fellowships at Case Western Reserve University, Department of Molecular Biology (1970 to 1971) and The Ohio State University, Department of Microbiology (1975 to 1976). She was an invited short term visiting professor at the University of Guelph, College of Veterinary Medicine, Canada (1987). She has served as a FAO consultant on neonatal diarrhea in animals in Latin America and the Carribean (1988) and chaired international workshops on this topic in Argentina and Peru. In 1989, she was the recipient of the Beecham Award for Research Excellence in the College of Veterinary Medicine, Ohio State University, Columbus.

Dr. Saif has been the recipient of grants from the National Institutes of Health, the U. S. Department of Agriculture, and private industry. She has served on peer review panels for the National Institutes of Health and the U. S. Department of Agriculture. She has published more than 100 research papers. Her current research interests are cultivation and characterization of enteric viruses including newly described viruses, and diagnosis and immunoprevention of enteric viral infections.

Kenneth W. Theil, Ph.D., is Associate Professor in the Food Animal Health Research Program, Department of Veterinary Preventive Medicine, at the Ohio Agricultural Research and Development Center of The Ohio State University in Wooster.

Dr. Theil received his B.S., M.S., and Ph.D. degrees in 1966, 1970, and 1972, respectively, from the Department of Microbiology, University of Maryland, College Park. From 1972 to 1975 he worked in the Virus Products Testing Section of the Bioquality Control Department of Norden Laboratories, Inc., in Lincoln, Nebraska. In 1976 he became a Research Associate in the Department of Veterinary Science, Ohio Agricultural Research and Development Center, and was appointed Assistant Professor in this department in 1977. He became Associate Professor in the Food Animal Health Program of the Department of Veterinary Preventive Medicine in 1986.

Dr. Theil is a member of the American Association for the Advancement of Science, American Association of University Professors, American Association of Veterinary Laboratory Diagnosticians, American Society for Microbiology, Conference of Research Workers in Animal Disease, and the honorary societies Phi Kappa Phi and Sigma Xi. He is also an associate member of the American Association of Avian Pathologists. He has been the recipient of research grants from the National Institutes of Health and the U. S. Department of Agriculture.

Dr. Theil is author or coauthor of over 70 papers, abstracts, and book chapters. In 1987 he was the recipient of the P. P. Levine Award for the best paper published in volume 30 of Avian Diseases. His current major research interests relate to enteric viruses, particularly rotavirus, that infect food-producing animals.

CONTRIBUTORS

David A. Benfield, Ph.D.
Associate Professor
Department of Veterinary Science
South Dakota State University
Brookings, South Dakota

Janice C. Bridger, Ph.D.
Department of Microbiology
Agricultural and Food Research Council
Insititute for Animal Health
Near Newbury, Berkshire, England

Kenny V. Brock, D.V.M., Ph.D.
Assistant Professor
Food Animal Health Research Program
Wooster, Ohio and
Department of Veterinary Preventive
Medicine
The Ohio State University
Columbus, Ohio

Harry Greenberg, M.D.
Chief
Division of Gasteroenterology
Associate Professor
Departments of Medicine
and Microbiology and Immunology
Stanford University Medical School
Stanford, California

Robert A. Heckert, Ph.D.
Research Associate
Food Animal Health Research Program
Wooster, Ohio and
Department of Veterinary Preventive
Medicine
The Ohio State University
Columbus, Ohio

Marian C. Horzinek, D.V.M., Ph.D.
Professor and Head
Department of Infectious Diseases and
Immunology
Institute of Virology
State University of Utrecht
Utrecht, The Netherlands

Alan J. Husband, Ph.D.
Associate Professor of Immunology
Faculty of Medicine
University of Newcastle
Newcastle, New South Wales, Australia

Daral J. Jackwood, Ph.D.
Assistant Professor
Food Animal Health Research Program
Ohio Agricultural Research and
Development Center
Wooster, Ohio and
Department of Veterinary Preventive
Medicine
The Ohio State University
Columbus, Ohio

Stephen S. Monroe, Ph.D.
Research Biologist
Division of Viral Diseases
Centers for Disease Control
Atlanta, Georgia

Linda J. Saif, Ph.D.
Associate Professor
Food Animal Health Research Program
Ohio Agricultural Research and
Development Center
Wooster, Ohio and
Department of Veterinary Preventive
Medicine
The Ohio State University
Columbus, Ohio

Marna M. Skaar, M.D.
Research Associate
Department of Gastroenterology
Stanford University Medical School
Stanford, California

David Snodgrass, D.V.M., Ph.D.
Head
Department of Microbiology
Moredun Research Institute
Edinburgh, Scotland

Kenneth W. Theil, Ph.D.
Associate Professor
Food Animal Health Research Program
Ohio Agricultural Research and Development Center
Wooster, Ohio and
Department of Veterinary Preventive Medicine
The Ohio State University
Columbus, Ohio

Ingrid Uhnoo, M.D., Ph.D.
Department of Infectious Diseases
University Hospital
Uppsala, Sweden

Göran Wadell, M.D., Ph.D.
Department of Virology
University of Umea
Umea, Sweden

Marianne Weiss, Ph.D.
Institute of Veterinary Virology
University of Berne
Berne, Switzerland

ACKNOWLEDGMENTS

We would like to acknowledge the supporting facilities at the OARDC and people connected with them who have contributed to several of the chapters contained in this book. We are greatly indebted to Hannah Gehman for her expert secretarial assistance and endless patience in the typing process. We thank Bob Whitmoyer, Elke Kretzscmar, and Fran Butts of the electron microscope lab for their assistance with the electron micrographs. Ken Chamberlain, Margaret Latta, and Cindy Gray assisted with the many photographs and the artwork. The OARDC librarian, Connie Britton, contributed time and effort in assisting us with the bibliographic sections of our chapters.

The book is dedicated to my husband, Yehia, my son, Justin, my parents, and various mentors throughout my career, including Drs. E. H. Bohl and E. M. Kohler. It is also dedicated to the excellent graduate students, post-doctorals, visiting scientists, and research assistants who have worked with me and contributed to the knowledge reviewed in this book.

L. J. S.

This book is dedicated to my wife, Beverly, and my sons, Kenneth and Karl, and my mother, Catherine.

K. W. T.

TABLE OF CONTENTS

I. Historical Perspective

Chapter 1

INTRODUCTION

Linda J. Saif

Diarrheal diseases have a worldwide impact on both human and animal health and are a major cause of morbidity and mortality. In Asia, Africa, and Latin America, 3 to 5 billion episodes of diarrhea and 5 to 10 million diarrheal-associated deaths occur in people annually, with children often the most severely affected.[1] Diarrheal diseases are also common in children in developed countries and are a frequent cause of patient visits to doctors.[2] The trend toward increasing numbers of infants and young children enrolled in day care centers, has also accentuated diarrheal diseases among this highly susceptible segment of the population in developed countries.[3]

In food-producing animals, diarrheal disease or "scours" is a common and significant problem among neonatal calves, pigs, and more recently, poultry.[4] Accurate morbidity, case fatality, and population rates are not readily available for animals affected with diarrheal diseases, particularly from developing countries. However, estimates of deaths in pigs and calves under 3 weeks of age, attributable to diarrheal disease, range up to 18%.[5,6] Economic losses due to neonatal diarrhea were estimated at $500 million annually for calves in North America alone[6] and $200 million annually for suckling pigs in the U. S. alone.[7] On a worldwide basis, this translates into significant economic losses in animal production. Such monetary losses resulting from mortality alone do not reflect other important economic losses due to increased costs for care and treatment, subsequent poor growth of some recovered animals, or increased susceptibility to secondary infections. The international trend toward intensive farm management practices including close confinement of large numbers of animals is likely to accentuate enteric infections among food-producing animals in the future.

In previous years, enteropathogenic strains of *Escherichia coli* were widely accepted as the major, if not the only, cause of infectious diarrhea in young animals and children. In 1945, transmissible gastroenteritis, subsequently shown to be caused by a coronavirus, was identified as the first infectious viral diarrheal disease of swine.[8] In the early 1970s Mebus and colleagues isolated in cell culture and reproduced diarrhea in calves using field rotavirus and coronavirus strains.[9,10] Morphologically similar rotaviruses were subsequently identified in young children with diarrhea in 1973[11] and in the young of many other species thereafter. Group A rotaviruses are now established as the most important cause of severe diarrhea in children less than 2 years of age in both developed and developing countries.[4,12] Rotaviruses as well as coronaviruses are also significant primary enteropathogens in food-producing animals. The sudden global appearance of canine parvovirus gastroenteritis in the late 1970s emphasizes the need for continued monitoring for emerging enteric pathogens in both man and animals.[13] Likewise, the more recent emergence of group B rotavirus infections associated with severe diarrhea in human adults and infants in China stresses the need for continuous vigilance against new enteric disease agents — particularly ones with zoonotic potential.[14]

At the present time, lack of cell culture systems and sensitive, widely available diagnostic assays have hindered detection and characterization of newly described enteric viruses, many of which are less readily visualized by electron microscopy or present in feces in low numbers. In spite of our present progress in cell culture growth and characterization of rotaviruses and coronaviruses in animals, efficacious vaccines to prevent rotavirus or coronavirus infections in animals are still not widely available,[15,16] and more basic studies of mechanisms of enteric viral pathogenesis and mucosal immunity are needed to define parameters important in protection.

Lack of efficacy of a bovine RIT rotavirus vaccine in infants also highlights the need for additional research on enteric viral vaccine approaches in children.[17]

The following discussion focuses on some of the major problems encountered in studies of enteric viruses and how these problems have been circumvented or resolved to attain our current level of understanding. Progress in the study of enteric viral infections of man and animals has been slow until recently. A major impediment was the failure of these viruses to replicate in conventional cell culture systems, thereby precluding the initial detection of the agents by using established cell culture procedures and propagation of virus for further studies.

The first breakthrough in the study of the enteropathogenic viruses came in the late 1960s through the 1970s when electron microscopy was used as a method to detect these fastidious viruses *directly* in feces or intestinal epithelial cells from animals and man.[9-11] This allowed detection and morphologic classification of the enteric viruses. It also provided a means of virus detection applicable to epidemiologic studies of enteric disease. Even now previously unrecognized enteric viruses continue to be identified directly in feces by electron microscopy. This has left the legacy that new enteric virus groups are frequently named for their distinctly recognized morphology — hence, rotavirus, astrovirus, torovirus, and coronavirus.

However, continued failure of many of these fastidious viruses to replicate *in vitro* still creates difficulties in availability of virus for further study and for the generation of simpler, more widely available, diagnostic tests. Although many enteric viruses are very restricted in their host range, such as Norwalk virus which has been recovered only from man, others, such as group A rotavirus have a wider host range. This property allowed adaptation of human rotaviruses to grow in alternative species thereby permitting amplification of the input virus and studies of viral pathogenesis and immunity. The only laboratory animals consistently susceptible to infection with heterologous group A (nonhuman) rotavirus strains are young mice. However, the age of susceptibility to rotaviral disease in this species is very limited, so no studies of protection against challenge are possible: in addition only specimens containing small amounts of virus can be obtained from mice. Another animal model which had been developed and used extensively in veterinary medicine is the gnotobiotic animal. The infectivity and pathogenicity of enteric viruses, including coronaviruses and rotaviruses, had previously been confirmed in the 1960s by studies using gnotobiotic pigs and calves. Such animals provide the additional advantage that the potential pathogenicity of a single agent can be determined free of the effects from enteropathogenic bacteria or other infectious microorganisms. Use of this model established the role of enteric viruses in diarrheal diseases, which were previously believed to be caused mainly by enteropathogenic bacteria. Passage of human group A rotavirus strains in gnotobiotic pigs or calves reproduced diarrhea and provided amplification of the virus in the inoculum.[18,19] Furthermore, passage of human group A rotavirus in an unnatural host (the gnotobiotic pig) may have facilitated its subsequent adaptation, reported shortly thereafter, to growth in cell culture.[20]

An additional factor which has been of major importance in the adaptation of many different enteric viruses to replication *in vitro* has been addition of proteolytic enzymes to the cell culture medium. This apparently mimics the conditions required for growth of these viruses in the intestinal tract. Enzymes such as trypsin or pancreatin have been most commonly used, but for some viruses, such as the porcine enteric calicivirus (PEC), the factor(s) present in intestinal contents required for replication have not been identified.[21] At this time, PEC replicates *in vitro* only in the continued presence of whole intestinal contents from uninfected gnotobiotic pigs. Recent progress in adaptation of fastidious human enteropathogenic adenoviruses to growth in cell culture (semipermissive) was achieved by use of 293 cells, human embryonic kidney cells immortalized by transfection with E1A and E1B regions of adenovirus.[4,22]

Continued success in cultivation of fastidious enteric viruses *in vitro,* and their adaptation to growth in cell lines is a key factor for epidemiologic, biochemical, and genetic analysis of these viruses and eventual vaccine development. Although enormous progress has been made in the study of human and animal group A rotaviruses and human enteropathogenic adenoviruses since their initial identification as recently as 15 to 20 years ago, other enteric viruses which have not been successfully adapted to serial propagation in cell culture, remain only poorly characterized. In addition, it is important to note that in spite of our progress in identification of new enteric viruses during the past decade, as many as 50% of the cases of acute, nonbacterial diarrhea in man and animals remain undiagnosed in some studies.[4] Thus, the field of viral gastroenteritis is likely to remain an important subject area for application of new and continued innovative research techniques in both human and veterinary medicine.

This book includes a description of each of the major families of enteropathogenic viruses found in man and animals and their properties. Also described in related chapters are comparative aspects of the *in vivo* replication and pathogenesis of enteric viruses, important characteristics of the intestinal immune system, and information regarding vaccination approaches for control of enteric viral infections in man and animals.

REFERENCES

1. **Walsh, J. A. and Warren, K. S.,** Selective primary health care. An interim strategy for disease control in developing countries, *N. Engl. J. Med.*, 301, 967, 1979.
2. **Dingle, J. H., Badger, G. F., and Jordan, W. S.,** Illness in the home: a study of 25,000 illnesses in a group of Cleveland families, The Press of Western Reserve University, Cleveland, Ohio, 1964, 19.
3. **Pickering, L. K., Bartlett, A. V., and Woodward, W. E.,** Acute infectious diarrhea among children in day care: epidemiology and control, *Rev. Infect. Dis.*, 8, 539, 1986.
4. **Tzipori, S.,** *Infectious Diarrhea in the Young,* Elsevier Science Publ., Amsterdam, The Netherlands, 1985.
5. **Vrbanac, I. M., Herceg, Z., Pavlovski, Naranac, V., Sviben, M., and Sobocanc, R.,** Rate of piglet losses according the days of suckling in a large pig farm. Proc. Intern. Pig Vet. Soc. Congress, Copenhagen, Denmark, June 30 to July 3, 1980, 92.
6. **House, J. A.,** Economic impact of rotavirus and other neonatal disease agents of animals, *JAVMA*, 173, 573, 1978.
7. **King, N. B.,** Contributions and needs of animal health and disease research, *Am. J. Vet. Res.*, 42, 1093, 1981.
8. **Doyle, L. P. and Hutchings, L. M.,** A transmissible gastroenteritis in pigs, *JAVMA*, 108, 257, 1946.
9. **Mebus, C. A., Kono, M., Underdahl, N. R., and Twiehaus, M. J.,** Cell culture propagation of neonatal calf diarrhea (scours) virus, *Can. Vet. J.*, 12, 69, 1971.
10. **Mebus, C. A., Stair, E. L., Rhodes, M. B., and Twiehaus, M. L.,** Neonatal calf diarrhea: propagation, attenuation and characteristics of a coronavirus-like agent, *Am. J. Vet. Res.*, 34, 145, 1973.
11. **Bishop, R. F., Davidson, G. P., Holmes, I. H., and Ruck, B. J.,** Virus particles in epithelial cells of duodenal mucosa from children with viral gastroenteritis, *Lancet,* 2, 1281, 1973.
12. **Black, R. E., Merson, M. H., Mizanur-Rahman, A. S. M., Yunus, A.R.M.A., Huq, I., Yolken, R. H., and Curlin, G. T.,** A 2 year study of bacterial, viral and parasitic agents associated with diarrhea in rural Bangladesh, *J. Infect. Dis.*, 142, 660, 1980.
13. **Carmichael, L. E. and Binn, L. N.,** New enteric viruses in the dog, *Adv. Vet. Sci. Comp. Med.*, 25, 1, 1981.
14. **Hung, T.,** Rotavirus and adult diarrhea, *Adv. Virus Res.*, 35, 193, 1988.
15. **Saif, L. J.,** Passive immunity to coronavirus and rotavirus infections in swine and cattle: enhancement by maternal vaccination, in *Infectious Diarrhea in the Young,* Tzipori, S., Ed., Elsevier Science Pub., Amsterdam, The Netherlands, 1985, 456.
16. **Acres, S. D. and Radostits, O. M.,** The efficacy of a modified live reovirus-like vaccine and an *E. coli* bacterin for prevention of acute neonatal diarrhea of beef calves, *Can. Vet. J.*, 17, 197, 1976.
17. **DeMol, P., Zissis, G., Butzler, J. P., Muwewingabo, A., and Andre, R. E.,** Failure of live attenuated oral rotavirus vaccine, *Lancet,* 2, 108, 1986.
18. **Middleton, P. J., Petrie, M., and Szymanski, M. T.,** Propagation of infantile gastroenteritis virus (orbi-group) in conventional and germfree piglets, *Infect. Immun.*, 12, 1276, 1975.
19. **Mebus, C. A., Wyatt, R. G., Sharpee, R. L., Sereno, M. M., Kalica, A. R., Kapikian, A. Z., and Twiehaus, M. J.,** Diarrhea in gnotobiotic calves caused by the reovirus-like agent of human infantile gastroenteritis, *Infect. Immun.*, 14, 471, 1976.

20. **Wyatt, R. G., James, W. D., Bohl, E. H., Theil, K. W., Saif, L. J., Kalica, A. R., Greenberg, H. B., Kapikian, A. Z., and Chanock, R. M.,** Human rotavirus type 2: cultivation *in vitro*, *Science*, 207, 189, 1980.
21. **Flynn, W. T. and Saif, L. J.,** Serial propagation of porcine enteric calicivirus-like virus in primary porcine kidney cell cultures, *J. Clin. Microbiol.*, 26, 206, 1988.
22. **Takiff, H. E., Strauss, S. E., and Garon, C. F.,** Propagation and *in vitro* studies of previously non-cultivable enteral adenoviruses in 293 cells, *Lancet*, 2, 832, 1981.

II. Comparative Properties and Pathogenicity of Enteric Viruses

Chapter 2

COMPARATIVE ASPECTS OF ENTERIC VIRAL INFECTIONS

Linda J. Saif

TABLE OF CONTENTS

I. DESCRIPTION AND PROPERTIES OF ENTEROPATHOGENIC VIRUSES

At least 9 different classes of enteropathogenic viruses comprising 6 known or proposed families have been described (Table 1). Most of these viruses have as a common characteristic their ability to cause cytolytic infections in villous and/or crypt enterocytes, thereby inducing villous atrophy and consequently a malabsorptive diarrhea. Although other viruses including enteroviruses, reoviruses, and hepatitis virus, may be commonly shed by some species in feces, most of these viruses have not been shown to cause diarrhea and do not replicate in either villous or crypt enterocytes, but infect intestinal lymphoid cells or the liver, respectively (Table 1).[1-4] An exception is reovirus types 1 and 3 which cause diarrhea in mice.[5,6] The remainder of this chapter will focus on the enteropathogenic viruses categorized as Type I or II which are associated with infections of the villous or crypt enterocytes, respectively.

Types I and II enteropathogenic viruses possess a number of both common and distinct characteristics. Type I viruses differ in their pathogenesis from Type II viruses, not only in intestinal cell tropism, but also in their failure to cause generalized infections, remaining highly localized in the intestinal tract. An exception may be certain strains of enteric adenoviruses which may cause generalized infections in some species (bovine).[7,8] In contrast, viruses which infect primarily crypt enterocytes (Type II) also cause generalized infections in addition to intestinal infection. At present, these include only parvoviruses and bovine virus diarrhea virus (BVDV).[9-15] Many of the viruses summarized in Table 1, particularly Type I viruses, have been identified only within the past 10 to 15 years. Progress in their characterization has been hampered by the inability to serially propagate many of these fastidious viruses in cell culture. Presently, serogroup B, E, F, and G rotaviruses and Norwalk virus and related viruses have remained refractory to cell culture adaptation. Within several of the other families only certain strains of viruses have been propagated recently.[16-22] Factors conducive to the *in vitro* replication of these viruses often mimicked *in vivo* conditions in the intestinal tract. These included use of proteolytic enzymes or other unidentified factors in intestinal contents to induce or enhance *in vitro* replication of enteric viruses.[16-25] Similarly, many rotaviruses required use of roller tubes for *in vitro* propagation, perhaps simulating the peristalsis and rapid turnover of nutrients within the intestinal tract.[21,26,27] The majority of the enteric viruses studied were shown to be stable to the effects of proteolytic enzymes and low pH, factors important for their survival in the intestine:[28] exceptions were rat group B rotavirus and attenuated strains of transmissible gastroenteritis virus (TGEV).[29,30] Most enteropathogenic viruses are heat labile at 56° for 30 min, except porcine Group A Gottfried rotavirus.[28] This property may partially explain the prevalence of viral diarrhea outbreaks during winter months.

Coronaviruses, Bredaviruses, and BVDV are enveloped viruses and are generally less stable in the environment and more susceptible to inactivation by many of the commonly used disinfectants. Their persistence in the environment may relate to a respiratory stage in their pathogenesis, as discussed in a subsequent section.[31,32,129] The nonenveloped viruses, particularly rotaviruses and parvoviruses, are extremely stable and less affected by many common disinfectants.

Presently, multiple serotypes have been described for only rotaviruses (approximately ten serotypes),[33] adenovirus (two serotypes),[34] Bredavirus (two serotypes)[35] and astrovirus (five serotypes).[17,36] Limited information is available regarding most of the other Type I viruses. Multiple serogroups of rotaviruses have recently been recognized which share common antigens within a serogroup but not between groups.[37,38] They now comprise at least seven antigenically distinct serogroups (A to G), some of which occur in several species of animals.[37,38]

All of the enteropathogenic viruses summarized in Table 1 have been associated with diarrhea in a variety of natural hosts, with the exception of Norwalk virus (a probable Calicivirus) and BVDV. Man appears to be the only susceptible host for Norwalk virus, while

TABLE 1
Classificaton and Characterization of Enteric Viruses According to Virus Family, Intestinal Cell Tropism and Enteropathogenicity

Family/virus	Size	Envelope	Nucleic acid	Infection of enterocyte: Villous	Crypt	Diarrhea
				Type I — villous enterocytes		
Coronaviridae/coronavirus	60—220 nm	+	ssRNA	+	±	+
Toroviridae/Bredavirus	70—120 nm	+	ssRNA	+	±	+
Reoviridae/Gp A rotavirus	55—70 nm	–	dsRNA	+	–	+
Non Gp A rotavirus	55—70 nm	–	dsRNA	+	–	+
Adenoviridae/adenovirus	70—90 nm	–	DNA	+	±	+
Caliciviridae/calicivirus	30—40 nm	–	ssRNA	+	–	+
Norwalk	25—32 nm	–	?	+	–	+
Astrovirus	28—30 nm	–	ssRNA	+	±	±
				Type II — crypt enterocytes		
Togaviridae/Bovine virus diarrhea	40—120 nm	+	ssRNA	–	+	+
Parvoviridae/Parvovirus DNA	18—26 nm	–	DNA	–	+	+
				Type III — intestinal lymphoid cells		
Picornaviridae/enterovirus	25—30 nm	–	ssRNA	–	–	–
Reoviridae/reovirus	60—80 nm	–	dsRNA	–	–/±	–/±
				Type IV — liver		
Hepadenaviridae/Hepatitis	42 nm	–	DNA	–	–	±

BVDV induces enteritis in cattle.[39,40] Parvovirus enteritis has been described only in cats, dogs, or calves but can infect other ruminants.[9-14] Rotaviruses and certain enteric coronaviruses which originate from different host species share common group antigens, and cross-species transmissions have been documented (Table 2).[25,33,41-44] However, most rotavirus or coronavirus strains seem to possess some degree of host preference (Table 2).[44] The zoonotic potential of rotaviruses and caliciviruses, in particular, is unclear.[37,41-43,50] No antigenic relationships have been demonstrated between astroviruses, caliciviruses, or parvoviruses (except for feline panleukopenia and canine parvovirus) and morphologically similar enteric viruses from other species.

II. CELL TROPISMS OF TYPE I ENTEROPATHOGENIC VIRUSES

Type I viruses are transmitted by the fecal-oral route resulting in direct infection of the absorptive villous epithelial cells via the luminal surface of these cells. Thus, incubation periods for these enteric viruses are usually very short, the viruses are excreted in feces in large numbers and spread to susceptible animals occurs rapidly. Whether certain of these viruses, particularly the enveloped viruses, can be transmitted by aerosol exposure, with virus initially replicating in the oropharynx leading to massive doses of viruses being swallowed, is unclear and needs to be investigated further. This may be the case for coronavirus infections, since experimental oral/intranasal (IN) or IN exposure of colostrum-deprived or gnotobiotic calves to bovine coronavirus led to coronavirus antigen, not only in the small intestine and colon, but also in nasal epithelial cells, nasal turbinates, and trachea.[31] In studies in our laboratory we have observed a similar infection of nasal epithelial cells by Bredavirus after oral/IN exposure of calves.[129] Replication of the coronavirus, transmissible gastroenteritis virus (TGE) also occurs in tissues

TABLE 2
Replication Sites of Enteric Coronavirus Strains in Pigs and Their Vertical and Longitudinal Distribution

	Primary site of replication							
	Vertical			Longitudinal				
Virus	Villous	(Location)	Crypt	D, J, I[e]	Colon	Extent	Primary site of lesion or villous atrophy (extent)	Diarrhea
Coronavirus								
Virulent TGE[a]	+	Entire	–	D, J, I	–	Continuous	D, J, I severe	Severe
Attenuated TGE[b]								
P115	+	Entire	–	J, I	–	Patchy	J, I (mild)	Mild-none
B1-350	+	Entire	–	J	–	Patchy	J (mild)	None
SPV	–	Lamina propria cells	–	J, I	–	NR[e]	NR	None
Canine[c]								
UDC-1	+	NR	+	I	–	Patchy	NR (slight)	None
Feline[c]								
FIPV	+	NR	–	J, I	–	Continuous	NR (severe)	Severe
Porcine epidemic diarrhea virus[d]	+	Sides/base	±	D, J, I	+	NR	J, I (moderate-severe)	Moderate

[a] Virulent Miller or B1-2 strains of transmissible gastroenteritis virus (TGE), based on data from Reference 45,46. All the viruses footnoted a to c are antigenically related.
[b] P-115= purdue strain of TGEV passaged 115 times in cell culture (Reference 45); B1 strain of TGEV passaged 350 times in cell culture (Reference 46); SPV-small plaque variant strain of TGEV (Reference 44,47).
[c] Canine coronavirus (strain UDC-1), Reference 44; FIPV = feline infectious peritonitis virus, Reference 44.
[d] Antigenically distinct coronavirus of swine, data from Reference 48,49.
[e] NR= not reported; D—duodenum; J—jejunum; I—ileum.

of the upper respiratory tract.[51] A viremia generally has not been reported for most Type I viruses, which would preclude spread to the respiratory tract via this mechanism. These findings suggest the possibility that more labile enveloped enteric viruses, such as coronavirus and perhaps Bredavirus, may undergo initial replication in the oropharynx contributing massive doses of virus to the small intestine upon swallowing virus-laden mucous. This mucous coating may also aid in stabilizing the virus during passage through the stomach.

The primary sites of vertical and longitudinal replication in the intestine for Type I viruses are summarized in Tables 2 to 4. The data summarized often represents a consensus of findings from several investigators and thus reflects multiple observations. Therefore, although some of these differences may reflect different stages in the development of lesions rather than differences in preferred sites of infection, they appear to be consistent observations for each virus type. The studies cited were reported using animals of various ages and euthanized at various times after the onset of diarrhea. Most studies were conducted in gnotobiotic or seronegative specific pathogen-free (SPF) animals, thereby eliminating the effects of differences in levels of preexisting immunity or concurrent infections on viral pathogenesis.

The major differences noted are discussed as follows. Most Type I enteric viral infections resulted in infection and lesions which were most extensive in the distal small intestine. In rotavirus and coronavirus infections, it has been reported that most infections begin in the proximal part of the small intestine and progress caudally to enterocytes in the jejunum and ileum, with the greatest percentage of infected cells and lesions present at these sites (Tables 3,4).[57,73] However, exceptions in this time course were noted including TGE virus[81] porcine epidemic diarrhea coronavirus (PEDV)[48,49] and lamb rotavirus,[74] in which ileal infections occurred almost simultaneously with infections in other regions. By comparison, all calicivirus infections and lesions were limited to the proximal portion of the small intestine, regardless of the species examined.[68-70] These observations were not altered in animals examined early or late after infection. Group B rotaviruses of swine also resulted in lesions in the duodenum in one study.[77]

Only enveloped viruses, including bovine coronaviruses and Bredavirus and porcine epidemic diarrhea virus, consistently induced infection and lesions in the colon and were the only Type I viruses other than bovine adenovirus to cause colonic lesions.[7,8,31,48,49,52,53,59-61] There were variable reports of rotavirus infections in the colon of certain species, but no colonic lesions were noted.[72,74,75] Only rotaviruses in serogroup B were associated with the occurrence of syncytia on the surface of villi.[29,78] This property of group B rotaviruses may be a pathognomonic lesion which distinguishes these viruses from other rotavirus serogroups. Although there is a report of syncytia associated with a fringed virus-like particle in calves, the causative agent was subsequently identified as a bovine group B rotavirus.[82] However, occurrence of syncytia in intestinal enterocytes has also been reported in association with coronavirus enteritis in mice.[58] All group B rotaviruses examined infected mature enterocytes at the tips of the villi and produced patchy foci of infection, mainly in the mid region of the small intestine.[29,77-79]

Two groups of Type I viruses resulted in infection or necrosis of Peyer's patches. These included bovine adenovirus and astrovirus.[7,8,60] A similar pattern of pathogenesis was not reported for adenoviruses or astroviruses from other species. Astroviruses, which do not cause diarrhea in calves, may replicate in intestinal M cells, the specialized epithelial cells on the surface of dome villi over Peyer's patches.[60] The M cells function in local antigen processing and presentation to the underlying lymphoid tissue. Whether this type of infection, by destroying dome epithelia, compromises the local gut immune system thereby predisposing the host to infection with other enteric pathogens, is unknown and requires further investigation. An interesting corollary may be noted between bovine astrovirus and reovirus infections of mice, most of which result in systemic but not enteric infections.[83] In types 1 and 3 reovirus infections, both viruses penetrate the ileal epithelial barrier by traversing M cells, but only type 1 reoviruses selectively adhere to M cells and not absorptive cells in suckling mice.[83] Subsequent studies suggested that after oral or intravenous inoculation of adult mice with 10^{10} PFU/ml of reovirus

TABLE 3
Replication Sites of Type I Enteropathogenic Viruses and Their Vertical and Longitudinal Distribution in the Intestine

	Primary site of replication								
	Vertical			Longitudinal					
Virus	Villous	(location)	Crypt	D, J, I[a]	Colon	Extent	Primary site of lesion or villous atrophy	Diarrhea	Ref.
Coronavirus									
Bovine	+	Entire	±	J, I	+	Continuous	J, I colon	Severe	31, 52, 53
Others	+	Tips/base, entire	–	J, I	–	Patchy-continuous	J, I	Moderate to severe	45, 54—58
Bredavirus	+	Sides/base	+	J, I	+	Patchy-continuous	J, I colon	Moderate	59—61
Adenovirus									
Bovine	±	Vascular endothelial cells	–	I	+	NR[b]	I, colon (necrosis)	Mild to severe	7, 8
Others	+	Sides/base	±	J, I	–	Patchy	J, I	Mild to none	62—64
Astrovirus									
Bovine	+	Dome[c]	–	J, I	–	NR	Peyer's patches	None	60
Others	+	Tips	–	D, J, I	–	Patchy	J, I	Mild	65—67
Calicivirus									
Human	+	NR	–	J	–	NR	J	Severe	68
Porcine	+	Tips/sides	–	D, J	–	Patchy-continuous	D, J	Moderate	69
Bovine	+	Base/entire	–	D, J	–	Patchy-continuous	D, J	Moderate	70

[a] D = duodenum; J = jejunum; I = ileum.
[b] NR = not reported.
[c] Dome — M cells of dome villi overlying Peyer's patches.

TABLE 4
Replication Sites of Rotaviruses and Their Vertical and Longitudinal Distribution in the Intestine

Virus	Primary site of replication						Primary site of lesion or villous atrophy (extent)	Diarrhea	Ref.
	Vertical			Longitudinal					
	Villous	(location)	Crypt	D, J, I[a]	Colon	Extent			
Group A rotavirus									
Mouse	+	Tips	–	J,I	±	Patchy-continuous	J, I	Mild	71, 72
Bovine	+	Apical half	–	J, I	–	Patchy-continuous	J, I	Moderate	73
Ovine	+	Apical half	–	J, I	+	Patchy-continuous	J, I	Mild-moderate	74
Porcine	+	Entire	–	D, J, I	±	Patchy-continuous	J, I	Mild-moderate	75, 76
Group B rotavirus									
Porcine	+	Tips	–	D, J, I	–	Patchy	D, J	Mild	77
Bovine	+	Tips	–	J, I	–	Patchy	J, syncytia	Mild	78, 79
Rodent	+	Tips	–	NR[b]	–	Patchy	NR, syncytia	Mild	29
Group C rotavirus									
Porcine	+	Entire	–	J, I	–	Patchy-continuous	J	Mild-severe	80

Note: See table 3 for footnotes.

type 1, uptake of virus occurred via M cells into the Peyer's patches.[6] After administration by either route, reovirus infected crypt cells in the ileum, where it was theorized that a receptor may exist on the basolateral membrane. Villous atrophy and crypt hyperplasia were noted, but no diarrhea was reported. All inoculated mice died, and generalized infections were apparent. No information was provided as to whether fecal-oral transmission of reovirus type 1 to other mice could occur under these circumstances. Thus reovirus type 1 may infect crypt cells not via the luminal surface, but by way of a "back door" mechanism.[6] Circumstantial evidence also exists that other enteropathogenic viruses, such as bovine adenoviruses, parvoviruses, and BVDV (discussed in a subsequent section), may also infect crypt enterocytes in a similar manner, either directly by a hematogenous route (adenovirus) or following infection of intestinal lymphoid cells.[7-15] Viruses which infect intestinal lymphoid cells as a primary site of replication, such as enteroviruses or reoviruses,[1-3] may then be disseminated hematogenously to secondary target tissues, including crypt enterocytes, as was observed following administration of high titers of reovirus type 1 to adult mice.[6]

There are no definite explanations to account for the villous enterocyte tropism of Type I viruses. Possible explanations include the role of cell receptors and various enzymes within the brush borders of the enzymatically mature villous epithelial cells. These enzymes may be required for viral uncoating or exposure of cell receptors, in a similar manner to the *in vitro* requirement for intestinal enzymes such as trypsin, chymotrysin, or pancreatin for replication of most rotaviruses, astroviruses, and caliciviruses in cell culture.[16-25] Additionally in the neonate, some viruses might be taken up in absorptive villous enterocytes nonspecifically via the tubulovacuolar system during the period of colostrum absorption. In regard to this latter point, studies in mice indicated that glucocorticoid treatment, which induces premature gut closure, decreased the susceptibility of suckling mice to rotavirus (EDIM) infection.[84]

III. PATHOGENICITY OF TYPE I VIRUSES

Type I viruses infect and destroy villous enterocytes resulting in villous atrophy and frequently fusion of adjacent villi. Pathologic lesions which may result in villous atrophy are shown schematically in Figure 1. Examples of the villous atrophy or fusion produced following infection with various Type I viruses are shown in the scanning electron micrographs in Figures 2 to 4. Loss and fusion (Figure 2) of the mature absorptive cells leads to reduced absorptive surfaces in the intestine. A malabsorptive, maldigestive diarrheal syndrome results, often accompanied by dehydration and death in cases of severe diarrhea. Unlike the secretory diarrhea associated with many enteric bacterial infections, the mechanism of fluid loss induced by enteric viruses relates to direct morphologic and functional damage of absorptive enterocytes.[85] Loss of mature absorptive cells results in loss of enzymes responsible for digestion of disaccharides causing maldigestion.[41,85-87] Glucose-sodium carrier and (Na^+K^+)-ATPase is diminished resulting in loss of sodium, potassium, chloride, bicarbonates, and water, with ensuing acidosis and malabsorption of nutrients.[41,85-87] The presence of undigested lactose induces further osmotic loss of fluids and increases bacterial fermentation, which in turn contributes to additional osmotic pressure in the intestine.[41,85-87] Finally, replacement of absorptive cells by enzymatically immature crypt cells, which retain their secretory capacity, as well as crypt hyperplasia also lead to increased secretion in the gut lumen.[85] Failure to correct these conditions in severely dehydrated children or animals by prompt administration of electrolytes may result in death.

In describing mechanisms of diarrheal pathogenesis, Moon drew a corollary between the severity of enteric viral infections and the extent of viral replication vertically in enterocytes lining both the sides and tips of the villi.[85] Whereas rotaviruses and astroviruses mainly infect cells on the apical half of the villi (the more enzymatically mature cells) producing a milder transient diarrhea (Tables 3 and 4, Figure 1, regions 1 and 2 and Figure 4a), coronaviruses infect

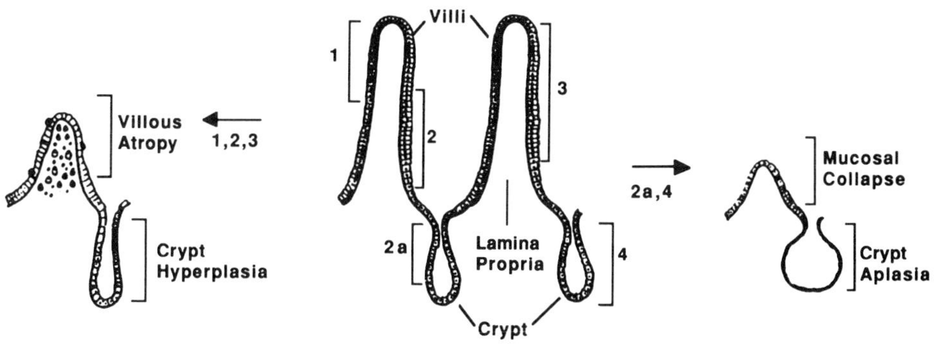

FIGURE 1. Schematic of small intestinal villi depicting vertical replication sites of enteropathogenic viruses in the villus (regions 1, 2, or 3) or crypts (2a, 4) of, and the pathologic lesions associated with such infections.

enterocytes throughout the length of the villi causing more pronounced villous atrophy and also diarrhea (Tables 2 and 3; Figure 1, region 3 and Figure 4 b,c). Epithelial cells on the villi tips constantly are being replaced by cells that proliferate in the crypts and migrate up the sides of the villi.[85,88] The turnover rate of these cells is slower in younger or gnotobiotic animals, leading to less rapid repair of villous atrophy, which may contribute to the enhanced susceptibility of neonates to these viral diarrheas.[89] The constant replacement of damaged villous enterocytes by cells originating in the crypts, which are refractory to further infection by several viruses,[55,73] suggests that in most animals these infections will be self-limiting. This may not be the case for all Type I viruses, however, since porcine epidemic diarrhea coronavirus appears capable of infecting regenerating cells.[48] In particular, Type I viruses which also replicate to a limited extent in crypt cells[7,8,52,59] may be able to initiate additional cycles of replication in the newly replaced enterocytes which have enzymatic profiles similar to crypt cells.[41] This may account for a longer persistence of such viruses in the intestinal tract.[7,8]

Besides the vertical stratification of virus replication sites within intestinal enterocytes, there also appears to be longitudinal differentiation of viral replication sites in the intestine (Tables 2 to 5). In general, viruses that replicate only in limited portions of the intestine, or that infect only scattered enterocytes, cause only mild or no villous atrophy and subsequently mild or no diarrhea. In comparison, viruses that replicate throughout the entire intestine or that infect a high percentage of enterocytes might be expected to cause more severe villous atrophy and diarrhea. For example, a virus such as astrovirus produces only a patchy immunofluorescence (indicative of only scattered infected cells), and in the calf, lamb, and pig this limited replication results in little or no villous atrophy or diarrhea (Table 3).[60,65-67] Many coronaviruses (TGEV and bovine), on the other hand, produce an almost continuous infection of entire villous enterocytes throughout the distal portion of the small intestine (Figure 4b, c) and colon (bovine coronavirus only).[31,45,46,52,54,55] These enteric coronaviruses are usually associated with a severe diarrheal syndrome in experimentally inoculated or naturally infected pigs and calves.[31,45,46,52-55] Bovine Bredavirus has a similar predilection for cells in the distal portion of the small intestine and colon, but it generally produces a more patchy infection and milder diarrheal syndrome.[59-61]

Rotaviruses (serogroups A to C) generally replicate and cause villous atrophy in the distal half of the small intestine, but not the colon (Table 4).[71,73,76-80] The group A and C rotaviruses usually result in a higher percentage of infected cells in the intestine compared with group B rotaviruses, which produce scattered foci of infection throughout the intestine, but mainly limited to the tips or sides of villi (Figure 4a).[29,77-79] Human, bovine and porcine caliciviruses replicate and produce villous atrophy only in the proximal half of the small intestine (Table 3; Figure 3a).[68-70]

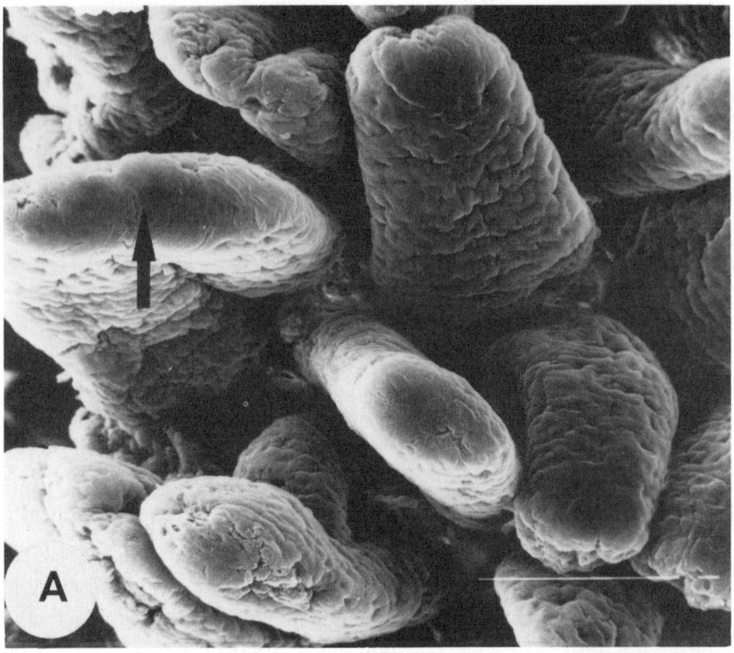

A

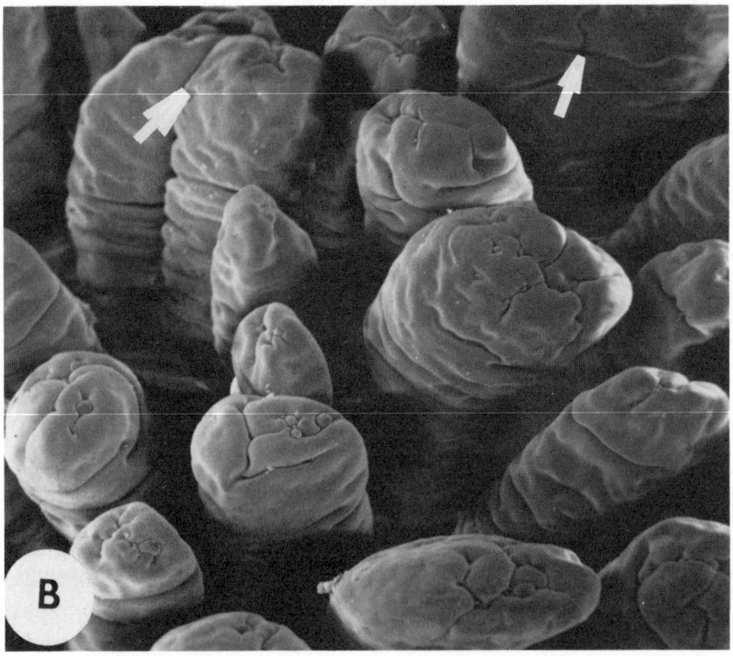

B

FIGURE 2. Scanning electron micrographs of the duodenum showing fused villi (arrows). (A) A gnotobiotic pig 7 days post-inoculation (DPI) with porcine enteric calicivirus.[69] (× 250.) (B) A gnotobiotic calf 3 DPI with bovine coronavirus.[31] (× 250.) Normal villi in the duodenum of a control uninoculated gnotobiotic pig are shown in Figure 3B.

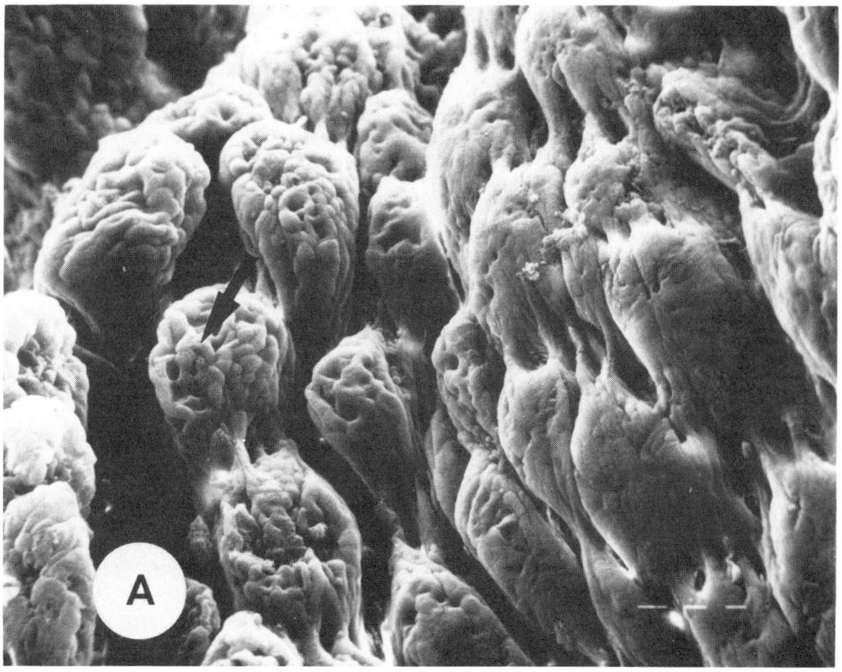

A

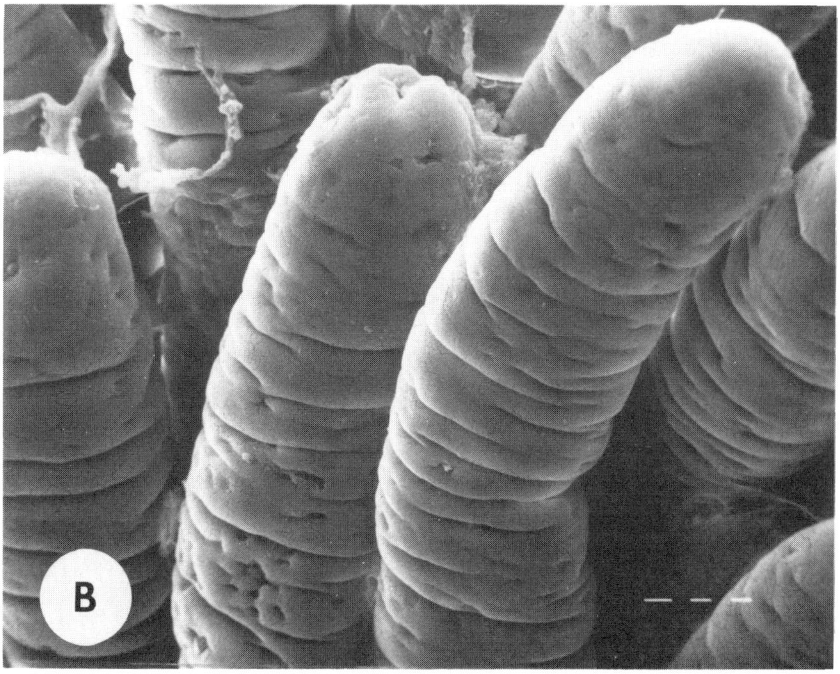

B

FIGURE 3. Scanning electron micrographs of the duodenum of gnotobiotic pigs. (A) Severely shortened, stump-like villi with distinct swollen irregular enterocytes (arrow) from a pig at 3 DPI with porcine enteric calicivirus.[69] (× 350.) (B) Long, finger-like villi containing tranverse farrows from a control uninfected pig, euthanized at the same age (22 days). (× 350.)

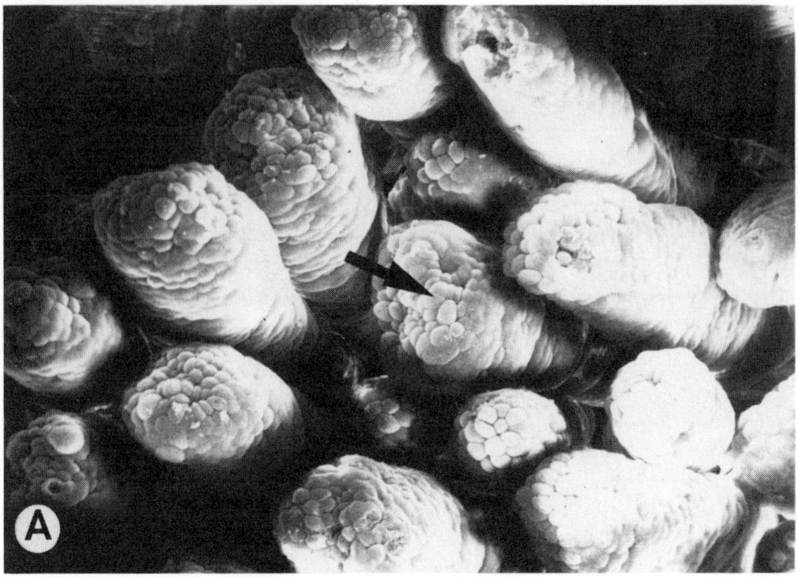

A

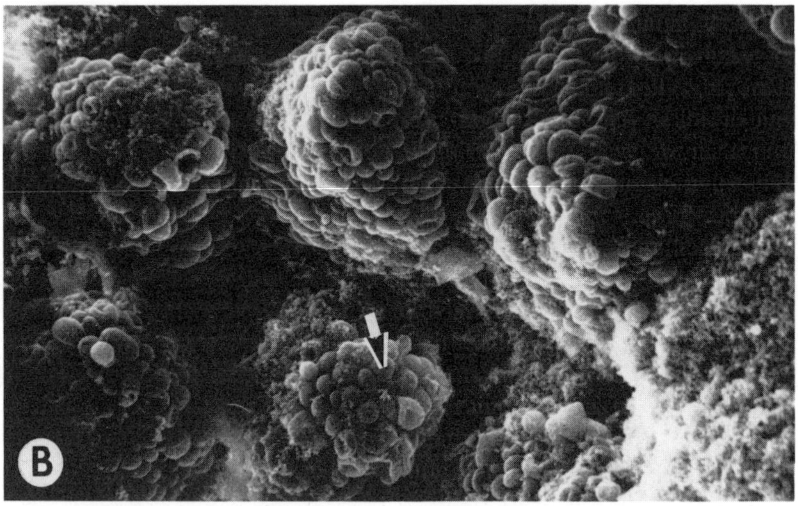

B

FIGURE 4. Scanning electron micrograph of the ileum showing various degrees of villous atrophy. (A) A gnotobiotic pig 1 DPI with porcine group B rotavirus showing villi of irregular length and with swollen enterocytes on the tips (arrow). (× 250.) (B) A gnotobiotic pig 1 DPI with the coronavirus transmissible gastroenteritis virus showing severe villous atrophy with stump-like villi covered with swollen, protruding cells. (×250.) (C) A gnotobiotic calf 4 DPI with bovine coronavirus showing severe villous atrophy. (× 60.) Normal villi are shown in Figure 3B for comparison.

The location of the viral lesion may also influence the severity of diarrhea. Because glucose and sodium absorption are the greatest in the proximal and middle section of the jejunum, extensive damage in this area may cause the most severe diarrhea.[81,87] However, sodium is also absorbed in the colon, which may compensate for electrolyte imbalances unless the colonic

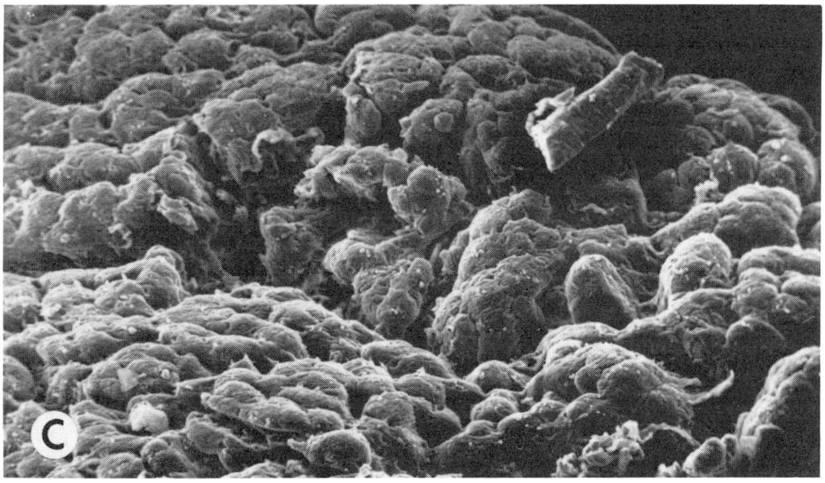

FIGURE 4C.

epithelium is also infected and severely damaged. This latter observation may account for the greater severity of diarrhea observed in calves infected with bovine coronavirus compared with calves challenged with bovine rotavirus.[90] Extensive infection of the ileum may also lead to impaired absorption of bile salts which then pass to the colon and stimulate further secretion.[85]

Use of neonatal gnotobiotic or colostrum-deprived SPF pigs has permitted researchers to study the comparative pathogenesis of various strains of enteric coronaviruses in a single susceptible species.[44-49] The principal findings from these studies are summarized in Table 2. All pigs were under 4 days old at the time of inoculation. Pigs were inoculated orally, with the exception of ones given canine or feline coronaviruses, which were inoculated via stomach tube. In a comparison of the pathogenicity of virulent and attenuated strains of TGEV, conclusions were as follows: (1) virulent strains of TGEV replicated throughout enterocytes of the entire villus and small intestine, inducing severe villous atrophy and diarrhea; (2) strains of TGEV attenuated by 115 to 350 passages in cell culture (P115 and B1-350) produced scattered foci of infection in villous enterocytes only in the jejunum or jejunum and ileum; and (3) only mild villous atrophy occurred in these regions and mild or no diarrhea was evident.[45,46] Thus, a reverse correlation existed between the degree of cell culture attenuation of TGEV and the extent of intestinal infection. Another attenuated strain of TGEV was produced by passage of TGEV in leukocyte cell cultures and selection of a small plaque variant (SPV).[47] This virus infected undefined cells in the lamina propia and did not cause diarrhea (Table 2). Further characteristics of the SPV have not been reported.

Passage of the antigenically related canine coronavirus and feline infectious peritonitis (FIP) coronavirus in pigs produced only slight villous atrophy with no diarrhea or severe villous atrophy with severe diarrhea, respectively.[44] In addition, although canine coronavirus replication was restricted to villous enterocytes in the natural host, in the pig, crypt enterocytes were also infected.[44,56]

Recently, a second antigenically unrelated enteric coronavirus (PEDV) was isolated from pigs.[48,49] This virus also differed from TGEV in its pathogenesis in pigs.[48,49] Differences observed included infection of cells at the base of the villus and occasional crypt cells, infection of cells in the colon, less severe villous atrophy and diarrhea than produced by TGEV infection, and possible infection of regenerating cells as indicated by a second peak of infection. Attributes of enteric coronavirus strains which influence their virulence, including regions and intensity of replication in the piglet intestinal tract, are unknown. The rate of viral replication may greatly

TABLE 5
Replication Sites of Type II Enteropathogenic Viruses and Their Vertical and Longitudinal Distribution in the Intestine

	Primary sites of replication								
	Vertical			Longitudinal					
	Villous	(Location)	Crypt	D, J, I[a]	Colon	Extent	Primary site of lesion or villous atrophy	Diarrhea	Ref.
Bovine virus									
Diarrhea	–	–	+	I	+	Patchy-continuous	NR	Mild-severe	15
Parvovirus									
Canine	–	–	+	D ,J, I	–	Patchy-continuous	D, J, I	Mild-severe	10—12
Feline	–	–	+	D, I	+	Patchy-continuous	D, J, I, colon	Mild-severe	13, 14
Bovine	–	–	+	D, J, I	+	Patchy-continuous	NR	Mild	9

Note: See Table 3 for footnotes.

influence the disease outcome. PEDV replicated more slowly in pigs as indicated by a longer incubation period and induced milder lesions and diarrhea than virulent TGEV. Viral mutants produced by *in vitro* attenuation or by adaptation in a heterologous host may replicate more slowly with fewer cells infected resulting in a longer incubation period.[44-46] Suboptimal viral doses may also lengthen the incubation period.[91] Both factors may permit development of host compensatory mechanisms, including development of local intestinal immunity.

As for cell tropisms, reasons for the predilections of different viruses for certain regions of the intestine are unclear. Physiologic factors such as different enzyme concentrations, pH, and presence of bile salts all may play a role. Studies have shown longitudinal differentiation in the small intestine with regard to absorption of nutrients and cell turnover, with both activities increased in the duodenum.[92] All these variables may influence virus stability, receptors, and replication. Immunologic factors such as regional variation in concentrations of local antibodies, cell-mediated immunity, or interferon also may influence intestinal viral replication sites.

IV. CELL TROPISMS OF TYPE II ENTEROPATHOGENIC VIRUSES

Viruses of the second type that infect mainly crypt enterocytes and intestinal lymphoid cells (Table 5) also are thought to be transmitted via the fecal-oral and respiratory routes. Studies of BVDV and feline and canine parvoviruses suggest the initial site of viral replication is in the upper respiratory tract (oropharyngeal tissue).[10-15] From there, virus may be disseminated hematogenously to other target tissues such as intestinal lymphoid tissues, crypt enterocytes, or other secondary sites either via a viremia, infected leukocytes, or both. The predilection of BVDV and parvovirus for crypt epithelial and intestinal lymphoid cells probably is related to the rapid proliferative rate of these cells since *in vivo* replication of these viruses requires a population of cells undergoing frequent mitosis.[11-15] As postulated previously for reovirus type 1, viruses which infect primarily crypt enterocytes may do so mainly by way of infection via the basolateral membrane of the crypt cell and not via the luminal surface.[6]

V. PATHOGENICITY OF TYPE II VIRUSES

As discussed by Moon, infection of crypt enterocytes by viruses also leads to severe villous atrophy and diarrhea with loss of crypt and consequently replacement villous epithelial cells (Figure 1, regions 2a, 4).[85] In severe cases, this results in a state of total mucosal collapse.[12,85] Replication of these viruses in lymphoid tissue may also lead to more severe disease because of interference with local immune responses and possible immunosuppression.[93,94]

A cause of some concern might be multiple infections that occur naturally with several Type I viruses which replicate in different regions of the intestine (Tables 3, 4; Figure 1),[95-97] or Type I and II viruses that destroy both villous and crypt enterocytes (Table 5, Figure 1).[98] Such combined infections presumably could lead to much greater morbidity and mortality than observed with either agent alone, and in one study mortality rates increased with the number of enteric viruses detected.[99] It is possible that infections with Type I viruses which damage the villous epithelium may predispose animals to infection with Type II viruses by inducing crypt hyperplasia thereby increasing crypt mitotic activity, an important factor in intestinal replication of Type II viruses. There are few experimental studies of the pathogenesis or possible additive or synergistic interactions of multiple enteric viral infections in animals, but detection of several viruses from a single diarrheic animal is common.[95-97,99,100-106] It is possible that immunosuppressive effects mediated by BVD virus infections may predispose calves to or enhance the severity of enteric viral infections in a manner similar to its accentuation of bovine respiratory disease.[94,107] Multiple infections of man or animals with enteric viruses and bacteria/protozoa are disease.[94,107] Multiple infections of man or animals with enteric viruses and bacteria/protozoa are also common[100-102,106,108] and the subject of enteric viral-bacterial/protozoal synergistic infections will be reviewed in detail in another chapter.

VI. HOST FACTORS WHICH INFLUENCE SUSCEPTIBILITY TO ENTEROPATHOGENIC VIRUSES

The severity of enteric viral infections under natural conditions is highly variable and is related to complex interactions among agent, host, and environment. Subclinical infections occur, particularly in adults, and it is often difficult to reproduce the disease in conventional animals. Subclinical infections probably occur when very limited numbers of cells are infected, leading to only mild intestinal damage and little disruption of absorption. Low virus infection rates may be due to low initial infective doses or avirulent virus strains allowing the host's specific and non specific immunity time to contain the infection. The host's immune status at the time of infection is also important. In the presence of adequate levels of antibody, infection can still occur, but the extent of replication is limited, resulting in little or no diarrhea.[109,110] Variables which influence immunity to enteropathogenic viruses will be reviewed in a later chapter. In general, for viruses which infect villous enterocytes via the luminal surface, local intestinal immunity appears to be necessary for protection as in TGEV and rotavirus infections.[54,109-111] If crypt enterocytes are infected, apparently via a hematogenous route, then humoral antibodies may provide sufficient protection as seems to be the case for canine and feline parvovirus infections.[112]

A. AGE, WEANING, INTESTINAL FLORA, AND DIET

Enteric viral infections commonly cause the most severe diarrhea in nursing or young animals or infants, but adults and animals of all ages are susceptible. In most cases infection of older animals or adults may be subclinical, but exceptions have been noted. Coronaviruses have been associated with outbreaks of winter dysentery in adult cattle.[113] Adenovirus, BVDV, rotavirus, and parvovirus infections are often reported in older or weaned (BVDV, adenovirus)[7,8,40] animals with the latter two infections (rotavirus and parvovirus) especially prominent in recently weaned pigs or dogs, respectively.[10-12,14] Although active immunity undoubtedly plays a role in the resistance of older animals, age may also influence maturation of intestinal cells.[89] The following observations related to epithelial cell kinetics are summarized in Table 6. Enterocytes proliferate in the crypts of the small intestine and become more enzymatically differentiated as they migrate to the tips of the villi where they are shed into the lumen.[85,88] As noted in Table 6, absorptive enterocytes regenerate more slowly in newborns than in older animals.[89] Newborn conventional piglets require 7 to 10 days to replace villous epithelium, whereas in 21-day-old pigs, only 2 to 4 days is required. Similar renewal times were reported in humans (4 to 6 days)[115] and 21-day-old calves and lambs (2 to 3 days).[116] These slower migration rates in younger hosts may increase the severity of Type I virus infections of villous enterocytes. Likewise it is possible that the aging process may impair rapid epithelial cell turnover rates, leading to the enhanced susceptibility of elderly adults or aged animals to enteric viral infections.[50,117] This possibility has not been addressed in any experimental studies.

The infectious dose of TGEV needed to infect a 6 month-old pig was 10^4 times that needed to infect a 2-day-old piglet and virus production in the slowly replaced, enzymatically differentiated cells of newborn pigs was higher than in older animals.[118,119] Additionally, Type I enteric viruses may accumulate in the tubulovacuolar system which is extensive in newborn pigs and mice prior to weaning, but diminishes in pigs older than 3 weeks, or mice after weaning. It was shown in mice that early maturation of this system by treatment with glucocorticoids, decreased the susceptibility of suckling mice to rotavirus (EDIM) infection.[84]

TABLE 6
Effect of Host Age on Epithelial Cell Kinetics

Pig age	Microbial status	Crypt to villous migration time[a] (d)
1	Conventional	7—10
21	Conventional	2—4
14	Conventional	4
14	Gnotobiotic	8

[a] Data from Moon, H. W., *J. Am. Vet. Med. Assoc.*, 172, 443, 1978.

On the other hand, more rapid enterocyte migration rates which occur in older animals (Table 6) may increase the severity of viral infections of crypt epithelial cells[86] (Type II viruses), since these viruses have a predilection for rapidly dividing cells.[9-15,85] As another example of factors influencing epithelial cell migration rates, feline parvovirus (panleukopenia) causes severe disease in conventional kittens but only mild disease in gnotobiotic kittens.[13] This latter observation in kittens and comparisons of gnotobiotic and conventional pigs (Table 6) reflects the influence of intestinal flora on enterocyte migration rates. Thus, the presence of intestinal flora increases the mitotic activity of crypt enterocytes and, consequently, epithelial cell migration rates in the intestine.[120] Therefore, the increased severity of parvovirus enteritis observed in recently weaned dogs may result from increased enterocytes migration rates due to both age of the animals as well as the dramatic changes which occur in the intestinal flora at weaning.[10-12] In addition, dietary stress at weaning, particularly alterations in dietary regimes, may also result in increased turnover of intestinal cells as occurs by starvation followed by an upsurge in feeding.[10,114,120,121] Such factors as dietary stress and alterations in intestinal flora may also reduce intestinal peristalsis, a nonspecific host factor important for resistance to gut colonization by infectious microorganisms.[114,121]

B. HORMONAL FACTORS

The effect of the hormones on intestinal epithelial migration rates or intestinal peristalsis has not been well studied. It has been postulated that the hormone enteroglucagon is a stimulator of intestinal mucosal growth. In support of this theory, blood levels were increased in calves after infection with enteropathogenic, but not nonenteropathogenic viruses.[122] Other indirect observations suggest hormones may be a factor in influencing host susceptibility to enteric viral infections. Pregnant or lactating seropositive animals have been shown to re-excrete enteric viruses, such as rotavirus or coronavirus, thereby serving as a possible source of virus for susceptible neonates.[84,123-126] It is possible pregnancy-associated hormones induce immunosuppressive factors which influence intestinal immunity and allow reinfection of the gut or re-excretion of enteric viruses. In addition, similar hormones may induce effects which reduce epithelial cell turnover rates, rendering the intestine more susceptible to reinfection by enteric viruses. More research is needed to investigate these possibilities.

VII. ENVIRONMENTAL FACTORS WHICH INFLUENCE HOST SUSCEPTIBILITY

There is a paucity of information on the influence of environmental factors on host susceptibility to enteric viral infections. Improvements in hygienic conditions, including thorough disinfection and cleansing and adequate ventilation in hospital nurseries or in animal-rearing facilities, will provide a reduction in viral exposures of neonates and young children or animals. In studies of rotaviral diarrhea in pigs weaned at 1 day of age, Lecce and colleagues reported an increased incidence and severity of diarrhea when pigs were weaned into a nursery with increasing levels of rotavirus contamination from previously weaned pigs.[91] Similar conditions may prevail in some heavily used hospital nurseries or in dairy or swine operations in which younger animals are constantly placed in close contact with older animals or in premises which are in constant and heavy use. Likewise, such conditions may foster concurrent exposure to other microbial agents which may also enhance the severity of enteric viral infections in the host.[114,121]

Poor nutrition and stress may also influence host susceptibility, perhaps by compromising immune responses, or affecting proliferation rates of enterocytes. Precise information on the role these factors play in host susceptibility to enteric viral infections is lacking.

In neonates and young animals in particular, cold stress or widely fluctuating ambient temperatures may influence the severity of diarrhea. Pigs challenged with TGEV and maintained in environments at high temperatures (30°C) remained free of disease, while pigs exposed to temperature extremes (from 30° to 4°C) after challenge developed severe diarrhea.[127] Although the authors speculated interference with local cell-mediated immunity played an important role in the temperature-related disease severity, other factors may also be important.[128]

In summary, a plethora of enteric viruses have been described, many only recently, which are capable of causing diarrhea in man or animals. However, their causative role in diarrhea syndromes under natural conditions may be difficult to assess due to the influence of other contributing factors. Such factors include ones discussed in this chapter such as viral dose, host, age, diet and nutrition, microbial flora, hormonal influences, and environmental factors. Other highly relevant factors such as immunity and concurrent microbial infections will be discussed in subsequent chapters. An understanding of the pathogenesis of enteric viral infections as reviewed in the present chapter is useful for devising strategies including appropriate methods of immunoprophylaxis for prevention and control.

REFERENCES

1. **Wenner, H. A., Beran, G. W., and Weder, A. A.,** Enteroviruses in swine. II. Studies on the natural history of infection and immunity, *Am. J. Vet. Res.*, 21, 958, 1960.
2. **Melnick, J. L.,** Enteroviruses: polioviruses, coxsackieviruses, echoviruses and newer enteroviruses, in *Virology*, Fields, B. N., Ed., Raven Press, New York, 1985, 739.
3. **Kasza, L.,** Isolation and characterization of a reovirus from pigs, *Vet. Rec.*, 87, 681, 1970.
4. **Hollinger, F. B. and Melnick, J. L.,** Features of viral hepatitis, I and II. in *Virology*, Fields, B. N., Ed., Raven Press, New York, 1985, 1417.
5. **Cook, I.,** Reovirus type 3 infection in laboratory mice, *Aust. J. Exp. Biol.*, 41, 651, 1963.
6. **Rubin, P. H., Eaton, M. A., and Anderson, A. O.,** Reovirus infection in adult mice: the virus hemagglutinin determines the site of intestinal disease, *Microbiol. Pathogenesis*, 1, 79, 1986.
7. **Thompson, K. G., Thompson, G. W., and Henry, J. N.,** Alimentary tract manifestations of bovine adenovirus infections, *Can. Vet. J.*, 22, 68, 1981.
8. **Orr, J. P.,** Necrotizing enteritis in a calf infected with adenovirus, *Can. Vet. J.*, 25, 72, 1984.
9. **Storz, J., Leary, J. J., Carlson, J. H., and Bates, R. C.,** Parvoviruses associated with diarrhea in calves, *JAVMA*, 173, 624, 1978.
10. **Carman, P. S. and Povey, R. C.,** Experimental challenge of dogs with canine parvovirus, *Can. J. Comp. Med.*, 436, 33, 1982.
11. **O'Sullivan, G., Durham, P. J. K., Smith, J. R., and Campbell, R. S. F.,** Experimentally induced severe canine parvoviral enteritis, *Aust. Vet. J.*, 61, 1, 1984.
12. **Macartney, L., McCandlish, I. A. P., Thompson, H., and Cornwell, H. J. C.,** Canine parvovirus enteritis I: clinical, haematological and pathological features of experimental infections, *Vet. Rec.*, 115, 201, 1984.

13. **Carlson, J. H., Scott, F. W., and Duncan, J. R.,** Feline panleukopenia. I. Pathogenesis in germfree and specific pathogen-free cats, *Vet. Pathol.,* 14, 79, 1977.
14. **Kahn, D. E.,** Pathogenesis of feline panleukopenia, *J. Am. Vet. Med. Assoc.,* 173, 628, 1978.
15. **Ohmann, H. B.,** Pathogenesis of bovine viral diarrhoea-mucosal disease: distribution and signficance of BVDV antigen in diseased calves, *Res. Vet. Sci.,* 34, 5, 1983.
16. **Lee, T. W. and Kurtz, J. B.,** Serial propagation of astrovirus in tissue culture with the aid of trypsin, *J. Gen. Virol.,* 57, 421, 1981.
17. **Woode, G. N., Kelso Gourley, N. E., Pohlenz, J. F., Liebler, E. M., Mathews, S. L., and Hutchinson, M. P.,** Serotypes of bovine astrovirus, *J. Clin. Microbiol.,* 22, 668, 1985.
18. **Cubitt, W. D. and Barrett, A. D. T.,** Propagation of human candidate calicivirus in cell culture, *J. Gen. Virol.,* 65, 1123, 1984.
19. **Flynn, W. T. and Saif, L. J.,** Serial propagation of porcine enteric calicivirus-like virus in primary porcine kidney cell cultures, *J. Clin. Microbiol.,* 26, 206, 1988.
20. **McNulty, M. S., Allen, G. M., Todd, D., McFerran, J. B., and McCracken, R. M.,** Isolation from chickens of a rotavirus lacking the rotavirus group antigen, *J. Gen. Virol.,* 55, 405, 1981.
21. **Saif, L. J., Terrett, L. A., Miller, K. L., and Cross, R. F.,** Serial propagation of porcine group C rotavirus (pararotavirus) in a continuous cell line and characterization of the passaged virus, *J. Clin. Microbiol.,* 26, 1277, 1988.
22. **Lamouliatte, F., Pasquier, P. D. U., Rossi, F., Laporte, J., and Loze, J. P.,** Studies on bovine Bredavirus, *Vet. Microbiol.,* 15, 261, 1987.
23. **Babiuk, L. A., Mohammed, K., Spence, L., Fauvel, M., and Petro, R.,** Rotavirus isolation and cultivation in the presence of trypsin, *J. Clin. Microbiol.,* 6, 610, 1977.
24. **Theil, K. W., Bohl, E. H., and Agnes, A. G.,** Cell culture propagation of porcine rotavirus (reovirus-like agent), *Am. J. Vet. Res.,* 38, 1765, 1977.
25. **Wyatt, R. G., Janes, W. D., Bohl, E. H., Theil, K. W., Saif, L. J., Kalica, A. R., Greenberg, H. B., Kapikian, A. Z., and Chanock, R. M.,** Human rotavirus type 2: cultivation *in vitro, Science,* 207, 189, 1980.
26. **Sato, K., Inaba, Y., Shinozaki, T., Fujii, R., and Matumoto, M.,** Isolation of human rotavirus in cell culture, *Arch. Virol.,* 9, 155, 1981.
27. **Urasawa, T., Urasawa, S., and Taniguchi, K.,** Sequential passages of human rotavirus in MA104 cells, *Microbiol. Immunol.,* 25, 1025, 1981.
28. **Terrett, L. A., Saif, L. J., Theil, K. W., and Kohler, E. M.,** Physicochemical characterization of porcine pararotavirus and detection of virus and viral antibodies using cell culture immunofluorescence, *J. Clin. Microbiol.,* 25, 268, 1987.
29. **Vonderfecht, S. L., Huber, A. C., Eiden, J., Mader, L. C., and Yolken, R. H.,** Infectious diarrhea in infant rats produced by a rotavirus-like agent, *J. Virol.,* 52, 94, 1984.
30. **Hess, R. G. and Bachmann, P. A.,** *In vitro* differentiation and pH sensitivity of field and cell culture attenuated strains of transmissible gastroenteritis virus (TGE), *Infect. Immun.,* 13, 1642, 1976.
31. **Saif, L. J., Redman, D. R., and Theil, K. W.,** Experimental coronavirus infections in calves: viral replication in the respiratory and intestinal tracts, *Am. J. Vet. Res.,* 47, 1426, 1986.
32. **Potgieter, L. N. D., McCracken, M. D., Hopkins, F. M., and Walker, R. D.,** Effect of the pneumopathogenicity of two strains of bovine viral diarrhea virus, *Am. J. Vet. Res.,* 46, 151, 1985.
33. **Hoshino, Y., Wyatt, R. G., Greenberg, H. B., Flores, I., and Kapikian, A. Z.,** Serotypic similarity and diversity of rotaviruses of mammalian and avian origin studied by plaque-reduction neutralization, *J. Infect. Dis.,* 149, 694, 1984.
34. **Uhnoo, I., Wadell, G., Svensson, L., and Johansson, M.,** Two new serotypes of enteric adenovirus causing infantile diarrhoea, *Dev. Biol. Stand.,* 53, 311, 1983.
35. **Woode, G. N., Saif, L. J., and Quesada, M.,** Comparative studies on three isolates of "Breda" virus of calves, *Am. J. Vet. Res.,* 46, 1003, 1985.
36. **Kurtz, J. B. and Lee, T. W.,** Human astrovirus serotypes, *Lancet,* 2, 1405, 1984.
37. **Saif, L. J. and Theil, K. W.,** Antigenically distinct rotaviruses of human and animal origin, *Proc. Infect. Diarrhea in the Young,* Tzipori, S., Ed., Elsevier, Amsterdam, The Netherlands, 1985, 208.
38. **Bridger, J. C.,** Novel rotaviruses in animals and man, in *Ciba Foundation Symp. 128, Novel Diarrhea Viruses,* Bock, G. and Whelan, J., Eds., J. Wiley & Sons, Chichester, U. K., 1987, 5.
39. **Kapikian, A. Z., Greenberg, H. B., Wyatt, R. G., Kalica, A. R., and Chanock, R. M.,** The Norwalk group of viruses-agents associated with epidemic viral gastroenteritis, in *Virus Infections of the Gastrointestinal Tract,* Tyrell, D. A. J. and Kapikian, A. Z., Eds., Marcel Dekker, New York, 1982, 147.
40. **Baker, J. C.,** Bovine viral diarrhea virus: a review, *J. Am. Vet. Med. Assoc.,* 190, 1449, 1987.
41. **Davidson, G. P., Gall, D. G., Butler, D. G., Petric, M., and Hamilton, J. R.,** Human rotavirus gastroenteritis induced in conventional piglets: intestinal structure and transport, *J. Clin. Invest.,* 60, 1402, 1977.
42. **Woode, G. N., Bridger, J. C., Jones, J. M., Flewett, T. H., Bryden, A. S., Davie, H. A., and White, G. B. B.,** Morphological and antigenic relationships between viruses (rotaviruses) from acute gastroenteritis of children, calves, piglets, mice and foals, *Infect. Immun.,* 14, 804, 1976.

43. **Tzipori, S. and Williams, I. H.,** Diarrhea in piglets inoculated with rotavirus, *Aust. Vet. J.*, 54, 188, 1978.
44. **Woods, R. D., Cheville, N. F., and Gallagher, J. E.,** Lesions in the small intestine of newborn pigs inoculated with porcine, feline, and canine coronaviruses, *Am. J. Vet. Res.*, 42, 1163, 1981.
45. **Frederick, G. T., Bohl, E. H., and Cross, R. F.,** Pathogenicity of an attenuated strain of TGE virus for newborn pigs, *Am. J. Vet. Res.*, 37, 165, 1976.
46. **Hess, R. G., Bachmann, P. A., and Hanichen, T.,** Versuche zur Entwicklung einer Immunprophylaxe gegen die Ubertragbare Gastroenteritis (TGE) der Schweine. I. Pathogenitat des Stammes B1 im Verlaufe von Serienpassagen, *Zentralbl. Veterinaermed. B*, 24, 753, 1977.
47. **Woods, R. D.,** Small plaque variant transmissible gastroenteritis virus, *J. Am. Vet. Med. Assoc.*, 173, 643, 1978.
48. **DeBouck, P., Pensaert, M., and Coussement, W.,** The pathogenesis of an enteric infection in pigs, experimentally induced by the coronavirus-like agent, CV777, *Vet. Microbiol.*, 6, 157, 1981.
49. **Pospischil, A., Hess, R. G., and Bachman, P. A.,** Light microscopy and ultrahistology of intestinal changes in pigs infected with epizootic diarrhea virus (EVD): comparison with transmissible gastroenteritis (TGE) virus and porcine rotavirus infections, *Zentralbl. Veterinaermed. B*, 28, 564, 1981.
50. **Humphrey, T.-J., Crwckshank, J. G., and Cubitt, W. D.,** An outbreak of calicivirus associated gastroenteritis in an elderly persons home. A possible zoonosis, *J. Hyg.*, 92, 293, 1984.
51. **Kemeny, L. J., Wiltsey, V. L., and Riley, J. L.,** Upper respiratory infection of lactating sows with TGE virus following contact exposure to infected piglets, *Cornell Vet.*, 65, 352, 1975.
52. **Mebus, C. A., Stair, E. L., Rhodes, M. B., and Twiehaus, M. J.,** Pathology of neonatal calf diarrhea induced by a coronavirus-like agent, *Vet. Pathol.*, 10, 45, 1973.
53. **Langpap, T. J., Bergeland, M. E., and Reed, D. E.,** Coronaviral enteritis of young calves: virologic and pathologic findings in naturally occurring infections, *Am. J. Vet. Res.*, 40, 1476, 1979.
54. **Hooper, B. E. and Haelterman, E. O.,** Concepts of pathogenesis and passive immunity in transmissible gastroenteritis of swine, *J. Am. Vet. Med. Assoc.*, 149, 1580, 1966.
55. **Pensaert, M., Haelterman, E. O., and Burnstein, T.,** Transmissible gastroenteritis of swine: virus-intestinal cell interactions. I. Immunofluorescence, histopathology and virus production in the small intestine through the course of infection, *Arch. Gesamte Virusforsch.*, 31, 321, 1970.
56. **Keenan, K. P., Jervis, H. R., Marchwicki, R. H., and Binn, L. N.,** Intestinal infection of neonatal dogs with canine coronavirus 1-71: studies by virologic, histologic, histochemical, and immunofluorescent techniques, *Am. J. Vet. Res.*, 37, 247, 1976.
57. **Gonder, E., Patel, B. L., and Pomeroy, B. S.,** Scanning electron, light and immunofluorescent microscopy of coronaviral enteritis of turkeys (bluecomb), *Am. J. Vet. Res.*, 37, 1435, 1976.
58. **Ishida, T., Taguchi, F., Lee, Y., Yamada, A., Tamura, T., and Fujiwara, K.,** Isolation of mouse hepatitis virus from infant mice with fatal diarrhea, *Lab. Anim. Sci.*, 28, 269, 1978.
59. **Pohlenz, J. F. L., Cheville, N. F., Woode, G. N., and Mokresh, A. H.,** Cell lesions in the intestinal mucosa of gnotobiotic calves experimentally infected with a new unclassified bovine virus (Breda virus), *Vet. Pathol.*, 21, 407, 417.
60. **Woode, G. N., Pohlenz, J. F., Kelso Gourley, N. E., and Fagerland, J. A.,** Astrovirus and breda virus infections of dome cell epithelium of bovine ileum, *J. Clin. Microbiol.*, 19, 623, 1984.
61. **Saif, L. J., Redman, D. R., Theil, K. W., Moorhead, P. D., and Smith, C. K.,** Studies of an enteric "Breda" virus in calves, Proc. Conf. Res. Workers in Animal Dis., Abstr. 236, Chicago, IL, November 9 to 10, 1981.
62. **Clemmer, D. I. and Ichinose, H.,** The cellular site of virus replication in the intestine of chicks infected with avian adenovirus, *Arch. Gesamte Virusforsch.*, 25, 277, 1968.
63. **Takeuchi, A. and Hoshimoto, K.,** Electron microscope study of experimental enteric adenovirus infection in mice, *Infect. Immun.*, 13, 569, 1976.
64. **Ducatelle, R., Coussement, W., and Hoorens, J.,** Sequential pathological study of experimental porcine adenovirus enteritis, *Vet. Pathol.*, 19, 179, 1982.
65. **Snodgrass, D. R., Angus, K. W., Gray, E. W., Menzies, J. D., and Paul, G.,** Pathogenesis of diarrhoea caused by astrovirus infection in lambs, *Arch. Virol.*, 60, 217, 1979.
66. **Saif, L. J., Bohl, E. H., Theil, K. W., Kohler, E. M., and Cross, R. F.,** 30 nm virus-like particles resembling astrovirus in intestinal contents of a diarrheic pig, Proc. Conf. Res. Workers in Animal Diseases, Abstr. 149, Chicago, IL, November 10 to 11, 1980.
67. **Phillips, A. D., Rice, S. J., and Walker-Smith, J. A.,** Astrovirus within human small intestinal mucosa, *Gut*, 23, A923, 1982.
68. **Schreiber, D. S., Blacklow, N. R., and Trier, J. S.,** The mucosal lesion of the proximal small intestine in acute infectious nonbacterial gastroenteritis, *N. Engl. J. Med*, 288, 1318, 1973.
69. **Flynn, W. T., Saif, L. J., and Moorhead, P. D.,** Pathogenesis of a porcine enteric calicivirus in gnotobiotic pigs, *Am. J. Vet. Res.*, 49, 819, 1988.
70. **Hall, G. A., Bridger, J. C., Brooker, B. E., Parsons, K. R., and Ormerod, E.,** Lesions of gnotobiotic calves experimentally infected with a calici-like (Newbury) agent, *Vet. Pathol.*, 21, 208, 215.

71. **Adams, W. R. and Kraft, L. M.,** Electron-microscopic study of the intestinal epithelium of mice infected with the agent of epizootic diarrhea of infant mice (EDIM virus), *Am. J. Pathol.*, 51, 39, 1967.
72. **Wilsnack, R. E., Blackwell, J. H., and Parker, J. C.,** Identification of an agent of epizootic diarrhoea of infant mice by immunofluorescent and complement-fixation tests, *Am. J. Vet. Res.*, 30, 1195, 1969.
73. **Mebus, C. A. and Newman, L. E.,** Scanning electron, light and immunofluorescent microscopy of intestine of gnotobiotic calf infected with reovirus-like agent, *Am. J. Vet. Res.*, 38, 553, 1977.
74. **Snodgrass, D. R., Angus, K. W., and Gray, E. W.,** Rotavirus infection in lambs: pathogenesis and pathology, *Arch. Virol.*, 55, 263, 1977.
75. **Theil, K. W., Bohl, E. H., Cross, R. F., Kohler, E. M., and Agnes, A. G.,** Pathogenesis of porcine rotaviral infection in experimentally-inoculated gnotobiotic pigs, *Am. J. Vet. Res.*, 39, 213, 1978.
76. **Pearson, G. R. and McNulty, M. S.,** Pathological changes in the small intestine of neonatal pigs infected with a pig reovirus-like agent (rotavirus), *J. Comp. Pathol.*, 87, 363, 1977.
77. **Theil, K. W., Saif, L. J., Moorhead, P. D., and Whitmoyer, R. E.,** Porcine rotavirus-like virus (Group B rotavirus): characterization and pathogenicity for gnotobiotic pigs, *J. Clin. Microbiol.*, 21, 340, 1985.
78. **Mebus, C. A., Rhodes, M. B., and Underdahl, N. R.,** Neonatal calf diarrhoea caused by a virus that induces villous epithelial cell syncytia, *Am. J. Vet. Res.*, 39, 1223, 1978.
79. **Saif, L. J., Theil, K. W., and Redman, D. R.,** Detection and pathogenicity of a bovine rotavirus-like agent, Abstr. 98, Conf. Res. Workers in Animal Dis., Chicago, IL, November 8 to 9, 1982.
80. **Bohl, E. H., Saif, L. J., Theil, K. W., Agnes, A. G., and Cross, R. F.,** Porcine pararotavirus: detection, differentiation from rotavirus and pathogenesis in gnotobiotic pigs, *J. Clin. Microbiol.*, 25, 812, 1982.
81. **Shepherd, R. W., Butler, D. G., Cutz, E., Gall, D. G., and Hamilton, J. R.,** The mucosal lesion in viral enteritis: extent and dynamics of the epithelial response to virus invasion in transmissible gastroenteritis in piglets, *Gastroenterology*, 76, 770, 1979.
82. **Vonderfecht, S. L., Eiden, J. J., Torres, A., Miskuff, R. L., Mebus, C. A., and Yolken, R. H.,** Identification of a bovine enteric syncytial virus as a nongroup A rotavirus, *Am. J. Vet. Res.*, 47, 1913, 1986.
83. **Wolf, J. L., Kauffman, R. S., Finberg, R., Dambrauskas, R., Fields, B. N., and Trier, J. S.,** Determinants of reovirus interaction with the intestinal M cells and absorptive cells of murine intestine, *Gastroenterology*, 85, 291, 1983.
84. **Wolf, J. L., Cukor, G., Blacklow, N. R., Dambrauskas, R., and Trier, J. S.,** Susceptibility of mice to rotavirus infection: effects of age and administration of corticosteroids, *Infect. Immun.*, 33, 565, 1981.
85. **Moon, H. W.,** Mechanisms in the pathogenesis of diarrhea: a review, *J. Am. Vet. Med. Assoc.*, 172, 443, 1978.
86. **Lewis, L. D. and Philips, R. W.,** Pathophysiologic changes due to coronavirus induced diarrhea in a calf, *J. Am. Vet. Med. Assoc.*, 173, 636, 1978.
87. **Bachman, P. A. and Hess, R. G.,** Comparative aspects of pathogenesis and immunity in animals, in *Virus Infections of the Gastrointestinal Tract*, Tyrrell, D. A. J. and Kapikıan, A. Z., Eds., Marcel Dekker, New York, 1985, 361.
88. **Lipkin, M.,** Proliferation and differentiation of gastrointestinal cells, *Physiol. Rev.*, 53, 891, 1973.
89. **Moon, H. W.,** Epithelial cell migration in the alimentary mucosa of the suckling pig, *Proc. Soc. Exp. Biol. Med.*, 137, 151, 1971.
90. **Saif, L. J.,** Development of nasal, fecal and serum isotype-specific antibodies in calves challenged with bovine coronavirus or rotavirus, *Vet. Immunol. Immunopathol.*, 17, 425, 1987.
91. **Lecce, J. G., King, M. W., and Dorsey, W. E.,** Rearing regimen producing piglet diarrhea (rotavirus) and its relevance to acute infantile diarrhea, *Science*, 199, 776, 1978.
92. **Baintner, K. and Beress, B.,** Longitudinal differentiation of the small intestine, *Nature (London)*, 215, 774, 1967.
93. **Johnson, P. W. and Mucoplat, C. C.,** Immunologic abnormalities in calves with chronic bovine viral diarrhea, *Am. J. Vet. Res.*, 34, 1139, 1973.
94. **Edwards, S. J., Wood, L., Hewitt-Taylor, C., and Drew, T. W.,** Evidence for an imunocompromising effect of bovine pestivirus on bovid herpesvirus 1 vaccination, *Vet. Res. Commun.*, 10, 302, 1986.
95. **Theil, K. W., Saif, L. J., Bohl, E. H., Agnes, A. G., and Kohler, E. M.,** Concurrent porcine rotaviral and transmissible gastroenteritis viral infections in a three-day-old piglet, *J. Am. Vet. Med. Assoc.*, 40, 719, 1979.
96. **Saif, L. J., Bohl, E. H., Theil, K. W., Cross, R. F., and House, J. A.,** Rotavirus-like, calicivirus-like, and 23-nm virus-like particles associated with diarrhea in young pigs, *J. Clin. Microbiol.*, 12, 105, 1980.
97. **Saif, L. J., Saif, Y. M., and Theil, K. W.,** Detection and pathogenicity of enteric viruses recovered from diarrheic turkeys: role of a rotavirus-like agent, Proc. XVII World's Poultry Congr., Helsinki, Finland, August, 1984, 539.
98. **Evermann, J. F., Stann, S., DiGiacomo, R. F., Bergstrom, P. K., McKeirnan, A. J., and Giddens, W. E.,** Epizootiologic and diagnostic features of canine diarrheal disease in high and low risk dog populations, in *Proc. Am. Vet. Lab. Diag.*, 25, 229, 1982.

99. **Van Opdenbosch, E., Wellemans, G. and Oudewater, J.,** De synergische werking van BVD-corona-and rotavirus in het neonatale diarree komplex: emkele infektieproeven by pasgeboren kalveren, *Vlaams Dier. Tijd.,* 50, 163, 1981.
100. **Acres, S. D., Saunders, J. R., and Radostits, O. M.,** Acute undifferentiated neonatal diarrhea of beef calves: the prevalence of enterotoxigenic *E. coli,* reo-like (rota) virus and other enteropathogens in cow-calf herds, *Can. Vet. J.,* 8, 113, 1977.
101. **Marsolais, G., Assaf, R., Montpetit, C., and Marois, P.,** Diagnosis of viral agents associated with neonatal calf diarrhea, *Can. J. Comp. Med.,* 42, 168, 1978.
102. **Morin, M., Larviere, S., Lallian, R., Begin, M., Roy, R., and Ethin, R.,** Neonatal calf diarrhea, pathology and microbiology of spontaneous cases in dairy herds and incidence of enteropathogens implicated as aetiological agents, in *Proc. of the Second Symp. on Neonatal Diarrhea,* VIDO, Saskatoon, Canada, 1978, 347.
103. **Bridger, J. C.,** Detection by electron microscopy of caliciviruses, astroviruses and rotavirus-like particles in the faeces of piglets with diarrhoea, *Vet. Rec.,* 107, 532, 1980.
104. **McNulty, M. S., Todd, D., Allan, G. M., McFerran, J. B., and Greene, J. A.,** Epidemiology of rotavirus infection in broiler chickens: recognition of four serogroups, *Arch. Virol.,* 81, 113, 1984.
105. **Saif, L. J.,** Enteric viruses in diarrheic turkey poults, *Avian Dis.,* 29, 798, 1985.
106. **Tzipori, S.,** *Infectious Diarrhea in the Young: Strategies for Control in Humans and Animals,* Elsevier Sci. Publ. B.V., The Netherlands, 1985.
107. **Potgieter, L. N. D., McCracken, M. D., Hopkins, F. M., and Walker, R. D.,** Effect of bovine viral diarrhea virus infection on the distribution of infectious bovine rhinotracheitis virus in calves, *Am. J. Vet. Res.,* 45, 687, 1984.
108. **Bolivar, R., Conklin, R. H., Vollett, J. J., Pickering, L. K., Dupont, H. L., Walters, D. L., and Kohl, S.,** Rotavirus in Traveler's diarrhea: study of an adult student population in Mexico, *J. Infect. Dis.,* 137, 324, 1978.
109. **Saif, L. J., Redman, D. R., Smith, K. L., and Theil, K. W.,** Passive immunity to bovine rotavirus in newborn calves fed colostrum supplements from immunized or non-immunized cows, *Infect. Immun.,* 41, 1118, 1983.
110. **Snodgrass, D. R. and Wells, P. W.,** The immunoprophylaxis of rotavirus infections in lambs, *Vet. Rec.,* 102, 146, 1978.
111. **Mebus, C. A., White, R. G., Bass, E. P., and Twiehaus, M. J.,** Immunity to neonatal calf diarrhea virus, *JAVMA,* 163, 880, 1973.
112. **Meunier, P. C., Cooper, B. J., Appel, M. J. G., Lanieu, M. E., and Slauson, D. O.,** Pathogenesis of canine parvovirus enteritis: sequential virus distribution and passive immunization studies, *Vet. Pathol.,* 22, 617, 1985.
113. **Saif, L. J., Redman, D. R., Brock, K. V., Kohler, E. M., and Heckert, R. A.,** Winter dysentery in adult dairy cattle: detection of coronavirus in the feces, *Vet. Rec.,* 123, 300, 1988.
114. **Leece, J. G., Clare, D. A., Balsbaugh, R. K., and Collier, D. N.,** Effect of dietary regime on rotavirus *E. coli* weanling diarrhea of pigs, *J. Clin. Microbiol.,* 17, 689, 1983.
115. **Shorter, R. G., Moertel, C. G., Titus, J. L., and Reitemeier, R. G.,** Cell kinetics in the jejunum and rectum of man, *Am. J. Dig. Dis.,* 9, 760, 1964.
116. **Moon, H. W. and Joel, D. D.,** Epithelial cell migration in the small intestine of sheep and calves, *Am. J. Vet. Res.,* 36, 187, 1975.
117. **Halvorsrud, J. and Orstavik, I.,** An epidemic of rotavirus-associated gastroenteritis in a nursing home for the elderly, *Scand. J. Infect. Dis.,* 12, 161, 1980.
118. **Moon, H. W., Kemeny, L. J., Lambert, G. L., Stark, S. L., and Booth, G. D.,** Age-dependent resistence to transmissible gastroenteritis of swine. III. Effects of epithelial cell kinetics on coronavirus production and atrophy of intestinal villi, *Vet. Pathol.,* 12, 434, 1975.
119. **Witte, K. H. and Walther, C.,** Age-dependent susceptibility of pigs to infection with the virus of transmissible gastroenteritis, *Proc. 4th Int. Congr. Pig Vet. Soc.,* Iowa State University, Ames, 1976, K3.
120. **Sprinz, H.,** Factors influencing intestinal cell renewal. A statement of principles, *Cancer,* 28, 71, 1971.
121. **Tzipori, S., Chandler, D., Smith, M., Makin, T., and Hennessy D.,** Factors contributing to postweaning diarrhea in a large intensive piggery, *Aust. Vet. J.,* 56, 274, 1980.
122. **Hall, G. A., Parsons, K. R., and Bridger, J. C.,** Plasma enteroglucagon and neurotensin levels in gnotobiotic calves infected with enteropathogenic and non-enteropathogenic viruses, *Res. Vet. Sci.,* 38, 99, 1985.
123. **Kraft, L. M.,** Studies on the etiology and transmission of epidemic diarrhea of infant mice, *J. Exp. Med.,* 106, 743, 1957.
124. **Benfield, D. A., Stotz, I., Moore, R., and McAdaragh, J. P.,** Shedding of rotavirus in feces of sows before and after farrowing, *J. Clin. Microbiol.,* 16, 186, 1982.
125. **Crouch, C. F., Bielefeldt Ohmann, H., Watts, R. C., and Babiuk, L. A.,** Chronic shedding of bovine enteric coronavirus antigen-antibody complexes by clinically normal cows, *J. Gen. Virol.,* 66, 1489, 1985.
126. **Collins, J. K., Riegel, C. A., Olson, J. D., and Fountain, A.,** Shedding of enteric coronavirus in adult cattle, *Am. J. Vet. Res.,* 48, 361, 1987.

127. **Shimizu, M., Shimizu, Y., and Kodama, Y.,** Effects of ambient temperatures on induction of transmissible gastroenteritis in feeder pigs, *Infect. Immun.*, 21, 747, 1978.
128. **Shimizu, M. and Shimizu, Y.,** Effects of ambient temperatures on clinical and immune responses of pigs infected with transmissible gastroenteritis virus, *Vet. Microbiol.*, 4, 109, 1979.
129. **Saif, L. J.,** unpublished data, 1988.

III. Nonenveloped Enteropathogenic Viruses

Chapter 3

GROUP A ROTAVIRUSES

Kenneth W. Theil

TABLE OF CONTENTS

I. INTRODUCTION

Viral etiologies for enteric illness were long suspected, but practically all rotaviral infections associated with diarrhea eluded detection until the last two decades. Rotaviruses ordinarily do not propagate in cell cultures used for virus isolation and went unrecognized until specimens from affected animals were examined by other techniques. The report of rotaviral-induced diarrhea in experimentally inoculated colostrum-deprived and gnotobiotic calves,[1] though presaged by studies on the EDIM agent,[2] evoked an intensive search for similar viruses causing gastroenteritis in other species. Within a few years, viruses resembling the bovine rotavirus and EDIM agent were found in stools of diarrheic infants[3,4] and the antigenic relatedness of these three viruses was confirmed.[5,6] Rotaviruses have since been identified as a common cause of enteric illness in many mammalian and avian species, and they are now recognized as the single most important etiologic agent of acute gastroenteritis leading to hospitalization of infants and young children.[7]

Rotaviruses, which superficially resemble reoviruses and orbiviruses, were once called reovirus-like or orbivirus-like viruses, even though they were antigenically unrelated to either.[6,8-10] The name rotavirus (from the Latin *rota,* a wheel) was proposed for these newly recognized viruses because the smooth outer capsid layer surrounding the spoke-like capsomeres of the inner capsid gave the virions the appearance of diminutive wheels in negatively stained preparations.[5] Further characterization revealed that rotaviruses had segmented double-stranded RNA genomes,[11,12] an attribute of the Reoviridae.[13] The genus rotavirus was officially established within the Reoviridae family in 1978.[14] Initial studies demonstrated that rotaviruses from diverse mammalian hosts shared a common antigen,[5,6,10] but later some isolates without this antigen were recovered from pigs and chickens.[15-18] Accordingly, it was proposed that rotaviruses be separated into groups, with group members sharing their own distinctive common antigen[19,20] and more recent studies have upheld this recommendation.[21-23] Currently, five groups, designated A through E, are recognized. The originally recognized rotaviruses belong to group A as do most rotavirus isolates.

The lack of cell culture procedures for isolation and propagation of rotaviruses hindered early investigations, but two circumstances offset this impediment and permitted the steady acquisition of new information. First, rotaviruses, which are relatively large and possess a distinctive morphology, often are shed in great numbers; fecal extracts can contain over 10^9 particles per milliliter.[24] Therefore, rotaviruses were detected easily in many specimens by electron microscopic examination.[8,9,25] Second, several mammalian group A rotaviruses were fortuitously adapted to cell culture propagation[9,26,27] thereby providing convenient antigens for serologic assays to detect rotavirus antibodies.

Cell culture propagation of group A rotavirus isolates is now routine and powerful new techniques are available that enable the precise characterization of rotaviruses and their infections. These changes have brought an explosive increase in published data on rotaviruses, making it impossible to exhaustively survey this vast literature within one review. This review, perforce, selectively focuses on some of the recent findings. Regrettably, it is not possible to acknowledge here many who have made early, substantive contributions during the difficult, formative years of rotavirus research. A more comprehensive historical perspective on this research is available in past reviews.[28-35]

II. MORPHOLOGY AND MORPHOGENESIS

Rotaviruses have a common morphology and two particle types occur in negatively stained preparations.[8,11,36-46] One particle type, 65 to 75 nm in diameter, has two concentric capsid layers with the thin outer capsid conferring a smooth edge upon the virion. The outer capsid is permeable to negative stains and the inner capsid structure is readily apparent within the complete particles. Rotaviruses with both capsid layers are called complete, smooth, or double-shelled particles. Individual complete rotaviruses particles surrounded by membrane envelopes are rarely reported,[36] although large complexes of virions are often associated with membranes.[47,48] The smaller particle type, 55 to 60 nm in diameter, has only the inner capsid and is called incomplete, rough, or single-shelled. The inner capsid consists of symmetrically arranged ring- or doughnut-shaped morphological units forming the spoke-like projections characteristic of the virion. The inner capsid surrounds a core which, if penetrated by the negative stain, appears hexagonal and about 40 nm in diameter. Thin-shelled, hexagonal-shaped cores with smooth surfaces, 40 to 45 nm in diameter, can be formed by treating purified rotavirus with calcium chloride, sodium thiocyanate, or trypsin-versene.[42,49,50] These particle types separate in cesium chloride density gradients and complete virus, incomplete virus, and core particles have densities of 1.36, 1.38, and 1.44 g/ml, respectively.[40,50-52]

The precise arrangement and structure of the capsid layers, particularly the inner one, are difficult to assess, due to the confounding effects produced by superimposed two-sided images of negatively stained virions. Numerous conflicting interpretations resulted.[53-56] To overcome moire pattern artifacts caused by this superimposition, virion structure was determined by examination of single-sided images obtained with platinum-coated, freeze-dried rotavirus preparations.[58] Under these conditions the inner capsid appeared to consist of 132 capsomers distributed in a skewed pattern with a T = 13 arrangement around five- and sixfold axes. The outer capsid layer appeared as a smooth shell perforated with holes regularly organized around five- and sixfold axes of symmetry. Studies on chemically disrupted virions concur with these findings except that the five- and six-coordinated axes were "holes" formed by smaller wedge-shaped trimeric subunits distributed in a skewed pattern with a T = 13 arrangement.[58]

Besides virions, several types of tubular structures, most of which exhibit hexagonally arrayed subunits, are found in fecal and cell cultures specimens containing group A rotaviruses.[10,37,43,59-63] Most of these structures range from 50 to 80 nm in diameter, and they react with virus specific antisera.[43,59,60,63] Their function is unknown, but they may represent structures involved in virus morphogenesis, products of aberrant capsid assembly, or accumulations of surplus capsid material.

The number of group A rotaviral structural polypeptides is unclear at present. Early studies often gave conflicting results and it was suggested that the complete virion contained from five to eight polypeptides.[64-68] Despite this uncertainty, good evidence now exists for at least six distinct polypeptides (designated VP1, VP2, VP3, VP4, VP6, and VP7) in the complete virion.[69] The VP4, which has only recently been resolved from VP3 in polyacrylamide gels, was formally designated VP3.[69] A seventh structural polypeptide, VP9, with a molecular weight of 27,000 (27 kDa), has been described,[68] but is not universally detected.[66,70] Even though polypeptide variation among group A rotaviruses recovered from different species has been noted,[71] the molecular weights of these six simian rotavirus SA11 structural polypeptides listed in Table 1 can serve as a useful guide. For example, these molecular weight values agree well with those reported for the structural polypeptides of a human group A rotavirus, although nine structural polypeptides were found in this instance.[72]

Core particles contain VP1, VP2, and VP3.[50,69,73] The major core polypeptide, VP2, binds RNA in a nucleotide sequence-independent manner and has a higher affinity for single-stranded RNA than for double-stranded RNA.[74] This suggests it may be involved in an early step of RNA replication. Moreover, VP2 is myristylated and the myristic acid moiety is linked covalently through an amide group.[75] This implies that VP2 may be a scaffolding protein which initiates encapsidation by condensing around a set of rotavirus 11 single-stranded RNA segments. The functions of VP1 and VP3 are undefined, but they may act as part of the virion-associated RNA polymerase. Group A rotaviruses contain an endogenous RNA polymerase that is activated by heat shock or calcium chelation,[76-81] the latter process converting complete virions to incomplete

TABLE 1
Structural Polypeptides of Simian Rotavirus SA11

Virion polypeptide	Location	Molecular weight	Percentage of virion protein	Approximate molar ratio
VP1	Core	125,000	2	1
VP2	Core	94,000	15	10
VP3	Core	88,000	0.5	0.35
VP4	Outer capsid	88,000	1.5	1
VP6	Inner capsid	41,000	51	78
VP7	Outer capsid	34,000	30	55

Adapted from Lui, M., Offit, P. A., and Estes, M. K., *Virology,* 163, 26, 1988.

ones. Core particles are necessary, but may not be sufficient, for RNA polymerase activity[50] as those produced by a high concentration of calcium chloride lacked enzymatic activity unless reassociated with VP6.

The inner capsid surrounding the core particle consists of VP6, the major internal structural polypeptide of the virion.[67,69,82,83] The VP6 contains the subgroup antigen specificity.[84-86] Like VP2, VP6 is myristlyated signifying that it too could serve as a scaffolding polypeptide during virus assembly.[75]

The outer capsid contains two polypeptides, a major one, VP7, and a minor one, VP4.[68,87] The VP9, if present, probably constitutes a very minor outer capsid polypeptide.[68,87] Early on it was shown that complete, but not incomplete, rotavirus virions, were agglutinated by concanavalin A, a lectin with D-mannose or D-glucose specificities.[88] This suggested that the outer capsid contained either of these sugar residues in terminal nonreducing positions. This finding was verified by the demonstration that rotavirus propagation in the presence of tunicamycin, an inhibitor of the synthesis of glycopeptides containing N-glycosidically linked oligosaccharides, reduced virus yield and prevented the proper assembly of the outer capsid polypeptides.[89] Additional studies using tunicamycin, radiolabeled glucosamine and mannose, enzymatic cleavage with endo-β-*N*-acetylglucosaminidase H, and concanavalin A, demonstrated that VP7 was a glycopolypeptide containing N-linked high mannose oligosaccharides.[68,72,90,91] Further enzymatic cleavage studies revealed that whereas some oligosaccharide moieties are exposed on the virion surface, others are in cryptic positions and less accessible to the enzyme.[92,93] The VP7 mediates virus attachment to the cell surface receptors but its oligosaccharides are not involved in this interaction.[94,95] This function is lost if the conformation of VP7 is disrupted by reducing the disulfide bonds but is unaltered if glycosylation is blocked by tunicamycin.[94] On the other hand, the minor outer capsid polypeptide, VP4, is not glycosylated[68,72] and is the hemagglutinin.[96] Proteolytic cleavage of VP4 enhances virus infectivity and permits the virus to replicate in cell cultures.[97,98] Although cleavage of VP4 by proteolytic enzymes is essential for viral infectivity, this polypeptide does not bind to cell surface receptors.[95] It appears that once cell attachment is mediated by VP7, the protease-cleaved VP4 initiates the selective internalization of the inner capsid components into the cell, a process necessary for infection.[95,96] Moreover, a recent study using newborn mice orally inoculated with bovine and simian SA11 rotaviruses has implicated VP4 as a significant determinant of gastrointestinal virulence.[99] This finding is corroborated by another study demonstrating that the fourth genome segments, which code for VP4, of group A rotaviruses recovered from asymptomatic newborn infants, are closely related to each other but not to fourth genome segments of virulent rotaviruses.[100] Genetic and monoclonal antibody studies have established that both VP4 and VP7 elicit neutralizing antibody production,[84,85,101-105] and monoclonal antibodies to either VP4 or VP7 provide passive protection against rotavirus-induced diarrhea in suckling mice.[106]

Rotavirus morphogenesis has been extensively studied.[37,63,107-114] Several studies suggest that rotaviruses gain entrance into the cell by receptor-mediated endocytosis and then move to the lysosomes[110,115] where they are presumably uncoated and the endogenous RNA polymerase activated. More recent studies show, however, that rotaviruses grown in the absence of trypsin enter cells via endocytosis and are converted cores by the lysosomes, but they fail to replicate.[116] Alternatively, infectious rotaviruses, pretreated with trypsin, were rapidly uncoated upon entry into the cell membrane and converted to nucleoids. Moreover, lysosomotropic drugs known to inhibit virus entry into cells by the endocytotic pathway, and cytochalasin B, which suppresses endocytosis, have little effect on the yield of rotavirus grown in the presence of trypsin.[117] This data supports the hypothesis that rotaviruses enter the cell by direct penetration of the nucleoid through the cell membrane. Rotavirus attachment to cell receptor sites is mediated by the major outer capsid glycosylated polypeptide, VP7, but proteolytic cleavage of the minor outer capsid polypeptide, VP4, is required before the inner capsid components are internalized by the cell.[94,95] Following infection, dense, granular, nonmembranous cytoplasmic inclusions, or viroplasms, appear and these are frequently situated near regions of the rough endoplasmic reticulum with dilated cisternae. The periphery of the viroplasm is usually surrounded with particles resembling incomplete virions, which acquire an outer capsid layer and an envelope by budding through ribosome-free regions of the endoplasmic reticulum. The envelope is a transient structure and is lost shortly after the virion penetrates the dilated cisternae. Eventually, the complete virions are released, often in membranous complexes,[48] by cell lysis. Tubular or filamentous structures are often detected within the cytoplasm and, sometimes, the nuclei of infected cells.[37,63,108,110,112,114]

Electron microscopic immunoperoxidase methods have provided a more precise understanding of rotavirus morphogenesis.[118] Hyperimmune antiserum to simian rotavirus SA11 reacted strongly with virus particles, membranes of virus-associated cisternae, and viroplasms. A general reaction with the cytoplasm, but not the nuclei, of infected cells was also noted. Monospecific antiserum to VP7, the major outer capsid glycoprotein, reacted with the virus particles and the virus-associated endoplasmic reticulum, but not with viroplasms. Monospecific antisera to VP2, the major core particle polypeptide, and VP6, the major inner capsid polypeptide, reacted strongly with viroplasms. Both antisera gave generalized reactions with the cytoplasm of infected cells, and the anti-VP6 serum reacted with incomplete virus particles. From these findings it was inferred that the inner capsid polypeptides are synthesized throughout the cytoplasm and accumulate in the viroplasms, whereas the outer coat glycoprotein is synthesized in the rough endoplasmic reticulum. The virus then acquires VP7 while budding into the cisternae.

Additional studies with simian rotavirus SA11 using colloidal gold and monospecific or monoclonal antibodies also show that VP3, VP6, and VP7 are present in complete virus particles that had been deenveloped.[119] Moreover, antiserum to VP7 reacted strongly with intracytoplasmic and intranuclear tubular structures, whereas antiserum to VP6 reacted weakly with the intracytoplasmic tubules. Antiserum to the product of genome segment 11, thought to be the precursor of VP9, a suspected minor structural polypeptide, reacted with viroplasms. Antiserum to a 35 kDa nonstructural polypeptide, NS35, synthesized by genome segment 8, reacted strongly with viroplasms, and it was speculated that this polypeptide may be a component of the viral RNA replicase. Antiserum to a 29 kDa nonstructural glycoprotein, NS29, reacted with the cytoplasmic side of the endoplasmic reticulum membrane and incomplete, nonenveloped virus particles nearest the viroplasms. Virus particles with intact membranes reacted poorly with this antiserum, whereas particles deenveloped with Triton X-100 reacted well. It was hypothesized that NS29 situated within the endoplasmic reticulum acts as a receptor for newly assembled, incomplete virus particles emerging from the viroplasms, thereby providing the mechanism for the budding process. Location of NS29 within the endoplasmic reticulum membrane is consistent with previous observations that it contains high mannose oligosaccharides.[68] Other studies substantiate the contention that NS29 aids the assembly of complete virions. For example, a simian SA11 rotavirus variant with a nonglycosylated VP7 undergoes normal morphogenesis.[120] If, however, this variant is grown in the presence of tunicamycin large numbers of enveloped virus particles, but not unenveloped particles, accumulate within the cisternae, suggesting that a glycosylated NS29 is structurally important for outer capsid assembly and virus maturation, even though it is later removed along with the envelope.

More recent studies indicate that calcium is also crucial for rotavirus maturation.[121,122] Cells infected with bovine rotavirus and maintained in calcium-free medium produce little, if any infective virus.[121] The calcium-dependent period of virus replication occurs during the log phase, and without calcium only incomplete virus particles are produced. Reduced levels of VP7 were detected in infected cells maintained in calcium-deficient medium. The VP7 produced without calcium is glycosylated, but is preferentially degraded within the cell.[122] Calcium stabilizes VP7 and is incorporated into double, but not single, capsid particles. Thus, calcium appears necessary for the final stages of rotavirus assembly and maturation. These observations help explain earlier findings of rotaviruses instability in the presence of chelating agents that bind calcium.[123-125]

III. PHYSICOCHEMICAL PROPERTIES

Great numbers of group A rotavirus particles are excreted in stools of infected animals, and large quantities of stool may be passed from a single diarrheic case. Such massive shedding of virus into the environment presents a considerable obstacle to any attempt at controlling the spread of infection. In addition, attempts to disinfect contaminated environmental surfaces are likely to be inpeded by the fact that rotaviruses are shed into stools as large clumps of membrane-associated complexes.[49,126] Virus sequestered within such aggregates may be protected from inactivation. This is particularly relevant considering the recent findings that the rotavirus minimum infective dose for adult humans and newborn pigs is equivalent to the minimum infective dose detected in cell cultures.[127,128] Therefore, information on group A rotavirus stability is of considerable practical importance.

Most studies demonstrate that group A rotaviruses are stable viruses. For instance, a human group A rotavirus in a 20% fecal suspension remained completely infective over a 7-month-period when maintained at 4°C or room temperature in the presence of magnesium or calcium ions.[129] However, it was much less stable at 37°C, as all infectivity was lost by 2 months; loss of infectivity coincided with loss of the outer capsid layer, a finding that is consistent with a previous observation that infectivity is associated with complete virions.[51] As noted previously, rotaviruses are unstable in the presence of chelating agents that bind calcium.[123-125]

Human group A rotavirus in feces on a nonporous, inanimate surface held at 20°C remained infective for 9 d or longer, depending upon the relative humidity;[128] at 4°C and high relative humidity, it remained infective through 45 d. In another study, infective human rotavirus persisted on a variety of surfaces for extended periods, but it did not survive long on paper currency.[130] Human group A rotavirus in fecal suspension remains infective for over 4 h on human fingerpads,[131] and infective group A rotavirus was isolated from the hands of a worker, and several environmental surfaces, in a day care center housing 6- to 18-month-old children during a nonoutbreak period.[132] Rotaviruses survive in aerosols for long periods, depending on the relative humidity and temperature,[133-135] and infective simian rotavirus SA11 survived 9 d in aerosols held at 20°C and midrange humidity.[134] Prolonged stability of rotavirus in aerosols is particularly interesting as early studies implied that airborne spread of murine rotavirus (EDIM agent) was likely.[2] Furthermore, some epidemics of human rotaviral gastroenteritis, characterized by high attack rates and rapid spread through populations with limited personal contact, also suggest respiratory transmission of the virus.[135,136]

Early, limited studies disclosed that commonly used disinfectants often were unable to completely inactivate rotavirus in suspensions containing fecal or organic matter.[137-139] Ethanol was found to be an effective disinfectant,[138] which is consistent with its observed ability to remove the outer capsid from the virion.[140] Under one fourth of the 76 commercial or noncommercial disinfectant formulations tested could completely inactivate >6 $\log_{10}$ of infectious human rotavirus suspended in the presence of fecal matter or tryptose phosphate broth.[141] Disinfection of human rotavirus from contaminated, nonporous inanimate surfaces is even more difficult, and formulations effective against virus in suspension are not always effective against virus on contaminated surfaces.[142] These findings, that many frequently used chemical disinfectants and antiseptics are unsuitable for rotavirus disinfection, are disconcerting and argue that judicious selection of these products is mandatory to prevent rotavirus transmission.

Group A rotaviruses are stable at pHs between 3 and 7,[27,123,124,128,140,143-148] but are unstable at pHs below 3.[123,128,146,148] This instability at low pH becomes pronounced at 37°C.[148] Rotavirus infectivity is also destroyed by gastric juice at low pH.[128,148] Rotavirus instability at low pH and in gastric juice is of practical significance and must be overcome to improve the efficacy of orally administered vaccines.[146] The seroconversion rate was increased in Finnish infants given milk formula to neutralize gastric acidity before oral vaccination with RIT 4237, a bovine rotavirus.[146]

One important environment in which enteric viruses may survive is wastewater or wastewater sludge. Rotavirus does not absorb well to aluminum hydroxide or activated sludge flocs,[149] thus, treatment processes effective in removing enteroviruses from wastewater cannot be relied upon to remove rotavirus. Experimental studies have also revealed that simian rotavirus SA11 is relatively stable in fresh and estaurine water[150] and rotavirus has been detected in domestic sewage.[151,152] As predicted, in contrast to enteroviruses, which were found in reduced levels in treated sewage, rotavirus concentration in treated sewage was essentially equal to the input titer detected in the raw sewage.[152] Moreover, at least one human rotavirus gastroenteritis outbreak has been associated with the consumption of inadequately chlorinated water containing coliform bacteria.[153] Because of these findings, and earlier ones demonstrating rotavirus resistance to a disinfectant containing chlorine,[137] there is considerable public health interest in determining if rotaviruses can be spread through the aquatic environment and if they are inactivated by common procedures used to treat or disinfect water. Results of several studies indicate that rotaviruses are rapidly inactivated by chlorine and therefore should pose no problem in the disinfection process used in water supply operations.[154,155]

Rotaviruses are resistant to inactivation by ether,[27,124,140,144,145,147] but their resistance to inactivation by chloroform is variable.[27,124,140,144,147] Rotaviruses are labile at 50°C in the presence of magnesium chloride,[124,144,147] but are stable in the presence of magnesium sulfate.[124]

IV. GENOME

The rotavirus genome consists of 11 discrete double-stranded RNA segments of various molecular weights.[11,12] These segments separate upon electrophoresis in polyacrylamide gels and are distributed among four size classes (Figure 1).[11,12,156-158] The four largest segments (segments 1 through 4) comprise the first class, two slightly smaller segments (segments 5 and 6) comprise the second class, a triplet of still smaller, closely migrating segments (segments 7, 8, and 9) make up the third class, and the fourth class consists of the two smallest segments (segments 10 and 11). The first molecular weight estimates for these segments were obtained by coelectrophoresis with reovirus double-stranded RNA.[11,12,159-162] These studies gave approximate molecular weights ranging from about 2.1×10^6 for the largest segment to 0.2×10^6 for the smallest, and an estimated total rotavirus genome molecular weight of about 11×10^6. However, the five smallest rotavirus genome segments migrate faster than the smallest reovirus genome segment, necessitating estimation of their molecular weights by extrapolation, a procedure that

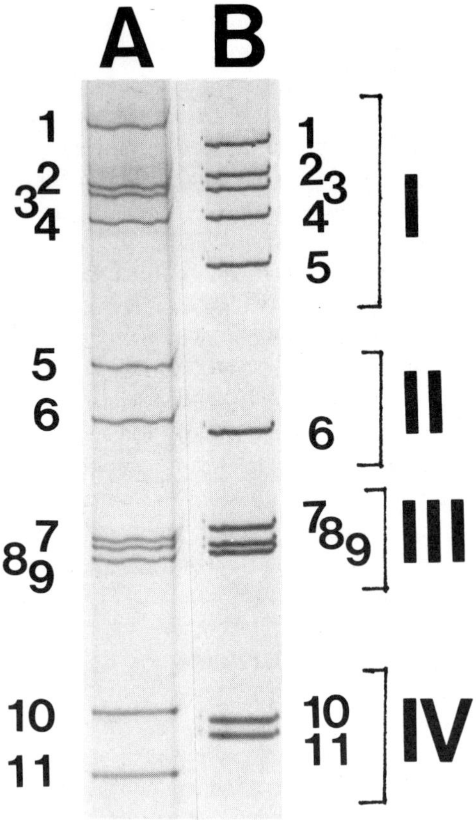

FIGURE 1. Genome electropherotypes of mammalian and avian group A rotaviruses in the same polyacrylamide gel slab. Migration is from top to bottom. Roman numerals on the right indicate the double-stranded RNA size classes. Numbers to the left designate segments of the mammalian group A rotavirus genome whereas the numbers to the right designate the segments of the avian group A rotavirus genome. Lane A: bovine group A rotavirus genome electropherotype; Lane B: turkey group A rotavirus genome electropherotype.

entails some risk. More recently, bovine rotavirus genome segment sizes were determined by electron microscopy using the relaxed replicative form of øX174 double-stranded DNA as an internal standard.[163] These estimates ranged from 2.25×10^6 to 0.42×10^6; the estimated molecular weight of the entire genome was 12.35×10^6. Molecular weight values obtained in this manner for segments 7, 8, 9, and 11 agreed closely (within 4%) with those predicted for these segments based on sequencing data.

Each rotavirus genome segment is base-paired its full length, lacks a 3′ terminal polyadenylate sequence,[164] and contains a capped 5′ structure, presumably on the plus strand.[81,164] Moreover, both ends of a segment contain a highly conserved region of 8 nucleotides.[164,166] Immediately internal to this region of absolute conservation lies another region, approximately 40 nucleotides long, in which the G-residue positions are conserved;[166] it is presumed that all residue positions in this region are also conserved. The cap structure and conserved sequence regions of these segments may serve as recognition sites for the initiation of messenger RNA synthesis by the virion transcriptase, RNA synthesis by the replicase, or they may participate in some way in virus core assembly. Full length messenger RNA transcripts are synthesized from each rotavirus genome segment.[165,167] Rotavirus genome segment plus strands migrate faster than their complementary minus strands on agarose-urea gels.[168] In this respect, rotavirus more closely resembles the cytoplasmic polyhedrosis virus than the reovirus. The explanation for this strand migration difference is unknown, but since all plus strands exhibit this behavior a common structural feature is suggested.

Each genome segment codes for a single polypeptide,[69,70,82-85,87,169-171] except genome segment 9, which contains two in-phase initiation codons and is functionally bicistronic.[172] Genome segments 1 through 4 code for the polypeptides VP1, VP2, VP3, and VP4, respectively.[68,70,82] VP1, VP2, and VP3 are core particle structural polypeptides with molecular weights of approximately 125, 94, and 88 kDa, respectively. The function of VP1 is unknown. As previously noted, VP2 is a myristylated polypeptide that binds to RNA, and may be involved in an early viral replication step.[74,75] The function of VP3 is unclear at present, but evidence suggests that it is involved in viral RNA synthesis. A simian rotavirus SA11 temperature-sensitive mutant with a lesion in genome segment 3 produces reduced amounts of single-stranded RNA and no double-stranded RNA at the nonpermissive temperature.[173] VP4 is a minor outer capsid polypeptide with a molecular weight of approximately 88K.[68] It is the hemaglutinin,[96] and it must be cleaved by a proteolytic enzyme (i.e., trypsin) before the virion will become infective.[97,98] The VP4 also elicits the production of neutralizing antibody.[101-104] Segment 5 codes on a nonstructural polypeptide with a molecular weight of 53 kDa (NS53).[87] Segment 6 codes for the major structural polypeptide of the virion, VP6, which is located in the inner capsid and has a molecular weight of about 41 kDa.[68,87] The VP6 is myristylated[75] and contains the subgroup antigen specificity.[84-86] Segments 7, 8, and 9 have similar electrophoretic mobilities, and consequently the coding assignments given to these segments vary with the rotavirus isolate. Depending on the isolate, segment 7, 8, or 9 can code for VP7, the major outer capsid glycosylated polypeptide with a molecular weight of about 34 kDa.[84,85,169-171] The VP7 also contains serotype antigen specificity and can evoke the production of neutralizing antibody.[84,85,101-104] As noted above, genome segment 9 of the simian rotavirus SA11 genome is bicistronic and codes for two similar, though not identical, VP7 glycopolypeptides.[172] It has been proposed that these two VP7s, which have molecular weights of 37 and 35.3 kDa, be designated VP7(1) and VP7(2), respectively. It is not known whether these two VP7s are functionally distinct. The other two segments in the 7, 8, 9 triplet code for nonstructural polypeptides with molecular weights of about 34 and 35 kDa (NS35 and NS34).[170] The function of NS34 is not known, but NS35 occurs in the viroplasm of infected cells and may be part of the viral RNA replicase.[119] Genome segment 10 codes for the precursor of a glycosylated nonstructural polypeptide with a molecular weight of 29 kDa (NS29)[87] that appears to be structurally important for the final stages of rotavirus morphogenesis.[120] Genome segment 11 encodes for a 27 kDa polypeptide believed to be VP9, a minor outer capsid component,[87] the function of which is unknown. In contrast, coding assignments determined for genome segments 10 and 11 of simian rotavirus SA11 in an earlier study[70] were inexplicably the reverse of those reported later.[87]

Polyacrylamide gel electrophoresis characterization of the group A rotavirus genome quickly lead to the recognition that isolates from different host species, and even different isolates from the same host species, were distinguishable by subtle variations in the migration rates of their genome segments.[155-159,161,174] Thus, the genome of each rotavirus strain produces a distinct migration pattern, called an electropherotype or genome electropherotype, by which that strain can be identified. These observations prompted the global application of the genome electropherotyping technique to epidemiologic studies at the molecular level on rotavirus infections. It was soon noted that some human group A rotaviruses, instead of having the expected genome electropherotypes, produced electropherotypes in which genome segments 10 and 11 apparently migrated at slower rates.[175] These isolates were said to possess “short” genome electropherotypes, and this “shortness” resulted from the genome segment 11 of these strains migrating slower than their normally migrating genome segment 10.[70] Although human group A rotaviruses with short genome electropherotypes occur worldwide, they are less frequently detected than isolates with long genome electropherotypes.[175-188] More recently, human group A rotaviruses with 10th genome segments possessing even slower migration rates than those in the short genome electropherotype have been isolated from diarrheic children in Indonesia.[189-191] These isolates were said to have "super-short" genome electropherotypes.

The genomes of group A rotaviruses from mammals other than man have not been studied extensively, but almost all are long genome electropherotypes.[12,156-161,174,192-204] So far, the only mammalian group A rotavirus with a short genome electropherotype were isolated from diarrheic rabbit in Japan.[205,206] Mammalian group A rotaviruses with supershort genome electropherotypes have been isolated from or detected in specimens from diarrheic rabbits[206,207] and calves[208-210] in the U.S. Lapine group A rotavirus short and supershort genome electropherotypes and a bovine group A rotavirus supershort genome electropherotype are due to the decreased migration of genome segment 11[206,208] similar to the situation previously noted with the human group A rotavirus short genome electropherotype.[70] Murine group A rotaviruses have genome electropherotypes in which the mobility of segment 11 is decreased so that it migrates near to segment 10.[192,211-213] Avian group A rotaviruses have genome electropherotypes resembling the murine group A rotavirus genome electropherotype, except that the avian group A rotavirus genome segment 5 also has decreased mobility and migrates near genome segment 4 (Figure 1).[214-218] Why short, supershort, and other distinct group A rotavirus genome electropherotypes exist and why some distinct genome electropherotypes are associated exclusively with certain host species (i.e., mice and birds) is not known, but in some instances these genome electropherotypes may indicate that the virus has undergone considerable adaptation to a specific host species.

Aside from the genome electropherotypes mentioned above, group A rotaviruses with unusual or grossly altered genome electropherotypes are occasionally detected.[219-224] In many cases, one of the smaller genome segments, 10 or 11, has greatly reduced mobility.[206,220,221,223,224] Group A rotaviruses recovered from immunodeficient children often exhibit considerable genomic heterogeneity with many extra segments occurring within the genome electropherotypes.[219,222] In one study, the additional larger genome segments had sequence homologies with smaller genome segments, indicating that they had arisen not from larger segments by deletions but from smaller ones by the formation of covalently linked concatemers.[219] Interestingly, rotavirus particles containing between 450 and 1,790 additional base pairs due to rearranged genomes were morphologically indistinguishable from normal virions,[222] although their density increased.[225] Thus, the rotavirus virion has considerable capacity to package additional genomic RNA, and the upper limit of this capacity remains to be determined. Surprisingly, despite substantial genomic rearrangements, many isolates synthesize normal, or nearly normal, polypeptides.[206,222-224] Because the rotavirus genome undergoes extensive rearrangement, and the virion can accommodate considerable extra genomic RNA, genome segment rearrangement may be a mechanism that influences the evolution of rotaviruses.[222]

Because of their segmented genomes, group A rotaviruses have the potential to form reassortants, which contain genome segments derived from each of two genetically distinct parental isolates, during mixed infections. Formation of such rotavirus reassortants *in vitro* has been well documented.[103,226-231] Efficient reassortant formation also occurs *in vivo* under experimental conditions, as mice simultaneously infected with simian rotavirus SA11 and rhesus rotavirus yield numerous reassortants.[232] These reassortants appear early in infection, suggesting that immune pressure does not induce their formation, and, interestingly, genome segments 3 and 5, which do not encode neutralizing antigens, segregated in a nonrandom manner. Moreover, emerging evidence indicates that group A rotaviruses may also undergo genetic reassortment in nature.[233-237]

Besides mutations, which occur frequently with RNA viruses, genome segment reassortment, and genome segment rearrangement, there is some recent evidence that group A rotaviruses might also undergo recombination. The major inner capsid polypeptide, VP6, of an equine group A rotavirus isolate (F1-14) bears both subgroup 1- and subgroup 2-specific eptiopes, instead of just one.[238] Moreover, the molecular weight of this VP6 (45 kDa) is greater than those of VP6 polypeptides (42 kDa) from other group A rotaviruses possessing only one subgroup specificity. It may be that the equine isolate obtained its additional subgroup specificity through recombination with the sixth genome segment of another isolate bearing the other subgroup-specific epitope. Obviously more research is needed to confirm this intriguing hypothesis.

V. ANTIGENIC RELATIONSHIPS

The antigenic complexity and diversity of group A rotaviruses is just becoming fully appreciated. Although it was initially documented that they shared a common antigen,[5,6,10] further antigenic analysis of most group A rotaviruses was difficult without suitable cell culture systems for their isolation and propagation. The first systematic attempt to antigenically characterize human and animal group A rotaviruses unexpectedly revealed that these viruses had at least two distinct antigenic specificities that were coded for by separate genes.[239] Another noteworthy observation was that animal group A rotaviruses shared subgroup antigen specificity with some human isolates, even though they were distinguishable by neutralization assays. Accordingly, it was proposed that group A rotavirus isolates be defined by at least two distinct specificities; serotype specificity designated rotaviral antigen reacting with neutralizing antibody, whereas subgroup specificity identified rotaviral antigen reacting with specific nonneutralizing antibody. Two subgroups, 1 and 2, were recognized. Subgroup specificity was associated with VP6, and most group A rotaviruses belonged to either subgroup 1 or subgroup 2;[86,239] some isolates, notably the murine and avian ones, did not possess either subgroup antigen.

Soon after subgroups were discovered, it was realized that human group A rotaviruses with subgroup 1 antigen had short genome electropherotypes, whereas those with subgroup 2 antigen had long genome electropherotypes.[240] This association of subgroup specificity with genome electropherotype was repeatedly verified with human group A rotaviruses immediately thereafter.[176,179,185,187,241,242] However, differences between subgroup 1 and 2 human group A rotaviruses extend beyond those in genome segments 10 and 11, as genetic hybridization studies show the two groups differ significantly in all genome segments.[243-245] Based on these genetic differences, it was proposed that at least two human group A rotavirus "families" exist, one including isolates with short genome electropherotypes and another including those with long genome electropherotypes. Human group A rotavirus isolates not genetically related to members of either family were likewise detected, implying the existence of more families.[245] The possibility of such additional families, or subgroups, was also suspected because some human and animal group A rotavirus isolates possess neither subgroup 1 nor subgroup 2 antigen.[86,239,246-249] Moreover, analysis of the reactivity patterns of human and animal group A rotavirus isolates with monclonal antibodies prepared against equine group A rotavirus (F1-14) VP6, which bears both subgroup specificities, further suggests that human group A rotaviruses may have two ancestral lineages:[238] one (subgroup II) with porcine-human lineage, and the other (subgroup I) with bovine-simian-human lineage. One of these monoclonal antibodies reacted with all mammalian group A rotaviruses tested, demonstrating a common VP6 epitope present in all mammalian group A rotaviruses. However, this monoclonal antibody, along with the others tested, failed to react with avian group A rotaviruses, indicating that these viruses have unique VP6 epitopes. Of further interest, genetic hybridization studies also provided evidence that gene reassortment between members of the two human group A rotavirus families probably occurs in nature, as some isolates with long genome electropherotypes partially cross-hybridized with RNA probes prepared from short genome electropherotype.[245]

Why human group A rotavirus short and long genome electropherotypes are associated with subgroup 1 and 2 specificities is difficult to understand, since the subgroup antigen is encoded for not by genome segment 10 or 11, but by genome segment 6. Recently, a human group A rotavirus with a long genome electropherotype and subgroup 1 specificity was generated *in vitro* by reassortment,[250] establishing that this association is not absolute. It was construed that isolates similar to this reassortant might occur in nature, and, indeed, human group A rotaviruses possessing novel combinations of subgroup specificities and genome electropherotypes are occasionally found in clinical specimens.[234,236,237]

Genome segment 6 nucleotide sequences of a subgroup 1 (simian rotavirus SA11) and a subgroup 2 (human rotavirus Wa) group A rotavirus are known. The inferred amino acid sequences of the two subgroup VP6 polypeptides are more than 90% conserved,[251] implying that variation within this major structural polypeptide responsible for subgroup specificity, is limited and probably constrained by the conformational requirements of virus assembly. Intriguing preliminary evidence supports the existence of an additional subgroup specific antigen in VP2, the major structural polypeptide of the virus core.[252]

Less is known about the subgroup specificities of animal group A rotaviruses, although most isolates have subgroup 1 antigen.[86,195,196,202,204,207,209,238,239] However, some porcine and one lapine group A rotavirus isolates have subgroup 2 antigen.[86,202,206,238] A correlation between genome electropherotype and subgroup specificity does not occur with animal group A rotaviruses. For examples, subgroup 1 and subgroup 2 porcine group A rotavirus isolates both have long genome electropherotypes,[86,197,238] whereas bovine group A rotavirus isolates with short and long genome electropherotypes have subgroup 1 antigen.[86,204,209,238,239] Some animal group A rotaviruses, mainly avian, equine, and murine isolates, cannot be subgrouped[86,238,239] suggesting that additional subgroups await definition.

The advent of routine cell culture procedures for group A rotavirus propagation enabled the development of a unified serotypic classification system for human and animal isolates. Initially, seven serotypes were recognized by plaque-reduction neutralization assays.[198] Serotypes 1 and 2 contain only human isolates, serotype 6 contains only bovine isolates, and serotype 7 contains only avian isolates. Serotype 3 encompasses the most diverse group of isolates and includes those from humans, vervet and rhesus monkeys, dogs,[253] cats,[254] pigs,[255] rabbits,[206] horses,[196] and mice.[213] Serotype 4 includes human and porcine isolates, whereas serotype 5 includes porcine and equine[195] isolates. Undoubtedly additional serotypes will be incorporated into this scheme as apparently new serotypic isolates from humans, and possibly bovines, have been recovered.[190,199,256,257]

This classification system clearly documented the shared serotypic specificity between human and animal group A rotaviruses as two serotypes (3 and 4) include isolates of human and animal origin. Most surprising, however, was the detection of an intertypic isolate, SB-1A from pigs, that was antigenically related to members of both serotypes 4 and 5 by reciprocal cross neutralization tests.[198] Shortly thereafter, another intertypic isolate, M37, which was antigenically related to both serotypes 1 and 4, was recovered from an asymptomatic neonate in Venezuela.[258] These confusing observations were ultimately reconciled by the demonstration that the SB-1A isolate has a VP4 antigenically related to serotype 5 and a VP7 antigenically related to serotype 4.[233] Similarly, M37 was shown to possess a VP4 antigenically related to serotype 4 and a VP7 antigenically related to serotype 1.[102,235] Because the neutralizing specificities present on the VP4 and VP7 can segregate independently, it is now recognized that any serotypic classification scheme describing the group A rotaviruses should be a binary system, similar to that employed for the influenza A viruses, in which the neutralization specificity of both VP4 and VP7 are indicated.

Group A rotavirus antigenic combinations involving subgroup and serotype specificities are quite complex. For example, serotype 3 contains an isolate with subgroup 1 specificity (simian rotavirus SA11), with subgroup 2 specificity (human rotavirus M), with undefined subgroup specificity (equine rotavirus H-2), and with both subgroup 1 and subgroup 2 specificities together (equine rotavirus F1-14).[238] Likewise, serotype 4 contains an isolate with subgroup 1 specificity (porcine rotavirus SB-2) with subgroup 2 specificity, (porcine rotavirus Gottfried), and with subgroup 2 specificity, and shared specificity with serotype 5 (porcine rotavirus SB-1A isolate). Undoubtedly, future studies will reveal additional combinations of antigenic specificities among group A rotavirus isolates.

It is extremely difficult to perform reliable cross protection studies with group A rotaviruses *in vivo*. Group A rotaviruses are ubiquitous, so animals are exposed to virus in the environment soon after birth. Moreover, these animals are likely to passively acquire antibodies against the virus before, or shortly after, birth via the dam's serum or colostrum. Therefore, *in vivo* cross protection studies are usually conducted with gnotobiotic animals, a costly, laborious process seldom done now that there are *in vitro* systems available. However, results of those group A rotavirus cross protection studies conducted in gnotobiotic animals generally confirm serotypic differences detected by *in vitro* methods.[197,256]

VI. ISOLATION AND CULTIVATION

Before routine serial propagation of group A rotavirus in cell culture was possible, isolation of these viruses associated with enteric illness was difficult and usually required the inoculation of gnotobiotic animals.[1,259,260] Early work with the EDIM agent (murine rotavirus) provides the single exception to this situation, and even so, elaborate precautions were required in this case to ensure reproducible results.[2] Moreover, initial studies involving human group A rotaviruses often included attempts to passage isolates in gnotobiotic animals to obtain sufficient virus and specific antiserum for additional investigations.[261-263] Because gnotobiotic animals, particularly calves and pigs, are costly and are not universally available, studies on most human and animal group A rotavirus isolates were severely limited.

Although several group A rotavirus isolates were adapted to serial propagation in conventional cell culture systems,[9,26,27,264-266] most isolates would not adapt to growth in cell culture. It is noteworthy that all but one of the isolates initially propagated in cell culture were from ruminants, mainly bovines, suggesting that isolates from these host species might be less fastidious in their growth requirements. A major impediment to group A rotaviruses investigations was eliminated by the discovery that pretreatment of the virus with proteolytic enzymes (i.e., trypsin), or the incorporation of proteolytic enzymes into the cell culture maintenance medium during virus infection, permitted the serial propagation of previously noncultivatable bovine and porcine isolates.[267,268] Subsequently, a gnotobiotic pig-passaged human group A rotavirus isolate was serially propagated in cell culture with the aid of trypsin treatment.[269] Soon thereafter, human group A rotaviruses were isolated directly from fecal specimens, without previous gnotobiotic animal passage, in embryonic rhesus monkey kidney cell line (MA104) monolayers with the aid of trypsin.[270,271] Now, use of MA104 cells, maintenance medium containing trypsin, and roller culture techniques are standard procedures for cell culture isolation of most group A rotaviruses. The use of trypsin treatments has enabled the routine cell culture isolation of avian,[216-218,272] bovine,[145,199,204,208,246,273] canine,[46,274,275] equine,[195,196,276,277] feline,[254] lapine,[147,205,207] murine,[212,213] ovine,[278] and porcine,[197,221,255,268,279,280] group A rotaviruses.

Group A rotaviruses form plaques in cell monolayers maintained under serum-free overlay medium containing trypsin,[281-283] but trypsin may also cause deterioration of the monolayer and obscure plaque formation. Fetal bovine serum, which is essentially devoid of antibody, cannot be included in the medium to maintain these monolayers since it contains anti-protease activity.[284] Chicken serum, which has little trypsin-inhibiting activity and presumably low levels of neutralizing antibody to mammalian rotaviruses, improves plaque formation by some mammalian group A rotaviruses when incorporated into the overlay medium.[285]

Although inclusion of trypsin into the cell culture medium is necessary for isolation of most group A rotaviruses, a few isolates are periodically recovered without the aid of trypsin. As noted above, several bovine group A rotaviruses were isolated without trypsin treatments. More recently, however, eight human group A rotavirus isolates from pediatric and adult patients with gastroenteritis in Italy were recovered in cell cultures without trypsin treatments.[286] Since these isolates possess cultural characteristics so remarkably different from those of other human group A rotaviruses, it will be of interest to know if they also differ in other significant biologic properties.

VII. PATHOGENESIS AND PATHOGENICITY

Although group A rotavirus induces diarrhea in conventional suckling mice, it is the ability of these viruses to infect and induce diarrhea in gnotobiotic calves,[1,256,287,291] lambs,[292-294] pigs,[259,295-299] and dogs[300-302] that provides the unequivocal evidence that they are primary etiologic agents of diarrhea in many species. Furthermore, human group A rotavirus isolates can infect and induce diarrhea in gnotobiotic calves, pigs and lambs.[261-263,303,304] In addition, group A rotaviruses have induced diarrhea or enteric illness in experimentally inoculated foals,[305] rabbits,[306] and turkeys.[307-309] Recent studies have also shown that human group A rotaviruses can infect orally inoculated volunteers and induce illness.[127,310]

Based on observations with gnotobiotic animals (which lack passive immunity and are therefore highly susceptible) inoculated with their homologous group A rotavirus, the following sequence of events occurs during infection. The virus infects the differentiated villous columnar epithelial cells (enterocytes) lining the small intestine and multiplies quickly. Virus replication rapidly leads to an extensive destruction and desquamation of enterocytes throughout the small intestine, and the villi become covered with immature, nondifferentiated cuboidal epithelium. Enterocyte loss may be so pronounced during the acute stage of infection that villi throughout the small intestine become temporarily shortened or even disappear (villous atrophy). Extensive enterocyte destruction causes an impairment of the normal digestive-absorptive processes occurring at the luminal surface of the villi, thereby inducing an acute, transitory, malabsorptive diarrhea. Epithelial cells within the crypts of the small intestinal mucosa then respond with increased mitotic activity in an attempt to quickly replace the depleted enterocyte population and a transitory crypt hyperplasia ensues. Diarrhea persists until sufficient villi within the small intestine are again covered with mature, differentiated enterocytes to permit resumption of the normal digestive-absorptive processes. Group A rotavirus infections in gnotobiotic animals are usually self-limiting and if infected gnotobiotes are allowed to survive, they usually return to normal by 7 to 10 d after infection.

Not all group A rotaviruses, however, are pathogenic for newborn colostrum-deprived or gnotobiotic animals, implying that there is likely to be considerable biologic variation among even those isolates recovered from the same host species. For example, an ovine group A rotavirus recovered from diarrheic lambs in Scotland induced diarrhea in experimentally inoculated day-old gnotobiotic lambs,[292] whereas another isolate from diarrheic lambs in Northern Ireland infected newborn colostrum-deprived lambs but did not induce diarrhea.[311] Several bovine group A rotavirus isolates have also been shown to be avirulent for gnotobiotic calves,[312] even though comparable amounts of both virulent and avirulent virus were excreted in the feces during infection. Of considerable interest was the finding that one avirulent bovine isolate, PP-1, was virulent for gnotobiotic pigs.[313] Antigenic analysis of this isolate suggests that it could be a natural hybrid between bovine and porcine group A rotaviruses. Obviously, more studies are needed to corroborate this interesting concept, or to affirm if this is another instance of a naturally occurring intertypic group A rotavirus.

Group A rotavirus infections of premature and full term newborn babies are often asymptomatic or mild.[314,315] Isolates from such cases, however, do not possess novel antigenic characteristics or belong exclusively to one serotype, as those recovered from asymptomatic neonates in four countries belong to one of the four previously recognized human group A rotavirus serotypes.[258] On the other hand, despite this serotypic polymorphism, genetic analysis of isolates recovered from asymptomatic neonates in nurseries indicates that their fourth genome segment is highly conserved compared to the fourth genome segment of community-derived isolates from symptomatic children.[100] This is an especially provocative finding in light of recent evidence implicating genome segment 4 as a significant determinant of virulence *in vivo*.[99] Group A rotaviruses from asymptomatic neonates may represent naturally occurring attentuated strains, and thus have potential as vaccine candidates.[258] Asymptomatic group A rotavirus infections occur commonly in pediatric patients,[316,317] but nothing is known about the strains associated with these cases as they are uncharacterized.

Several studies have attempted to determine if differences in clinical manifestations exist in infants infected with subgroup 1 or subgroup 2 group A rotaviruses. One study on Guatemalan children revealed that diarrhea and dehydration were more strongly associated with subgroup 2 rotavirus infections,[318] whereas another study, conducted in Venezuela, found no differences in fever or vomiting between infections with viruses of either subgroups, but duration of illness was longer in children infected with subgroup 2 group A rotaviruses.[176] A later investigation in Sweden revealed that fever with temperature exceeding 39°C was more frequently associated with subgroup 1 group A rotavirus infections.[184] Although diarrhea and vomiting occurred at similar rates in patients infected with either subgroup, clinical signs were more pronounced in those infected with subgroup 2 group A rotavirus. More recently, a study conducted in the Republic of South Africa, disclosed that fever and temperatures exceeding 39°C were more frequent in children infected with subgroup 2 group A rotaviruses, whereas vomiting was more pronounced in those infected with subgroup 1 isolates;[319] diarrhea, dehydration and duration of illness, however, were similar in infections with either subgroup. Based on the findings of these studies, it is difficult to conclude that infection with group A rotaviruses belonging a particular subgroup is consistently associated with specific clinical manifestations. However, it must be noted that these observations are based on only a limited number of investigations and more studies should be performed to substantiate these earlier findings.

In some instances, but not all, the age of the host at infection can determine if clinical illness results. Mice, for example, become resistant to infection with group A rotavirus as they approach weaning age.[320] This increased resistance parallels a reduction in virus-specific receptor sites on the enterocytes lining the small intestine.[321] Administration of cortisone acetate to suckling mice, which induces partial premature intestinal maturation, leads to a decreased susceptibility to group A rotavirus infection.[320] Thus, in mice, susceptibility to group rotavirus A rotavirus infection appears to be age associated and correlates with the presence of virus-specific receptor sites on the enterocytes. Experiments using ligated segments of the small intestine have also demonstrated an increased resistance of enterocytes to group A rotavirus infection in pigs.[322] By the time pigs are 2 d old, enterocytes in the upper small intestine become resistant to group A rotavirus infection, and this increased resistance correlated with the loss in ability of the pigs to transport macromolecules through the enterocytes. By 8 d, enterocytes in the lower small intestine are also resistant. Obviously, this resistance in pigs cannot be absolute, since group A rotavirus infections commonly occur in weaned pigs 3 weeks of age or older.[260,323] In contrast, a virulent bovine group A rotavirus isolate induced diarrhea in gnotobiotic calves up to at least 116 d of age.[312] Similarly, turkey group A rotavirus induced more severe clinical signs in birds 112 d of age than in young poults.[309]

VIII. EPIDEMIOLOGY

Group A rotaviruses are widely distributed in nature and a common cause of enteric illness in the young of many mammalian and avian species. Reports documenting the medical and veterinary significance of these viruses, and the clinical aspects of their infections, have been the subject of numerous previous reviews[28-35,324-327] and will not be reviewed here.

Although group A rotavirus infections primarily occur in the young, evidence has accumulated over recent years that older individuals may become infected.[328] Experimental studies with adult volunteers orally inoculated with human group A rotaviruses have clearly documented that these viruses are capable of infecting and inducing clinical illness in some individuals.[127,310] Group A rotavirus infections associated with gastroenteritis in adults have been detected in hospital outpatients and admissions,[329,330] in bone-marrow-transplant recipients undergoing cytotoxic chemotherapy,[331] in patients in geriatric wards or nursing homes,[332-334] in homosexual men with symptomatic human immunodeficiency virus infections,[335] in patients with travellers' diarrhea,[336,337] and members of isolated populations.[339,340] Many studies demonstrate that young adults often acquire group A rotavirus infections by close contact with pediatric cases,[341-347] but based on serologic data obtained in a recent study, it was concluded that group A rotaviruses are maintained in the community by chronically infected older individuals who periodically infect children under 10 years old who, in turn, spread the infection to young adults.[340] Spread of group A rotavirus from adults to newborns was also suggested by findings that sows shed virus around farrowing time, thereby assuring that the nursery environment is contaminated at a time when susceptible hosts will be available.[348]

Group A rotaviruses presumbly spread by the fecal-oral route, and numerous experimental studies with group A rotavirus infections have confirmed this.[292,293,296-313] Nonetheless, group A rotaviruses are relatively stable in aerosols,[134] and the patterns of several human group A rotaviruses gastroenteritis epidemics have suggested respiratory transmission of the virus.[135,136] If the virus is spread by aerosols, it apparently does not cause upper respiratory tract illness as patients in one study with both respiratory and enteric illness commonly had dual infections with group A rotavirus and a respiratory virus.[349]

Numerous studies have shown that under experimental conditions group A rotaviruses recovered from one host species can infect, and often induce diarrhea in, susceptible animals of another species.[261-263,303,304,350-361] Consequently, there is considerable interest in determining if cross-species infections occur in nature with group A rotaviruses. Genetic hybridization studies between group A rotaviruses of human and animal origin show that generally the genomes of human and animal isolates have a relatively low order of homology, whereas those of isolates from the same host species have a relatively high degree of homology.[194,362] These genetic sequence divergences among group A rotaviruses from different hosts argue that these viruses will likely have restricted host ranges. Moreover, one comprehensive epidemiologic study of Panamanian cattle ranchers provided data indicating that children, not farm animals, are the most important reservoir for intrafamilial rotavirus infections.[363] Thus to date, there is no convincing evidence to indicate that frequent cross-species transmission of group A rotavirus occurs in nature. On the other hand, the high incidence of group A rotavirus infections in colostrum-deprived neonates of many different exotic animal species reared in a zoo nursery provides suggestive evidence that under some circumstances cross-species infections occur.[364]

During the last decade, epidemiologic studies[175,177-188,200-202,204] using the genome electropherotyping technique have revealed that an extraordinary number of distinct group A rotavirus strains may be present, either simultaneously or sequentially, within the population under investigation. At the time it was not known if this extensive genomic diversity reflected antigenic diversity. More recently, the human group A rotavirus serotypes prevalent within several different populations over relatively short time periods were analyzed and in each study 3 or 4 serotypes were detected.[185-187,365] Although distinct serotypes of animal group A rotaviruses exist,[197-199,255,256] it remains to be determined if several serotypes are simultaneously prevalent within individual herds. The frequent circulation of several human group A rotavirus serotypes over short time intervals, however, does indicate that any effective vaccination program will likely have to provide protection against more than one serotype.

IX. IMMUNITY

Group A rotaviruses primarily, if not exclusively, infect the epithelium lining the small intestinal villi; thus immunity to this infection revolves around protecting the enterocyte. Since group A rotaviruses are stable, ubiquitous viruses, essentially all infections occur in hosts possessing some passively acquired immunity. Numerous studies in animals and humans have established that rotavirus neutralizing antibody within the small intestinal lumen at the time of exposure to virus provides protection from clinical illness or infection.[263,366-375] In nature, this antibody is provided by the dam via colostrum and milk while the newborn suckles. Unfortunately, group A rotavirus antibody titers decline markedly during the transition from colostrum to milk so that by a few days after birth the intestinal tract may no longer be protected from high doses of infective virus.[374-376] Whereas repeated natural exposures of the dam's intestinal tract to group A rotavirus induces production of neutralizing antibodies, the concentration of these antibodies normally secreted in the milk can seldom assure the offspring of protection for more than a few days.[377-379] Intramuscular vaccination of the dam with group A rotavirus during gestation induces higher antibody levels in the colostrum and milk for prolonged periods and thereby affords better protection against infection or clinical illness to those animals ingesting these secretions.[368,370-375,380] Recent experimental studies indicate that oral immunization of dams protected suckling mice against diarrhea only when the immunizing and challenge viruses were of the same serotype.[381] Thus, under natural conditions, passive immunity to group A rotavirus infection is often likely to be serotype-specific. While most of the passive protection within the gut to group A rotavirus infections stems from the antibodies present in the ingested milk, certain evidence suggests that some circulating passively-acquired antibodies, previously absorbed from the colostrum through the intestinal lining, may be transfered back into the intestinal lumen, thereby providing additional protection as milk antibody levels decline.[266,382] However, this contrasts with experimental findings in mice which demonstrated that circulating neutralizing antibodies of high titer did not protect against an oral challenge with simian group A rotavirus SA11.[383] Moreover, milk may contain other protective factors besides neutralizing antibodies; it has recently been shown that protection against group A rotavirus infections in breat-fed human neonates depends on the presence of high levels of both antibody and trypsin inhibitors in the milk.[384]

Cross-protection studies in gnotobiotic animals with group A rotaviruses have given conflicting results that await clarification. In most instances, gnotobiotic animals orally or intranasally inoculated with group A rotaviruses have active immunity at the surface of the small intestinal mucosa that confers serotype-specific protection.[197,256,385] However, serotype-specific neutralization is not always observed, as gnotobiotic calves exposed *in utero* to bovine group A rotavirus developed heterologous neutralizing antibody and were resistant shortly after birth to diarrheal disease induced by a serotypically unrelated human group A rotavirus.[386,387] Similarly, infection of gnotobiotic calves with an avirulent bovine rotavirus provided protection against a virulent isolate poorly related to it by neutralization tests.[388] These latter findings suggest that mechanisms other than neutralizing antibody may have a role in providing protective immunity at the mucosal surface.

How long protective active immunity lasts at the mucosal surface is unknown at present. In one study, neonates that had experienced an asymptomatic group A rotavirus infections, while not immune to reinfection, were protected against clinically severe disease during the next 3 years.[389] Serologic data from an epidemiologic study suggests that humans possessing serum antibodies to group A rotavirus have long lasting protection against clinically severe rotavirus infection.[390] Obviously, individuals with serum-neutralizing antibodies are not always protected from clinical disease, as volunteers with serum-neutralizing antibodies became infected and ill following oral exposure to human group A rotavirus.[127,310] In one study, serum antibody titers did not correlate with protection from infection or illness,[127] whereas in the other study prechallenge serum antibody did correlate with resistance to diarrhea and shedding of virus.[310] Since it is the antibody present at the surface of the enterocytes that provides protection, a better means of determining its concentration is needed. The recent finding that the concentration of IgA to group A rotavirus in the feces of young children has a high predictive accuracy for estimating duodenal IgA is of interest.[391] Such a test will probably be of great value in the future for determining the active immunity status of individuals.

The role each component of the immune system contributes to protection against group A rotavirus infections is unclear at present. Children with X-linked agammaglobulinemia, severe combined immunodeficiency, or cellular immunodeficiency may become chronically infected with group A rotaviruses.[219,222,392] Experimental infection of congenitally athymic suckling mice with murine group A rotavirus results in a typical, self-limiting infection indicating that T-lymphocytes are not required for recovery from a primary rotavirus infection.[393] On the other hand, group A rotaviruses are commonly detected in diarrheic stools collected from patients with symptomatic human immunodeficiency virus infections[335] suggesting that T-lymphocyte depletion prevents a normal termination of this infection. These observations may be reconciled by more recent findings that both serotype-specific and broadly cross-reactive cytotoxic T-lymphocytes are generated after primary rotavirus inoculation.[394] Thus, cytotoxic T-lymphocytes obtained after oral inoculation may be more important in providing heterotypic protection later during reinfection than in recovery from a primary infection with group A rotaviruses.

Obviously, many points concerning active and passive immunity to group A rotavirus infections in humans and animals require further elucidation. A better understanding of the factors inducing homotypic and heterotypic immunity to these infections will have immense practical value.

X. DIAGNOSIS

The vast majority of young mammals and birds possess passively acquired antibody when infected, so diagnosis of group A rotavirus infections is limited to detecting virus particles, virus antigens, or virus nucleic acids in specimens. By the time clinical signs are manifest, group A rotavirus replication within the small intestine has been sufficiently extensive to disrupt the normal digestion and absorption processes. At this time, the small intestinal mucosa contains numerous infected enterocytes and the intestinal contents and feces will contain large quantities of virus. However, as the disease progresses, infected enterocytes rapidly exfoliate and they, along with the virus, are lost with the diarrheal stool. Therefore, the optimal specimens for group A rotavirus diagnosis should be collected as near the onset of diarrhea as possible. This is not a great problem for veterinary cases, since the practitioner has some control over which affected animals to examine. In humans, however, many specimens are obtained from gastroenteritis cases of sufficiently long duration that medical attention is being sought for dehydration.

The specimens collected are largely determined by the host species involved. For humans and large, valuable animals, the specimens are mostly feces or rectal swabs. For smaller farm animals, such as pigs or birds, live animals are often submitted and intestinal sections and contents are collected. Some of the most commonly used diagnostic assays include direct and immune negative stain electron microscopic examination of fecal or intestinal extracts, immunofluorescent staining of small intestinal sections or smears, immunofluorescent staining of inoculated cell cultures, enzyme-linked immunosorbent assays of fecal extracts or rectal swabs, latex agglutination assays of fecal extracts, and detection of virus nucleic acid in feces by polyacrylamide gel electrophoresis. The choice of assays used is determined largely by the requirements of the laboratory which must take into account a number of factors including the equipment available, the number of specimens to be tested, the time required to conduct the test, the cost per test, and the sensitivity and specificity of the test.

Negative stain electron microscopic examination was one of the first methods used to detect group A rotavirus in fecal extracts.[3,4,8,25,259,274,290,311,323] This technique still remains among the most sensitive for detecting group A rotaviruses and has often been used as the gold standard with which to compare other assays. This test also has the advantage of permitting the visualization of other viruses that could be present in the specimen. However, it does not lend itself well to the examination of large numbers of specimens and it requires equipment not universally available. Further, it does not distinguish between group A rotaviruses and other rotaviruses. This latter deficiency can be overcome by reacting the virus with specific antibody prior to negative staining (immune electron microscopy).[43,260,264] Immune electron microscopic examination is more sensitive than direct negative stain electron microscopy for detecting virus,[43,396] but it also requires longer to perform.

Immunofluorescent staining of small intestinal sections and smears has often been used to detect group A rotavirus infection in animals.[260,287,288,305,395] This test requires special equipment and specific antisera. Its main disadvantages are that it requires the sacrifice of the animal, it doesn't lend itself the examination of numerous specimens, and it may not detect infected enterocytes in animals that are in the later stages of infection. In addition, procedures for immunofluorescent staining of inoculated cell cultures are available that permit the easy examination of numerous specimens, usually rectal swabs, in a relatively short time period.[397] Besides the advantage of not having to sacrifice the animal, this technique can detect a few infective virus particles that would be missed by other assays. Some specimens, however, may be cytotoxic and require testing at higher dilutions, thereby reducing sensitivity. Moreover, this test relies on infective virus, so it will be relatively insensitive for detecting virus in those specimens that contain a preponderance of particles already neutralized by milk antibodies.

Enzyme-linked immunosorbent assays (ELISAs) have been developed for the detection of group A rotaviruses in human and animal specimens,[318,398-400] and they are generally at least as sensitive as direct negative-staining electron microscopic examination.[396,398-400] These tests are relatively easy to perform, often don't require elaborate equipment, and can be adapted to the testing of numerous specimens in short periods of time. In fact, these tests have become the standard method in many laboratories throughout the world for detecting group A rotaviruses. Numerous ELISAs for the detection of group A rotaviruses in pediatric stools are commercially available.[401-404] Commercial ELISAs, though developed to detect human isolates, have also been used to detect group A rotaviruses in veterinary specimens.[395,405,406] These tests compared favorably with direct negative stain electron microscopic examination with bovine and equine specimens, but fared less well with porcine and turkey specimens. Lower sensitivity with turkey specimens might be explained by the findings that turkeys are frequently infected with group D rotaviruses.[215,327,407]

Latex agglutination assays were developed for the detection of group A rotaviruses in human stools; early studies indicated that these assays were as sensitive as direct negative stain electron microscopic examination of specimens[408] or slightly less sensitive than a commercial enzyme-linked immunosorbent assay.[409] These assays are rapid, simple to perform, require little equipment, and are well suited for the laboratory that only occasionally examines specimens for group A rotavirus. Latex agglutination assays for detecting group A rotaviruses in human stools are now commercially available and, though comparative tests have given conflicting results on their performance, they are highly specific and generally comparable or slightly less sensitive than direct negative stain electron microscopic examination or commercial enzyme-linked immunosorbent assays.[402,404,410-413] When used with veterinary specimens from calves, pigs, and turkeys, two commercial latex agglutination assays, though highly specific, were found to have low sensitivities.[406]

Although it lacks serologic specificity, detection of group A rotavirus nucleic acid by genome electropherotyping in polyacrylamide gels has been used recently with clinical specimens from humans and animals. This technique compares favorably with direct and immune negative stain electron microscopic examination of specimens[407,414] and enzyme-linked immunosorbent assays.[414-416] Besides detecting group A rotaviruses, this technique can provide additional epidemiologic information on the number of distinct strains that are prevalent in a population, and it can also detect non-group A rotaviruses in specimens. Because the genome electropherotyping technique lacks specificity, definitive classification of a virus detected by this assay must be done serologically with specific antisera.

XI. PREVENTION AND CONTROL

Group A rotaviruses are widespread pathogens that are shed into the environment in extraordinary numbers via the feces of infected humans and animals. They are quite stable and retain their infectivity for months in fecal suspensions at room temperature[129] and on nonporous surfaces at low temperature and high humidity.[128] Moreover, they survive on fingers for hours.[131] Commonly used disinfectants and antiseptics are frequently inefficacious in completely inactivating group A rotavirus infectivity.[137-142] As noted previously, the group A rotavirus minimum infective dose for humans and pigs is very low, equivalent to a single cell culture infective dose.[127,128] Apparently, chronically infected asymptomatic adults can periodically shed small quantities of group A rotavirus, thereby serving as reservoirs for susceptible younger individuals.[340,348] Because of these facts, eradication of group A rotaviruses will never be accomplished, and consequently, exposure to virus is inevitable. Thus, the practical objective becomes reducing exposure of young, susceptible hosts to virus and, if possible, enhancing their passive and active immunities so when infection does occur it will be less likely to result in clinical illness.

Aside from asymptomatic infections of human neonates, group A rotavirus infections are more severe in younger animals.[2,417] Relatively simple management procedures can be used to significantly reduce the spread of group A rotavirus infection among susceptible, young individuals. Experimental studies with pigs show that infection occurs at increasingly earlier ages as the level of group A rotavirus contamination increases within the environment;[417] the severity of clinical signs increased as the age at infection decreased. Thorough cleaning and fumigation of the pig nursery successfully interrupted this cycle of increasingly severe diarrhea. Similar observations are made under field conditions, as the cycle of group A rotavirus infection in calves was interrupted by cleaning and disinfecting the calf house.[418] Such management procedures also appear to be beneficial in the hospital setting. Disinfection and fumigation was reported to be the only way of eradicating group A rotavirus from the hospital nursery.[419] In addition, group A rotaviruses are among the agents identified in stools collected from diarrheic infants and toddlers in day-care centers.[420] Handwashing by children and staff in day-care centers after toilet activities and before eating reduced the incidence of diarrhea by half, suggesting that simple handwashing programs will probably prevent at least some diarrhea in the day-care center setting. Prevention or reduction of group A rotavirus infections among day-care center children would be significant, since many epidemiologic studies indicate that the young child is probably the index case as well as the principal agent of intrafamilial spread of virus.[341-347]

As previously mentioned, passive immunity to group A rotavirus infection in young animals depends predominantly upon the presence of neutralizing antibody within the gastrointestinal tract at the time of exposure to virus.[263,366-375] This antibody is provided by the dam to the nursing offspring through the colostrum and milk and is often called lactogenic immunity. Feeding animals and humans colostrum supplements containing neutralizing antibodies to group A rotavirus for extended periods can have a protective effect against clinically severe infections.[369,370,372] Enhancing and prolonging lactogenic immunity has been primarily of major interest to veterinary medicine as most farm animals are reared under intensive management conditions in which newborns are exposed to potential enteropathogens at an early age. Numerous studies conducted on enhancing this lactogenic immunity have been extensively reviewed elsewhere.[373-375,422] Although a modified-live bovine group A rotavirus vaccine is commercially available for intramuscular inoculation of pregnant cows, it did not significantly enhance antibody titers in the colostrum or milk of vaccinates.[423] In another study, intramuscular immunization of pregnant cows with an adjuvanted combined bovine group A rotavirus and enterotoxigenic *Escherichia coli* vaccine led to a significantly decreased incidence of diarrhea in calves born to vaccinates.[380] In a third study, a commercially available combined, inactivated, adjuvanted bovine group A rotavirus and *Escherichia coli* K99 vaccine given intramuscularly to pregnant cows resulted in a decreased incidence of group A rotavirus shedding and diarrhea in calves fed colostrum and milk from vaccinates.[424] A modified-live porcine group A rotavirus vaccine for oral or intramuscular inoculation of pregnant swine is also commercially available in the U.S. and reportedly enhances the lactogenic immunity provided to suckling pigs.[425]

Efficacious active oral immunization of nursing animals to group A rotaviruses is likely to be hindered by both antibodies in colostrum and milk and the simultaneous exposure of vaccinates to virulent group A rotavirus contaminating the environment. An attenuated, modified-live bovine group A rotavirus vaccine for oral immunization of calves is commercially available and its efficacy was established by protection studies in colostrum-deprived calves.[379] Its efficacy under field conditions, however, has been controversial.[426-430] Similarly, a modified-live porcine group A rotavirus vaccine for oral inoculation of suckling pigs is commercially available,[431] but the efficacy of such a vaccine could not be confirmed in one study.[432]

Group A rotavirus vaccines for the active immunization of infants and young children are not yet available commercially, although several candidate vaccines are in the preliminary stages of clinical evaluations for safety and efficacy. Many preliminary studies and strategies regarding these potential human vaccines have been reviewed in depth elsewhere.[433-436] One vaccine candidate, RIT 4237, is a derivative of the Lincoln isolate of bovine group A rotavirus used in a commercially available vaccine.[434] This vaccine gave high protection rates against clinically significant rotavirus diarrhea when given orally to Finnish infants,[434,437] but did not protect Gambian infants against severe group A rotavirus diarrhea.[438] Current evidence suggests that this vaccine provides protection against human group A rotavirus serotype 1.[437] Another candidate vaccine strain, rhesus group A rotavirus MMU 18006 belongs to serotype 3.[433] This virus is highly immunogenic,[439] but it also causes more side effects.[440] This vaccine gave 100% protection against severe rotavirus diarrheal episodes to orally vaccinated Venezuelan infants, and preliminary evidence suggests that this protection was against group A rotavirus serotype 3.[441] Although these initial trials indicate that these vaccines have promise, considerably more investigation is necessary before commercial vaccines will be available for widespread use.

The use of antiviral compounds against group A rotavirus infections has not been studied extensively. Two studies revealed that ribavirin had little effect on murine group A rotavirus infections in suckling mice.[211,442] The inactivity of ribavirin within the gut might be explained by the presence of guanosine inhibitors. Interestingly, clioquinol, an 8-hydroxyquinoline derivative and the active ingredient of Entero-Vioform (Ciba Pharmaceutical Co.) did significantly reduce murine group A rotavirus infection of suckling mice when administered at 12-h intervals after inoculation.[443] It was speculated that clioquinol may inactivate the virus by chelating divalent cations. Bismuth subsalicylate, which is effective in treating some infectious diarrheas, was found to have no direct effect on group A rotavirus viability *in vitro*.[444] On the other hand, the antiviral compound MDL 20,610 was active against group A rotaviruses *in vitro*.[445] It is unlikely that interferon will be effective in treating group A rotavirus infections *in vivo* as cell culture studies indicate that the virus is only weakly susceptible to its action.[446] Further studies on potential antiviral compounds for the treatment of group A rotavirus infections *in vivo* would be considerable interest.

REFERENCES

1. **Mebus, C. A., Underdahl, N. R., Rhodes, M. B., and Twiehaus, M. J.,** Calf diarrhea (scours) reproduced with a virus from a field outbreak, *Univ. Neb. Res. Bull.*, 233, 1, 1969.
2. **Kraft, L. M.,** Observations on the control and natural history of epidemic diarrhea of infant mice (EDIM), *Yale J. Biol. Med.*, 31, 121, 1958.
3. **Flewett, T. H., Bryden, A. S., and Davies, H.,** Virus particles in gastroenteritis, *Lancet,* 1, 1497, 1973.
4. **Bishop, R. F., Davidson, G. P., Holmes, I. H., and Ruck, B. J.,** Detection of a new virus by electron microscopy of faecal extracts from children with acute gastorenteritis, *Lancet,* 1, 149, 1974.
5. **Flewett, T. H., Bryden, A. S., Davies, H., Woode, G. N., Bridger, J. C., and Derrick, J. M.,** Relation between viruses from acute gastroenteritis of children and newborn calves, *Lancet,* 2, 61, 1974.
6. **Kapikian, A. Z., Kim, H. Y., Wyatt, R. G., Rodriguez, W. J., Ross, S., Cline, W. L., Parrot, R. H., and Chanock, R. M.,** Reoviruslike agent in stools, association with infantile diarrhea and development of serologic tests, *Science,* 185, 1049, 1974.
7. **Kapikian, A. Z., Wyatt, R. G., Greenberg, H. B., Kalica, A. R., Kim, H. W., Brandt, C. D., Rodriguez, W. J., Parrot, R. H., and Chanock, R. M.,** Approaches to immunization of infants and young children against gastroenteritis due to rotaviruses, *Rev. Infect. Dis.*, 2, 459, 1980.
8. **Flewett, T. H., Davies, H., Bryden, A. S., and Robertson, M. J.,** Diagnostic electron microscopy of faeces. II. Acute gastroenteritis associated with reovirus-like particles, *J. Clin. Pathol.*, 27, 608, 1974.
9. **Fernelius, A. L., Ritchie, A. E., Classick, L. G., Norman, J. O., and Mebus, C. A.,** Cell culture adaptation and propagation of a reovirus-like agent of calf diarrhea from a field outbreak in Nebraska, *Arch. Gesamte Virusforsch.*, 37, 114, 1972.
10. **Woode, G. N., Bridger, J. C., Jones, J. M., Flewett, T. H., Bryden, A. S., Davies, H. A., and White, G. B. B.,** Morphological and antigenic relationships between viruses (rotaviruses) from acute gastroenteritis of children, calves, piglets, mice, and foals, *Infect. Immun.*, 14, 804, 1978.
11. **Newman, J. F. E., Brown, F., Bridger, J. C., and Woode, G. N.,** Characterisation of a rotavirus, *Nature (London),* 258, 631, 1975.
12. **Rodger, S. M., Schnagl, R. D., and Holmes, I. H.,** Biochemical and biophysical characteristics of diarrhea viruses of human and calf origin, *J. Virol.*, 16, 1229, 1975.
13. **Joklik, W. K.,** The members of the family reoviridae, in *The Reoviridae,* Joklik, W. K., Ed., Plenum Press, New York, 1983, chap. 1.
14. **Matthews, R. E. F.,** The classification and nomenclature of viruses. Summary of results of meetings of the International Committee on Taxonomy of Viruses, The Hague, September, 1978, *Intervirology,* 11, 133, 1979.
15. **Bridger, J. C.,** Detection by electron microscopy of caliciviruses, astroviruses, and rotavirus-like particles in the faeces of piglets with diarrhoea, *Vet. Rec.*, 107, 532, 1980.
16. **McNulty, M. S., Allan, G. M., Todd, D., McFerran, J. B., and McCracken, R. M.,** Isolation from chickens of a rotavirus lacking the rotavirus group antigen, *J. Gen. Virol.*, 55, 405, 1981.
17. **Saif, L. J., Bohl, E. H., Theil, K. W., Cross, R. F., and House, J. A.,** Rotavirus-like, calicivirus-like, and 23-nm virus-like particles associated with diarrhea in young pigs, *J. Clin. Microbiol.*, 12, 105, 1980.
18. **Chasey, D., Bridger, J. C., and McCrae, M. A.,** A new type of atypical rotavirus in pigs, *Arch. Virol.*, 89, 235, 1986.
19. **Pedley, S., Bridger, J. C., Brown, J. F., and McCrae, M. A.,** Molecular characterization of rotaviruses with distinct group antigens, *J. Gen. Virol.*, 64, 2093, 1983.
20. **Pedley, S., Bridger, J. C., Chasey, D., and McCrae, M. A.,** Definition of two new groups of atypical rotaviruses, *J. Gen. Virol.*, 67, 131, 1986.
21. **Eiden, J., Vonderfecht, S., Theil, K., Torres-Medina, and Yolken, R. H.,** Genetic and antigenic relateness of human and animal strains of antigenically distinct rotaviruses, *J. Infect. Dis.*, 154, 972, 1986.
22. **Bremont, M., Cohen, J., and McCrae, M. A.,** Analysis of the structural polypeptides of a porcine group C rotavirus, *J. Virol.*, 62, 2183, 1988.

23. **Yolken, R. H., Arango-Jaramillo, S., Eiden, J., and Vonderfecht, S.,** Lack of genomic reassortment following infection of infant rats with group A and group B rotaviruses, *J. Infect. Dis.*, 158, 1120, 1988.
24. **Chrystie, I. L., Totterdell, B., Baker, M. J., Scopes, J. W., and Banatvala, J. E.,** Rotavirus infections in a maternity unit, *Lancet*, 2, 79, 1975.
25. **Flewett, T. H.,** Electron microscopy in the diagnosis of infectious diarrhea, *J. Am. Vet. Med. Assoc.*, 173, 538, 1978.
26. **Malherbe, H., Harwin, R., and Ulrich, M.,** The cytopathic effects of vervet monkey viruses, *S. African Med. J.*, 37, 407, 1963.
27. **Malherbe, H. H. and Strickland-Chomley, M.,** Simian virus SA11 and the related O agent, *Arch. Gesamte Virusforsch.*, 22, 238, 1967.
28. **Hodes, H. L.,** American Pediatric Society presidential address, *Ped. Res.*, 10, 201, 1976.
29. **McNulty, M. S.,** Rotaviruses, *J. Gen. Virol.*, 40, 1, 1978.
30. **Flewett, T. H. and Woode, G. N.,** The rotaviruses. Brief review, *Arch. Virol.*, 57, 1, 1978.
31. **Kapikian, A. Z., Yolken, R. H., Wyatt, R. G., Kalica, A. R., Chanock, R. M., and Kim, H. W.,** Viral diarrhea. Etiology and control, *Am. J. Clin. Nutr.*, 31, 2219, 1978.
32. **Wyatt, R. G., Kalica, A. R., Mebus, C. A., Kim, H. W., London, W. T., Chanock, R. M., and Kapikian, A. Z.,** Reovirus-like agents (rotaviruses) associated with diarrheal illness in animals and man, *Perspect. Virol.*, 10, 121, 1978.
33. **Holmes, I. H.,** Viral gastroenteritis, *Prog. Med. Virol.*, 25, 1, 1979.
34. **Blacklow, N. R. and Cukor, G.,** Viral gastroenteritis, *N. Eng.. J. Med.*, 304, 397, 1981.
35. **Estes, M. K., Palmer, E. L., and Obijeski, J. F.,** Rotaviruses: a review, *Curr. Top. Microbiol. Immunol.*, 105, 123, 1983.
36. **Els, H. J. and Lecatsas, G.,** Morphological studies on simian virus S. A. 11 and the "related" O agent, *J. Gen. Virol.*, 17, 129, 1972.
37. **Holmes, I. H., Ruck, B. J., Bishop, R. F., and Davidson, G. P.,** Infantile enteritis viruses: morphogenesis and morphology, *J. Virol.*, 16, 937, 1975.
38. **Martin, M. L., Palmer, E. L., and Middleton, P. J.,** Ultrastructure of infantile gastroenteritis virus, *Virology*, 68, 146, 1975.
39. **Rodger, S. M., Schnagl, R. D., and Holmes, I. H.,** Biochemical and biophysical characteristics of diarrhea viruses of human and calf origin, *J. Virol.*, 16, 1229, 1975.
40. **Bridger, J. C. and Woode, G. N.,** Characterization of two particle types of calf rotavirus, *J. Gen. Virol.*, 31, 245, 1976.
41. **Snodgrass, D. R., Smith, W., Gray, E. W., and Herring, J. A.,** A rotavirus in lambs with diarrhoea, *Res. Vet. Sci.*, 20, 113, 1976.
42. **Palmer, E. L., Martin, M. L., and Murphy, F. A.,** Morphology and stability of infantile gastroenteritis virus: comparison with reovirus and bluetongue virus, *J. Gen. Virol.*, 35, 403, 1977.
43. **Saif, L. J., Bohl, E. H., Kohler, E. M., and Hughes, J. H.,** Immune electron microscopy of transmissible gastroenteritis virus and rotavirus (reovirus-like agent of swine), *Am. J. Vet. Res.*, 38, 13, 1977.
44. **Todd, D. and McNulty, M. S.,** Biochemical studies on a reovirus-like agent (rotavirus) from lambs, *J. Virol.*, 21, 1215, 1977.
45. **Hoshino, Y., Baldwin, C. A., and Scott, F. W.,** Isolation and characterization of feline rotavirus, *J. Gen. Virol.*, 54, 312, 1981.
46. **Hoshino, Y., Wyatt, R. G., Scott, F. W., and Appel, M. J.,** Isolation and characterization of a canine rotavirus, *Arch. Virol.*, 72, 113, 1982.
47. **Soler, C., Musalem, C., Lorono, M., and Espejo, R. T.,** Association of viral particles and viral proteins with membranes in SA11-infected cells, *J. Virol.*, 44, 983, 1982.
48. **Williams, F. P.,** Membrane-associated viral complexes observed in stools and cell culture, *Appl. Environ. Microbiol.*, 50, 523, 1985.
49. **Almeida, J. D., Bradburne, A. F., and Wreghitt, T. G.,** The effect of sodium thiocyanate on virus structure, *J. Med. Virol.*, 4, 269, 1979.
50. **Bican, P., Cohen, J., Charpilienne, A., and Scherrer, R.,** Purification and characterization of bovine rotavirus cores, *J. Virol.*, 43, 1113, 1982.
51. **Elias, M. M.,** Separation and infectivity of two particle types of human rotavirus, *J. Gen. Virol.*, 37, 191, 1977.
52. **Rodger, S. M., Schnagel, R. D., and Holmes, I. A.,** Further biochemical characterization, including the detection of surface glycoproteins, of human, calf, and simian rotaviruses, *J. Virol.*, 24, 91, 1977.
53. **Martin, M. L., Palmer, E. L., and Middleton, P. J.,** Ultrastructure of infantile gastroenteritis virus, *Virology*, 68, 146, 1975.
54. **Kogasaka, R., Akihara, M., Horino, K., Chiba, S., and Nakao, T.,** A morphological study of human rotavirus, *Arch. Virol.*, 61, 41, 1979.
55. **Stannard, L. M. and Schoub, B. D.,** Observations on the morphology of two rotaviruses, *J. Gen. Virol.*, 37, 435, 1977.

56. **Esparza, J. and Gil, F.,** A study on the ultrastructure of human rotavirus, *Virology,* 91, 141, 1978.
57. **Roseto, A., Escaig, J., Delain, E., Cohen, J., and Scherrer, R.,** Structure of rotaviruses as studied by the freeze-drying technique, *Virology,* 98, 471, 1979.
58. **Ludert, J. E., Gil, F., Liprandi, F., and Esparza, J.,** The structure of the rotavirus inner capsid studied by electron microscopy of chemically disrupted particles, *J. Gen. Virol.,* 67, 1721, 1986.
59. **Kimura, T. and Murakami, T.,** Tubular structures associated with acute nonbacterial gastroenteritis in young children, *Infect. Immun.,* 17, 157, 1977.
60. **Kimura, T.,** Immuno-electron microscopic study on the antigenicity of tubular structures associated with human rotavirus, *Infect. Immun.,* 33, 611, 1981.
61. **Palmer, E. and Martin, M. L.,** Further observations on the ultrastructure of human rotaviruses, *J. Gen. Virol.,* 62, 105, 1982.
62. **Chasey, D. and Labram, J.,** Electron microscopy of tubular assemblies associated with naturally occurring bovine rotavirus, *J. Gen. Virol.,* 64, 863, 1983.
63. **Kimura, T. and Hase, A.,** Three different forms of tubular structures associated with the replication of bovine rotavirus in a tissue culture system, *Arch. Virol.,* 92, 165, 1987.
64. **Kalica, A. R. and Theodore, T. S.,** Polypeptides of simian rotavirus (SA-11) determined by a continuous polyacrylamide gel electrophoresis method, *J. Gen. Virol.,* 43, 463, 1979.
65. **Matsuno, S. and Mukoyama, A.,** Polypeptides of bovine rotavirus, *J. Gen. Virol.,* 43, 309, 1979.
66. **Espejo, R., Lopez, S., and Arias, C.,** Structural polypeptides of simiam rotavirus SA11 and the effect of trypsin, *J. Virol.,* 37, 156, 1981.
67. **Novo, E. and Esparza, J.,** Composition and topography of structural polypeptides of bovine rotavirus, *J. Gen. Virol.,* 56, 325, 1981.
68. **Ericson, B. L., Graham, D., Y., Mason, B. B., and Estes, M. K.,** Identification, syntheis, and modifications of simian rotavirus SA11 polypeptides in infected cells, *J. Virol.,* 42, 825, 1982.
69. **Lui, M., Offit, P. A., and Estes, M. K.,** Identification of the simian rotavirus SA11 genome segment 3 product, *Virology,* 163, 26, 1988.
70. **Dyall-Smith, M. and Holmes, I. H.,** Gene coding assignments of the rotavirus double-stranded RNA segments 10 and 11, *J. Virol.,* 38, 1099, 1981.
71. **Thouless, M. E.,** Rotavirus polypeptides, *J. Gen. Virol.,* 44, 187, 1977.
72. **Svennson, L.,** Identification of an outer capsid glycoprotein of human rotavirus by concanavalin A, *J. Gen. Virol.,* 65, 2183, 1984.
73. **Helmberger-Jones, M. and Patton, J. T.,** Characterization of subviral particles in cells infected with simian rotavirus SA11, *Virology,* 155, 655, 1986.
74. **Boyle, J. F. and Holmes, K. V.,** RNA-binding proteins of bovine rotavirus, *J. Virol.,* 58, 561, 1986.
75. **Clark, B. and Desselberger, U.,** Myristylation of rotavirus proteins, *J. Gen. Virol.,* 69, 2681, 1988.
76. **Cohen, J.,** Ribonucleic acid polymerase activity associated with purified calf rotavirus, *J. Gen. Virol.,* 36, 395, 1977.
77. **Hruska, J. F., Notter, M. F. D., Menegus, M. A., and Steinhoff, M. C.,** RNA polymerase associated with human rotaviruses in diarrhea stools, *J. Virol.,* 26, 544, 1978.
78. **Cohen, J., Laporte, J., Charpilienne, A., and Scherrer, R.,** Activation of rotavirus RNA polymerase by calcium chelation, *Arch. Virol.,* 60, 177, 1979.
79. **Bernstein, J. M. and Hruska, J. F.,** Characterization of RNA polymerase products of Nebraska calf diarrhea virus and SA11 rotavirus, *J. Virol.,* 37, 1071, 1981.
80. **Spencer, E. and Aria, M. L.,** *In vitro* transcription catalyzed by heat-treated human rotavirus, *J. Virol.,* 40, 1, 1981.
81. **Spencer, E. and Garcia, B. I.,** Effect of S-adenosylmethionine on human rotavirus RNA synthesis, *J. Virol.,* 52, 188, 1984.
82. **Mason, B. B., Graham, D. Y., and Estes, M. K.,** *In vitro* transcription and translation of simian rotavirus SA11 gene products, *J. Virol.,* 33, 1111, 1980.
83. **Smith, M. L., Lazdins, I., and Holmes, I. H.,** Coding assignments of double-stranded RNA segments of SA 11 rotavirus established by *in vitro* translation, *J. Virol.,* 33, 976, 1980.
84. **Kalica, A. R., Greenberg, H. B., Wyatt, R. G., Flores, J., Sereno, M. M., Kapikian, A. Z., and Chanock, R. M.,** Genes of human (strain Wa) and bovine (UK) rotaviruses that code for neutralization and subgroup antigens, *Virology,* 112, 385, 1981.
85. **Greenberg, H. B., Flores, J., Kalica, A. R., Wyatt, R. G., and Jones, R.,** Gene coding assignments for growth restriction, neutralization and subgroup specificities of the W and DS-1 strains of rotavirus, *J. Gen. Virol.,* 64, 313, 1983.
86. **Greenberg, H., McAuliffe, V., Valdesus, J., Wyatt, R., Flores, J., Kalica, A., Hoshino, Y., and Singh, N.,** Serological analysis of the subgroup protein of rotavirus, using monoclonal antibodies, *Infect. Immun.,* 39, 91, 1983.

87. **Mason, B. B., Graham, D. Y., and Estes, M. K.,** Biochemical mapping of the simian rotavirus SA11 genome, *J. Virol.,* 46, 413, 1983.
88. **Cohen, J., Maget-Dana, R., Roche, A.-C., and Monsigny, M.,** Calf rotavirus: detection of outer capsid glycoproteins by lectins, *FEBS Lett.,* 87, 26, 1978.
89. **Sabara, M., Babiuk, L. A., Gilchrist, J., and Misra, V.,** Effect of tunicamycin on rotavirus assembly and infectivity, *J. Virol.,* 43, 1082, 1983.
90. **Aria, C. R., Lopez, S., and Espejo, R.,** Gene protein products of SA11 simian rotavirus genome, *J. Virol.,* 41, 42, 1982.
91. **Killen, H. M. and Dimmock, N. J.,** Identification of a neutralization-specific antigen of a calf rotavirus, *J. Gen. Virol.,* 62, 297, 1982.
92. **Kouvelos, K., Petric, M., and Middleton, P. J.,** Oligosaccharide composition of calf rotavirus, *J. Gen. Virol.,* 65, 1159, 1984.
93. **Kouvelos, K., Petric, M., and Middleton, P. J.,** Comparison of bovine, simian, and human rotavirus structural glycoproteins, *J. Gen. Virol.,* 65, 1211, 1984.
94. **Sabara, M., Gilchrist, J. E., Hudson, G. R., and Babiuk, L. A.,** Preliminary characterization of an epitope involved in neutralization and cell attachment that is located on the major bovine rotavirus gylcoprotein, *J. Virol.,* 53, 58, 1985.
95. **Fukuhara, N. Yoshie, O., Kitaoka, S., and Konno, T.,** Role of VP3 in human rotavirus internalization after target cell attachment via VP7, *J. Virol.,* 62, 2209, 1988.
96. **Matsuno, S. and Inouye, S.,** Purification of an outer capsid glycoprotein of neonatal calf diarrhea virus and preparation of its antisera, *Infect. Immun.,* 39, 155, 1983.
97. **Estes, M. K., Graham, D. Y., and Mason, B. B.,** Proteolytic enhancement of rotavirus infectivity: molecular mechanisms, *J. Virol.,* 39, 879, 1981.
98. **Kalica, A. R., Flores, J., and Greenberg, H. B.,** Identification of the rotaviral gene that codes for hemagglutination and protease-enhanced plaque formation, *Virology,* 125, 194, 1983.
99. **Offit, P. A., Blavat, G., Greenberg, H. B., and Clark, F. C.,** Molecular basis of rotavirus virulence: role of gene segment 4, *J. Virol.,* 57, 46, 1986.
100. **Flores, J., Midthun, K., Hoshino, Y., Green, K., Gorziglia, M., Kapikian, A. Z., and Chanock, R. M.,** Conservation of the fourth gene among rotaviruses recovered from asymptomatic newborn infants and its possible role in attenuation, *J. Virol.,* 60, 972, 1986.
101. **Greenberg, H. B., Valdesuso, J., Van Wyke, K., Midthun, K., Walsh, M., McAuliffe, V., Wyatt, R. G., Kalica, A. R., Flores, J., and Hoshino, Y.,** Production and preliminary characterization of monoclonal antibodies directed at two surface proteins of rhesus rotavirus, *J. Virol.,* 47, 267, 1983.
102. **Hoshino, Y., Sereno, M. M., Midthun, K., Flores, J., Kapikian, A. Z., and Chanock, R. M.,** Independent segregation of two antigenic specificities (VP3 and VP7) involved in neutralization of rotavirus infectivity, *Proc. Natl. Acad. Sci.,* 82, 8701, 1985.
103. **Offit, P., and Blavat, G.,** Identification of the two rotavirus genes determining neutralization specificities, *J. Virol.,* 57, 376, 1986.
104. **Shaw, R. D., Vo, P. T., Offit, P. A., Coulson, B. S., and Greenberg, H. B.,** Antigenic mapping of the surface proteins of rhesus rotavirus, *Virology,* 155, 434, 1986.
105. **Gorziglia, M., Aguirre, Y., Hoshino, Y., Esparza, J., Blumemtals, I., Askaa, J., Thompson, M., Glass, R. I., Kapikian, A. Z., and Chanock, R. M.,** VP7 serotype-specific glycoprotein of OSU porcine rotavirus: coding assignment and gene sequence, *J. Gen. Virol.,* 67, 2445, 1986.
106. **Offit, P. A., Shaw, R. D., and Greenberg, H. B.,** Passive protection against rotavirus-induced diarrhea by monoclonal antibodies to surface proteins VP3 and VP7, *J. Virol.,* 58, 700, 1986.
107. **McNulty, M. S., Curran, W. L., and McFerran, J. B.,** The morphogenesis of a cytopathic bovine rotavirus in Madin-Darby bovine kidney cells, *J. Gen. Virol.,* 33, 503, 1976.
108. **Saif, L. J., Theil, K. W., and Bohl, E. H.,** Morphogenesis of porcine rotavirus in porcine kidney cell cultures and intestinal epithelial cells, *J. Gen. Virol.,* 39, 205, 1978.
109. **Esparza, J., Gorziglia, M., Gil, F., and Romer, H.,** Multiplication of human rotavirus in cultured cells: an electron microscopic study, *J. Gen. Virol.,* 47, 461, 1980.
110. **Altenburg, B. C., Graham, D. Y., and Estes, M. K.,** Ultrastructural study of rotavirus replication in cultured cells, *J. Gen. Virol.,* 46, 75, 1980.
111. **Carpio, M. M., Babiuk, L. A., Misra, V., and Blumenthal, R. M.,** Bovine rotavirus-cell interactions: effect of virus infection on cellular integrity and macromolecular synthesis, *Virology,* 114, 86, 1981.
112. **Suzuki, H., Kutsuzawa, T., Konno, T., Ebina, T., and Ishida, N.,** Morphogenesis of human rotavirus type 2 Wa strain in MA104 cells, *Arch. Virol.,* 70, 33, 1981.
113. **Petrie, B. L., Graham, D. Y., and Estes, M. K.,** Identification of rotavirus particle types, *Intervirology,* 16, 20, 1981.

114. **Suzuki, H., Konno, T., Kitaoka, S., Sato, T., Ebina, T., and Ishida, N.,** Further observations on the morphogenesis of human rotavirus in MA 104 cells, *Arch. Virol.,* 79, 147, 1984.
115. **Quan, C. M. and Doane, F. W.,** Ultrastructural evidence for the cellular uptake of rotavirus by endocytosis, *Intervirology,* 20, 223, 1983.
116. **Suzuki, H., Kitaoka, S., Konno, T., Sato, T., and Ishida, N.,** Two modes of human rotavirus entry in MA 104 cells, *Arch. Virol.,* 85, 25, 1985.
117. **Fukuhara, N., Yoshie, O., Kitaoka, S., Konno, T., and Ishida, N.,** Evidence for endocytosis-independent infection by human rotavirus, *Arch. Virol.,* 97, 93, 1987.
118. **Petrie, B. L., Graham, D. Y., Hanssen, H., and Estes, M. K.,** Localization of rotavirus antigens in infected cells by ultrastructural immunocytochemistry, *J. Gen. Virol.,* 63, 457, 1982.
119. **Petrie, B. L., Greenberg, H. B., Graham, D. Y., and Estes, M. K.,** Ultrastructural localization of rotavirus antigens using colloidal gold, *Virus Res.,* 1, 133, 1984.
120. **Petrie, B. L., Estes, M. K., and Graham, D. Y.,** Effects of tunicamycin on rotavirus morphogenesis and infectivity, *J. Virol.,* 46, 270, 1983.
121. **Shahrabadi, M. S. and Lee, P. W. K.,** Bovine rotavirus maturation is a calcium-dependent process, *Virology,* 152, 298, 1986.
122. **Shahrabadi, M. S., Babiuk, L. A., and Lee, P. W. K.,** Further analysis of the role of calcium in rotavirus morphogenesis, *Virology,* 158, 103, 1987.
123. **Palmer, E. L., Martin, M. L., and Murphy, F. A.,** Morphology and stability of infantile gastroenteritis virus: comparison with reovirus and bluetongue virus, *J. Gen. Virol.,* 35, 403, 1977.
124. **Estes, M. K., Graham, D. Y., Smith, E. M., and Gerba, C. P.,** Rotavirus stability and inactivation, *J. Gen. Virol.,* 43, 403, 1979.
125. **Shirley, J. A., Beards, G. M., Thouless, M. E., and Flewett, T. H.,** The influence of divalent cations on the stability of human rotavirus, *Arch. Virol.,* 67, 1, 1981.
126. **Narang, H. K. and Codd, A. A.,** Frequency of preclumped virus in routine fecal specimens from patients with acute nonbacterial gastroenteritis, *J. Clin. Microbiol.,* 13, 982, 1981.
127. **Ward, R. L., Bernstein, D. I., Young, E. C., Sherwood, J. R., Knowlton, D. R., and Schiff, G. M.,** Human rotavirus studies in volunteers: determination of infectious dose and serological response to infection, *J. Infect. Dis.,* 154, 871, 1986.
128. **Graham, D. Y., Dufour, G. R., and Estes, M. K.,** Minimal infective dose of rotavirus, *Arch. Virol.,* 92, 261, 1987.
129. **Moe, K. and Shirley, J. A.,** The effects of relative humidity and temperature on the survival of human rotavirus in faeces, *Arch. Virol.,* 72, 179, 1982.
130. **Sattar, S. A, Lloyd-Evans, N., and Springthorpe, V. S.,** Institutional outbreaks of rotavirus diarrhoea: potential role of fomites and environmental surfaces as vehicles for virus transmission, *J. Hyg.,* 96, 277, 1986.
131. **Ansari, S. A., Sattar, S. A., Springthorpe, V. S., Wells, G. A., and Tostowaryk, W.,** Rotavirus survival on human hands and transfer of infectious virus to animate and nonporous inanimate surfaces, *J. Clin. Microbiol.,* 26, 1513, 1988.
132. **Keswick, B. H., Pickering, L. K., DuPont, H. L., and Woodward, W. E.,** Survival and detection of rotaviruses on environmental surfaces in day care centers, *Appl. Environ. Microbiol.,* 46, 813, 1983.
133. **Moe, K. and Harper, G. J.,** The effect of relative humidity and temperature on the survival of bovine rotavirus in aerosol, *Arch. Virol.,* 76, 211, 1983.
134. **Sattar, S. A., Ijaz, M. K., Johnson-Lussenburg, M., and Springthorpe, V. S.,** Effect of relative humidity on the airborne survival of rotavirus SA11, *Appl. Environ. Microbiol.,* 47, 879, 1984.
135. **Foster, S. O., Palmer, E. L., Gary, W. L., Martin, M. L., Herrmann, K. L., Beasley, P., and Sampson, J.,** Gastroenteritis due to rotavirus in an isolated Pacific Island Group: an epidemic of 3,439 cases, *J. Infect. Dis.,* 141, 32, 1980.
136. **Santosham, M. Yolken, R. H., Wyatt, R. G., Bertrando, R., Black, R. E., Spira, W. M., and Sack, B.,** Epidemiology of rotavirus diarrhea in a prospectively monitored American Indian population, *J. Infect. Dis.,* 152, 778, 1985.
137. **Snodgrass, D. R. and Herring, J. A.,** The action of disinfectants on lamb rotavirus, *Vet. Rec.,* 101, 81, 1977.
138. **Tan, J. A. and Schnagl, R. D.,** Inactivation of a rotavirus by disinfectants, *Med. J. Aust.,* 1, 19, 1981.
139. **Sattar, S. A., Raphael, R. A., Lochman, H., and Springthorpe, V. S.,** Rotavirus inactivation by chemical disinfectants and antiseptics used in hospitals, *Can. J. Microbiol.,* 29, 1464, 1983.
140. **Bishai, F. R., Blaskovic, P., and Goodwin, D.,** Physicochemical properties of Nebraska calf diarrhea virus hemagglutin, *Can. J. Microbiol.,* 24, 1425, 1978.
141. **Springthorpe, V. S., Grenier, J. L., Lloyd-Evans, N., and Sattar, S. A.,** Chemical disinfection of human rotaviruses: efficacy of commercially-available products in suspension tests, *J. Hyg.,* 97, 139, 1986.
142. **Lloyd-Evans, N., Springthorpe, V. S., and Sattar, S. A.,** Chemical disinfection of human rotavirus-contaminated inanimate surfaces, *J. Hyg.,* 97, 163, 1986.

143. **Welch, A. B.**, Purification, morphology and partial characterization of a reovirus-like agent associated with neonatal calf diarrhea, *Can. J. Comp. Med.*, 35, 195, 1971.
144. **Welch, A. B. and Thompson, T. L.**, Physicochemical characterization of a neonatal calf diarrhea virus, *Can. J. Comp. Med.*, 37, 295, 1972.
145. **Ihara, T., Samejima, T., Kuwahara, H., and Tajima, M.**, Isolation of new serotypes of bovine rotavirus, *Arch. Virol.*, 78, 145, 1983.
146. **Vesikari, T., Isolauri, E., D'Hondt, E., Delem, A., and Andre, F. E.**, Increased "take" rate of oral rotavirus vaccine in infants after milk feeding, *Lancet*, 2, 700, 1984.
147. **Castrucci, G., Ferrari, M., Frigeri, F., Cilli, V., Perucca, L., and Donelli, G.**, Isolation and characterization of cytopathic strains of rotavirus from rabbits, *Arch. Virol.*, 83, 99, 1985.
148. **Weiss, C. and Clark, H. F.**, Rapid inactivation of rotaviruses by exposure to acid buffer or acidic gastric juice, *J. Gen. Virol.*, 66, 2725, 1985.
149. **Farrah, S. R., Goyal, S. M., Gerba, C. P., Conklin, R. H., and Smith, E. M.**, Comparison between adsorption of poliovirus and rotavirus by aluminum hydroxide and activated sludge flocs, *Appl. Environ. Microbiol.*, 35, 360, 1978.
150. **Hurst, C. J. and Gerba, C. P.**, Stability of simian rotavirus in fresh and estuarine water, *Appl. Environ. Microbiol.*, 39, 1, 1980.
151. **Steinmann, J.**, Detection of rotavirus in sewage, *Appl. Environ. Microbiol.*, 41, 1043, 1981.
152. **Smith, E. M. and Gerba, C. P.**, Development of a method for detection of human rotavirus in water and sewage, *Appl. Environ. Microbiol.*, 43, 1440, 1982.
153. **Sutmoller, F., Azeredo, R. S., Lacerda, M. D., Barth, O. M., Pereira, H. G., Hoffer, E., and Schatzmayr, H. G.**, An outbreak of gastroenteritis caused by both rotavirus and Shigella sonnei in a private school in Rio de Janeiro, *J. Hyg.*, 88, 285, 1982.
154. **Berman, D. and Hoff, J. C.**, Inactivation of simian rotavirus SA11 by chlorine, chlorine dioxide, and monochloramine, *Appl. Environ. Microbiol.*, 48, 317, 1984.
155. **Vaughn, J. M., Chen, Y.-S., and Thomas, M. Z.**, Inactivation of human and simian rotaviruses by chlorine, *Appl. Environ. Microbiol.*, 51, 391, 1986.
156. **Kalica, A. R., Garon, C. F., Wyatt, R. G., Mebus, C. A., Van Kirk, D. H., Chanock, R. M., and Kapikian, A. Z.**, Differentiation of human and calf reoviruslike agents associated with diarrhea using polyacrylamide gel electrophoresis of RNA, *Virology*, 74, 86, 1976.
157. **Kalica, A. R., Sereno, M. M., Wyatt, R. G., Mebus, C. A., Chanock, R. M., and Kapikian, A. Z.**, Comparison of human and animal rotavirus strains by gel electrophoresis of viral RNA, *Virology*, 87, 247, 1978.
158. **Kalica, A. R., Wyatt, R. G., and Kapikian, A. Z.**, Detection of differences among human and animal rotaviruses, using analysis of viral RNA, *J. Am. Vet. Med. Assoc.*, 173, 531, 1978.
159. **Schnagl, R. D. and Holmes, I. H.**, Characteristics of the genome of human infantile enteritis virus (rotavirus), *J. Virol.*, 19, 267, 1976.
160. **Todd, D. and McNulty, M. S.**, Characterization of pig rotavirus RNA, *J. Gen. Virol.*, 33, 147, 1976.
161. **Verly, E. and Cohen J.**, Demonstration of size variation of RNA segments between different isolates of calf rotavirus, *J. Gen. Virol.*, 35, 583, 1977.
162. **Barnett, B. B., Egbert, L. N., and Spendlove, R. S.**, Characterisitics of neonatal calf diarrhea virus ribonucleic acid, *Can. J. Comp. Med.*, 42, 46, 1978.
163. **Rixon, F., Taylor, P., and Desselberger, U.**, Rotavirus RNA segments sized by electron microscopy, *J. Gen. Virol.*, 65, 233, 1984.
164. **Imai, M., Akatani, K., Ikegami, N., and Furuichi, Y.**, Capped and conserved terminal structures in human rotavirus genome double-stranded RNA segments, *J. Virol.*, 47, 125, 1983.
165. **McCrae, M. A. and McCorquodale, J. G.**, Molecular biology of rotaviruses. V. Terminal structure of viral RNA species, *Virology*, 126, 204, 1983.
166. **Clarke, I. N. and McCrae, M. A.**, The molecular biology of rotaviruses. VI. RNA species-specific terminal conservation in rotaviruses, *J. Gen. Virol.*, 64, 1877, 1983.
167. **Flores, J., Myslinski, J., Kalica, A. R., Greenberg, H. B., Wyatt, R. G., Kapikian, A. Z., and Chanock, R. M.**, *In vitro* transcription of two human rotaviruses, *J. Virol.*, 43, 1032, 1983.
168. **Patton, J. T. and Stacy-Phipps, S.**, Electrophoretic separation of the plus and minus strands of rotavirus SA11 double-stranded RNAs, *J. Virol. Meth.*, 13, 185, 1986.
169. **Flores, J., Greenberg, H. B., Myslinski, J., Kalica, A. R., Wyatt, R. G., Kapikian, A. Z., and Chanock, R. M.**, Use of transcription probes for genotyping rotavirus reassortants, *Virology*, 121, 288, 1982.
170. **Kantharidis, P., Dyall-Smith, M. L., and Holmes, I. H.**, Completion of the gene coding assignments of SA11 rotavirus: gene products of segments 7, 8, and 9, *J. Virol.*, 48, 330, 1983.

171. **Ward, R. L., Knowlton D. R., Schiff, G. M., Hoshino, Y., and Greenberg, H. B.,** Relative concentration of serum neutralizing antibody to VP3 and VP7 proteins in adults infected with a human rotavirus, *J. Virol.*, 62, 1543, 1988.
172. **Chan, W.-K., Penaranda, M. E., Crawford, S. E., and Estes, M. K.,** Two glycoproteins are produced from the rotavirus neutralization gene, *Virology*, 151, 243, 1986.
173. **Gombold, J. L., Estes, M. K., and Ramig, R. F.,** Assignment of simian rotavirus SA11 temperature-sensistive mutant groups B and E to genome segments, *Virology*, 309, 1985.
174. **Rodger, S. M. and Holmes, I. H.,** Comparison of the genomes of simian, bovine, and human rotaviruses by gel electrophoresis and detection of genomic variation among bovine isolates, *J. Virol.*, 30, 1979.
175. **Rodger, S. M., Bishop, R. F., Birch, C., McLean, B., and Holmes, I. H.,** Molecular epidemiology of human rotaviruses in Melbourne, Australia, from 1973 to 1979, as determined by eletrophoresis of genome ribonucleic acid, *J. Clin. Microbiol.*, 13, 272, 1981.
176. **White, L., Perez, I., Perez, M., Urbina, G., Greenberg, H., Kapikian, A., and Flores, J.,** Relative frequency of rotavirus subgroups 1 and 2 in Venezuelan children with gastroenteritis as assayed with monoclonal antibodies, *J. Clin. Microbiol.*, 19, 516, 1984.
177. **Albert, M. J., Soenarto, Y., and Bishop, R. F.,** Epidemiology of rotavirus diarrhea in Yogyakarta, Indonesia, as revealed by electrophoresis of genome RNA, *J. Clin. Microbiol.*, 16, 731, 1982.
178. **Albert, M. J., Bishop, R. F., and Shann, F. A.,** Epidemiology of rotavirus diarrhea in the Highlands of Papua, New Guinea, in 1979, as revealed by electrophoresis of genome RNA, *J. Clin. Microbiol.*, 17, 162, 1983.
179. **Kutsuzawa, T., Konno, T., Suzuki, H., Ebina, T., and Ishida, N.,** Two distinct electrophoretic migration patterns of RNA segments of human rotaviruses prevalent in Japan in relation to their serotypes, *Microbiol. Immunol.*, 26, 271, 1982.
180. **Spencer, E., Avendano, F., and Araya, M.,** Characteristics and analysis of electropherotypes of human rotavirus isolated in Chile, *J. Infect. Dis.*, 148, 41, 1983.
181. **Nakagomi, O., Nakagomi, T., Oyamada, H., and Suto, T.,** Relative frequency of human rotavirus subgroups 1 and 2 in Japanese children with acute gastroenteritis, *J. Med. Virol.*, 17, 29, 1985.
182. **Cash, P., Freebain, E., Brown, T., and Reid, T. M. S.,** Molecular epidemiology of human rotavirus, *J. Hyg.*, 96, 265, 1986.
183. **Tam, J. S., Kum, W. W. S., Lam, B., Yeung, C. Y., and Ng, M. H.,** Molecular epidemiology of human rotavirus infection in children in Hong Kong, *J. Clin. Microbiol.*, 23, 660, 1986.
184. **Uhnoo, I. and Svensson, L.,** Clinical and epidemiological features of acute infantile gastroenteritis associated with human rotavirus subgroups 1 and 2, *J. Clin. Microbiol.*, 23, 551, 1986.
185. **Urasawa, T., Urasawa, S., Chiba, Y., Taniguchi, K., Kobayashi, N., Mutanda, L. N., and Tukei, P. M.,** Antigenic characterization of rotaviruses isolated in Kenya from 1982 to 1983, *J. Clin. Microbiol.*, 25, 1891, 1987.
186. **Brown, D. W. G., Mathan, M. M., Mathew, M., Martin, R., Beards, G. M., and Mathan, V. I.,** Rotavirus epidemiology in Vellore, South India: group, subgroup, serotype, and electropherotype, *J. Clin. Microbiol.*, 26, 2410, 1988.
187. **Georges-Courbot, M. C., Beraud, A. M., Beards, G. M., Campbell, A. D., Gonzalez, J. P., Georges, A. J., and Flewett, T. H.,** Subgroups, serotypes, and electropherotypes of rotavirus isolated from children in Bangui, Central African Republic, *J. Clin. Microbiol.*, 668, 1988.
188. **Sethi, S. K., Olive, D. M., Strannegard, O. O., and Al-Nakib, W.,** Molecular epidemiology of human rotavirus infections based on genome segment variations in viral strains, *J. Med. Microbiol.*, 26, 249, 1988.
189. **Hasegawa, A., Inouye, S., Matsuno, S., Yamaoka, K., Eko, R., Suharyono,** Isolation of human rotaviruses with a distinct RNA electrophoretic pattern from Indonesia, *Microbiol. Immunol.*, 28, 719, 1984.
190. **Matsuno, S., Hasegawa, A., Mukoyama, A., Inouye, S.,** A candidate for a new serotype of human rotavirus, *J. Clin. Microbiol.*, 54, 623, 1985.
191. **Albert, M. J., Unicomb, L. E., and Bishop, R. F.,** Cultivation and characterization of human rotaviruses with "super short" RNA patterns, *J. Clin. Microbiol.*, 25, 183, 1987.
192. **Smith, M. and Tzipori, S.,** Gel electrophoresis of rotavirus RNA derived from six different animal species, *Aust. J. Exp. Biol. Med. Sci.*, 57, 583, 1979.
193. **Rodger, S. M., Holmes, I. H., and Studdert, M. J.,** Characteristics of the genomes of equine rotaviruses, *Vet. Microbiol.*, 5, 243, 1980.
194. **Schroeder, B. A., Street, J. E., Kalmakoff, J., and Bellamy, A. R.,** Sequence relationships between the genome segments of human and animal rotavirus strains, *J. Virol.*, 43, 379, 1982.
195. **Hoshino, Y., Wyatt, R. G., Greenberg, H. B., Kalica, A. R., Flores, J., and Kapikian, A. Z.,** Isolation and characterization of an equine rotavirus, *J. Clin. Microbiol.*, 18, 585, 1983.
196. **Hoshino, Y., Wyatt, R. G., Greenberg, H. B., Kalica, A. R., Flores, J., and Kapikian, A. Z.,** Isolation, propagation, and characterisation of a second equine rotavirus serotype, *Infect. Immun.*, 41, 1031, 1983.
197. **Bohl, E. H., Theil, K. W., and Saif, L. J.,** Isolation and serotyping of porcine rotaviruses and antigenic comparison with other rotaviruses, *J. Clin. Microbiol.*, 19, 105, 1984.

198. **Hoshino, Y., Wyatt, R. G., Greenberg, H. B., Flores, J., and Kapikian, A. Z.,** Serotypic similarity and diversity of rotaviruses of mammalian and avian origin as studied by plaque reduction neutralization, *J. Infect. Dis.*, 149, 694, 1984.
199. **Ojeh, C. K., Snodgrass, D. R., and Herring, A. J.,** Evidence for serotypic variation among bovine rotaviruses, *Arch. Virol.*, 79, 161, 1984.
200. **Bellinzoni, R. C., Matton, N., La Torre, J. L., and Scodeller, E. A.,** Incidence of rotavirus in beef herds in Argentina, *Res. Vet. Sci.*, 42, 257, 1987.
201. **Fijtman, N. L., Barrandeguy, E. M., and Schudel, A. A.,** Variations and persistency of electropherotypes of bovine rotavirus field isolates, *Arch. Virol.*, 96, 275, 1987.
202. **Liprandi, F., Garcia, D., Botero, L., Gorziglia, M., Cavazza, M.-E., Perez-Schael, I., and Esparza, J.,** Characterization of rotaviruses isolated from pigs with diarrhoea in Venezuela, *Vet. Microbiol.*, 13, 35, 1987.
203. **Pocock, D. H.,** Characterisation of rotavirus isolates from sub-clinically infected calves by genome profile analysis, *Vet. Microbiol.*, 13, 27, 1987.
204. **Theil, K. W. and McCloskey, C. M.,** Molecular epidemiology and subgroup determination of bovine group A rotaviruses associated with diarrhea in dairy and beef calves, *J. Clin. Microbiol.*, 27, 126, 1989.
205. **Sato, K., Inaba, Y., Miura, Y., Tokuhisa, S., and Matumoto, M.,** Isolation of lapine rotavirus in cell cultures, *Arch. Virol.*, 71, 267, 1982.
206. **Tanaka, T. N., Conner, M. E., Graham, D. Y., and Estes, M. K.,** Molecular characterization of three rabbit rotavirus strains, *Arch. Virol.*, 98, 253, 1988.
207. **Thouless, M. E., DiGiacomo, R. F., and Neuman, D. S.,** Isolation of two lapine rotaviruses: characterization of their subgroup, serotype, and RNA electropherotypes, *Arch. Virol.*, 89, 161, 1986.
208. **Paul, P. M., Lyoo, Y. S., Woode, G. N., Zheng, S., Greenberg, H. B., Matsui, S., Schwartz, K. J., and Hill, H. T.,** Isolation of a bovine rotavirus with a "super-short" RNA electrophoretic pattern from a calf with diarrhea, *J. Clin. Microbiol.*, 26, 2139, 1988.
209. **Theil, K. W. and McCloskey, C. M.,** Partial characterization of a bovine group A rotavirus with a short genome electropherotype, *J. Clin. Microbiol.*, 26, 1094, 1988.
210. **Allen, S., Mitchell, J., Jones, W., and Quinn, M.,** A novel bovine rotavirus electropherotype from outbreaks of neonatal diarrhea in Utah beef herds, *J. Vet. Diagn. Invest.*, 1, 74, 1989.
211. **Smee, D. F., Sidwell, R. W., Clark, S. M., Barnett, B. B., and Spendlove, R. S.,** Inhibition of rotaviruses by selected antiviral substances: mechanisms of viral inhibition and in vivo activity, *Antimicrob. Agents. Chemother.*, 21, 66, 1982.
212. **Tajima, T., Suzuki, E., Ushijima, H., Araki, K., Kim, B., Shinozaki, T., and Fujii, R.,** Isolation of murine rotavirus in cell cultures, *Arch. Virol.*, 82, 119, 1984.
213. **Greenberg, H. B., Vo, P. T., and Jones, R.,** Cultivation and characterization of three strains of murine rotavirus, J. Virol., 57, 585, 1986.
214. **Todd, D., McNulty, M. S., and Allan, G. M.,** Polyacrylamide gel electrophoresis of avian rotavirus RNA, *Arch. Virol.*, 63, 87, 1980.
215. **Saif, L. J., Saif, Y. M., and Theil, K. W.,** Enteric viruses in diarrheic turkey poults, *Avian Dis.*, 29, 798, 1985.
216. **Yason, C. V. and Schat, K. A.,** Isolation and characterization of avian rotaviruses, *Avian Dis.*, 29, 499, 1985.
217. **Kang, S. Y., Nagaraja, K. V., and Newman, J. A.,** Electropherotypic analysis of rotaviruses isolated from turkeys, *Avian Dis.*, 30, 794, 1986.
218. **Theil, K. W., Reynolds, D. L., and Saif, Y. M.,** Isolation and serial propagation of turkey rotaviruses in a fetal rhesus monkey kidney (MA104) cell line, *Avian Dis.*, 30, 93, 1986.
219. **Pedley, S., Hundley, F., Chrystie, I., McCrae, M. A., and Desselberger, U.,** The genomes of rotaviruses isolated from chronically infected immunodeficient children, *J. Gen. Virol.*, 65, 1141, 1984.
220. **Besselaar, T. G., Rosenblatt, A., and Kidd, A. H.,** Atypical rotavirus from South African neonates, *Arch. Virol.*, 87, 327, 1986.
221. **Bellinzoni, R. C., Mattion, N. M., Burrone, O., Gonzalez, A., La Torre, J. L., and Scodeller, E. A.,** Isolation of group A swine rotavirus displaying atypical electropherotypes, *J. Clin. Microbiol.*, 25, 952, 1987.
222. **Hundley, F., McIntyre, M., Clark, B., Beards, G., Wood, D., Chrystie, I., and Desselberger, U.,** Heterogeneity of genome rearrangements in rotaviruses isolated from a chronically infected immunodeficient child, *J. Virol.*, 61, 3365, 1987.
223. **Pocock, D. H.,** Isolation and characterization of two group A rotaviruses with unusual genome profiles, *J. Gen. Virol.*, 68, 653, 1987.
224. **Mattion, N., Gonzalez, S. A., Burrone, O., Bellinzoni, R., La Torre, J. L., and Scodeller, E. A.,** Rearrangement of genomic segment 11 in two swine rotavirus strains, *J. Gen. Virol.*, 69, 695, 1988.
225. **McIntyre, M., Rosenbaum, V., Rappold, W., Desselberger, M., Wood, D., and Desselberger, U.,** Biophysical characterization of rotavirus particles containing rearranged genomes, *J. Gen. Virol.*, 68, 2961, 1987.
226. **Matsuno, S., Hasegawa, A., Kalica, A. R., and Kono, R.,** Isolation of a recombinant between simian and bovine rotaviruses, *J. Gen. Virol.*, 48, 253, 1980.

227. **Greenberg, H. B., Kalica, A. R., Wyatt, R. G., Jones, R. W., Kapikian, A. Z., and Chanock, R. M.,** Rescue of noncultivatable human rotavirus by gene reassortment during mixed infection with ts mutants of a cultivatable bovine rotavirus, *Proc. Natl. Acad. Sci.*, 78, 420, 1981.
228. **Garbarg-Chenon, A., Bricout, F., and Nicolas, J.-C.,** Study of genetic reassortment between two human rotaviruses, *Virology*, 139, 358, 1984.
229. **Midthun, K., Greenberg, H. B., Hoshino, Y., Kapikian, A. Z., Wyatt, R. G., and Chanock, R. M.,** Reassortant rotaviruses as potential live rotavirus vaccine candidates, *J. Virol.*, 53, 949, 1985.
230. **Graham, A., Kudesia, G., Allen, A. M., and Desselberger, U.,** Reassortment of human rotavirus possessing genome rearrangements with bovine rotavirus: evidence for host cell selection, *J. Gen. Virol.*, 68, 115, 1987.
231. **Gargarg-Chenon, A., Bricout, F., and Nicolas, J.-C.,** Serological characterization of human reassortant rotaviruses, *J. Virol.*, 59, 510, 1986.
232. **Gombold, J. L. and Ramig, R. F.,** Analysis of reassortment of genome segments in mice mixedly infected with rotaviruses SA11 and RRV, *J. Virol.*, 57, 1986.
233. **Hoshino, Y., Sereno, M. M., Midthun, K., Flores, J., Chanock, R. M., and Kapikian, A. Z.,** Analysis by plaque reduction neutralization assay of intertypic rotaviruses suggests that gene reassortment occurs in vivo, *J. Clin. Microbiol.*, 25, 290, 1987.
234. **Kitaoka, S., Nakagomi, Fukuhara, N., Hoshino, Y., Suzuki, H., Nakagomi, O., Kapikian, A. Z., Ebina, T., Konno, T., and Ishida, N.,** Serologic characteristics of a human rotavirus isolate, AU-1, which has a "long" RNA pattern and subgroup I specificity, *J. Med. Virol.*, 23, 351, 1987.
235. **Midthun, K., Valdesuso, J., Hoshino, Y., Flores, J., Kapikian, A. Z., and Chanock, R. M.,** Analysis by RNA-RNA hybridization assay of intertypic rotaviruses suggests that gene reassortment occurs *in vivo*, *J. Clin. Microbiol.*, 25, 295, 1987.
236. **Nakagomi, O., Nakagomi, T., Hoshino, Y., Flores, J., and Kapikian, A. Z.,** Genetic analysis of a human rotavirus that belongs to subgroup I but has an RNA pattern typical of subgroup II Human rotaviruses, *J. Clin. Microbiol.*, 25, 1159, 1987.
237. **Matsuno, S., Mukoyama, A., Hasegawa, A., Taniguch, K., and Inouye, S.,** Characterization of a human rotavirus strain which is possibly a naturally-occuring reassortant virus, *Virus Res.*, 10, 167, 1988.
238. **Hoshino, Y., Gorziglia, M., Valdesuso, J., Askaa, J., Glass, R. I., and Kapikian, A. Z.,** An equine rotavirus (Fl-14 strain) which bears both subgroup I and subgroup II specificities on it VP6, *Virology*, 157, 488, 1987.
239. **Kapikian, A. Z., Cline, W. L., Greenberg, H. B., Wyatt, R. G., Kalica, A. R., Banks, C. E., James, H. D., Flores, J., and Chanock, R. M.,** Antigenic characterization of human and animal rotaviruses by immune adherence hemagglutination assay (IAHA): evidence for distinctness of IAHA and neutralization antigens, *Infect. Immun.*, 33, 1981.
240. **Kalica, A. R., Greenberg, H. B., Espejo, R. T., Flores, J., Wyatt, R. G., Kapikian, A. Z., and Chanock, R. M.,** Distinctive ribonucleic acid patterns of human rotavirus subgroups 1 and 2, *Infect. Immun.*, 33, 958, 1981.
241. **Beards, G. M.,** Polymorphism of genomic RNAs within rotavirus serotypes and subgroups, *Arch. Virol.*, 74, 65, 1982.
242. **Thouless, M. E., Beards, G. M., and Flewett, T. H.,** Serotyping and subgrouping of rotavirus strains by the ELISA test, *Arch. Virol.*, 73, 219, 1982.
243. **Flores, J. Perez, I., White, L., Perez, M., Kalica, A. R., Marquina, R., Wyatt, R. G., Kapikian, A. Z., and Chanock, R. M.,** Genetic relatedness among human rotaviruses as determined by RNA hybridization, *Infect. Immun.*, 37, 648, 1982.
244. **Street, J. E., Croxson, M. C., Chadderton, W. F., and Bellamy, A. R.,** Sequence diversity of human rotavirus strains investigated by Northern blot hybridization analysis, *J. Virol.*, 43, 369, 1982.
245. **Flores, J., Perez-Schael, I., Boeggeman, E., White, L., Perez, M., Purcell, R., Hoshino, Y., Midthun, K., Chanock, R. M., and Kapikian, A. Z.,** Genetic relatedness among human rotaviruses, *J. Med. Virol.*, 17, 135, 1985.
246. **Lambert, J.-P., Marissens, D., Marbehant, P., and Zissis, G.,** Prevalence of subgroup 1, 2, and 3 rotaviruses in Belgian children suffering from acute diarrhea (1978—1981), *J. Med. Virol.*, 11, 31, 1983.
247. **Tufvesson, B.,** Detection of a human rotavirus strain different from types 1 and 2 — a new subgroup? Epidemiology of subgroups in a Swedish and an Ethiopian community, *J. Med. Virol.*, 12, 111, 1983.
248. **Aboudy, Y., Shif, I., Zilberstein, I., and Gotlieb-Stematsky, T.,** Use of polyclonal and monoclonal antibodies and analysis of viral RNA in the detection of unusual group A human rotaviruses, *J. Med. Virol.*, 25, 351, 1988.
249. **Svensson, L., Grahnquist, L., Pettersson, C.-A., Grandien, M., Stintzing, G., and Greenberg, H. B.,** Detection of human rotaviruses which do not react with subgroup I- and II-specific monoclonal antibodies, *J. Clin. Microbiol.*, 26, 1238, 1988.
250. **Urasawa, S., Urasawa, T., and Taniguchi, K.,** Genetic reassortment between two human rotaviruses having different serotype and subgroup specificities, *J. Gen. Virol.*, 67, 1551, 1986.

251. **Both, G. W., Siegman L. J., Bellamy, A. R., Ikegami, N., Shatkin, A. J., and Furuichi, Y.**, Comparative sequence analysis of rotavirus genomic segment 6 — the gene specifying viral subgroups 1 and 2, *J. Virol.*, 51, 97, 1984.
252. **Taniguchi, K., Urasawa, T., and Urasawa, S.**, Reactivity patterns to human rotavirus strains of a monoclonal antibody against VP2, a component of the inner capsid of rotavirus, *Arch. Virol.*, 87, 135, 1986.
253. **Hoshino, Y., Wyatt, R. G., Greenberg, H. B., Kalica, A. R., Flores, J., and Kapikian, A. Z.**, Serological comparison of canine rotavirus with various simian and human rotaviruses by plaque reduction neutralization and hemagglutination inhibition tests, *Infect. Immun.*, 41, 169, 1983.
254. **Birch, C. J., Heath, R. L., Marshall, J. A., Lui, S., and Gust, I. D.**, Isolation of feline rotaviruses and their relationship to human and simian isolates by electropherotype and serotype, *J. Gen. Virol.*, 66, 2731, 1985.
255. **Nagesha, H. S. and Holmes, I. H.**, New porcine rotavirus serotype antigenically related to human rotavirus serotype 3, *J. Clin. Microbiol.*, 26, 171, 1988.
256. **Woode, G. N., Kelso, N. E., Simpson, T. F., Gaul, S. K., Evans, L. E., Babiuk, L.**, Antigenic relationships among some bovine rotaviruses: serum neutralization and cross-protection in gnotobiotic calves, *J. Clin. Microbiol.*, 18, 358, 1983.
257. **Clark, H. F., Hoshino, Y., Bell, L. M., Groff, J., Hess, G., Bachman, P., and Offit, P. A.**, Rotavirus isolate WI61 representing a presumptive new human serotype, *J. Clin. Microbiol.*, 25, 1757, 1987.
258. **Hoshino, Y., Wyatt, R. G., Flores, J., Midthun, K., and Kapikian, A. Z.**, Serotypic characterization of rotaviruses derived from asymptomatic human neonatal infections, *J. Clin. Microbiol.*, 21, 425, 1985.
259. **Woode, G. N., Bridger, J., Hall, G. A., Jones, J. M., and Jackson, G.**, The isolation of reovirus-like agents (rotaviruses) from acute gastroenteritis of piglets, *J. Med. Microbiol.*, 9, 203, 1976.
260. **Bohl, E. H., Kohler, E. M., Saif, L. J., Cross, R. F., Agnes, A. G., and Theil, K. W.**, Rotavirus as a cause of diarrhea in pigs, *J. Am. Vet. Med. Assoc.*, 172, 458, 1978.
261. **Torres-Medina, A., Wyatt, R. G., Mebus, C. A., Underdahl, N. R., and Kapikian, A. Z.**, Diarrhea caused in gnotobiotic piglets by the reovirus-like agent of human infantile gastroenteritis, *J. Infect. Dis.*, 133, 22, 1976.
262. **Mebus, C. A., Wyatt, R. G., Sharpee, R. L., Sereno, M. M., Kalica, A. R., Kapikian, A. Z., and Twiehaus, M. J.**, Diarrhea in gnotobiotic calves caused by the reovirus-like agent of human infantile gastroenteritis, *Infect. Immun.*, 14, 471, 1976.
263. **Snodgrass, D. R., Madeley, C. R., Wells, P. W., and Angus, K. W.**, Human rotavirus in lambs: infection and passive protection, *Infect. Immun.*, 16, 268, 1977.
264. **Bridger, J. C., and Woode, G. N.**, Neonatal calf diarrhoea: identification of a reovirus-like (rotavirus) agent in faeces by immunofluorescence and immune electron microscopy, *Br. Vet. J.*, 131, 258, 1975.
265. **L'Haridon, R. and Scherrer, R.**, Culture *in vitro* du rotavirus associe aux diarrhees neotnatles du veau, *Ann. Rech. Veter.*, 7, 373, 1976.
266. **McNulty, M. S., Allan, G. M., and McFerran, J. B.**, Isolation of a cytopathic calf rotavirus, *Res. Vet. Sci.*, 21, 114, 1976.
267. **Babiuk, L. A., Mohammed, K., Spence, L., Fauvel, M., and Petro, R.**, Rotavirus isolation and cultivation in the prescence of trypsin, *J. Clin. Microbiol.*, 6, 610, 1977.
268. **Theil, K. W., Bohl, E. H., and Agnes, A. G.**, Cell culture propagation of porcine rotavirus (reovirus-like agent), *Am. J. Vet. Res.*, 38, 1765, 1977.
269. **Wyatt, R. G., James, W. D., Bohl, E. H., Theil, K. W., Saif, L. J., Kalica, A. R., Greenberg, H. B., Kapikian, A. Z., and Chanock, R. M.**, Human rotavirus type 2: cultivation in vitro, *Science*, 207, 189, 1980.
270. **Sato, K., Inaba, Y., Shinozaki, T., Fujii, R., and Matumoto, M.**, Isolation of human rotavirus in cell cultures, *Arch. Virol.*, 69, 155, 1981.
271. **Urasawa, T., Urasawa, S., and Taniguchi, K.**, Sequential passages of human rotavirus in MA-104 cells, *Microbiol. Immunol.*, 25, 1025, 1981.
272. **McNulty, M. S., Allan, G. M., Todd, D., and McFerran, J. B.**, Isolation and cell culture propagation of rotaviruses from turkeys and chickens, *Arch. Virol.*, 61, 13, 1979.
273. **Theodoridis, A., Prozesky, L. and Els, H. J.**, The isolation and cultivation of calf rotavirus in the Republic of South Africa, *Onderstepoort J. Vet. Res.*, 46, 65, 1979.
274. **England, J. J. and Poston, R. P.**, Electron microscopic identification and subsequent isolation of a rotavirus from a dog with fatal neonatal diarrhea, *Am. J. Vet. Res.*, 41, 782, 1980.
275. **Fulton, R. W., Johnson, C. A., Pearson N. J., and Woode, G. N.**, Isolation of a rotavirus from a newborn dog with diarrhea, *Am. J. Vet. Res.*, 42, 841, 1981.
276. **Imagawa, H., Ando, Y., Sugiura, T., Wada, R., Hirasawa, K., and Akiyama, Y.**, Isolation of foal rotavirus in MA-104 cells, *Bull. Equine Res. Inst.*, 18, 119, 1981.
277. **Gillespie, J., Kalica, A., Conner, M., Schiff, E., Barr, M., Holmes, D., and Frey, M.**, The isolation, propagation and characterization of tissue-cultured equine rotaviruses, *Vet. Microbiol.*, 9, 1, 1984.
278. **Makabe, T., Komaiwa, H., Kishi, Y., Yataya, K., Imagawa, H., Sato, K., and Inaba, Y.**, Isolation of bovine rotavirus in cell cultures, *Arch. Virol.*, 83, 123, 1985.

279. **Fukusho, A., Shimizu, Y., and Ito, Y.**, Isolation of cytopathic porcine rotavirus in cell roller culture in the presence of trypsin, *Arch. Virol.*, 69, 49, 1981.
280. **Paul, P. S., Lyoo, Y. S., Andrews, J. J., and Hill, H. T.**, Isolation of two new serotypes of porcine rotavirus from pigs with diarrhea, *Arch. Virol.*, 100, 139, 1988.
281. **Matsuno, S., Inouye, S., and Kono, R.**, Plaque assay of neonatal calf diarrhea virus and the neutralizing antibody in human sera, *J. Clin. Microbiol.*, 5, 1, 1977.
282. **Ramia, S. and Sattar, S. A.**, Simian rotavirus SA-11 plaque formation in the presence of trypsin, *J. Clin. Microbiol.*, 10, 609, 1979.
283. **Smith, E. M., Estes, M. K., Graham, D. Y., and Gerba, C. P.**, A plaque assay for the simian rotavirus SA11, *J. Gen. Virol.*, 43, 513, 1979.
284. **Graham, D. Y. and Estes, M. K.**, Proteolytic enhancement of rotavirus infectivity: biologic mechanisms, *Virology*, 101, 432, 1980.
285. **Twist, E. M., Kolonich, K., and Rubin, D. H.**, Propagation of rotaviruses in the presence of chicken serum, *J. Gen. Virol.*, 65, 1207, 1984.
286. **Agliano, A. M., Rossi, A., and Sanna, A.**, Isolation from faecal specimens of new strains of human rotavirus primarily cytopathic for stationary cell cultures without trypsin, *Arch. Virol.*, 84, 119, 1985.
287. **Mebus, C. A., Stair, E. L., Underdahl, N. R., and Twiehaus, M. J.**, Pathology of neonatal calf diarrhea induced by a reo-like virus, *Vet. Pathol.*, 8, 490, 1971.
288. **Mebus, C. A., Stair, E. L., Rhodes, M. B., Underdahl, N. R., and Twiehaus, M. J.**, Calf diarrhea of viral etiology, *Ann. Rech. Vet.*, 4, 71, 1973.
289. **Stair, E. L., Mebus, C. A., Twiehaus, M. J., and Underdahl, N. R.**, Neonatal calf diarrhea. Electron microscopy of intestines infected with a reovirus-like agent, *Vet. Pathol.*, 10, 155, 1973.
290. **Woode, G. N., Bridger, J. C., Hall, G., and Dennis, M. J.**, The isolation of a reovirus-like agent associated with diarrhea in colostrum-deprived calves in Great Britain, *Res. Vet. Sci.*, 16, 102, 1974.
291. **Mebus, C. A. and Newman, L. E.**, Scanning electron, light, and immunofluorescent microscopy of intestine of gnotobiotic calf infected with reovirus-like agent, *Am. J. Vet. Res.*, 38, 553, 1977.
292. **Snodgrass, D. R., Herring, J. A., and Gray, E. W.**, Experimental rotavirus infection in lambs, *J. Comp. Pathol.*, 86, 637, 1976.
293. **Snodgrass, D. R., Angus, K. W., and Gray, E. W.**, Rotavirus infection in lambs: pathogenesis and pathology, *Arch. Virol.*, 55, 263, 1977.
294. **Snodgrass, D. R., Ferguson, A., Allan, F., Angus, K. W., and Mitchell, B.**, Small intestinal morphology and epithelial cell kinetic in lamb rotavirus infections, *Gastroenterology*, 76, 477, 1979.
295. **Crouch, C. F. and Woode, G. N.**, Serial studies of virus multiplication and intestinal damage in gnotobiotic piglets infected with rotavirus, *J. Med. Microbiol.*, 11, 325, 1978.
296. **Theil, K. W., Bohl, E. H., Cross, R. F., Kohler, E. M., and Agnes, A. G.**, Pathogenesis of porcine rotaviral infection in experimentally inoculated gnotobiotic pigs, *Am. J. Vet. Res.*, 39, 213, 1978.
297. **Tzipori, S. and Williams, I. H.**, Diarrhoea in piglets inoculated with rotavirus, *Aust. Vet. J.*, 54, 188, 1978.
298. **McAdaragh, J. P., Bergeland, M. E., Meyer, R. C., Johnshoy, M. W., Stotz, I. J., Benfield, D. A., and Hammer, R.**, Pathogenesis of rotaviral enteritis in gnotobiotic pigs: a microscopic study, *Am. J. Vet. Res.*, 41, 1572, 1980.
299. **Torres-Medina, A. and Underdahl, N. R.**, Scanning electron microscopy of intestine of gnotobiotic piglets infected with porcine rotavirus, *Can. J. Comp. Med.*, 44, 403, 1980.
300. **Johnson, C. A., Fulton, R. W., Henk, W. G., and Snider, T. G.**, Inoculation of neonatal gnotobiotic dogs with a canine rotavirus, *Am. J. Vet. Res.*, 44, 1682, 1983.
301. **Johnson, C. A., Snider, T. G., Fulton, R. W., and Cho, D.**, Gross and light microscopic lesions in neonatal gnotobiotic dogs inoculated with a canine rotavirus, *Am. J. Vet. Res.*, 44, 1687, 1983.
302. **Johnson, C. A., Snider, T. G., Henk, W. G., and Fulton, R. W.**, A scanning and transmission electron microscopic study of rotavirus-induced intestinal lesions in neonatal gnotobiotic dogs, *Vet. Pathol.*, 23, 443, 1986.
303. **Torres-Medina, A., Wyatt, R. G., Mebus, C. A., Underdahl, N. R., and Kapikian, A. Z.**, Patterns of shedding of human reovirus-like agent in gnotobiotic newborn piglets with experimentally-induced diarrhea, *Intervirology*, 7, 250, 1976.
304. **Mebus, C. A., Wyatt, R. G., and Kapikian, A. Z.**, Intestinal lesions induced in gnotobiotic calves by the virus of human infantile gastroenteritis, *Vet. Pathol.*, 14, 273, 1977.
305. **Kanitz, C. L.**, Identification of an equine rotavirus as a cause of neonatal foal diarrhea, Proc. 22nd Annual Conv. Am. Assoc. Equine Pract., Dallas, Texas, November 28 to December 1, 1976, 155.
306. **Thouless, M. E., DiGiacomo, R. F., Deeb, B. J., and Howard, H.**, Pathogenicity of rotavirus in rabbits, *J. Clin. Microbiol.*, 26, 943, 1988.
307. **Yason, C. V. and Schat, K. A.**, Pathogenesis of rotavirus infection in turkey poults, *Avian Pathol.*, 15, 421, 1986.

308. **Yason, C. V., Summers, B. A., and Schat, K. A.,** Pathogenesis of rotavirus infection in various age groups of chickens and turkeys: pathology, *Am. J. Vet. Res.*, 48, 927, 1987.
309. **Yason, C. V. and Schat, K. A.,** Pathogenesis of rotavirus infection in various age groups of chickens and turkeys: clinical signs and virology, *Am. J. Vet. Res.*, 48, 977, 1987.
310. **Kapikian, A. Z., Wyatt, R. G., Levine, M. M., Yolken, R. H., VanKirk, D. H., Dolin, R., Greenberg, H. B., and Chanock, R. M.,** Oral administration of human rotavirus to volunteers: induction of illness and correlates of resistance, *J. Infect. Dis.*, 147, 95, 1983.
311. **McNulty, M. S., Allan, G. M., Pearson, G. R., McFerran, J. B., Curran, W. L., and McCracken, R. M.,** Reovirus-like agent (rotavirus) from lambs, *Infect. Immun.*, 14, 1332, 1976.
312. **Bridger, J. C. and Pocock, D. H.,** Variation in virulence of bovine rotaviruses, *J. Hyg.*, 96, 257, 1986.
313. **Bridger, J. C. and Brown, J. F.,** Antigenic and pathogenic relationships of three bovine rotaviruses and a porcine rotavirus, *J. Gen. Virol.*, 65, 1151, 1984.
314. **Totterdell, B. M., Chrystie, I. L., and Banatvala, J. E.,** Rotavirus infections in a maternity unit, *Arch. Dis. Childhood*, 51, 924, 1976.
315. **Van Renterghem, L., Borre, P., and Tilleman, J.,** Rotavirus and other viruses in the stool of premature babies, *J. Med. Virol.*, 5, 137, 1980.
316. **Champsaur, H., Questiaux, E., Prevot, J., Henry-Amar, M., Goldszmidt, D., Bourjouane, M., and Bach, C.,** Rotavirus carriage, asymptomatic infection, and disease in the first two years of life. I. Virus shedding, J. *Infect. Dis.*, 149, 667, 1984.
317. **Burke, V., Gracey, M., and Masters, P.,** Rotavirus in children, *J. Infect. Dis.*, 152, 646, 1985.
318. **Yolken, R. H., Wyatt, R. G., Zissis, G., Brandt, C. D., Rodriguez, W. J., Kim, H. W., Parrott, R. H., Urrutia, J. J., Mata, L., Greenberg, H. B., Kapikian, A. Z., and Chanock, R. M.,** Epidemiology of human rotavirus types 1 and 2 as studied by enzyme-linked immunosorbent assay, *N. Engl. J. Med.*, 299, 1156, 1978.
319. **Steele, A. D., Bos, P., and Alexander, J. J.,** Clinical features of acute infantile gastroenteritis associated with human rotavirus subgroups I and II, *J. Clin. Microbiol.*, 26, 2647, 1988.
320. **Wolf, J. L., Cukor, G., Blacklow, N. R., Dambrauskas, R., and Trier, J. S.,** Susceptibility of mice to rotavirus infection: effects of age and administration of corticosteroids, *Infect. Immun.*, 33, 565, 1981.
321. **Riepenhoff-Talty, M., Lee, P.-C., Carmody, P. J., Barrett, H. J., and Ogra, P. L.,** Age-dependent rotavirus-enterocyte interactions, *Proc. Soc. Exp. Biol. Med.*, 170, 146, 1982.
322. **Kirstein, C. G., Clare, D. A., and Lecce, J. G.,** Development of resistance of enterocytes to rotavirus in neonatal, agammaglobulinemic piglets, *J. Virol.*, 322, 567, 1985.
323. **Lecce, J. G. and King, M. W.,** Role of rotavirus (reo-like) in weanling diarrhea of pigs, *J. Clin. Microbiol.*, 8, 454, 1978.
324. **Kurstak, E., Kurstak, C., Van Den Hurk, J., and Morisset, R.,** Animal Rotaviruses, in *Comparative Diagnosis of Viral Diseases*, Kurstak, E. and Kurstak, C., Eds., Academic Press, New York, 1981.
325. **Holmes, I. H.,** Rotaviruses, in *The Reoviridae*, Joklik, W. K., Ed., Plenum Press, New York, 1983, chap. 8.
326. **Woode, G. N.,** Porcine rotavirus infection, in *Diseases of Swine*, 6th ed., Leman, A. D., Straw, B., Glock, R. D., Mengeling, W. L., Penny, R. H. C., and Scholl, E., Eds., Iowa State University Press, Ames, 1986, chap. 28.
327. **Saif, Y. M., Theil, K. W., Reynolds, D. L., and Saif, L. J.,** Enteric viruses of turkeys, in *Advances in Turkey Science*, Butterworth, London, 1989, chap. 13.
328. **Hardy, D. B.,** Epidemiology of rotaviral infection in adults, *Rev. Infect. Dis.*, 9, 461, 1987.
329. **Von Bonsdorff, C.-H., Hovi, T., Makela, P., and Morttinen, A.,** Rotavirus infections in adults in association with acute gastroenteritis, *J. Med. Virol.*, 2, 21, 1978.
330. **Echeverria, P., Blacklow, N. R., Cukor, G. G., Vibulbandhitkit, S., Changchawalit, S., and Boonthai, P.,** Rotavirus as a cause of severe gastroenteritis in adults, *J. Clin. Microbiol.*, 18, 663, 1983.
331. **Yolken, R. H., Bishop, C. A., Townsend, T. R., Bolyard, E. A., Bartlett, J., Santos, G. W., and Sarai, R.,** Infectious gastroenteritis in bone-marrow-transplant recipients, *N. Engl. J. Med.*, 306, 1009, 1982.
332. **Cubitt, W. D. and Holzel, H.,** An outbreak of rotavirus infection in a long-stay ward of a geriatric hospital, *J. Clin. Pathol.*, 33, 306, 1980.
333. **Halvorsrud, J. and Orstavik, I.,** An epidemic of rotavirus-associated gastroenteritis in a nursing home for the elderly, *Scand. J. Infect. Dis.*, 12, 161, 1980.
334. **Marrie, T. J., Lee, S. H. S., Faulkner, R. S., Ethier, J., and Young, C. H.,** Rotavirus infection in a geriatric population, *Arch. Intern. Med.*, 142, 1982.
335. **Cunningham, A. L., Grohman, G. S., Harkness, J., Law, C., Marriot, D., Tindall, B., and Cooper, D. A.,** Gastrointestinal viral infections in homosexual men who were symptomatic and seropositive for human immunodeficiency virus, *J. Infect. Dis.*, 158, 386, 1988.
336. **Bolivar, R., Conklin, R. H., Vollet, J. J., Pickering, L. K., DuPont, H. L., Walters, D. L., and Kohl, S.,** Rotavirus in travelers' diarrhea: study of an adult student population in Mexico, *J. Infect. Dis.*, 137, 324, 1978.

337. **Vollet, J. J., Ericsson, C. D., Gibson, G., Pickering, L. K., DuPont, H. L., Kohl, S., and Conklin, R. H.,** Human rotavirus in an adult population with travelers' diarrhea and its relationship to the location of food consumption, *J. Med. Virol.,* 4, 81, 1979.
338. **Ryder, R. W., Oquist, C. A., Greenberg, H., Taylor, D. N., Orskov, F., Orskov, I., Kapikian, A. Z., and Sack, R. B.,** Travelers' diarrhea in Panamanian tourists in Mexico, *J. Infect. Dis.,* 144, 442, 1981.
339. **Linhares, A. C., Pinheiro, F. P., Freitas, R. B., Gabbay, Y. B., Shirley, J. A., and Beards, G. M.,** An outbreak of rotavirus diarrhea among a nonimmune, isolated South American Indian community, *Am. J. Epidemiol.,* 113, 703, 1981.
340. **Holdaway, M. D., Kalmakoff, J., Todd, B. A., and Jennings, L. C.,** Rotavirus infection in a small community, *J. Med. Virol.,* 15, 389, 1985.
341. **Von Bonsdorff, C.-H., Hove, T., Makela, P., Hove, L., Tevalvoto-Aarnio,** Rotavirus associated with acute gastroenteritis in adults, *Lancet,* 2, 423, 1976.
342. **Kim, H. W., Brandt, C. D., Kapikian, A. Z., Wyatt, R. G., Arrobio, J. O., Rodriguez, W. J., Chanock, R. M., and Parrott, R. H.,** Human reovirus-like agent infection. Occurrence in adult contacts of pediatric patients with gastroenteritis, *J. Am. Med. Assoc.,* 238, 404, 1977.
343. **Rodriguez, W. J., Kim, H. W., Brandt, C. D., Yolken, R. H., Richard, M., Arrobio, J. O., Schwartz, R. H., Kapikian, A. Z., Chanock, R. M., and Parrott, R. H.,** Common exposure outbreak of gastroenteritis due to type 2 rotavirus with high secondary attack rate within families, *J. Infect. Dis.,* 140, 353, 1979.
344. **Wenman, W. M., Hinde, D., Feltham, S., and Gurwith, M.,** Rotavirus infection in adults. Results of a prospective study, *N. Engl. J. Med.,* 301, 303, 1979.
345. **Grimwood, K., Abbott, G. D., Fergusson, D. M., Jennings, L. C., and Allan, J. M.,** Spread of rotavirus within families: a community based study, *Br. Med. J.,* 287, 575, 1983.
346. **Monto, A. S., Koopman, J. S., Longini, I. M., and Isaacson, R. E.,** The Tecumseh study. XII. Enteric agents in the community, 1976—1981, *J. Infect. Dis.,* 148, 284, 1983.
347. **Galil, A., Antverg, R., Katzir, G., Zentner, B., Margalith, M., Friedman, M. G., Sarov, B., and Sarov, I.,** Involvement of infants, children, and adults in a rotavirus gastroenteritis outbreak in a kibbutz in southern Israel, *J. Med. Virol.,* 18, 317, 1986.
348. **Benfield, D. A., Stotz, I., Moore, R., and McAdaragh, J. P.,** Shedding of rotavirus in feces of sows before and after farrowing, *J. Clin. Microbiol.,* 16, 186, 1982.
349. **Brandt, C. D., Kim, H. W., Rodriguez, W. J., Arrobio, J. O., Jefferies, B. C., and Parrott, R. H.,** Simultaneous infections with different enteric and respiratory tract viruses, *J. Clin. Microbiol.,* 23, 177, 1986.
350. **Bridger, J. C., Woode, G. N., Jones, J. M., Flewett, T. H., Brysden, A. S., and Davies, H.,** Transmission of human rotaviruses to gnotobiotic piglets, *J. Med. Microbiol.,* 8, 565, 1975.
351. **Middleton, P. J., Petric, M., and Szymanski, M. T.,** Propagation of infantile gastroenteritis virus (orbi-group) in conventional and germfree piglets, *Infect. Immun.,* 12, 1975.
352. **Hall, G. A., Bridger, J. C., Chandler, R. L., and Woode, G. N.,** Gnotobiotic piglets experimentally infected with neonatal calf diarrhoea reovirus-like agent (rotavirus), *Vet. Pathol.,* 13, 197, 1976.
353. **Wyatt, R. G., Sly, D. L., London, W. T., Palmer, A. E., Kalica, A. R., Van Kirk, D. H., Chanock, R. M., and Kapikian, A. Z.,** Induction of diarrhea in colostrum-deprived newborn rhesus monkeys with the human reovirus-like agent of infantile gastroenteritis, *Arch. Virol.,* 50, 17, 1976.
354. **Mitchell, J. D., Lambeth, L. A., Sosula, L., Murphy, A., and Albrey, M.,** Transmission of rotavirus gastroenteritis from children to a monkey, *Gut,* 18, 156, 1977.
355. **Majer, M., Behrens, F., Weinmann, E., Mauler, R., Maass, G., Baumeister, H. G., Luthardt, T.,** Diarrhea in newborn cynomolgus monkeys infected with human rotavirus, *Infection,* 6, 3, 1978.
356. **Tzipori, S. and Makin, T.,** Propagation of human rotavirus in young dogs, *Vet. Microbiol.,* 3, 55, 1978.
357. **Tzipori, S. R., Makin, T. J., and Smith, M. L.,** The clincal response of gnotobiotic calves, pigs, and lambs to inoculation with human, calf, pig, and foal rotavirus isolates, *Aust. J. Exp. Biol. Med. Sci.,* 58, 309, 1980.
358. **Schwers, A., Hoyois, P., Chappuis, G., Dagenais, L., and Pastoret, P. P.,** Propagation of bovine rotavirus by cats and dogs, *Ann. Rech. Vet.,* 13, 303, 1982.
359. **Schwers, A., Dagenais, L., Chappuis, G., Pastoret, P.-P., and Calberg-Bacq, C.-M.,** Propagation of bovine rotavirus by young dogs, *J. Comp. Pathol.,* 93, 135, 1983.
360. **Offit, P. A., Clark, H. F., Kornstein, M. J., and Plotkin, S. A.,** A murine model for oral infection with a primate rotavirus (simian SA11), *J. Virol.,* 51, 233, 1984.
361. **Gouvea, V. S., Alencar, A. A., Barth, O. M., De Castro, L., Fialho, A. M., Araujo, H. P., Majerowicz, S., and Pereira, H. G.,** Diarrhea in mice infected with a human rotavirus, *J. Gen. Virol.,* 67, 577, 1986.
362. **Flores, J., Hoshino, Y., Boeggeman, E., Purcell, R., Chanock, R. M., and Kapikian, A. Z.,** Genetic relatedness among animal rotaviruses, *Arch. Virol.,* 87, 273, 1986.
363. **Ryder, R. W., Yolken, R. H., Reeves, W. C., and Sack, R. B.,** Enzootic bovine rotavirus is not a source of infection in Panamanian cattle ranchers and their families, *J. Infect. Dis.,* 153, 1139, 1986.
364. **Baumeister, B. M., Castro, A. E., McGuire-Rodgers, S. J., and Ramsey, E. C.,** Detection and control of rotavirus infections in zoo animals, *J. Am. Vet. Med. Assoc.,* 183, 1252, 1983.

365. **Flores, J., Taniguchi, K., Green, K., Perez-Schael, I., Garcia, D., Sears, J., Urasawa, S., and Kapikian, A. Z.,** Relative frequencies of rotavirus serotypes 1, 2, 3, and 4, in Venezuelan infants with gastroenteritis, *J. Clin. Microbiol.*, 26, 2092, 1988.
366. **Snodgrass, D. R. and Wells, P. W.,** The immunoprophylaxis rotavirus infections in lambs, *Vet. Rec.*, 102, 146, 1978.
367. **Snodgrass, D. R. and Wells, P. W.,** Passive immunity in rotaviral infections, *J. Am. Vet. Med. Assoc.*, 173, 565,1978.
368. **Snodgrass, D. R., Fahey, K. J., Wells, P. W., Campbell, I., and Whitelaw, A.,** Passive immunity in calf rotavirus infections: maternal vaccination increases and prolongs immunoglobulin G1 antibody secretion in milk, *Infect. Immun.*, 28, 1980.
369. **Bridger, J. C. and Brown, J. F.,** Development of immunity to porcine rotavirus in piglets protected from disease by bovine colostrum, *Infect. Immun.*, 31, 906, 1981.
370. **Snodgrass, D. R., Stewart, J., Taylor, J., Krauth, F. L., and Smith, M. L.,** Diarrhoea in dairy calves reduced by feeding colostrum from cows vaccinated with rotavirus, *Res. Vet. Sci.*, 32, 70, 1982.
371. **Saif, L. J., Redman, D. R., Smith, K. L., and Theil, K. W.,** Passive immunity to bovine rotavirus in newborn calves fed colostrum supplements from immunized or nonimmunized cows, *Infect. Immun.*, 41, 1118, 1983.
372. **Ebina, T., Sato, A., Umezu, K., Ishida, N., Ohyama, S., Oizumi, A., Aikawa, K., Katagari, S., Katsushima, N., Imai, A., Kitaoka, S., Suzuki, H., and Konno, T.,** Prevention of rotavirus infection by oral administration of cow colostrum containing antihumanrotavirus antibody, *Med. Microbiol. Immunol.*, 174, 177, 1985.
373. **Saif, L. J. and Smith, K. L.,** Enteric viral infections of calves and passive immunity, *J. Dairy Sci.*, 68, 206, 1985.
374. **Saif, L. J.,** Passive immunity to coronavirus and rotavirus infections in swine and cattle: enhancement by maternal vaccination, in *Infectious Diarrhoea in the Young*, Tzipori, S., Ed., Elsevier Science Publ., Amsterdam, The Netherlands, 1985.
375. **Saif, L. J. and Jackwood, D. J.,** Enteric virus vaccines: theroretical considerations, current status and future approaches, in *Viral Diarrheas in Man and Animals*, Saif, L. J. and Theil, K. W., Eds., CRC Press, Boca Raton, FL, chap. 14.
376. **Woode, G. N., Jones, J. J., and Bridger, J.,** Levels of colostral antibodies against neonatal calf diarrhoea virus, *Vet. Rec.*, 97, 148, 1975.
377. **Kraft, L. M.,** Responses of the mouse to the virus of epidemic diarrhea of infant mice. Neutralizing antibodies and carrier state, *Lab. Anim. Care*, 11, 125, 1961.
378. **Schoub, B. D., Prozesky, O. W., Lecatsas, G., and Oosthuizen, R.,** The role of breast-feeding in the prevention of rotavirus infection, *J. Med. Microbiol.*, 11, 25, 1977.
379. **Mebus, C. A., White, R. G., Bass, E. P., and Twiehaus, M. J.,** Immunity to neonatal calf diarrhea virus, *J. Am. Vet. Med. Assoc.*, 163, 880, 1973.
380. **Snodgrass, D. R.,** Evaluation of a combined rotavirus and enterotoxigenic *Escherichia coli* vaccine in cattle, *Vet. Rec.*, 119, 39, 1986.
381. **Offit, P. A. and Clark, H. F.,** Maternal antibody-mediated protection against gastroenteritis due to rotavirus in newborn mice is dependent on both serotype and titer of antibody, *J. Infect. Dis.*, 152, 1152, 1985.
382. **Besser, T. E., Gay, C. C., McGuire, T. C., and Everman, J. F.,** Passive immunity to bovine rotavirus infection associated with transfer of serum antibody into the intestinal lumen, *J. Virol.*, 2238, 1988.
383. **Offit, P. A. and Clark, H. F.,** Protection against rotavirus-induced gastroenteritis in a murine model by passively acquired gastrointestinal but not circulating antibodies, *J. Virol.*, 54, 58, 1985.
384. **Jayashree, S., Bhan, M. K., Kumar, R., Bhandari, N., and Sazawai, S.,** Protection against neonatal rotavirus infection by breast milk antibodies and trypsin inhibitors, *J. Med. Virol.*, 26, 333, 1988.
385. **Gaul, S. K., Simpson, T. F., Woode, G. N., and Fulton, R. W.,** Antigenic relationships among some animal rotaviruses: virus neutralization in vitro and cross-protection in piglets, *J. Clin. Microbiol.*, 16, 495, 1982.
386. **Wyatt, R. G., Mebus, C. A., Yolken, R. H., Kalica, A. R., James, H. D., Kapikian, A. Z., and Chanock, R. M.,** Rotaviral immunity in gnotobiotic calves: heterologous resistance to human virus induced by bovine virus, *Science*, 203, 1979.
387. **Wyatt, R. G., Kapikian, A. Z., and Mebus, C. A.,** Induction of cross-reactive serum neutralizing antibody to human rotavirus in calves after in utero administration of bovine rotavirus, *J. Clin. Microbiol.*, 18, 505, 1983.
388. **Bridger, J. C. and Oldham, G.,** Avirulent rotavirus infections protect calves from disease with and without inducing high levels of neutralizing antibody, *J. Gen. Virol.*, 68, 2311, 1987.
389. **Bishop, R. F., Barnes, G. L., Cipriani, E., and Lund, J. S.,** Clinical immunity after neonatal rotavirus infection. A prospective longitudinal study in young children, *N. Engl. J. Med.*, 309, 72, 1983.
390. **Ryder, R. W., Singh, N., Reeves, W. C., Kapikian, A. Z., Greenberg, H. B., and Sack, R. B.,** Evidence of immunity induced by naturally acquired rotavirus and Norwalk virus infection on two remote Panamanian Islands, *J. Infect. Dis.*, 151, 99, 1985.
391. **Grimwood, K., Lund, J. C. S., Coulson, B. S., Hudson, I. L., Bishop, R. F., and Barnes, G. L.,** Comparison of serum and mucosal antibody responses following severe acute rotavirus gastroenteritis in young children, *J. Clin. Microbiol.*, 26, 732, 1988.

392. **Saulsbury, F. T., Winkelstein, J. A., and Yolken, R. H.,** Chronic rotavirus infection in immunodeficiency, *J. Pediatr.*, 97, 61, 1980.
393. **Eiden, J., Lederman, H. M., Vonderfecht, S., and Yolken, R.,** T-cell-deficient mice display normal recovery from experimental rotavirus infection, *J. Virol.*, 57, 706, 1986.
394. **Offit, P. A. and Dudzik, K. L.,** Rotavirus-specific cytotoxic T lymphocytes cross-react with target cells infected with different rotavirus serotypes, *J. Virol.*, 62, 127, 1988.
395. **Benfield, D. A., Stotz, I. J., Nelson, E. A., and Groon, K. S.,** Comparison of a commercial enzyme-linked immunosorbent assay with electron microscopy, fluorescent antibody, and virus isolation for the detection of bovine and porcine rotavirus, *Am. J. Vet. Res.*, 45, 1998, 1984.
396. **Brandt, C. D., Kim, H. W., Rodriguez, W. J., Thomas, L., Yolken, R. H., Arrobio, J. O., Kapikian, A. Z., Parrott, R. H., and Chanock, R. M.,** Comparison of direct electron microscopy, immune electron microscopy, and rotavirus enzyme-linked immunosorbent assay for detection of gastroenteritis viruses in children, *J. Clin. Microbiol.*, 13, 976, 1981.
397. **Bohl, E. H., Saif, L. J., Theil, K. W., Agnes, A. G., and Cross, R. F.,** Porcine pararotavirus: detection, differentiation from rotavirus, and pathogenesis in gnotobiotic pigs, *J. Clin. Microbiol.*, 15, 312, 1982.
398. **Ellens, D. J., de Leeuw, P. W., Straver, P. J., and van Balken, J. A. M.,** Comparison of five diagnostic methods for the detection of rotavirus antigens in calf faeces, *Med. Microbiol. Immunol.*, 166, 157, 1978.
399. **Ellens, D. J., and de Leeuw, P. W.,** Enzyme-linked immunosorbent assay for diagnosis of rotavirus infections in calves, *J. Clin. Microbiol.*, 6, 530, 1977.
400. **Robertson, D. M., Harrison, M., Hosking, C. S., Adams, L. C., Bishop, R. F.,** Rapid diagnosis of rotavirus infection: comparison of electron microscopy and enzyme linked immunosorbent assay (Elisa), *Aust. Paediatr. J.*, 15, 229, 1979.
401. **Chernesky, M., Castriciano, S., Mahony, J., Spiewak, M., and Schaefer, L.,** Ability of TESTPAK ROTAVIRUS enzyme immunoassay to diagnose rotavirus gastroenteritis, *J. Clin. Microbiol.*, 26, 2459, 1988.
402. **Dennehy, P. H., Gauntlett, D. R., and Tente, W. E.,** Comparison of nine commercial immunoassays for the detection of rotavirus in fecal specimens, *J. Clin. Microbiol.*, 26, 1630, 1988.
403. **Marchlewicz, B., Spiewak, M., and Lampinen, J.,** Evaluation of Abbott TESTPACK ROTAVIRUS with clinical specimens, *J. Clin. Microbiol.*, 26, 2456, 1988.
404. **Thomas, E. E., Puterman, M. L., Kawano, E., and Curran, M.,** Evaluation of seven immunoassays for detection of rotavirus in pediatric stool samples, *J. Clin. Microbiol.*, 26, 1189, 1988.
405. **Conner, M. E., Gillespie, J. H., Schiff, E. I., and Frey, M. S.,** Detection of rotavirus in horses with and without diarrhea by electron microscopy and rotazyme test., *Corn. Vet.*, 73, 280, 1983.
406. **Goyal, S. M., Rademacher, R. A., and Pomeroy, K. A.,** Comparison of electron microscopy with three commercial tests for the detection of rotavirus in animal feces, *Diagn. Microbiol. Infect. Dis.*, 6, 249, 1987.
407. **Theil, K. W., Reynolds, D. L., and Saif, Y. M.,** Comparison of immune electron microscopy and genome electropherotyping techniques for detection of turkey rotaviruses and rotaviruslike viruses in intestinal contents, *J. Clin. Microbiol.*, 23, 695, 1986.
408. **Sanekata, T., Yoshida, Y., and Okada, H.,** Detection of rotavirus in faeces by latex agglutination, *J. Immunol. Meth.*, 41, 377, 1981.
409. **Hughes, J. H., Tuomari, A. V., Mann, D. R., and Hamparian, V. V.,** Latex immunoassay for rapid detection of rotavirus, *J. Clin. Microbiol.*, 20, 441, 1984.
410. **Miotti, P. G., Eiden, J., and Yolken, R. H.,** Comparative efficiency of commercial immunoassays for the diagnosis of rotavirus gastroenteritis during the course of infection, *J. Clin. Microbiol.*, 22, 693, 1985.
411. **Pai, C. H., Shahrabadi, M. S., and Ince, B.,** Rapid diagnosis of rotavirus gastroenteritis by a commercial latex agglutination test, *J. Clin. Microbiol.*, 22, 846, 1985.
412. **Sanders, R. C., Campbell, A. D., and Jenkins, M. F.,** Routine detection of human rotavirus by latex agglutination: comparison with enzyme-linked immunosorbent assay, electron microscopy and polyacrylamide gel electrophoresis, *J. Virol. Meth.*, 13, 285, 1986.
413. **Brandt, C. D., Arndt, C. W., Evans, G. L., Kim, H. W., Stallings, E. P., Rodriguez, W. J., and Parrott, R. H.,** Evaluation of a latex test for rotavirus detection, *J. Clin. Microbiol.*, 25, 1800, 1987.
414. **Herring, A. J., Inglis, N. F., Ojeh, C. K., Snodgrass, D. R., and Menzies, J. D.,** Rapid diagnosis of rotavirus infection by direct detection of viral nucleic acid in silver-stained polyacrylamide gels, *J. Clin. Microbiol.*, 16, 473, 1982.
415. **Pacini, D. L., Brady, M. T., Budde, C. T., Connell, M. J., Hamparian, V. V., and Hughes, J. H.,** Polyacrylamide gel electrophoresis of RNA compared with polyclonal- and monoclonal-antibody-based enzyme immunoassays for rotavirus, *J. Clin. Microbiol.*, 26, 194, 1988.
416. **Kasempimolporn, S., Louisirirothchanakul, S., Sinarachatanant, P., and Wasi, C.,** Polyacrylamide gel electrophoresis and silver staining for detection of rotavirus in stools from diarrheic patients in Thailand, *J. Clin. Microbiol.*, 26, 158, 1988.
417. **Lecce, J. D., King, M. W., and Dorsey, W. E.,** Rearing regimen producing piglet diarrhea (rotavirus) and its relevance to acute infantile diarrhea, *Science*, 199, 776, 1978.

418. **McNulty, M. S. and Logan, E. F.,** Longitudinal survey of rotavirus infection in calves, *Vet. Rec.*, 113, 333, 1983.
419. **Flewett, T. H.,** Rotavirus in the home and hospital nursery, *Br. Med. J.*, 287, 568, 1983.
420. **Bartlett, A. V., Moore, M., Gary, G. W., Starko, K. M., Erben, J. J., and Meredith, B. A.,** Diarrheal illness among infants and toddlers in day care centers. I. Epidemiology and pathogens, *J. Pediatr.*, 107, 495, 1985.
421. **Black, R. E., Dykes, A. C., Anderson, K. E., Wells, J. G., Sinclair, S. P., Gary, G. W., Hatch, M. H., and Gangarosa, E. J.,** Handwashing to prevent diarrhea in day-care centers, *Am. J. Epidemiol.*, 113, 445, 1981.
422. **Crouch, C. F.,** Vaccination against enteric rota and coronaviruses in cattle and pigs: enhancement of lactogenic immunity, *Vaccine*, 3, 284, 1985.
423. **Saif, L. J., Smith, K. L., Landmeier, B. J., Bohl, E. H., Theil, K. W., and Todhunter, D. A.,** Immune response of pregnant cows to bovine rotavirus immunization, *Am. J. Vet. Res.*, 45, 49, 1984.
424. **McNulty, M. S. and Logan, E. E.,** Effect of vaccination of the dam on rotavirus infection in young calves, *Vet. Rec.*, 120, 250, 1987.
425. **Fitzgerald, G. R., Welter, M. W., and Welter, C. J.,** Evaluating the performance of a porcine rotavirus vaccine, *Vet. Med.*, 81, 188, 1986.
426. **Mebus, C. A., White, R. G., Stair, E. L., Rhodes, M. B., and Twiehaus, M. J.,** Neonatal calf diarrhea: results of a field trial using a reo-like virus vaccine, *Vet. Med. Sm. Anim. Clin.*, 67, 173, 1972.
427. **Acres, S. D. and Radostits, O. M.,** The efficacy of a modified live reo-like virus vaccine and an *E. coli* bacterin for prevention of acute undifferentiated neonatal diarrhea of beef calves, *Can. Vet. J.*, 8, 197, 1976.
428. **Blackmer, P. E.,** A practitioner's experience with experimental reo-coronavirus calf diarrhea vaccine, *Vet. Med. Sm. Anim. Clin.*, March, 351, 1976.
429. **Thurber, E. T., Bass, E. P., and Beckenhauser, W. H.,** Field trial evaluation of a reo-coronavirus calf diarrhea vaccine, *Can. J. Comp. Med.*, 41, 131, 1977.
430. **De Leeuw, P. W., Ellens, D. J., Talmon, F. P., Zimmer, G. N., and Kommerij, R.,** Rotavirus infections in calves: efficacy of oral vaccination in endemically infected herds, *Res. Vet. Sci.*, 29, 142, 1980.
431. **Fitzgerald, G. R., Welter, M. W., and Welter, C. J.,** Effect of porcine rotavirus vaccination on postweaning weight gains in baby pigs, *Mod. Vet. Pract.*, 67, 609, 1986.
432. **Hoblet, K. H., Saif, L. J., Kohler, E. M., Theil, K. W., Bech-Nielsen, S., and Stitzlein, G. A.,** Efficacy of an orally administered modified live porcine origin rotavirus vaccine against postweaning diarrhea in pigs, *Am. J. Vet. Res.*, 47, 1697, 1986.
433. **Kapikian, A. Z., Midthun, K., Hoshino, Y., Flores, J., Wyatt, R. G., Glass, R. I., Askaa, J., Nakagomi, O., Nakagomi, T., Chanock, R. M., Levine, M. M., Clements, M. L., Dolin, R., Wright, P. F., Belshe, R. B., Anderson, E. L., and Potash, L.,** Rhesus rotavirus: a candidate vaccine for prevention of human rotavirus disease, in *Vaccines 85*, Lerner, R. A., Chanock, R. M., and Brown, F., Eds., Cold Spring Harbor Laboratory, Cold Spring Harbor, NY, 1985, 357.
434. **Vesikari, T., Isolauri, E., Andre, F. E., d'Hondt, E., Delem, A., Zissis, G., Beards, G., and Flewett, T. H.,** Protection of infants against human rotavirus diarrhea by the RIT 4237 live attenuated bovine rotavirus vaccine, in *Vaccines 85*, Lerner, R. A., Chanock, R. M., and Brown, F., Eds., Cold Spring Harbor Laboratory, Cold Spring Harbor, NY, 1985, 369.
435. **Kapikian, A. Z., Flores, J., Hoshino, Y., Glass, R. I., Midthun, K., Gorziglia, M. and Chanock, R. M.,** Rotavirus: the major etiologic agent of severe infantile diarrhea may be controllable by a "Jennerian" approach to vaccination, *J. Infect. Dis.*, 153, 815, 1986.
436. **Chanock, R. M., Murphy, B. R., Collins, P. L., Coelingh, K. V. W., Olmsted, R. A., Snyder, M. H., Spriggs, M. K., Prince, G. A., Moss, B., Flores, J., Gorziglia, M., and Kapikian, A. Z.,** Live viral vaccines for respiratory and enteric tract disease, *Vaccine*, 6, 1988.
437. **Vesikari, T., Isolauri, E., d'Hondt, E., Delem, A., Andre, F. E., and Zissis, G.,** Protection of infants against rotavirus diarrhoea by RIT 4237 attentuated bovine rotavirus strain vaccine, *Lancet*, 1, 977, 1984.
438. **Hanlon, P., Hanlon, L., Marsh, V., Byass, P., Shenton, F., Hassan-King, M., Jobe, O., Sillah, H., Hayes, R., M'Boge, B. H., Whittle, H. C., and Greenwood, B. M.,** Trial of an attentuated bovine rotavirus vaccine (RIT 4237) in Gambian infants, *Lancet*, 1, 1342, 1987.
439. **Anderson, E. L., Belshe, R. B., Bartram, J., Crookshanks-Newman, Chanock, R. M., and Kapikian, A. Z.,** Evaluation of rhesus rotavirus vaccine (MMU 18006) in infants and young children, *J. Infect. Dis.*, 153, 823, 1986.
440. **Perez-Schael, I., Gonzalez, M., Daoud, N., Perez, M., Soto, I., Garcia, D., Daoud, G., Kapikian, A. Z., and Flores, J.,** Reactogenicity and antigenicity of the rhesus rotavirus vaccine in Venezuelan children, *J. Infect. Dis.*, 155, 334, 1987.
441. **Flores, J., Perez-Schael, I., Gonzalez, M., Garcia, D., Perez, M., Daoud, N., Cunto, W., Chanock, R. M., and Kapikian, A. Z.,** Protection against severe rotavirus diarrhoea by rhesus rotavirus vaccine in Venezuelan infants, *Lancet*, 1, 882, 1987.
442. **Schoub, B. D. and Prozesky, O. W.,** Antiviral activity of ribavirin in rotavirus gastroenteritis of mice, *Antimicrob. Agents Chemother.*, 12, 543, 1977.

443. **Bednarz-Prasad, A. J. and John, E. I.,** Effect of clioquinol, an 8-hydroxyquinoline derivative, on rotavirus infection in mice, *J. Infect. Dis.*, 148, 613, 1983.
444. **Ward, R. L., Sander, D. S., and Knowlton, D. R.,** In vitro activities of bismuth salts against rotaviruses and other enteric viruses, *Antimicrob. Agents Chemother.*, 27, 306, 1985.
445. **Kenny, M. T., Dulworth, J. K., Bargar, T. M., Torney, H. L., Graham, M. C., and Manelli, A. M.,** In vitro antiviral activity of the 6-substituted 2-(3′,4′-dichlorophenoxy)-2H-pyranol(2,3-b)pyridines MDL 20,610, MDL 20,646, and MDL 20,957, *Antimicrob. Agents Chemother.*, 30, 516, 1986.
446. **La Bonnardiere, C., De Vaureix, C., L'Haridon, R., and Scherrer, R.,** Weak susceptibility of rotavirus to bovine interferon in calf kidney cells, *Arch. Virol.*, 64, 167, 1980.

Chapter 4

NONGROUP A ROTAVIRUSES

Linda J. Saif

TABLE OF CONTENTS

I. INTRODUCTION

Rotaviruses are classified as a genus within the family Reoviridae, based on characteristic morphology and a genome containing 11 segments of dsRNA within a double-shelled capsid.[1,2] Early studies using serologic techniques such as immunofluorescence (IF), immune electron microscopy (IEM), and ELISA established that all rotaviruses, irrespective of their host origin share a common group antigen(s) located on the inner capsid layer.[1,2]

However, recently viruses were identified which resembled rotavirus in morphology, size, number of dsRNA genome segments, and tropism for villous enterocytes, but which were antigenically distinct. These viruses, which will be referred to collectively as nongroup A rotaviruses (nonA RV), were first identified in diarrheic swine[3,4] and were subsequently identified in poultry,[5-7] calves,[8-10] man,[11-18] lambs,[9,19] rats,[20] and ferrets.[21] The nonA RV have been referred to previously as pararotaviruses (PaRV), rotavirus-like viruses (RVLV), atypical, antigenically distinct, and novel rotaviruses. The conventional rotaviruses which possess a common group antigen will be called Group (Gp) A or conventional rotaviruses.

Although nonA RV possess 11 segments of dsRNA, their genome electrophoretic migration patterns differ from one another and from conventional (GpA) rotaviruses.[10,22] At least seven distinct rotavirus groups have been proposed based on serologic (serogroup) and electropherotype (electropherogroup) analyses of prototype viruses which include: GpA (A/conventional rotaviruses); GpB (B/porcine NIRD-1);[4,23] GpC (C/porcine Cowden);[3,24] GpD (D/avian 132);[5-7] GpE (E/porcine DC-9);[25] GpF (F/avian A4);[7,26] and GpG (G/avian 555)[7,26] rotaviruses.[10,22,27,28] Each different nonA RV will be subsequently referred to by either their individual or group names as listed above. The nonA RV which are similar to electropherogroups A to G above, but which have not yet been serogrouped will be designated by their electropherogroup or referred to as unclassified nonA RV.

Group A rotaviruses are widespread and important causes of diarrheal disease in man and animals; however the incidence and significance of nonA RV infections is unclear. While in comparison to GpA rotaviruses, their detection is still rare[9-15,18,29] (except in poultry and lambs),[7,8,19,30] nonA RV (GpB) have been associated with epidemics of diarrhea predominantly among adults in China.[16,17,31-33] Moreover, the prevalence of antibodies to nonA RV is high among adults of several species.[33-39] The nonA RV also are a possible source of confusion in diagnosis of rotavirus infections based solely on electron microscopy (EM).

II. MORPHOLOGY AND MORPHOGENESIS

All nonA RV detected to date have been morphologically indistinguishable from GpA rotaviruses. Both single- and double-shelled capsid particles occur with diameters ranging from 54 to 65 nm and 67 to 75 nm, respectively,[3-25] which are similar to the sizes reported for GpA rotavirus particles.[1,2] One report suggested that nonA RV particles were slightly larger than rotaviruses.[23] A difference consistently noted between nonA RV and rotaviruses was the prevalence of core-like particles, 48 to 52 nm in diameter, in preparations of GpB rotaviruses.[16,17,23,40] Additionally, the large numbers of virus particles often seen in feces after acute infections with GpA rotaviruses are rarely seen following infections with nonA RV, particularly GpB rotaviruses.[20,34,40]

In our laboratory, using identical negative staining EM conditions (3% phosphotungstic acid [PTA], pH 7.0)[41] for all virus isolates, the GpA porcine and GpB porcine and bovine rotaviruses were predominately single shelled[40] whereas the GpA bovine, GpC porcine[3,24] and GpD and tentative GpF turkey rotaviruses were primarily double shelled.[30] An electron micrograph of negatively stained rotavirus particles, representing this observation for rotavirus serogroups A to D is depicted in Figure 1. Others have reported that different negative staining EM conditions may influence nonA RV morphology for GpB and E rotaviruses: use of PTA often led to appearance of single-shelled and core-like forms whereas with ammonium molybdate or uranyl acetate a higher proportion of particles retained their double-shelled morphology.[25,42-44]

The morphogenesis of nonA RV in small intestinal villous epithelial cells has been characterized in several species infected with GpB (rats and calves),[20,34,45] GpC (pigs),[83] GpD (chickens),[5] GpE (pigs),[25] and two unclassified nonA RV, one from pigs[46] and the second from lambs.[47] The morphogenesis resembled that described for GpA rotaviruses with viroplasm present within the rough endoplasmic reticulum and both enveloped and unenveloped virus particles (about 76 to 80 nm and 40 to 57 nm in diameter, respectively) evident. Shown in Figure 2 are GpC rotavirus particles in a villous epithelial cell from a gnotobiotic pig infected 18 h previously with the porcine GpC Cowden rotavirus. Results were similar in pigs and lambs infected with unclassified nonA RV.[46,47] In addition in these latter two studies, as in studies with GpB rat and calf rotavirus,[20,34,45] epithelial cells at the villous tips were often fused to form giant syncytia.

III. *IN VITRO* CULTIVATION

Rotaviruses in Gps A, B and C and one ungrouped nonA RV are capable of at least limited replication in a rhesus monkey kidney cell line (MA104).[36,37,47,48] However only GpA rotaviruses,[1,2,48] one avian GpD[5] rotavirus and one porcine GpC rotavirus (Cowden)[49,50] have been adapted to serial passage in cell culture. The only nonA RV adapted to serial propagation in a continuous cell line (MA104) is the porcine GpC rotavirus.[50] Procedures which were successful in serially propagating the GpC and D rotaviruses in cell culture were similar to ones initially used to propagate GpA rotaviruses[1,2] and included use of primary or MA104 cells, addition of proteolytic enzymes to the maintenance medium (trypsin or pancreatin),[5,49,50] and use of roller tubes (GpC rotavirus).[49,50] Use of proteolytic enzymes in the cell cultures obscured the ability to detect CPE, so GpC viral infectivity was monitored by immunofluorescent staining of infected cell monolayers. After 16 passages, titers of 10^7 fluorescent focus units were evident for GpC rotavirus, and typical GpC electropherotypes were obtained with dsRNA extracted from the cell-culture passaged virus.[49,50] Similar procedures have been unsuccessful for adapting GpB or non-porcine GpC rotaviruses to cell culture. Treatment with pancreatin or trypsin of field strains of porcine GpB or C rotaviruses centrifuged onto washed monolayers of MA104 cells resulted in primary infection of MA104 cells as demonstrated by immunofluorescent staining, although no cytopathic changes were evident.[36,37,47] Using this procedure it may be possible to detect GpB and C rotaviruses from fecal specimiens of infected pigs. However, the serial cultivation of GpC rotavirus in MA104 cells should provide a source of viral antigen for detecting GpC rotavirus antibodies in swine or other species or development of other diagnostic tests.[49,50] The porcine GpB rotaviruses were less consistent in their ability to cause primary infections in MA104 cells, but in contrast to other known rotaviruses, GpB rotaviruses frequently induced multinucleate syncytia.[36] Similar *in vitro* results were reported for a nonA RV from lambs.[47] These *in vitro* findings are consistent with *in vivo* observations of syncytia formation with other confirmed GpB rotaviruses including rat rotavirus[20] and a previously unclassified virus from calves which was confirmed as a bovine GpB rotavirus.[45,51] An unclassified nonA RV from swine caused similar changes in vivo.[46]

IV. PHYSICOCHEMICAL PROPERTIES

Most GpA rotaviruses with the exception of simian rotavirus SA11 are stable at pH 3.[2] The GpA rotaviruses are stable to ether but are inactivated when heated at 56°C,[2] with the exception of the Gottfried strain of porcine rotavirus.[37] Two strains of porcine GpC rotaviruses had similar

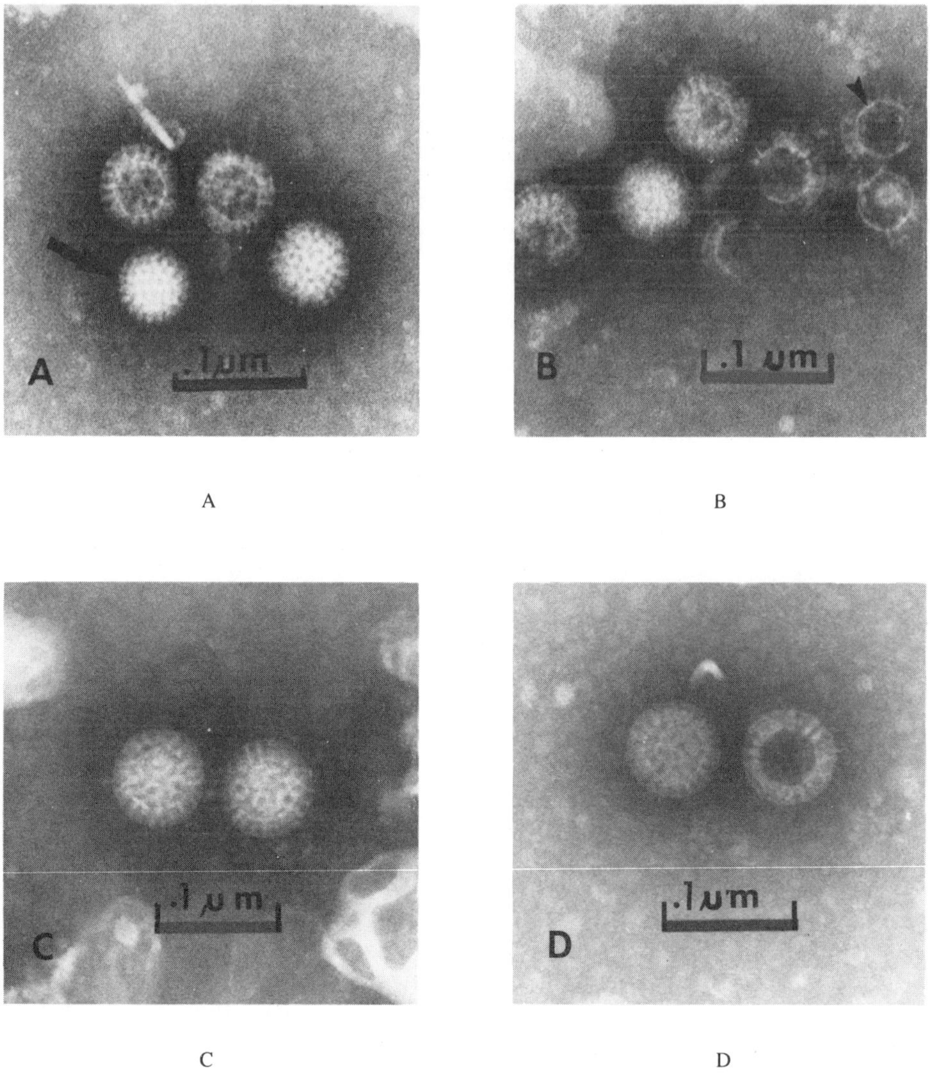

FIGURE 1. Electron micrograph of single (1A, arrow) and double-shelled rotavirus particles: (A) NCDV bovine rotavirus (GpA); (B) porcine RVLV (GpB); (C) porcine pararotavirus (GpC); and (D) turkey RVLV (GpD). Rotavirus core particles, typical of those seen with GpB rotaviruses are evident in 1B (arrow head).

properties to GpA rotaviruses: they were inactivated when heated at 56°C for 30 min, but were stable to ether and at pH 3.[37] This acid stability of most rotaviruses is in contrast to studies with a rat GpB rotavirus which was labile at pH 3.[20]

The GpA rotaviruses have buoyant densities in CsCl of 1.36 g/cm^3 and 1.38 g/cm^3 for double-shelled and single-shelled particles, respectively.[1,2] Similar densities were obtained for these two particle types from a human[18] and porcine GpC rotavirus[84] and a rat GpB rotavirus.[20] Little information is available concerning densities of other nonA RV. Only preliminary data is available on the numbers and molecular weights of nonA RV structural polypeptides and no information is available on the nonstructural viral proteins of nonA RV. Espejo and colleagues[18] observed six major proteins from purified human GpC rotavirus, ranging in molecular weight from 35 kilodaltons (kDa) to 93 kDa. By analogy with GpA rotaviruses, they concluded that proteins with molecular weights of 93 and 44 kDa might correspond to the major protein classes

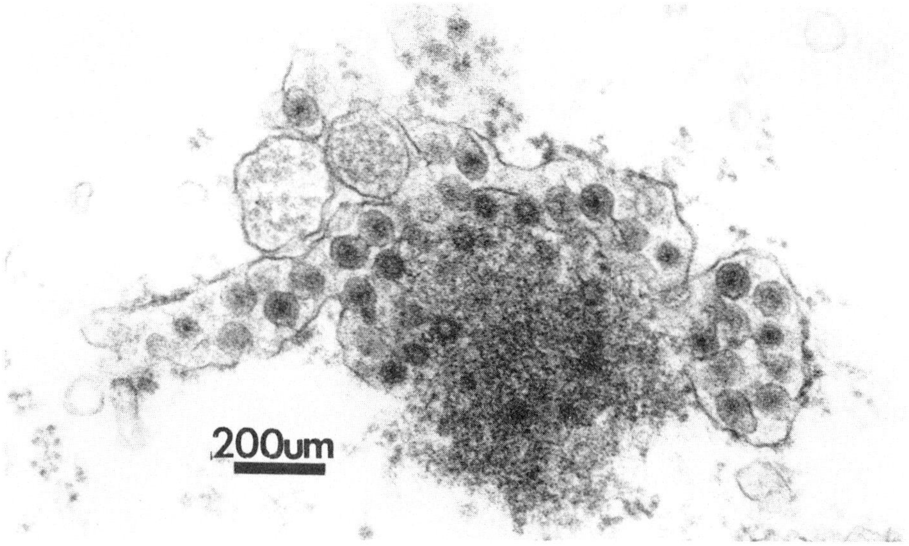

FIGURE 2. Electron micrograph of a viroplasm within the cytoplasm of a villous epithelial cell from a gnotobiotic pig infected 18 h previously with porcine GpC Cowden rotavirus. Both double-shelled (x = 78 nm) and single-shelled (x = 50 nm) particles are evident within the dilated cisterna of rough endoplasmic reticulum. Two budding particles are indicated by arrowheads.

in the virus inner capsid (VP2 and 6, respectively), whereas the other four proteins could be located in the outer capsid. Similar results were reported by Bremont and co-workers[52] following analysis of the structural proteins of a porcine GpC rotavirus. They identified six major structural proteins with migration patterns very similar to GpA rotaviruses and relative molecular weights ranging from 125 to 28 kDa. Two of the six proteins were associated with the outer shell (52 and 39 kDa) as indicated by the ability of EDTA to strip them from the virion. The 39 kDa protein was shown to be a glycoprotein based on sensitivity to endo-B-*N*-acetylglucosaminidase F, and thus may be comparable to the VP7 (37 kDa protein) of GpA rotaviruses. However, structural differences between human GpA and C rotaviruses have been reported.[53] These include removal of the inner and outer shells of GpC rotavirus by treatment with EDTA or EGTA in the presence of absence of Mg^{2+}: similar treatment of GpA rotaviruses removed only the outer shell. GpC rotavirus particles were not affected by treatment with either 1.5 *M* $CaCl_2$ or trypsin, both of which disrupt double-shelled particles of GpA rotaviruses. Such results suggest Ca^{2+} may play an important role in maintaining the association of the GpC viral core with the inner and outer protein shells. Whether nonA RV hemagglutinate erythrocytes as do GpA bovine and simian rotaviruses,[2] is uncertain, but investigators have reported that a GpB rotavirus from a Chinese infant agglutinated erythrocytes of rhesus monkeys.[54]

V. GENOME

Although genomic heterogeneity exists among strains of GpA rotaviruses, the overall electrophoretic pattern of the 11 dsRNA segments is characteristic. It is often described as 4-2-3-2 based on the division of genome segments into four size regions (I to IV).[55,56] Segments range in molecular weight from about 0.2 to 2.2×10^6 Da.[1,2] The nonA RV differ from GpA rotaviruses primarily in genome segments 7, 8, or 9 within region III. All nonA RV detected to date irrespective of species or origin, lack the tight 7-8-9 triplet characteristic of all GpA rotaviruses

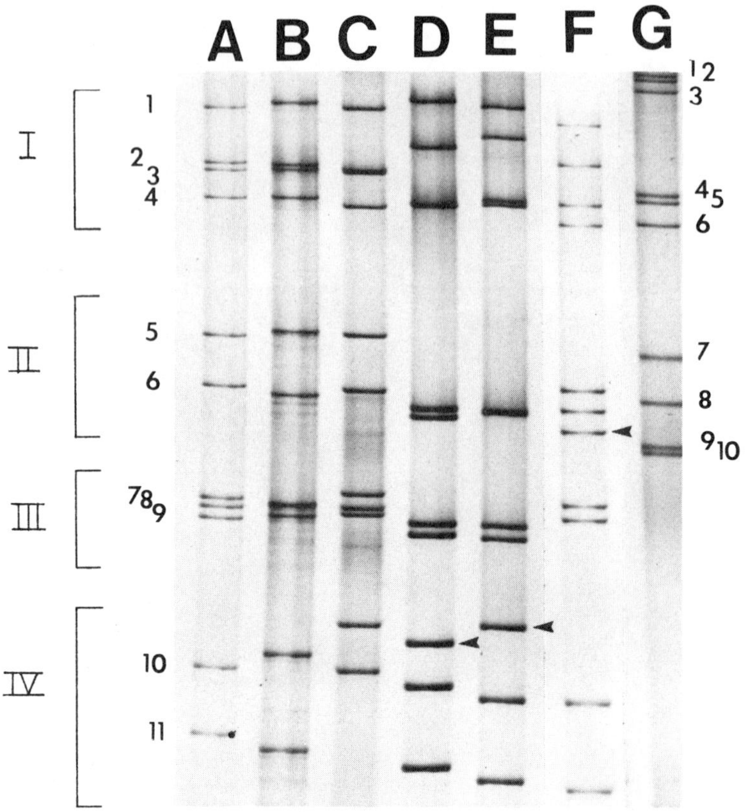

FIGURE 3. Comparison of mammalian rotavirus dsRNA genome electropherotypes in polyacrylamide gel electrophoresis (PAGE). Migration is from top to bottom. Roman numerals on the left designate dsRNA size classes,[55] while numbers indicate segments of reference rotavirus or reovirus genomes (lanes A or G). Lanes (left to right): A and C, human GpA rotaviruses, Wa and DS-1 strains, respectively; B, bovine GpA rotavirus; D and E, porcine and bovine GpB Ohio rotaviruses, respectively; F, porcine GpC Cowden rotavirus; and G, reovirus. Arrows denote bands displaced from the 7-8-9 triplet position of group A rotaviruses. (PAGE courtesy of Dr. K. W. Theil.)

examined.[10,26-27] Most nonA RV fall within electropherogroups (B to G) based on distinctive electrophoretic patterns.[10,26-27] An exception to this is GpE which more closely resembles the electropherotypes reported for GpB rotaviruses.[25] Preliminary information from our studies[10] and others[9] suggests most rotaviruses with electropherotypes resembling Gps B to D coincide closely with rotavirus serogroups (B to D) determined by serologic data.[22,27,28] Major migrational differences for viruses within the same group were conserved, but minor variations similar to those seen for GpA rotaviruses were evident. RNA electropherotypes of nonA RV from our laboratory, representing each rotavirus electropherogroup (A to D,F) are shown in Figure 3 (mammalian rotavirus electropherogroups) and Figure 4 (avian rotavirus electropherogroups).

These electrophoretic alterations can be further categorized by the method of Lourenco[55] and others,[56] as based on division of genome segments into four size regions (I to IV) as illustrated in the schematic in Figure 5. The GpB and E rotaviruses consistently had a faster migrating segment 9, while GpC, D, and F rotaviruses had a slower migrating segment 7 than GpA rotaviruses. Accordingly, GpB and E rotaviruses have different 4-2-2-3 patterns, GpC a 4-3-2-2 pattern and GpD and F different 5-2-2-2 patterns. This schematic was assembled from published electropherotypes, but suffers from failure to depict RNA electropherotypes run in a

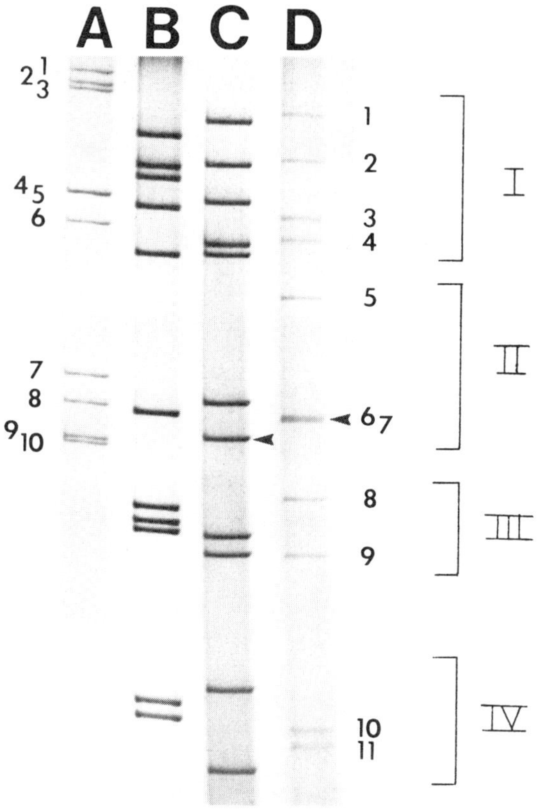

FIGURE 4. Comparison of avian rotavirus dsRNA genome electropherotypes in polyacrylamide gel electrophoresis (PAGE). Migration is from top to bottom. Roman numerals on the right designate dsRNA size classes,[55] while numbers indicate segments of a reovirus (lane A) or electropherogroup F (lane D) rotavirus genome. Lanes (left to right): A, reovirus; B, GpA turkey rotavirus; C, GpD turkey rotavirus (RVLV); and D, tentative GpF turkey rotavirus (ATR). Arrows denote bands displaced from the 7-8-9 triplet position of group A rotaviruses. (PAGE courtesy of Dr. K. W. Theil.)

single gel under uniform conditions. Nevertheless it suggests most of the nonA RV described in the literature can be tentatively classified by electropherotyping within the electropherogroups depicted. Coelectrophoresis of RNA from GpB human and rat rotaviruses with RNA from other GpB rotaviruses is needed to more accurately define their electropherotypes. All other human nonA RV described to date appear to fall within electropherogroup C,[9-15,18,57-61] with electropherotypes similar to porcine GpC rotavirus (Figures 3 and 5).[10,24] The GpB rotaviruses include Ohio bovine and porcine viruses (Figures 3 and 5)[10,40] as well as porcine, bovine and ovine viruses from the UK.[9,19,23] The GpD rotaviruses include Ohio turkey rotavirus (RVLV) (Figure 4)[10,30] and the chicken 132 rotavirus from Ireland (Figure 5).[5] The only GpE rotavirus reported to date is from swine (Figure 5).[25] A GpF rotavirus has been detected in chickens and by electropherotyping in turkeys (Figures 4 and 5).[7,26,62] The chicken Gp D, F and G rotaviruses are antigenically distinct from one another and from serogroups A, B, C, and E.[7,28] One additional distinct avian electropherogroup has have been identified, but it was not established if this virus was antigenically related to those previously reported.[26]

Because variations in electropherotypes do not always reflect antigenic variation among GpA rotaviruses,[56] electropherogrouping can be only a tentative means of assigning nonA RV

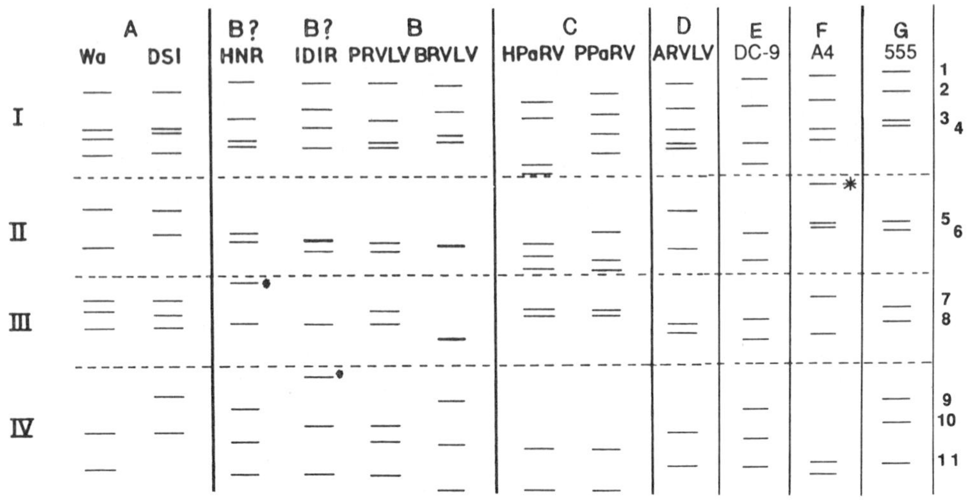

FIGURE 5. Schematic of RNA electropherotypes of serogroup A to G rotaviruses. The RNA patterns shown are representative of the following rotaviruses: human GpA (Wa, DS1);[1,2] human and rat GpB (HNR, IDIR);[16,20] porcine and bovine GpB (Ohio P and B RVLV);[9,10,23,40] human and porcine GpC (H PaRV; Ohio P PaRV);[3,11-15,24,57] avian GpD (turkey-A RVLV);[5,30] GpE (Porcine DC-9);[25] avian GpF (A4);[62] and avian GpG (555).[7] Roman numerals on the left designate dsRNA size classes, asterisks denote bands whose placement is unclear, and numbers on the right indicate segments of a reference rotavirus genome (lane G, electropherogroup G).

to groups. The final definition of distinct rotavirus groups should rely on serologic comparisons (serogroups).

Other methods utilized for detailed comparison of rotavirus genomes include hybridization techniques[2,56,57,63,64] and oligonucleotide fingerprinting of individual RNA segments.[22,27] This latter test has confirmed the electropherotypic differences between the porcine GpB (NIRD-1) rotavirus, porcine GpC (Cowden) rotavirus, avian GpD (132) rotavirus, porcine GpE (DC-9) rotavirus, and porcine GpA (OSU) rotavirus.[27] However, because of the greater complexity of the fingerprinting technique and lack of absolute sequence conservation at the termini of GpB and E rotaviruses,[27] this technique is less suitable for routine genomic comparisons of the nonA RV.

A technique of greater ease and more widespread diagnostic application for comparison of rotavirus genomes is nucleic acid cross-hybridization. To date, only limited cross-hybridization studies have been conducted between the RNAs of the various rotavirus groups, largely due to the limited availability of dsRNA, particularly from human nonA RV isolates. In three studies where dot-blot or Northern-blot hybridizations have been done to assess nucleic acid relationships among rotavirus groups, the following results were reported.[57,63,64] In the first study, a ^{32}P-labeled dsRNA genomic probe prepared from a human Australian rotavirus which was antigenically and electrophoretically similar to other GpC rotaviruses, cross-hybridized with two porcine GpC rotaviruses (Cowden and Strain 18) and a human Brazilian GpC rotavirus.[57] Only a slight cross-reaction was noted with GpA rotavirus RNAs. In a second study, a genomic RNA probe made against the Chinese GpB human rotavirus cross-hybridized with the NIRD-1 GpB porcine rotavirus.[63] In the third study, ^{32}P-labeled cDNA probes were prepared from GpA (SA11) rotavirus, GpB (rat and bovine) rotaviruses, and GpC (Cowden) porcine rotavirus.[64] The GpA probe hybridized in Northern blots to mammalian GpA rotaviruses, but not to GpB (rat, bovine and porcine) rotavirus, GpC (Cowden) porcine rotavirus or an avian nonA RV. Probes to GpB rotaviruses (rat and bovine) hybridized to all GpB rotaviruses examined (rat, bovine and porcine) and to segment 1 of GpA rotavirus, but not to GpC rotavirus. Hybridization was also observed in dot-blots between the Chinese human GpB rotavirus and the rat rotavirus cDNA probe. Probes to GpC rotavirus hybridized only to GpC rotavirus and to segment 1 of GpA rotaviruses, but not to GpB rotaviruses.

The *in vitro* transcription of a human GpC rotavirus has been described recently by Jashes and colleagues.[53] Different procedures were required to activate the human GpC rotavirus associated RNA-dependent RNA polymerase. Whereas temperatures were similar to those used for GpA rotaviruses, the polymerase of GpC rotavirus was not activated by chelating agents or by thermal shock as in GpA rotaviruses, although the enzyme was similar to those described for GpA rotaviruses. RNA synthesis occurred only when all four ribonucleoside triphosphates and Mg^{2+} were present during thermal shock. Analysis of *in vitro* transcripts revealed 11 RNA species with migration patterns similar to that of the genomic RNA.

VI. ANTIGENIC RELATIONSHIPS

A. *IN VITRO*

The availability of small numbers of nonA RV isolates, often in limited quantities, and failure of most nonA RV to serially replicate in cell culture[23,24,36,47] has hampered investigation of antigenic relationships and precluded use of many conventional serologic tests for establishing these relationships. The following assays have been used most frequently to detect rotavirus common group antigens and thereby establish rotavirus serogroups: immune electron microscopy (IEM); immunofluorescence (IF); and ELISA.[10,28] The antigen source for these tests was from feces, intestinal contents, or small intestinal enterocytes of infected animals. Antisera for these tests were prepared in gnotobiotic or SPF animals or were paired sera from infected individuals. The antigenic relationships among GpA rotaviruses and nonA RV from different species, investigated at our facilities and by others are summarized in Table 1.

Our data concur with findings of other investigators who also reported lack of reciprocal cross-reactivity between GpA to D rotaviruses[9,10,22,27] and more recently GpE rotaviruses.[27] Other putative GpB antigenically related rotaviruses include the UK bovine D522 and ovine E1101 which show a one-way cross-reaction with NIRD-1 porcine GpB rotavirus.[9] Thus, in summarizing the results in Table 1, occurrence of serogroup B rotaviruses have been confirmed in swine, cattle, man, rats and possibly sheep. Serogroup C rotaviruses have been detected only in swine and man. Serogroup E rotaviruses were identified only in swine, and serogroup D, F, and G rotaviruses were reported only in poultry.[7,26,28,30]

B. *IN VIVO*

In our studies, the Ohio bovine GpB rotavirus infected and caused profuse diarrhea in young gnotobiotic (GN) piglets.[10,65] Reciprocal cross-protection experiments were conducted between the Ohio bovine and porcine GpB rotaviruses in eight GN piglets. No cross-protection against either diarrhea or virus shedding was evident between the two GpB rotaviruses. Thus, although these two viruses share a common group antigen(s) by IF and IEM (Table 1), they may represent distinct serotypes within GpB.[65]

VII. DIAGNOSIS

Because of the failure of most nonA RV to serially propagate in cell culture, diagnosis has relied mainly on similar techniques initially used to detect nonreplicating *(in vitro)* GpA rotaviruses. The differentiation of nonA RV from GpA rotaviruses remains a problem in attempts to diagnose these viruses by routine negative staining EM techniques, since nonA RV are morphologically indistinguishable from GpA rotaviruses. Detection of samples which are positive for rotavirus by EM, but negative for GpA rotaviruses by IF, ELISA or other serologic

TABLE 1
Antigenic Relationships of Conventional and Nongroup A Rotaviruses Determined by Reciprocal Immunofluorescence and/or Immune Electron Microscopy[a]

	Rotavirus serogroups													
	Gp A	Gp B							Gp C				Gp D		Gp E
		Swine		Cattle			Human	Rat	Swine		Humans		Chicken	Turkey	Swine
Serogroup Antiserum	Many spp	NIRD-1	Ohio	Ohio	UK D522	Mebus	Chinese	IDIR	Cowden	UK D238	Australia	Brazil	132	RVLV	DC-9
A/many spp	+	–	–	–	–	–	–	–	–	–	–	–	–	–	–
B/Porcine NIRD-1	–	+	+	NT	–	NT	+	NT	–	–	NT	NT	–	NT	–
B/Porcine Ohio	–	+[b]	+	+	NT	+	+	+	–	NT	NT	NT	NT	–	NT
B/Bovine D522	–	NT	+	+	+[c]	+	+	+	–	NT	NT	NT	NT	–	NT
B/Human Chinese	–	+	+	+	NT	+	+	+	–	NT	NT	NT	NT	NT	NT
B/Rat IDIR	–	NT	+	+	NT	+	+	+	–	NT	NT	NT	NT	NT	NT
C/Porcine Cowden	–	–	–	–	–	–	–	–	+	+	+	+	–	–	–
C/Human Aust.	–	NT	–	–	NT	NT	NT	NT	+	NT	+	NT	NT	NT	NT
C/Human Brazil	–	NT	NT	NT	NT	NT	NT	NT	–	NT	+	+	NT	NT	NT
C/Human U.K.	–	NT	NT	NT	NT	NT	NT	NT	+	NT	+	NT	NT	NT	NT
D/Chicken 132	–	–	NT	NT	–	NT	NT	NT	–	–	NT	NT	+	+	–
D/Turkey Ohio RVLV	–	NT	–	–	NT	NT	NT	NT	–	NT	NT	NT	+	+	NT
E/Porcine DC-9	–	–	NT	NT	NT	NT	NT	NT	–	NT	NT	NT	–	NT	+

Data in this table were compiled from information in References 7, 9, 10, 25—28, 30, 42, 57, 63 and 64. Rotavirus strains isolated in this laboratory include porcine and bovine Ohio GpB, bovine Mebus GpB, porcine Cowden GpC, and turkey RVLV GpD.

Bridger, J. C., personal communication, 1986.

Snodgrass, D. R., personal communication, 1986.

tests which detect groupA rotavirus-specific antigens is suggestive of the presence of nonA RV. Further tests must then be done to distinguish the serogroups of nonA RV. Alternatively, IEM or solid phase IEM (SPIEM)[66] using rotavirus group-specific antisera may be used to distinguish rotavirus serogroups.[3,4,7,8,10,24,30,40,42,62,63] Both ELISA and IF tests can be used to detect nonA RV by using specific hyperimmune antisera or monoclonal antibodies which react with group-specific antigens common to each virus serogroup.[5,7,8,20,22-25,27,28,36,37,40,42,57,67-70] Although no researchers have analyzed the comparative sensitivity and specificity of the various methods for detection of nonA RV, preliminary data using these various assays in a time sequence study of GpB rotavirus infections in infant rats, suggested ELISA and SPIEM were the most sensitive methods for detecting virus in intestinal specimens.[71] Negative-stain EM and electropherotyping were the two least sensitive methods, and detected virus only at 1 DPI. The transient detection of GpB rotavirus in feces or gut contents from experimentally challenged rats[71] or pigs[83] suggest that researchers relying on the latter two techniques may underestimate the role of nonA RV in diarrhea in humans and animals. Recently, monoclonal antibodies have been prepared against animal nonA RV which are reactive with a group-specific antigen of GpB rotaviruses[70] or porcine GpC rotaviruses.[72] Availability of these immunoreagents for use in ELISA or IF assays should greatly facilitate diagnosis of nonA RV infections in animals and humans. Infected gut sections or, where possible, cell cultures or virus antigens extracted from feces have been successfully used as antigen sources for detection of antibodies which cross-react with viruses within each serogroup.[35-39,73,74]

RNA electropherotyping is also useful in detecting nonGpA rotaviruses which lack the tight 7,8,9 triplet. The migration patterns of rotavirus dsRNA segments, although slightly different for each rotavirus isolate are somewhat characteristic for each group of rotaviruses.[9,10,22,26,27,30,62] Thus they may provide tentative information for grouping nonA RV isolates, pending serogroup definition by serological data. RNA electropherotyping may be especially useful for detecting mixed infections of GpA and/or nonA RV as evident by additional RNA bands in the gels.

As discussed previously, nucleic acid hybridization probes have been used successfully to differentiate rotavirus groups.[57,63,64] However, the sensitivity and specificity of these probes for routine detection of nonA RV from fecal specimens has not been determined.

VIII. EPIDEMIOLOGY

A. INFECTIONS IN MAN

The nonA RV have been found in association with human enteric viral infections in the following geographic locations; North, Central and South America, Europe, Australia, China and Japan.[11-18,31-34,53,54,57-61,63] The GpB rotaviruses have been identified from human beings only in China and the U.S.,[16,17,34,54,63] while human GpC or electropherotypically GpC rotaviruses have been reported to occur in the other regions.[11-15,18,53,57-61]

The nonA RV in man were recognized initially by their atypical electropherotypes or by positive EM diagnosis combined with negative immunologic assays using antisera for GpA rotaviruses. In studies conducted on feces from hospitalized infants with diarrhea, detection of nonA RV was infrequent compared with detection of GpA rotaviruses. Human nonA RV (electropherogroup C) were found in only 0.25 to 7% of rotavirus samples tested by EM and/or electropherotyping.[11-14,18] These viruses were detected from children ranging in age from 4 to 48 months.[11-14,18,58-61]

An electropherotypically and antigenically distinct human rotavirus (electropherogroup B) was detected from mainly adults with acute epidemic diarrhea in nation-wide disease outbreaks which occurred in 1982 to 1983 in China. The virus was associated with a disease incidence of 5 to 31%.[16,17,31-33] Local outbreaks continue to occur and it has now been established that these viruses are serotypically related to other serogroup B rotaviruses. It was proposed that this disease was spread initially by contaminated water and later by the fecal-oral route.[16] The pattern of this epidemic, whereby all age groups (but especially adults) were clinically infected, differed significantly from infections caused by GpA rotaviruses, which are often subclinical in adults. This may be correlated with a low prevalence of antibody to GpB rotavirus, compared with a high prevalence of antibody to GpA rotavirus in adults.[38,57,73,74] More recently, an outbreak of mild diarrhea was described in China among 2- to 8-day-old hospitalized infants.[54] Electropherogroup B rotaviruses were identified in 67% of the fecal specimens and one isolate was shown to be antigenically related to the Chinese GpB adult diarrhea rotavirus characterized previously. Antibodies to GpB rotavirus have also been detected in swine and rats in China, but it is unknown whether these antibodies reflect exposure to human GpB rotaviruses or antigenically related viruses which occur in these species.[33,74]

A virus antigenically related to a GpB rotavirus, known to cause enteric disease in rats, was found to be associated with diarrhea in about 38% (6/16) of subjects examined (adults and children) who had gastroenteritis not attributed to GpA rotaviruses or other enteric viruses.[34] However, at least two of these subjects had direct or indirect contact with rats experimentally infected with the rat GpB rotavirus and all but one specimen came from individuals associated with the hospital or hospital personnel. Recognition of GpB rotaviruses or antibodies in man, rats and pigs, and GpC rotaviruses from only man and pigs has raised questions about the possible zoonotic significance of these nonA RV. However, as with GpA rotaviruses, there is as yet no direct evidence for the natural occurrence of cross-species infection or disease induced by nonA RV.

Limited serological investigations have been done in attempts to further assess the prevalence of nonA RV in human populations. In many of these studies antigenically-related rotaviruses from animals were used as a source of virus or virus antigen. Antibodies were assayed using IEM, IF, counterimmunoelectrophoresis (CIE) or ELISA tests. The following results were reported for prevalence of antibodies to nonA RV in humans (Table 2). Using human or porcine GpB rotavirus purified from stools as the antigen, the percentage of sera from adults positive for GpB rotavirus antibodies was 3 to 10% in the U.K.;[28,38] 12 to 41% in China;[33] 5 to 18% in Thailand and Hong Kong, respectively;[33,73] 0 to 15% in Australia,[33,73] 2 to 13% in the U.S. and Canada; and 10% in Kenya.[33,73] No sera from children in the above studies were seropositive for GpB rotavirus.[28,73] Using rat GpB rotavirus as the antigen source in ELISA, the prevalence of antibodies in human sera from the U.S. was 51% and 88% positive for children and adults, respectively.[34] In Mexico, 42% of adult sera and no sera from children were positive for antibodies to human GpC rotavirus by IEM.[18] Using IF with porcine Cowden GpC rotavirus as antigen, none of children's sera and 11% of adult sera from the U.K. were positive for GpC rotavirus antibodies.[28] These very limited surveys of antibody prevalence to nonA RV in human sera support the virus detection data which suggests that infection of humans with GpC rotaviruses is rare, while infection with GpB rotaviruses may be more prevalent, but only in certain geographical areas. There are a number of possible explanations for these findings, however, which may result in lower detection rates for nonA RV than their true incidence. These include the observations that nonA RV are often shed in lower numbers as intact particles in stools, are shed for a shorter duration and are less stable than GpA rotaviruses; this would often preclude their detection by EM or electropherotyping which successfully detect GpA rotaviruses.[10,20,34,71,75] Second, these infections may be more transient than GpA rotavirus infections, thereby evoking lower antibody responses, less readily detected by the lower sensitivity of assays currently available for non-cell culture adapted viruses. Third, many studies of nonA RV antibody prevalence in human populations have relied on the use of heterologous nonA RV for antigen; this may be less sensitive than techniques using homologous antigen.[42] Finally, nonA RV infection may occur together with GpA infections but in lower titer, thus precluding their detection and differentiation from GpA. Besides possible differences in detection sensitivities for GpA and nonA RV, the nonA RV may have different patterns of infection. They may evoke

TABLE 2
Prevalence of Antibody to Three Rotavirus Serogroups (A, B, C,) in Human Sera

Source	Country	No. of sera tested	Year(s) of sera collection	Rotavirus serogroup/virus[a] % positive		
				A/bovine	**B/porcine**	**C/porcine**
Adults[b]	U.K.	38	pre-1984	95	3	11
Children[b]	U.K.	15	1982	93	0	0
					B/human	
Adults[c]	U.K.	100	1986	ND[a]	10	ND
Veterinarians[c]	U.K.	110	1983—1984	ND	4	ND
Adults[d]	China	249	NR[a]	ND	12—41	ND
	Hong Kong	34	NR[a]	ND	18	ND
	Australia	40	NR[a]	ND	15	ND
	U.S.	202	NR[a]	ND	10	ND
	Canada	40	NR[a]	ND	13	ND
				A/SA11	**B/human**	
Adults[e]	U.S.	107	1984—86	97	2	ND
Children[e]	U.S.	57	1984—86	88	0	ND
Adults[e]	Australia	10	1981—83	100	0	ND
	Canada	10	1983—85	90	10	ND
	Kenya	10	1985	100	10	ND
	Thailand	20	1984—85	100	5	ND
Adults[f]	U.S.	33	NR	ND	88	ND
Children[f]	U.S.	35	NR	ND	51	ND
					A/SA11	**C/human**
Adults[g]	Mexico	12	NR	100	ND	42
Children[g]	Mexico	5	NR	ND	ND	0

[a] Virus antigens used included A/bovine U.K. or A/SA11; B/porcine NIRD-1 or B/human Chinese; C/porcine Cowden or C/human Mexican; ND = not determined.
[b] Data from Reference 28.
[c] Data from Reference 38.
[d] Data from Reference 33.
[e] Data from Reference 73.
[f] Data from Reference 34.
[g] Data from Reference 18.

milder, transient diarrhea among children and thus be less frequently detected among children hospitalized for acute diarrhea.

B. INFECTIONS IN OTHER ANIMALS

Infections with nonA RV have been reported in swine, poultry, cattle, sheep, rats and ferrets in the U.S., South America and Europe.[3-10,19-30,40,46,51,67,75,76] Serologic evidence exists for occurence of GpB rotavirus infections in rats and swine in the U.S. and China and sheep and cattle from the U.K. and U.S., but not from China (Table 3).[25,28,33,35-38,77] Although antibodies to GpC rotaviruses were reported in cattle in the U.K., no GpC rotaviruses have been identified in this species (Table 3).[28]

The most extensive studies of antibody prevalence to nonA RV have been conducted in swine and poultry,[35-39,77] the two species from which the greatest numbers of nonA RV have been identified. A comparison of antibody prevalence to nonA RV among different age groups of swine in the U.S. and U.K. is in Table 4.[25,35-37,77] While antibodies to GpB rotavirus were less prevalent in the U.S., the incidence of antibodies to GpA and C rotaviruses were similar in the U.S. and U.K. Antibodies to all groups of nonA RV were less prevalent than antibodies to GpA rotaviruses and less prevalent in younger compared with older swine.

TABLE 3
Prevalence of Antibody to Three Rotavirus Serogroups (A, B, C) in Sera from Animals

Source	Country	No. of sera tested	Year(s) of sera collection	Rotavirus serogroup/virus[a] % positive A/bovine	B/porcine	C/porcine
Cattle[b]	U.K.	118	1973—1983	91	20	24
Sheep[b]	U.K.	50	1981	43	19	0
					B/human	
Cattle[c]	U.K.	59	1983	ND[a]	71	ND
Sheep[c]	U.K.	11	1983	ND[a]	91	ND
Pigs[c]	U.K.	67	1983	ND[a]	97	ND
					B/human	
House Rats[d]	China	60	NR	ND	47	ND
Wistar Rats[d]	China	35	NR	ND	17	ND
Pigs[d]	China	202	NR	ND	36	ND
Cattle/Sheep/Horses[d]	China	46/29/5	NR	ND	0	ND

[a] Virus antigens used included A/bovine U.K.; B/porcine NIRD-1 or B/human Chinese; C/porcine Cowden; ND = not determined.
[b] Data from Reference 28.
[c] Data from Reference 38.
[d] Data from Reference 33.

The widespread prevalence of antibodies to nonA RV in swine, particularly GpC in the U.S. and U.K., suggests these latter viruses are enzootic. In the U.S. at least, the serologically less prevalent GpB rotaviruses may potentially occur as epizootics similar to GpB rotavirus epidemics in adults in China. The nonA RV may play a role in the etiology of porcine diarrhea under field conditions, as these viruses were shown to do experimentally in gnotobiotic pigs.[10,23,24,40,77] At present, the detection rate for nonA RV is low compared with that for GpA rotaviruses. Only 5% of porcine rotaviruses identified from cases of piglet diarrhea in one diagnostic lab in the U.K. in 1983 were nonA RV.[29] In a study conducted on diarrheal feces collected from Brazilian swine, 21% were positive for GpA rotavirus, but only 1% were positive for an electropherogroup B rotavirus.[75] However, improved diagnostic techniques for nonA RV may alter this detection rate in the future. As for human nonA RV, such factors as instability of the viruses, lower levels, and shorter duration of virus shedding greatly influence virus detection rates.

Although data regarding natural infections with nonA RV is limited, in our studies 12 outbreaks of diarrhea occurred among nursing and weaning pigs from which GpC rotaviruses were detected by IEM. Four cases were from 3- to 10-day-old pigs; five from 11- to 27-day-old pigs; and the remaining three from weaned pigs (3 to 7 days post-weaning). All the samples from weaned pigs and two of the other cases contained other enteric viruses including rotaviruses, caliciviruses, or small round viruses. GpB rotaviruses were identified from pigs from six outbreaks of diarrhea using IEM and RNA electropherotyping. Four cases occurred in 10- to 28-day-old nursing pigs and two were from weaning pigs. GpA rotaviruses were also present in two of six samples.

A comparative study of the pattern of GpA and C rotavirus excretion in pigs from three closed swine breeding herds was conducted in Belgium using ELISA tests for virus detection.[67] Both GpA and C rotaviruses were enzootic in the herds tested. Peak GpA rotavirus excretion occurred at 4 to 5 weeks of age, and GpC rotavirus was rarely detected before 3 weeks of age except in one herd. A second period of GpA rotavirus excretion occurred in many nursing piglets 3 to 4 weeks after the first, but pigs weaned at 6 to 10 weeks of age rarely excreted virus after weaning.

TABLE 4
Prevalence of Antibody (% Positive) to Four Different Rotavirus Serogroups in Swine

Antisera origin (Country)	Rotavirus serogroups											
	GpA			GpB			GpC			GpE		
	3—8 weeks	10—26 weeks	Adult	3—8 weeks	10—26 weeks	Adult	3—8 weeks	10—26 weeks	Adult	3—8 weeks	10—26 weeks	Adult
U.S.[a]	100	100	100	17	NT	24	86	59	100	NT	NT	NT
U.K.[b]	100	100	97	70	92	92	58	86	79	0	58	61

[a] Data on GpA and C rotavirus antibodies from Reference 37 and GpB rotavirus antibodies from Reference 36.
[b] Data on GpA, B and C rotavirus antibodies from Reference 35 and GpE rotavirus antibodies from Reference 25.

In contrast, most pigs weaned at 4 weeks of age shed GpC rotavirus at weaning, but weaning diarrhea was not observed. About 50% of fecal samples from suckling pigs with diarrhea contained GpA rotavirus. GpC rotavirus excretion was associated with subclinical infections in most older pigs in two herds (3 to 10 weeks of age) and with diarrhea in 70% of virus positive younger pigs (1 to 4 weeks of age) in the third herd.

Only in poultry, is there presently additional serologic and antigenic data regarding the incidence of nonA RV. In a survey of antibody prevalence to GpA and D rotaviruses in broiler breeder chickens, antibodies to each virus occurred with similar frequencies (63% and 70% positive for GpA and GpD antibodies, respectively).[39] Similarly in 10- to 21-day-old diarrheic turkey poults, a rotavirus-like virus (RVLV), subsequently identified as a turkey GpD rotavirus, was as prevalent (detected in 60% of flocks) as GpA rotaviruses (detected in 58% of flocks).[30]

In a subsequent study comparing prevalence of electropherogroup A and D rotaviruses in diarrheic turkeys, 69% of rotaviruses detected were electropherogroup A and 31% were electropherogroup D.[78] In another survey, RVLV (electropherogroup D) were detected most frequently in 3- to 4-week-old turkey poults and occurred in 67% of diseased flocks but only 26% of normal flocks. By comparison, GpA rotaviruses were detected in 22% of diseased flocks, but 26% of normal flocks.[79] A high percentage of turkey flocks examined (73%) had mixed infections with RVLV and other enteric viruses including GpA rotaviruses,[78] reoviruses, astroviruses, adenoviruses, or enteroviruses.[30,79] The turkey RVLV is electropherotypically and antigenically similar to a GpD (A/132) chicken rotavirus.[85] It is antigenically unrelated to rotaviruses in Gps A, B, and C.[5,30] Its role as a cause of diarrhea in turkeys is not defined, due to the presence of multiple enteric viruses in poults with enteritis and failure to separate these viruses for pathogenicity studies.[6]

Recently in a longitudinal survey of broiler chickens, two other serogroups of nonA RV besides GpA and D rotaviruses were identified. They were designated A4 (GpF) and 555 (GpG).[7,26] Recent studies suggest that these two avian rotaviruses are antigenically distinct from serogroup A to E rotaviruses and thus may be classified as new serogroups F and G, respectively.[28] Results of this longitudinal survey in chickens further indicated that rotaviruses antigenically and RNA electropherotypically similar to GpA and D rotaviruses were prevalent in birds at 22 to 29 days of age. Rotaviruses with dsRNA electropherotypically similar to A4 (GpF) were prevalent at 9 to 16 days of age, whereas between 43 to 50 days of age, the 555 (GpG) dsRNA electropherotypes predominated. Thus both GpA rotavirus and nonA RV infections occurred in waves with each wave of infection lasting about 1 week. However, different patterns of rotavirus shedding were observed in another longitudinal survey of broiler flocks.[26] A second nonA RV designated atypical rotavirus was recently described in diarrheic turkey poults and is electropherotypically similar to the chicken A4, serogroup F rotavirus (Figure 4).[62] As assessed by electropherotyping it was detected in the lowest prevalence (4%) from 1- to 12-week-old turkey poults. An atypical rotavirus with an electropherotype similar to serogroup F avian rotavirus was also described in 7- to 10-day-old chicks with diarrhea in Argentina.[76] The antigenic relationship of these electropherogroup F rotaviruses to the avian serogroup F rotavirus has not been analyzed.

Only limited information is available on the prevalence of nonA RV infections in cattle, sheep and rats. In one report based on electropherotyping, nonA RV infections were more prevalent in lambs in England than GpA rotavirus infections.[19] Only GpB rotaviruses have been identified in rats and there are no reports of group A or C rotavirus infections in this species.[33,34] Although antibodies to GpC rotavirus have been reported in cattle,[28] no electropherogroup C rotaviruses have been isolated from cattle or sheep.

Limited data are also available on the ability of nonA RV from one species to cross-infect other species. Preliminary experiments suggest a similar situation may occur as with GpA rotaviruses, with not all nonA RV capable of cross-infection between species.[80] Both UK and

TABLE 5
Comparative Pathogens of Porcine Nongroup A Rotaviruses in Gnotobiotic Pigs and Infectivity in MA104 Cells

Rotavirus sserogroup /strain	Infection of gut enterocytes		Villous atrophy		Diarrhea in pigs[a]		Infect MA104 cells	
	+/–	1° region	+/–	1° region	GN	Conv'l	+/–	Syncytia
A/OSU, Gottfried	+	Villous sides and tips	+	J,I	+	+	+	–
B/Ohio N-338	+	Villous tips	±	D,J or I	+ (Transient)	+	+[b]	+
C/Cowden	+	Villous sides and tips	+	J,I	+	+	+	–

[a] GN = gnotobiotic; Conv'l = conventional nursing or weaning pigs.
[b] Replicate upon initial but not serial passage.

Ohio bovine Gp B rotaviruses infected gnotobiotic (GN) lambs.[9,86] A lamb tentative GpB rotavirus infected and caused diarrhea in GN piglets, but three human GpC rotaviruses did not.[9,57] Similarly, a human GpB rotavirus failed to replicate in GN piglets in one study.[42] Ohio bovine and porcine GpB rotaviruses produced reciprocal cross-infections and mild diarrhea in GN pigs or calves.[65]

IX. PATHOGENESIS AND PATHOGENICITY

The pathogenesis of nonA RV infections has been described in chickens, pigs, calves, and rats.[5,6,8,10,17,23-25,40,45-47,50,71,77] However, little information is available concerning the pathogenesis of human nonA RV infections. Infection of SPF chickens with (132) chicken GpD rotavirus[24,50,77] or GN piglets with Cowden porcine GpC rotavirus resulted in infection of villous enterocytes and diarrhea resembling that produced by GpA rotavirus. Infection of GN piglets with porcine GpB Ohio[40] or GpE DC-9 rotaviruses,[25] GN calves with bovine GpB Ohio or Mebus rotaviruses,[8-10] and rats with rat GpB rotavirus[20] resulted in an acute transitory diarrhea; infection was often restricted to discrete foci of enterocytes near the villous tips or the luminal one-third of the villi. In our studies, infection of GN pigs with GpB or C rotaviruses or GN calves with GpB rotaviruses induced a more rapid onset of diarrhea and led to shedding of fewer virus particles for a shorter period of time than GpA rotavirus infections.[83] The comparative pathogenesis of porcine nonA RV in GN piglets is summarized in Table 5. Profuse watery yellow stools followed by brown liquid stools containing milk curds occurred in most GN piglets inoculated orally with GpA or nonA RV. In some litters, stools were milky white resembling undigested milk. Diarrhea, accompanied by anorexia was usually observed 24 to 36 h post-exposure (PE) to GpA rotaviruses.[81] In GpB in particular, but often also in GpC rotavirus infections, the onset of diarrhea and anorexia usually occurred more rapidly (by 12 to 24 h PE).[24,40] Occasionally, vomiting was observed in GpB rotavirus-infected piglets and upon necropsy the stomach was often greatly distended with milk. Following experimental inoculation, GpB rotaviruses induced an acute transient diarrhea which persisted about 3 d, but caused no mortality in pigs exposed at 5 to 6 d of age.[40] By comparison, GpA and C rotaviruses caused dehydration and death in most GN pigs exposed when <6 d of age.[24,81] Pigs >6 d old when exposed generally developed transient diarrhea which persisted 3 to 7 days, but little mortality ensued.

Four nonA RV, including two GpB rotaviruses (one from calves[45] and one from rats[20]), and two unclassified nonA RV (one from lambs,[47] and one from pigs[46]) differed from other rotaviruses in their ability to induce syncytia formation in villous enterocytes. Although the syncytia-forming virus from calves was initially described as a 100 nm fringed particle,[45] recent studies in our laboratory,[84] and another laboratory[51] have confirmed that the only virus detected from fecal or intestinal smear samples of infected calves is a GpB rotavirus. This virus is electropherotypically similar and antigenically related to porcine and bovine GpB rotaviruses in our studies (Table 1), and to a rat GpB rotavirus in serologic and nucleic acid hybridization studies.[51,64] A recent observation from two laboratories indicated the Ohio porcine GpB rotavirus[36] and the unclassified nonA RV from lambs[47] also induced syncytia formation in MA104 infected monolayers (Table 5). Whether syncytia formation is confined to nonA RV or GpB rotaviruses in particular, requires further investigation.

X. IMMUNITY

The lack of information on the epidemiology of nonA RV infections and lack of routine serologic assays to detect antibodies to nonA RV has greatly limited studies of immunity to nonA RV. The low prevalence of antibodies in human sera to the Chinese GpB rotavirus, except for areas in China where outbreaks occurred, suggests that many adults and children may be at risk of acquiring these infections.[33,74] Serologic studies in Baltimore, MD indicated that infection of individuals with an agent antigenically related to rat GpB rotavirus was common.[34] A corresponding increase in seropositives with increasing age up to 20 years, suggests that although infections may arise in children, they may also occur in older age groups. Further studies are needed to access the incidence, prevalence and clinical significance of nonA RV in the human population and to provide additional information on immunity to these infections.

The low prevalence of antibodies to GpB rotavirus in swine in one region of the U.S.[36] compared to the higher prevalence in U.K. herds[35] also suggests the possibility of regional variations in the incidence or patterns of infection with GpB rotaviruses in swine. By comparison, antibodies to GpC rotavirus were common in adult swine in both the U.S. and U.K., suggesting GpC rotavirus infections may be enzootic, similar to the situation with GpA rotaviruses.[35,37] As noted in our studies, GpC rotaviruses were encountered in association with diarrhea in nursing pigs in 75% of the diagnostic cases studied (12 positive samples) and with weaning pigs in the remaining cases.[77] If GpC rotaviruses are substantiated to be a frequent cause of diarrhea in nursing pigs, as commonly shown for GpA rotaviruses, then methods to induce or enhance passive immunity may be important in protection. Enhancement of antibody titers in milk by vaccination of the pregnant or lactating dam may be one means of accomplishing this objective, as was shown for GpA rotaviruses.[82] However, it should be noted that such vaccinations did not actually prevent rotavirus infections in nursing pigs, but primarily delayed their onset and association with diarrhea.

XI. PREVENTION AND CONTROL

There is little information on methods to prevent or control infections with nonA RV. Further knowledge about the severity of these infections, their epidemiology, and methods for serial propagation of the viruses in cell culture is important for possible future vaccine development.

The posssibility of waterborne spread of the GpB rotavirus in China suggests that proper sanitation may play an important role in controlling or preventing epidemics.[16] Detection of antibodies and antigenically related rotaviruses in animals such as swine and rats poses unanswered questions about the zoonotic potential of the nonA RV. The present low incidence of antibodies to nonA RV in several human populations compared with a higher incidence among swine and poultry, raises the possibility that nonA RV may be emerging infections in human beings, or may have entirely different epidemiologic patterns among human beings.

Studies of GpC rotaviruses in swine including antibody prevalence and virus shedding have suggested these viruses, like GpA rotaviruses are enzootic.[35,37,67,77] Pigs infected at an earlier age (1 to 4 weeks) with GpC rotavirus had a higher incidence of clinical diarrhea than pigs infected at later ages. Thus, factors which influence passive immunity may be important in controlling the severity of these infections. These include failure of pigs to nurse or the sow to provide milk shortly after birth, or high viral doses from a heavily contaminated environment may exceed protective levels of milk antibodies. It is likely that factors useful in controlling spread and severity of GpA rotavirus infections among man and other animals will also be useful for controlling infections with nonA RV.

REFERENCES

1. **Estes, M. K., Palmer, E. L., and Obijeski, J. F.,** Rotaviruses: a review, *Curr. Top. Microbiol. Immunol.*, 105, 123, 1983.
2. **Holmes, I. H.,** Rotaviruses, in *The Reoviridae*, Joklik, W. K., Ed., Plenum Press, New York, 1983, 359.
3. **Saif, L. J., Bohl, E. H., Theil, K. W., Cross, R. F., and House, J. A.,** Rotavirus-like, calicivirus-like, and 23-nm virus-like particles associated with diarrhea in young pigs, *J. Clin. Microbiol.*, 12, 105, 1980.
4. **Bridger, J. C.,** Detection by electron microscopy of caliciviruses, astroviruses and rotavirus-like particles in the faeces of piglets with diarrhoea, *Vet. Rec.*, 107, 532, 1980.
5. **McNulty, M. S., Allen, G. M., Todd, D., McFerran, J. B., and McCracken, R. M.,** Isolation from chickens of a rotavirus lacking the rotavirus group antigen, *J. Gen. Virol.*, 55, 405, 1981.
6. **Saif, L. J., Saif, Y. M., and Theil, K. W.,** Detection and pathogenicity of enteric viruses recovered from diarrheic turkeys: role of a rotavirus-like agent, Proc. XVII World's Poultry Congress, Helsinki, Finland, August, 1984, 539.
7. **McNulty, M. S., Todd, D., Allan, G. M., McFerran, J. B., and Greene, J. A.,** Epidemiology of rotavirus infection in broiler chickens: recognition of four serogroups, *Arch. Virol.*, 81, 113, 1984.
8. **Saif, L. J., Theil, K. W., and Redman, D. R.,** Detection and pathogenicity of an enteric bovine rotavirus-like agent (RVLA), Abstr. 98, Proc. Conf. of Res. Workers Anim. Dis., Chicago, IL, 1982.
9. **Snodgrass, D. R., Herring, A. J., Campbell, I., Inglis, J. M., and Hargreaves, F. D.,** Comparison of atypical rotaviruses from calves, piglets, lambs and man, *J. Gen. Virol.*, 65, 909, 1984.
10. **Saif, L. J. and Theil, K. W.,** Antigenically distinct rotaviruses of human and animal origin, in *Proc. Infectious Diarrhoea in the Young: Strategies for Control in Humans and Animals*, Tzipori, S., Ed., Elsevier Science Publ., Amsterdam, The Netherlands, 1985, 208.
11. **Rodger, S. M., Bishop, R. F., and Holmes, I. H.,** Detection of a rotavirus-like agent associated with diarrhea in an infant, *J. Clin. Microbiol.*, 16, 724, 1982.
12. **Buitenwerf, J., Muilwijk Van-Alphen, M., and Schoap, G. J. P.,** Characterization of rotaviral RNA isolated from children with gastroenteritis in two hospitals in Rotterdam, *J. Med. Virol.*, 12, 71, 1983.
13. **Dimitrov, D. H., Estes, M. K., Rangelova, S. M., Shindarov, L. M., Melnick, J. L., and Graham, D. Y.,** Detection of antigenically distinct rotaviruses from infants, *Infect. Immun.*, 41, 523, 1983.
14. **Nicolas, J. C., Cohen, J., Fortier, B., Lourenco, M. H., and Bricout, F.,** Isolation of a human pararotavirus, *Virology*, 124, 181, 1983.
15. **Pereira, H. G., Leite, J. P. G., Azeredo, R. S., de Farias, V., and Sutmoller, F.,** Atypical rotavirus detected in a child with gastroenteritis in Rio de Janeiro, Brazil, Memorias Instituto Oswaldo Cruz, 78, 245, 1983.
16. **Hung, R., Wang, C., Fang, Z., Chou, Z., Chang, X., Liong, X., Chen, G., Yo, H., Chao, T., Ye, W., Den S., and Chang, W.,** Waterborne outbreak of rotavirus diarrhoea in adults in China caused by a novel rotavirus, *Lancet*, 1, 1139, 1984.
17. **Bai, Z., Wu, M., Shen, S., Fang, Y., Sun, Y., Wu, M., Zhao, Y., Zheng, Y., and Li, W.,** Isolation of human pararotavirus in faeces of adults with epidemic diarrhoea in Lanzhou, *Chin. J. Microbiol. Immunol.*, 4, 277, 1984.
18. **Espejo, R. T., Puerto, F., Soler, C., and Gongalez, N.,** Characterization of a human pararotavirus, *Infect. Immun.*, 44, 112, 1984.

19. **Chasey, D. and Banks, J.**, The commonest rotaviruses from neonatal lamb diarrhoea in England and Wales have atypical electropherotypes, *Vet. Rec.*, 115, 326, 1984.
20. **Vonderfecht, S. L., Huber, A. C., Eiden, J., Mader, L. C., and Yolken, R. H.**, Infectious diarrhea in infant rats produced by a rotavirus-like agent, *J. Virol.*, 52, 94, 1984.
21. **Torres-Medina, A.**, Isolation of an atypical rotavirus causing diarrhea in neonatal ferrets, *Lab. Anim. Sci.*, 37, 167, 1987.
22. **Pedley, S., Bridger, J. C., Brown, J. F., and McCrae, M. A.**, Molecular characterization of rotaviruses with distinct group antigens, *Virology*, 64, 2093, 1983.
23. **Bridger, J. C., Clarke, I. N., and McCrae, M. A.**, Characterization of an antigenically distinct porcine rotavirus, *Infect. Immun.*, 35, 1058, 1982.
24. **Bohl, E. H., Saif, L. J., Theil, K. W., Agnes, A. A., and Cross, R. F.**, Porcine pararotavirus: detection, differentiation from rotavirus, and pathogenesis in gnotobiotic pigs, *J. Clin. Microbiol.*, 15, 312, 1982.
25. **Chasey, D., Bridger, J. C., and McCrae, M. A.**, A new type of atypical rotavirus in pigs, *Arch. Virol.*, 89, 235, 1986.
26. **Todd, D. and McNulty, M. S.**, Electrophoretic variation of avian rotavirus RNA in polyacrylamide gels, *Avian Pathol.*, 15, 149, 1986.
27. **Pedley, S., Bridger, J. C., Chasey, D., and McCrae, M. A.**, Definition of two new groups of atypical rotaviruses, *J. Gen. Virol.*, 67, 131, 1986.
28. **Bridger, J. C.**, Novel rotaviruses in animals and man, in *Ciba Foundation Symp. 128, Novel Diarrhea Viruses*, Bock, G. and Whelan, J., Eds., John Wiley & Sons, Chichester, U.K., 1987, 5.
29. **Chasey, D. and Davies, P.**, Atypical rotaviruses in pigs and cattle, *Vet. Rec.*, 114, 16, 1984.
30. **Saif, L. J.**, Enteric viruses in diarrheic turkey poults, *Avian Dis.*, 29, 798, 1985.
31. **Wang, S., Cai, R., Chen, J., Li, R., and Jiang, R.**, Etiological studies of the 1983 and 1984 outbreaks of epidemic diarrhea in Guangxi, *Intervirology*, 24, 140, 1985.
32. **Su, C., Wu, Y., Shen, H., Wang, D., Chen, Y., Wu, D., He, L., and Yang, Z.**, An outbreak of epidemic diarrhoea in adults caused by a new rotavirus in Anhiu province of China in the summer of 1983, *J. Med. Virol.*, 19, 167, 1986.
33. **Hung, T., Chen, G., Wang, C., Fan, R., Yong, R., Chang, J., Dan, R., and Ng, M. H.**, Seroepidemiology and molecular epidemiology of the Chinese rotavirus, in *Ciba Foundation Symp. 128, Novel Diarrhea Viruses*, Bock, G. and Whelan, J., Eds., John Wiley & Sons, Chichester, U.K., 1987, 49.
34. **Eiden, J., Vonderfecht, S., and Yolken, R.**, Evidence that a novel rotavirus-like agent of rats can cause gastroenteritis in man, Lancet, 2, 8, 1985.
35. **Bridger, J. C. and Brown, J. F.**, Prevalence of antibody to typical and atypical rotaviruses in pigs, *Vet. Rec.*, 116, 50, 1985.
36. **Theil, K. W. and Saif, L. J.**, In vitro detection of porcine rotavirus-like virus (group B rotavirus) and its antibody, *J. Clin. Microbiol.*, 21, 844, 1985.
37. **Terrett, L. A., Saif, L. J., Theil, K. W., and Kohler, E. M.**, Physiochemical characterization of porcine pararotavirus and detection of virus and viral antibodies using cell culture immunofluorescence, *J. Clin. Microbiol.*, 25, 268, 1987.
38. **Brown, D. W. G., Beards, G. M., Chen, G. M., and Flewett, T. H.**, Prevalence of antibody to group B (atypical) rotavirus in humans and animals, *J. Clin. Microbiol.*, 25, 316, 1987.
39. **McNulty, M. S., Allan, G. M., and McFerran, J. B.**, Prevalence of antibody to conventional and atypical rotaviruses in chickens, *Vet. Rec.*, 114, 219, 1984.
40. **Theil, K. W., Saif, L. J., Moorhead, P. D., and Whitmoyer, R. E.**, Porcine rotavirus-like virus (group B rotavirus): characterization and pathogenicity for gnotobiotic pigs, *J. Clin. Microbiol.*, 21, 340, 1985.
41. **Saif, L. J., Bohl, E. H., Kohler, E. M., and Hughes, J. H.**, Immune electron microscopy of TGE virus and rotavirus (reovirus-like agent) of swine, *Am. J. Vet. Res.*, 38, 13, 1977.
42. **Nakata, S., Estes, M. K., Graham, D. Y., Loosle, R., Hung, T., Wang, S., Saif, L. J., and Melnick, J. L.**, Antigenic characterization and ELISA detection of adults diarrhea rotaviruses, *J. Infect. Dis.*, 154, 448, 1986.
43. **Suzuki, H., Chen, G. M., Hung, T., Beards, G. M., Brown, D. W. G., and Flewett, T. H.**, Effects of two negative staining methods on the Chinese atypical rotavirus, *Arch. Virol.*, 94, 305, 1987.
44. **Nakata, S., Petrie, B. L., Calomeni, E. P., and Estes, M. K.**, Electron microscopy procedure influences detection of rotaviruses, *J. Clin. Microbiol.*, 25, 1902, 1987.
45. **Mebus, C. A., Rhodes, M. B., and Underdahl, M. S.**, Neonatal calf diarrhea caused by a virus that induces villous epithelial cell syncytia, *Am. J. Vet. Res.*, 39, 1223, 1978.
46. **Askaa, J. and Bloch, B.**, Infection in piglets with a porcine rotavirus-like virus. Experimental inoculation and ultrastructural examination, *Arch. Virol.*, 80, 291, 1984.
47. **Chasey, D. and Banks, J.**, Replication of atypical ovine rotavirus in small intestine and cell culture, *J. Gen. Virol.*, 67, 567, 1986.

48. **Theil, K. W., Bohl, E. H., and Agnes, A. G.,** Cell culture propagation of porcine rotavirus (reovirus-like agent), *Am. J. Vet. Res.*, 38, 1765, 1977.
49. **Terrett, L. A. and Saif, L. J.,** Serial propagation of porcine group C rotavirus (pararotavirus) in primary porcine kidney cell culture, *J. Clin. Microbiol.*, 25, 1316, 1987.
50. **Saif, L. J., Terrett, L. A., Miller, K. L., and Cross, R. F.,** Serial propagation of porcine group C rotavirus (pararotavirus) in a continuous cell line and characterization of the passaged virus, *J. Clin. Microbiol.*, 26, 1277, 1988.
51. **Vonderfecht, S. L., Eiden, J. J., Torres, A., Miskuff, R. L., Mebus, C. A., and Yolken, R. H.,** Identification of a bovine enteric syncytial virus as a nongroup A rotavirus, *Am. J. Vet. Res.*, 47, 1913, 1986.
52. **Bremont, M., Cohen, J., and McCrae, M. A.,** Analysis of the structural polypeptides of a porcine group C rotavirus, *J. Virol.*, 62, 2183, 1988.
53. **Jashes, M., Sandino, A. M., Faundez, G., Avendano, L. F., and Spencer, E.,** *In vitro* transcription of human pararotavirus, *J. Virol.*, 57, 183, 1986.
54. **Dai, G., Sun, M., Liu, S., Ding, X., Chen, Y., Wang, L., Du, D., Zhao, G., Su, Y., Li, J., Xu, W., Li, T., and Chen, X.,** First report of an epidemic of diarrhea in human neonates involving the new rotavirus and biological characteristics of the epidemic virus strain (KMB/R85), *J. Med. Virol.*, 22, 365, 1987.
55. **Lourenco, M. H., Nicolas, J. C., Cohen, J., Scherrer, R., and Bricout, F.,** Studies of human rotavirus genome by electrophoresis: attempt of classification among strains isolated in France, *Ann. Virol.*, 132, 161, 1981.
56. **Estes, M. K., Graham, D. Y., and Dimitrov, D. H.,** The molecular epidemiology of rotavirus gastroenteritis, *Prog. Med. Virol.*, 29, 1, 1984.
57. **Bridger, J. C., Pedley, S., and McCrae, M. A.,** Group C rotaviruses in humans, *J. Clin. Microbiol.*, 23, 760, 1986.
58. **Arista, S., Giovannelli, L., and Titone, L.,** Detection of an antigenically distinct human rotavirus in Palermo, Italy, *Ann. Inst. Pasteur,* 1368, 229, 1985.
59. **Sorrentino, A., Scodeller, E. A., Bellinzoni, R., Muchinik, G. R., and LaTone, J. L.,** Detection of an atypical rotavirus associated with diarrhoea in Chaco, Argentina, *Trans. R. Soc. Trop. Med. Hyg.*, 80, 120, 1986.
60. **Sziics, G., Kende, M., and Uj, M.,** Atypical human rotaviruses in Hungary, *Ann. Inst. Pasteur,* 138, 391, 1987.
61. **Oseto, M., Yamashita, Y., Takagi, K., Inoue, H., Ishimaru, Y., and Kobayashi, N.,** Four year survey of diarrhea due to rotavirus in Matsuyama City, Abstr., U.S.-Japan Panel on Viral Disease, Tokyo, Japan, July 17 to 22, 1988.
62. **Theil, K. W., Reynolds, D. L., and Saif, Y. M.,** Comparison of immune electron microscopy and genome electropherotyping techniques for detection of turkey rotaviruses and RVLV in intestinal contents, *J. Clin. Microbiol.*, 23, 695, 1986.
63. **Chen, G., Hung, T., Bridger, J. C., and McCrae, M. A.,** Chinese adult rotavirus is a group B rotavirus, *Lancet,* 2, 1123, 1985.
64. **Eiden, J., Vonderfecht, S., Theil, K., and Yolken, R.,** Antigenically distinct rotaviruses: genetic and antigenic relatedness of human and animal strains, *J. Infect. Dis.*, 154, 972, 1986.
65. **Saif, L. J., Redman, D. R., Mengel, J., and Estes, M. K.,** Porcine and bovine group B rotaviruses: antigenic relationships and cross-infectivity in pigs and calves, Abstr., U.S.-Japan Cooperative Med. Sci. Prog., Viral Diseases Panel, Monterey, CA, December 1 to 4, 1987.
66. **Gerna, G., Passarani, N., Sarasini, A., and Battaglia, M.,** Characterization of serotypes of human rotavirus strains by solid-phase immune electron microscopy, *J. Infect. Dis.*, 152, 1143, 1985.
67. **Debouck, P., Callebaut, P., and Pensaert, M.,** The pattern of rotavirus and pararotavirus excretions in pigs in closed swine herds, in *Proc. 4th Int. Symp. Neonatal Diarrhea,* Acres, S. D., Ed., VIDO, Saskatchewan, Canada, 1984, 77.
68. **Vonderfecht, S. L., Miskuff, R. L., Eiden, J. J., and Yolken, R. H.,** Enzyme immunoassay inhibition assay for the detection of rat rotavirus-like agent in intestinal and fecal specimens obtained from diarrheic rats and humans, *J. Clin. Microbiol.*, 22, 726, 1985.
69. **Wang, C. A., Yang, R. J., Liang, X. G., et al.,** Development of ELISA for detection of adult diarrhoea rotavirus antigen, *Chin. J. Virol.*, 2, 248, 1986.
70. **Yolken, R., Wee, S. B., Eiden, J., Kinney, J., and Vonderfecht, S.,** Identification of a group-reactive epitope of group B rotavirus recognized by monoclonal antibody and application to the development of a sensitive immunoassay for viral characterization, *J. Clin. Microbiol.*, 26, 1853, 1988.
71. **Vonderfecht, S. L., Eiden, J. J., Miskuff, R. L., and Yolken, R. H.,** Kinetics of intestinal replication of group B rotavirus and relevance to diagnostic methods, *J. Clin. Microbiol.*, 26, 216, 1988.
72. **Ojeh, C. K., Saif, L. J., and Kang, S. Y.,** Production and characterization of monoclonal antibodies to porcine group C rotavirus, Abstr. 328, Conf. Res. Workers in Animal Dis., Chicago, IL, November 14 to 15, 1988.
73. **Nakata, S., Estes, M. K., Graham, D. Y., Wang, S., Gary, G. W., and Melnick, J. L.,** Detection of antibody to group B adult diarrhea rotaviruses in humans, *J. Clin. Microbiol.*, 25, 812, 1987.

74. **Hung, T., Fan, R., Wang, C., Chen, G., Chow, D., Chang, J., McCrae, M. A., Wang, W., Se, W., Dan, R., and Mon, H. N.,** Seroepidemiology of adult rotavirus, *Lancet,* 2, 325, 1985.
75. **Sigolo De San Juan, C., Bellinzoni, R. C., Mattion, N., LaTorre, J., and Schodeller, E. A.,** Incidence of group A and atypical rotviruses in Brazilian pig herds, *Res. Vet. Sci.,* 41, 270, 1986.
76. **Bellinzoni, R., Mattion, N., Vallejos, L., LaTorre, J. L., and Schoeller, E. A.,** Atypical rotavirus in chickens in Argentina, *Res. Vet. Sci.,* 43, 130, 1987.
77. **Saif, L. J.,** Typical and atypical rotaviruses of swine, in Proc. George A. Young Conf., Lincoln, NB, August 5 to 6, 1985, 73.
78. **Kang, S. Y., Nagaraja, K. U., and Newman, J. A.,** Electropherotypic analysis of rotavirus isolated form turkeys, *Avian Dis.,* 30, 797, 1986.
79. **Reynolds, D. L., Saif, Y. M., and Theil, K. W.,** A survey of enteric viruses of turkey poults, *Avian Dis.,* 31, 89, 1987.
80. **Tzipori, S., Chandler, D., and Smith, M.,** The clinical manifestation and pathogenesis of enteritis associated with rotavirus and ETEC infections in domestic animals, *Prog. Food Nutri. Sci.,* 7, 193, 1983.
81. **Bohl, E. H., Kohler, E. M., Saif, L. J., Cross, R. F., Agnes, A. G., and Theil, K. W.,** Rotavirus as a cause of diarrhea in pigs, *J. Am. Vet. Med. Assoc.,* 172, 458, 1978.
82. **Saif, L. J.,** Passive immunity to coronavirus and rotavirus infections in swine and cattle: Enhancement by maternal vaccination, in *Proc. Infectious Diarrhoea in the Young, Strategies for Control in Humans and Animals,* Tzipori, S., Ed., Elsevier Science Publ., Amsterdam, The Netherlands, 1985, 456.
83. **Saif, L. J.,** unpublished data, 1988.
84. **Saif, L. J.,** unpublished data, 1986.
85. **McNulty, M. S. and Saif, L. J.,** unpublished data, 1988.
86. **Saif, L. J.,** unpublished data, 1983.

Chapter 5

ENTERIC ADENOVIRUSES OF MAN

Lennart Svensson, Ingrid Uhnoo, and Göran Wadell

TABLE OF CONTENTS

I. INTRODUCTION

Human adenoviruses were first detected in human adenoids by Rowe et al. in 1953.[1] Adenovirus can infect various species of mammals, birds, and amphibians.[2] There are 47 known serotypes of human adenovirus[3] classified into six subgenera (A to F). Most of the serotypes have a predilection for lymphoid tissue, hence their name. They can persistently infect these organs and are also associated with outbreaks of respiratory disease and keratoconjunctivitis.

The relationship between adenoviruses and acute infantile gastroenteritis has been questioned for many years. Since many adenovirus serotypes are excreted and detected in stools, it has been assumed that they would have a possible role in infantile diarrhea.[4,5] However, earlier epidemiologial studies generally found as many adenovirus isolations in the stools of controls as from those with diarrhea.[6] The failure to correlate adenovirus isolations from stool with clinical symptoms shows that adenoviruses should not be identified as the cause of medical illness just because they can be cultured from the stool of an individual with disease.

By the introduction of electron microscopy (EM) for examination of stool specimens, Flewett et al. observed in 1975[5] a previously unrecognized type of adenovirus that that could not grow in cell culture. These adenoviruses, which have been designated uncultivable,[7] enteric,[8,9] or fastidious[10] adenoviruses, have emerged as the prime pathogen among adenoviruses causing infantile gastroenteritis.[7,9,11,12] A potential medical importance of the enteric adenoviruses (EA) in acute diarrhea began to emerge when several groups reported that adenoviruses which were visualized by EM, but which could not be isolated in tissue culture systems, were more highly associated with gastroenteritis than those adenoviruses which could be isolated.[7,9,13] In several studies, enteric adenoviruses (EA) have been found in 4 to 12% of stool specimens from infants and young children with acute diarrhea.[7,13-17]

After rotavirus, the EA are now recognized as the second most commonly identified agent in stools of infants and young children with viral gastroenteritis. The inability of EA to grow *in vitro* initially hampered classification by accepted serological methods, such as neutralization and hemagglutination inhibition techniques. Alternative methods such as enzyme-linked immunosorbent assay (ELISA), SDS-polyacrylamide gel electrophoresis, and DNA restriction enzyme analysis were therefore employed for characterization. Two distinct serotypes, Ad 40 and Ad 41, belonging to subgenera F have now been identified.[18,19]

II. MORPHOLOGY AND CLASSIFICATION

Adenoviruses are nonenveloped icosahedral viruses with a diameter of 80 nm (Figure 1). The virion is composed of at least 10 different structural proteins and a linear 33 to 45 kilobase pair DNA molecule. The virus capsid is composed of 252 subunits (capsomers); of these 240 (hexons) are symmetrically arranged so that each hexon is surrounded by six capsomers. The 12 corners of the virion each contain a capsomer called a penton. Each of the pentons contains a base and a fiber projecting from the base.

By tradition, adenoviruses have been classified by their hemagglutination[20] and oncogenic[21] properties. Adenoviruses belonging to subgenus A are highly oncogenic and induce tumors in newborn hamsters within a few months. Adenoviruses belonging to subgenus B and E are weakly oncogenic and induce tumors in a few animals. Neither Ad 40 nor Ad 41 induces tumors in newborn hamsters.[22,23] Non-oncogenic adenoviruses that transform rat cells *in vitro* are classified into subgenera C and D.

Since hemagglutination and oncogenicity groupings are both based on properties representing minor regions (6% or less)[22] of the genome, Wadell et al.[22,23,24] studied the adenovirus relationships based on polypeptide profiles. By SDS-polyacrylamide gel electrophoresis it was found that the internal structural polypeptides V, VI, VII, and VIII are evolutionarily conserved and do not vary in size between serotypes of the same subgenus. Distinctly different sizes were

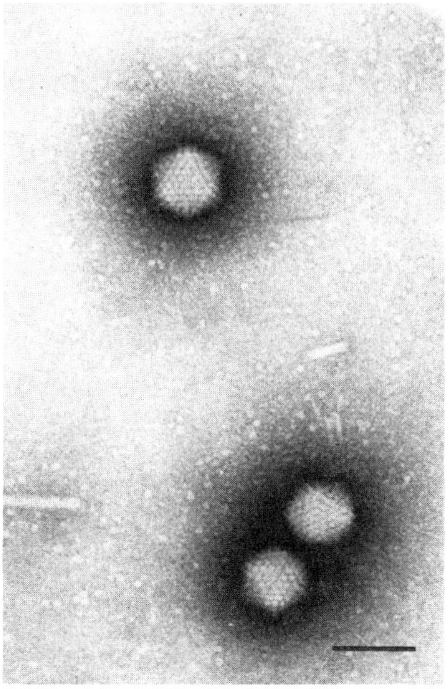

FIGURE 1. Electron micrograph of enteric adenovirus found in a stool specimen from a child with acute gastroenteritis. Bar represents 100 nm.

observed when members of different subgenera were compared. The polypeptide pattern could consequently be used to classify the 39 serotypes of adenovirus into 5 subgenera.[22,24]

By use of restriction enzyme analysis, Wadell et al.[22,23] compared the restriction endonuclease pattern of enteric adenovirus with adenoviruses from other subgenera (Figure 2), and found that Ad 40 and Ad 41 displayed unique restriction profiles with no resemblance to the established serotypes.

A subgenus is defined as DNA homology of more than 50% between members within a subgenus and less than 20% between members of different subgenera. Based on DNA homologies between 31 adenovirus serotypes, Green et al.[25] identified five groups designated A to E (Table 1). Because of the low sequence homologies (less than 23%) between members of different subgenera, it is unlikely that recombination occurs between adenoviruses of different subgenera. The established subgenera appear therefore to serve as recombination barriers, and, in fact, no intermediate strains with parent strains from two different subgenus have been reported. The human adenoviruses are newly classified into six subgenera, A to F, containing 41 serotypes (Table 1). As will be shown below, tropism and genetic variability varies between adenoviruses of the six subgenera.

Subgenus A — This consists of Ad 12, Ad 18, Ad 31. All three members can be isolated from the stool of healthy humans. Ad 31 has frequently been isolated from children with diarrhea, but a definitive relation remains to be established.

Subgenus B — Two distinct clusters of DNA homology can be seen after DNA restriction enzyme analysis of the members of subgenus B. The first group consists of Ad 3, Ad 7, and Ad 21, which account for 33% of all typed adenovirus isolates reported by WHO. They primarily cause outbreaks or respiratory disease, but can also cause a more generalized infection. The second group contains Ad 11, Ad 34, and Ad 35 which cause persistent infections in the urinary tract.

Subgenus C — Ad 1, Ad 2, Ad 5, and Ad 6 represent nearly 60% of all adenovirus isolates

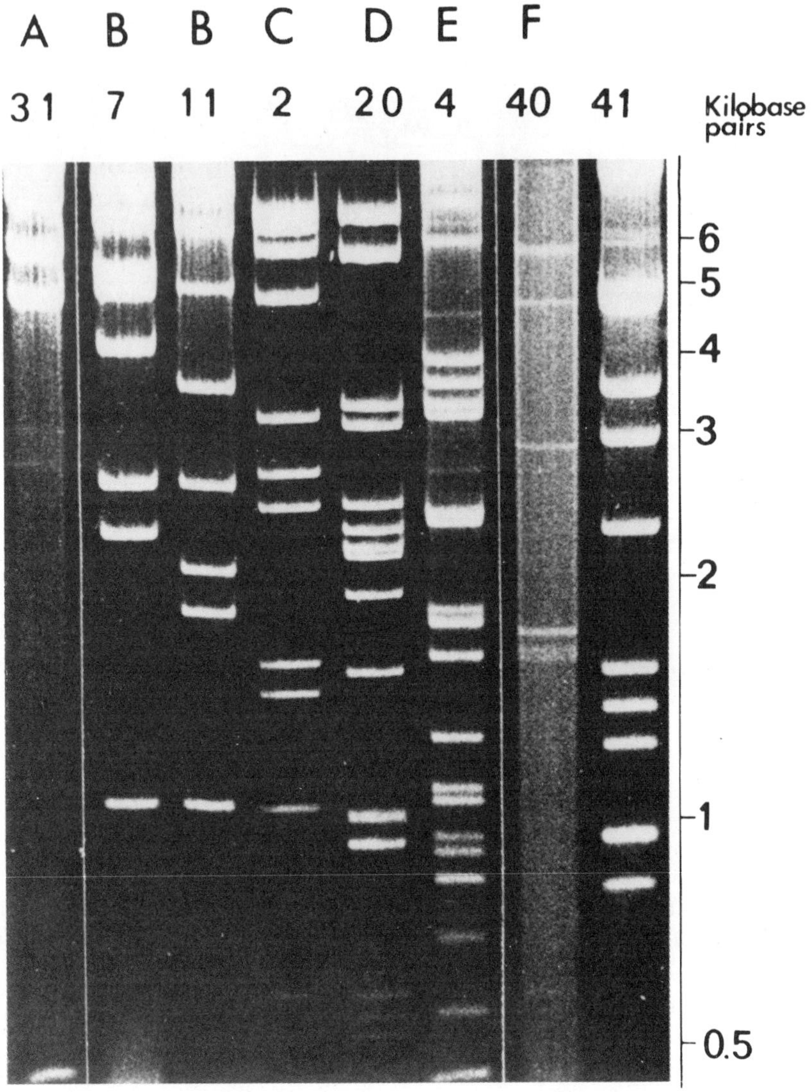

FIGURE 2. Enzymatic cleavage of adenovirus DNA from subgenera A to F by the restriction enzyme *Sma*I. (From Wadell, G., *Curr. Top. Microbiol Immunol.*, 100, 191, 1984. With permission.)

reported by WHO. They display a DNA homology of 98% in pairwise comparison. They show tropism for lymphoid tissue in which they can persist for years.

Subgenus D — This family consists of 23 established serotypes and five potential new serotypes.[3] They show a predilection for infecting the eye. The most important members of subgenus D are serotypes 8 and 19 which are associated with epidemic keratoconjuctivitis.

Subgenus E — Based on sequence homology, restriction endonuclease cleavage pattern and protein profiles, subgenus E contains only subgenus type 4. Ad 4 has been associated with both the epidemic follicular conjunctivitis and respiratory disease. When the Ad 4 and Ad 4a genome types are analyzed by restriction enzyme analysis only 45% of the DNA fragments comigrate. Such a pronounced genetic difference within one serotype is unique for Ad 4.[26]

Subgenus F — This subgenus contains the two enteric adenovirus serotypes 40 and 41, represented by the prototype strains Dugan and Tak, respectively. Both prototype strains were isolated in Holland from children with diarrhea.[18] By liquid hybridization von Loon et al.[27]

TABLE 1
Properties of Human Adenovirus Serotypes of Subgenera A-F

		DNA						
		Homology %						
Subgenus	Serotype	Intrageneric	Intergeneric	G + C (%)	No of *Sma*I[a] fragments	Hemagglutination[b] pattern	Oncogenicity in newborn hamsters	Tropism/ symptoms
A	12, 18, 31	48—69	8—20	48	4—5	IV	High (tumors in most animals in 4 months	Cryptic enteric infection
B:1[c]	3, 7, 16, 21,	88—94	9—20	51	8—10	I	Weak (tumors in a few animals in 14—18 months	Respiratory disease; persistent infections of kidney
B:2	14, 11, 34, 35							
C	1, 2, 5, 6,	99—100	10—16	58	10—12	III	Nil	Respiratory disease persists in lymphoid tissue
D	8, 9, 10, 13, 15, 17, 19, 20, 22—30, 32, 33, 36, 37, 38, 39	94—99	4—17	58	14—18	II	Nil	Keratoconjuctivitis
E	4		4—23	58	16—19	III	Nil	Conjunctivitis; respiratory disease
F	40, 41	62—69,	15—22	52	9—12	IV	Nil	Infantile diarrhea

[a] The restricted DNA fragments were analyzed on 0.8 to 1.2% agarose slab gels. DNA fragments smaller than 400 bp were not resolved.

[b] Complete agglutination of monkey erythrocytes; II, complete agglutination of rat erythrocytes; III, partial agglutination of rat erythrocytes (fewer receptors); IV, agglutination of rat erythrocytes discernible only after addition of heterotypic antisera.

[c] Members of subgenus B are divided into two clusters of DNA homology based on pronounced differences in DNA restriction sites.

Modified from Wadell, G., *Curr. Top. Microbiol. Immunol.*, 110, 191, 1984.

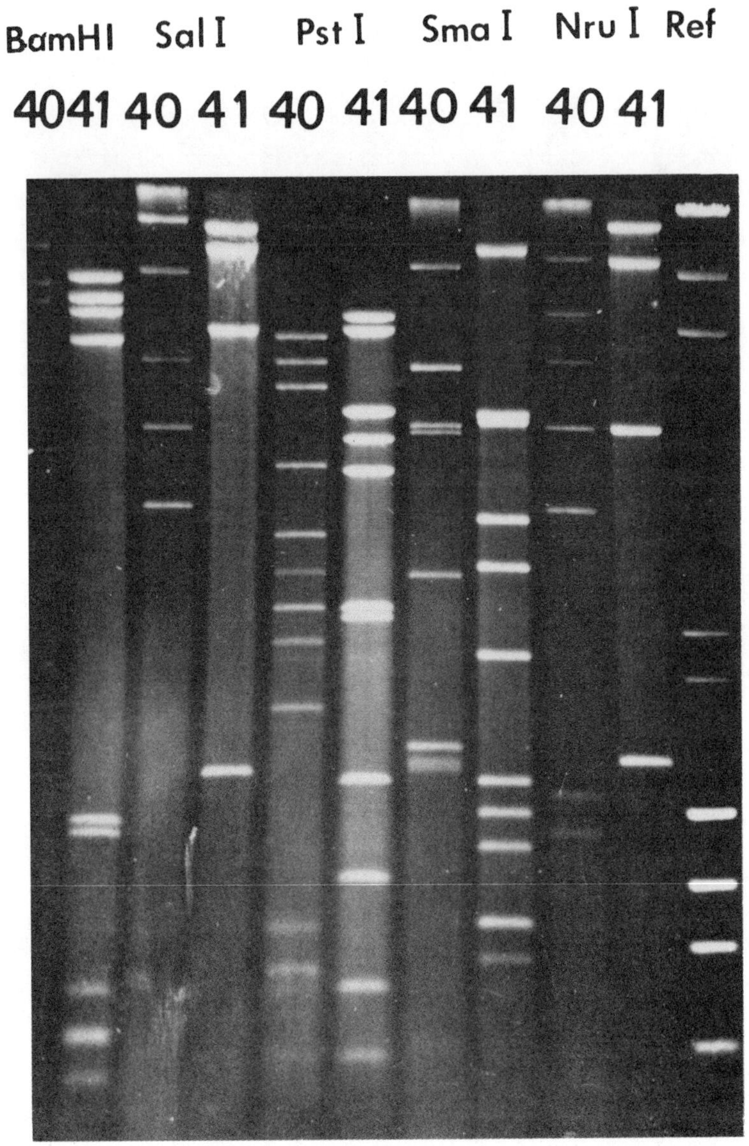

FIGURE 3. DNA restriction patterns of the enteric adenovirus Ad 40 and 41 (subgenus F) obtained with restriction endonucleases *Bam*HI, *Sal*I, *Pst*I, *Sma*I, and *Nru*I. Lambda DNA and ϕX174 digested with *Hind*III and *Hind*II, respectively, were used as size references. (From Wadell, G., Allard, A., Evander, M., and Li, Q.-G., *Chem. Scripta*, 26B, 325, 1986. With permission.)

determined the DNA homology between Ad 40 and Ad 41 to be 62 to 69%. This degree of homology is not very high, but together with the other characteristics in common, such as their restricted host cell range, their association with gastroenteritis and their cross-reactivity in immunological tests have classified Ad 40 and Ad 41 in subgenus F. Restriction enzyme analyses have revealed that only 18 of 177 DNA restriction fragments of Ad 40 and Ad 41 comigrate (Figures 3, 4)[11,23]

III. *IN VITRO* CULTIVATION

All gastroenteritis viruses, including enteric adenoviruses, share the property of being

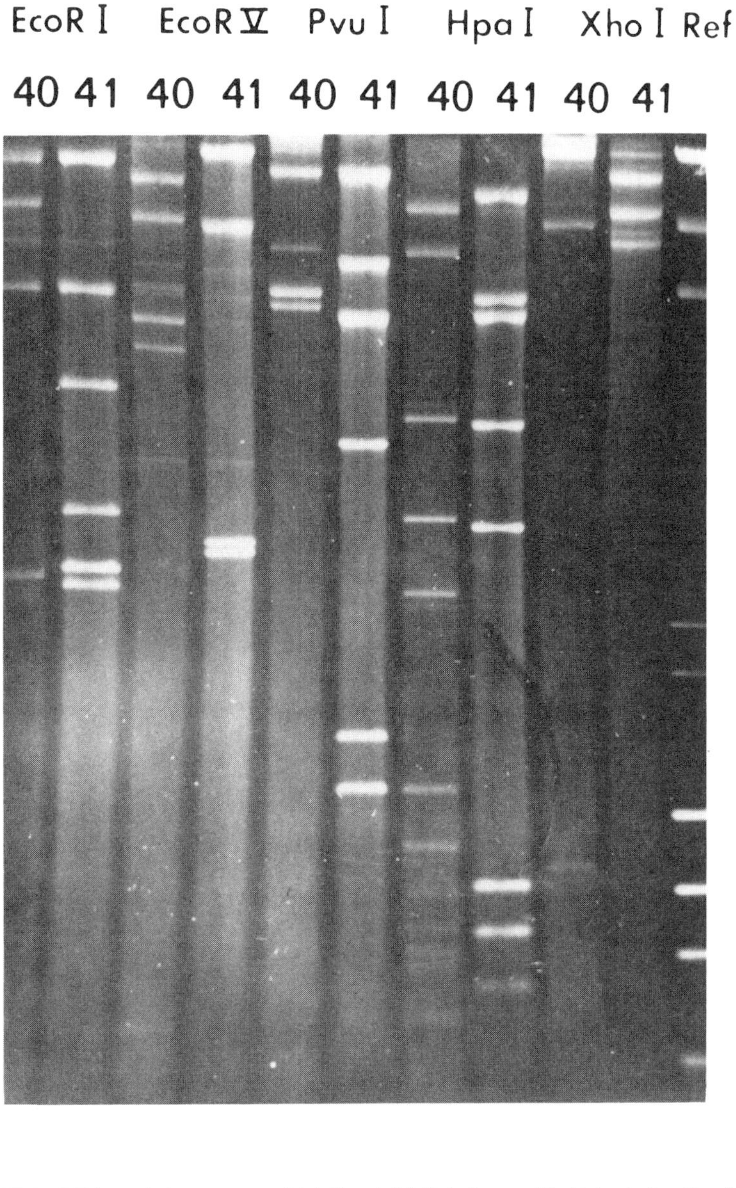

FIGURE 4. DNA restriction patterns of Ad 40 and Ad 41 (subgenus F) obtained after digestion with *Eco*RI, *Eco*RV, *Pvu*I, *Hpa*I, and *Xho*I. The size reference is described under Figure 3. (From Wadell, G., Allard, A., Evander, M., and Li, Q.-G., *Chem. Scripta,* 26B, 325, 1986. With permission.)

initially noncultivable *in vitro* systems. Whereas most established adenovirus serotypes can be cultivated and serially passaged in human embryonic kidney, HeLa, A 549, and KB cells, the enteric adenoviruses only undergo an abortive infection in these cells. The reason for the lack of growth of enteric adenoviruses in HeLa and KN cells has been investigated by Takiff and Strauss.[28] Their results suggest an early replication block in the growth of the virus. The enteric adenoviruses grow efficiently, however, in Graham 293 cells, an Ad 5-transformed human embryonic kidney cell line.[29,30] It has been suggested that the 293 cells enable growth of enteric adenoviruses by providing early (E1a and E1b) viral functions.[27] Enteric adenoviruses have also been shown to grow in tertiary cynomolgous monkey kidney cells, tCMK,[18] Chang Conjunctival cells,[10] HT29,[11] and Hep-2 cells.[18,31] The growth of characteristics of enteric adenovirus strains in Chang conjunctival, tCMK, Hep-2, HeLa, HDF, and Graham 293 cells has been found to be dependent on several parameters, including viral strain, cell type, subline and other still undefined conditions.[18] It has also been found that some viruses replicate in these cells without producing CPE. Whereas incorporation of trypsin in the medium and rolling of tCMK cultures enhances virus growth, Graham 293 and Chang cells produce CPE more readily when incubated in stationary culture.

IV. GENOME

All adenoviruses contain a double-stranded DNA genome of about 23×10^6 Da. The adenovirus genome has two unusual features. The 5′-end of each DNA strand is covalently attached to a protein[33,34] and the terminal DNA sequences of either strand are inverted duplicates of each other.[35]

Genetic variability of enteric adenovirus. — By analyzing 26 strains of Ad 40 from South Africa, Canada, and Europe, Kidd et al.[36] identified 6 distinct genome types using restriction endonucleases. Of these strains 73% were found to be identical to the characteristic Hovi strain.[22] In another study, Kidd et al.[37] analyzed 15 Ad 41 strains from Europe, Canada, and South Africa. The Sma-I restriction pattern revealed that all were similar to the original restriction pattern of Ad 41.[11] By employing 10 restriction endonucleases, four distinct genome types could be demonstrated among the South African strains, whereas the 10 Canadian and European strains could not be distinguished from each other.

V. PHYSICAL ORGANIZATION OF THE GENOME OF ENTERIC ADENOVIRUSES

Mapping of recognition sites for restriction endonuclease *Bam*HI, *Bgl*I, *Bst*EII, *Eco*RI, *Hind*III, *Hpa*I, *Kpn*I, *Nru*I, *Pst*I, *Pvu*I, *Sac*I, *Sal*I, *Sma*I, or *Xho*I has been performed in the Ad 40 or Ad 41 genome.[27,38,39] The E1 region of both Ad 40 and Ad 41 has been sequenced and compared to the corresponding regions of both Ad 12 (subgenus A), Ad 7 (subgenus B), and Ad 5 (subgenus C), and Ad 4 (subgenus E). A12S and A13S mRNA could be translated in the E1a region. In Ad 41 and 13S mRNA corresponds to a 251 amino acid protein. Within this protein three conserved sequences CS1–CS3 were found. CS1 is hydrophilic and only 10 of the 16 conserved amino acids are shared by Ad 40. CS2 is highly acidic and the most conserved domain. CS3 is more hydrophobic with a structure compatible with that of zinc finger protein required for efficient transcription and transactivation. Only 18 of 23 conserved amino acids are shared by Ad 41.[40] The E1a promotor of Ad 40 has only a limited *cis* acting activity and E1a has a considerably lower capacity to induce morphological transformation than the corresponding regions of Ad 5 and Ad 12.[41,42] Analysis of the DNA sequence of the Ad 41 hexon gene revealed the absence of a highly charged stretch of 32 amino acids that in the Ad 2 hexon is exposed on the surface of the Ad 2 capsid. It has has been suggested that this region is involved in the pH-induced conformation changes within the adenovirus capsid.[42a]

VI. BIOLOGICAL AND ANTIGENIC SPECIFICITIES OF STRUCTURAL PROTEINS

The structural proteins of adenoviruses possess several biological and antigenic specificities. All human adenovirus serotypes share a cross-reacting group antigen that is exposed on the inner surface of the hexon. In addition, the hexons (peptide II) carry intersubgenus, intrasubgenus, and type specificities.[43] The vertex capsomers (peptide III) carry group, intersubgenus and intrasubgenus specificities.[44] The fibers (peptide IV) carry intrasubgenus, intersubgenus, and type specificities. The regions of type specificity on the hexon and on the fiber both give rise to neutralizing antibodies.[45] The receptor-binding sites and the hemagglutinin are both present on the fiber. The adenovirus hemagglutinin can be identified with rat and rhesus monkey red blood cells.[20] Isolated pentons that are produced in excess are toxic and cause cytopathogenic effects on cells.[46]

The definition of a serotype relies on distinctive antigen determinants that are able to induce a neutralizing of type-specific antibody response. The serotypes are therefore based on biological properties; however, the gene products carrying those antigenic regions represent only a minor part of the viral genome. This is illustrated by the cross-reaction in neutralization tests between Ad 4 and Ad 16, members of subgenera E and B, respectively. The enteric Ad 40 and Ad 41 cannot be distinguished from each other by neutralization test in Graham 293 cells or Chang Conjunctival, but in tCMK cells;[18] nor can they be distinguished by hemagglutination inhibition test.[18]

VII. EPIDEMIOLOGY

Information on the distribution of enteric adenoviruses has been gained from sero-epidemiological surveys, the study of outbreaks, and prospective studies. Kidd et al.[47] conducted a serological survey of enteric adenovirus infection using sera from both developed and developing countries. More than one third of the sera from the U.K., New Zealand, Hong Kong, and Gambia contained neutralizing antibodies against Ad 40 and Ad 41, whereas all sera from Guatamala were negative. In another study, significant titers of enteric adenovirus-specific antibodies were found in pooled sera from West Germany and The Netherlands.[18] In addition, enteric adenoviruses have been found in the stools of young children with acute gastroenteritis in Europe, Asia, and Latin and North Americas.[18] These studies indicate that infection with enteric adenoviruses is common and widespread in both industrialized and developing countries.

Enteric adenoviruses have been associated with outbreaks of diarrhea in a long-stay children's ward[5] and an R.A.F. camp,[48] both in the U.K. The causative adenovirus was later identified as Ad 40.[12] Chiba et al.[49] reported on an outbreak of Ad 40-associated diarrhea in an orphanage in Sapporo, Japan.

Enteric adenoviruses have been detected in the stools of infants and young children with acute gastroenteritis in several prospective studies conducted in the developed countries.[12,13,16,17,50-52] The incidence of infection varied between 7 to 17%, a frequency second only to that of rotaviruses. It is, however, vital to distinguish between enteric adenoviruses and other adenoviruses, since some types, i.e., Ad 1, Ad 2, and Ad 5, may be shed in the stools for years after the primary infection without symptoms. In a Swedish study using enteric adenovirus-specific assays, Ad 40 and Ad 41 were found in 7.9% of 416 children with diarrhea.[5] Approximately 60% of the isolates in this study were identified at enteric adenoviruses. These findings are in accordance with those of other studies, and it is clear that enteric adenoviruses are the most important pathogens causing diarrhea among the adenoviruses. During this 1-year survey in Sweden, the two enteric adenovirus types were detected at equal frequency, whereas in a study from the Netherlands extending over several years, a temporal shift was seen, with one type predominating in 1981 (72% Ad 40 vs. 28% Ad 41) and the other in 1984 to 1985 (20% Ad 40 vs. 80% Ad 41).[53] During a 9-year study of adenovirus associated with diarrhea in children in Washington, D.C., Ad 40 and/or Ad 41 were detected in every calendar month, suggesting that the enteric adenoviruses might be endemic in this area.[54]

Few studies have been published on the incidence of adenovirus gastroenteritis in developing countries. Leite et al.[55] found an infection rate of 2% (14 of 746) for enteric adenovirus during a 2-year study in Rio de Janerio, Brazil, and Kidd et al.[56] reported a 6.5% (40 of 616) rate in a 7-month study done in Johannesburg, South Africa. A study in Thailand found that Ad 40 and

Ad 41 as well as nonenteric adenoviruses were significantly correlated with diarrhea. Two percent (11 of 1114) of children with gastroenteritis and 0.6% (6 of 947) of healthy controls were positive for enteric adenoviruses.[57] However, in a 1-year survey in India, the detection rate for enteric adenoviruses was similar in patients with diarrhea and in healthy controls, whereas nonenteric adenovirus types were significantly associated with gastroenteritis.[58] This data indicate that enteric adenoviruses are causes of diarrhea in these populations, but they may be proportionately less important than they are in the industrialized countries.

In contrast to the winter peak of rotavirus infection, no characteristic seasonal variation has been observed with enteric-associated gastroenteritis. In most studies, Ad 40 and Ad 41 were found throughout the year,[16,17,54] whereas in Sweden[12] and South Africa[56] a higher occurrence of enteric adenoviruses infection was noted during the summer months.

VIII. PATHOGENESIS

Enteric adenoviruses are excreted in large amounts up to 10^{11} particles per gram of feces at the acute stage of the disease.[9] This indicates that these viruses actively multiply in the gastrointestinal tract. Adenovirus particles have been demonstrated in the duodenal mucosa in a fatal case of Ad 41 gastroenteritis.[59] In this case, crystalline arrays of virus particles were observed in the small intestine mucosa, which is suggestive of their replication. The observation by Uhnoo et al.[12] that 70% of paired sera from children with enteric adenovirus infections displayed a type-specific sero-response measured by hemagglutination inhibition supports the notion that enteric adenoviruses are definite pathogens of diarrhea. Respiratory symptoms are common in children with established infections, whereas only a limited number of children with enteric adenovirus gastroenteritis exhibit signs of respiratory tract involvement.[12,49] Enteric adenoviruses have been reported to be associated with lactose and gluten intolerance in a few cases.[12,60] A possible role of human adenovirus in the pathogenesis of celiac disease has been suggested by Kagnoff et al.[61] These researchers screened 1498 proteins for their homology to gliadin and noted that the Ad 12 E1B protein contained a region of 12 amino acid residues that included eight identities and a hydrophilic identical pentapeptide. The native antibodies from rats carrying an Ad 12-transformed baby rat kidney cell tumor reacted both with gliadin and with a synthetic heptapeptide of gliadin from the region of homology. The significance of this molecular mimicry has to be evaluated by determining the conservation of this peptide in the early proteins of other adenoviruses infecting the gut, and by estimating the relative frequency of infection by the adenoviruses in question in the population of healthy and sick HLA-B8, DR3, DR7, and DC3 individuals.

IX. CLINICAL CHARACTERISTICS

The most prominent feature of enteric adenovirus infection is diarrhea. Other symptoms associated with the disease include vomiting, low grade fever, and dehydration. The course of the illness is usually mild, but can be more severe and a fatal case of Ad 41 infection has been reported.[59] Enteric adenoviruses mainly affect infants and young children and have only sporadically been found in adults. The incubation period is approximately 7 to 8 d and virus excretion in the stool lasts 10 to 14 d.[12]

Uhnoo et al.[12] carried out a detailed clinical study of 56 children with adenovirus gastroenteritis of whom 33 had enteric adenovirus infection. The clinical characteristics of childhood diarrhea associated with different types of adenovirus are shown in Table 2. The mean duration of diarrhea in children with Ad 40 and Ad 41 infection was 8.6 and 12.2 d, respectively. Prolonged diarrhea was common, particularly in association with Ad 41, with one third of the patients having symptoms for 14 d or more. Vomiting and fever were mild and lasted for a median of 2 d. Dehydration was mostly mild and isotonic in nature. One third of the patients required hospitalization. Upper respiratory symptoms were found in 21% of the patients. On the

TABLE 2
Clinical Characteristics of 55 Children with Gastroenteritis Due to Adenovirus

Type of infection	Total no. of patients	No. (%) with indicated clinical findings: Diarrhea	Mean duration (d)	Vomiting	Fever: 37.5—38.9°C	Fever: >39°C	Total no. with fever	Abdominal pain	5% dehydration or more	Respiratory symptoms
Ad 40	14	14 (100)	8.6	11 (79)	7 (50)	1 (7)	8 (57)	1 (7)	2 (14)	3 (21)
Ad 41	19	18 (95)	12.2	15 (79)	7 (37)	1 (6)	8 (42)	7 (37)	3 (16)	4 (21)
Established adenovirus	14	14 (100)	6.2	7 (50)	5 (36)	9 (64)	14 (100)	5 (36)	2 (14)	11 (79)
Untyped adenovirus	8	8 (100)	4.5	6 (75)	2 (25)	3 (38)	5 (63)	4 (50)	1 (13)	2 (25)

From Uhnoo, I., Wadell, G., Svensson, L., and Johansson, M., *J. Clin. Microbiol.*, 20, 365, 1984. With permission.

TABLE 3
Clinical Features in 393 Children with Acute Gastroenteritis in Relation to Enteropathogens detected in the Stools

	Groups of patients				
	1 Rotavirus	2 Enteric adenovirus	3 Bacteria	4 Bacteria and virus	5 No pathogens
N =	168	32	42	16	135
Diarrhea	164 (98)	31 (97)	42 (100)	16 (100)	132 (98)
Diarrhea >10 times daily	36 (21)	7 (22)	15 (36)	3 (19)	27 (20)
Vomiting	146 (87)	25 (78)	18 (43)***	15 (94)	72 (53)***
Vomiting >5 times daily	62 (37)	3 (9)**	3 (7)***	5 (31)	19 (14)***
Fever	141 (84)	14 (44)***	29 (69)*	15 (94)	83 (61)***
Fever >39°C	71 (42)	1 (3)***	18 (43)	7 (44)	45 (33)
Abdominal pain	31 (18)	8 (25)	21 (50)***	4 (25)	40 (30)*
Blood present in stools	2 (1)	1 (3)	17 (41)***	1 (6)	14 (10)***
Mucus present in stools	28 (17)	6 (19)	11 (26)	1 (6)	34 (25)
Respiratory symptoms	56 (33)	6 (19)	16 (38)	8 (50)	57 (42)
Admission to hospital	65 (39)	9 (28)	16 (38)	8 (50)	33 (24)*

Note: Values given are numbers and percentages; *p <0.05, **p <0.01, ***p <0.001 (p values denote significant differences by x2 test between the rotavirus group and each of the other groups).

From Uhnoo, I., Olding-Stenkvist, E., and Kreuger, A., *Arch. Dis. Child.*, 61, 732, 1986. With permission of the British Medical Association.

other hand, other adenoviruses caused a different clinical picture with diarrhea of shorter duration (mean 6.2 d), higher fever, and significantly increased occurrence of respiratory symptoms (79%).

The clinical characteristics also differed from those of rotavirus and bacterial infections (Tables 3 and 4).[83] In comparison with rotavirus, enteric adenoviruses caused a milder disease with less intense vomiting, fever, and dehydration, which made the parents seek medical advice later in the course of the illness. The duration of the diarrhea was significantly prolonged in association with enteric adenoviruses (mean 10.2 d) as compared to rotaviruses (mean 5.9 d).

In other studies, EA gastroenteritis has been characterized as mild and of short duration[5] as well as more severe and with a protracted course.[58,62] In one investigation, enteric adenoviruses were found to be associated with respiratory symptoms in 93% of 27 hospitalized children,[50] whereas in several other studies only a very low frequency has been observed.[5,49]

At follow-up of the 33 children with enteric adenovirus infection in the study by Uhnoo et al.[12] three children showed signs of lactose intolerance from 5 to 7 months, and 1 child did not tolerate gluten-containing food for 9 months. These observations suggest a possible role for enteric adenovirus in the etiology of chronic diarrhea and malnutrition in children.

X. DIAGNOSIS

Adenoviruses (established serotypes) were first detected in stools by routine cell culture techniques. However, enteric adenoviruses as well as other gastroenteritis viruses show the property of being initially noncultivable by conventional *in vitro* systems. Electron microscopy (EM) was therefore initially the only method of choice. A drawback of using EM for detection

TABLE 4
Clinical Course of Gastroenteritis in Children Infected with Different Enteropathogens

	Groups of patients				
	1 Rotavirus	2 Enteric adenovirus	3 Bacteria	4 Bacteria and virus	5 No pathogens
N =	168	32	42	16	135
Symptoms before hospital contact	2.9 (0.16)	5.3 (0.75)***	5.4 (0.59)***	3.0 (0.50)	3.9 (0.25)***
Diarrhea	5.9 (0.28)	10.8 (1.71)***	14.1 (2.18)***	8.4 (1.70)**	8.0 (0.57)***
Vomiting	2.5 (0.10)	3.2 (0.80)	2.1 (0.34)*	2.1 (0.24)	2.1 (0.16)
Fever	2.2 (0.12)	2.4 (0.35)	3.3 (0.39)**	2.5 (0.40)	2.5 (0.16)
Hospital stay	2.4 (0.19)	3.6 (1.18)	3.6 (1.20)	2.6 (0.56)	2.8 (0.48)

Note: Values are mean (SEM) duration of each symptom (days); *$p < 0.05$, **$p < 0.01$, ***$p < 0.001$ (p values denote significant differences by Mann Whitney U test between the rotavirus group and each of the other groups).

From Uhnoo, I., Olding-Stenkvist, E., and Kreuger, A., *Arch. Dis. Child.*, 61, 732, 1986. With permission of the British Medical Association.

of adenovirus is that no serotype specificity can be obtained. Several immunological and molecular methods have subsequently been established for identification of Ad 40 and Ad 41.

Solid phase immune electron microscopy (SPIEM) has been used to identify cross-reacting epitopes between Ad 40 and Ad 41.[63] More recently, Wood and Baily[64] established an immune electron microscopic method based on hyperimmune sera for identification of Ad 40 and Ad 41. Jacobson et al.[8] established an immunoelectroosmophoresis method that could differentiate Ad 40 from the established serotypes.

Several ELISA assays have been established for specific detection of Ad 40 and Ad 41. Two different approaches using either adsorbed polyclonal or monoclonal antibodies have been tried. Johansson et al.[65,66] adsorbed anti Ad 41 and Ad 41 virion sera with soluble adenovirus antigen from other subgenera to establish Ad 40- and Ad 41-specific ELISAs. In both assays, the capture antibody was a group-specific antibody, whereas the detection antibody was an adsorbed type 40 or 41 specific antibody. Singh-Naz and Naz[67] employed hybridoma technology to develop monoclonal antibodies against Ad 40 and Ad 41. Two of the monoclonals recognized a 17-kDa protein which probably is polypeptide VII. One of the two monoclonals cross-reacted between Ad 40 and Ad 41. Herman et al.[68] also developed monoclonal antibodies for diagnosis of EA. Neutralizing monoclonal antibodies specific for Ad 40 and Ad 41 have been developed by J. C. de Jong (personal communication).

In addition to immunological assays, several molecular biology-based methods have been described. Stolhanske et al.[69] described a nucleic acid hybridization assay on nitrocellulose filters utilizing radioactive probes from Ad 2 DNA and cloned *Bam*HI G fragment of Ad 41. Similar dot-blot hybridization tests have been described by Takiff et al.[70] who used probes from *Bgl*II and *Eco*RI fragments of Ad 40 and Ad 41 and by Kidd et al.[71] who employed radioactive probes from *Pst*I fragment of Ad 40 and Ad 41. A simplified hybridization system utilizing samples directly from stools rather than extracted specimens has been described by Hammond et al.[72] Niel et al.[73] have described described a peroxidaseprobe-labeled hybridization test for identification of Ad 40 and Ad 41.

DNA restriction enzyme has been found to be a useful technique for identification of adenoviruses.[23,74-76] Since the analysis can be done directly on viral DNA extracts from stools, it is suitable for identification of enteric adenoviruses. The restriction pattern obtained can be compared with the library of DNA restriction fragment patterns of all 41 adenovirus prototypes that has been reported by Adrian et al.[77]

Isolation of enteric adenoviruses can be performed in tCMK[18] or in 293 cells.[18] The isolation efficiency can be monitored with an assay based on fluorescence of infected cells or by restriction endonuclease analysis. It has to be noted that the growth of enteric adenoviruses is only semipermissive in 293 cells and that "blind" passages may allow overgrowth of other serotypes. Enteric adenovirus can be distinguished from the established serotypes by serum neutralization. Whereas enteric adenoviruses cross-reach extensively in hemagglutination inhibition tests, they can be distinguished from each other by serum neutralization tests on monkey kidney cells.

XI. IMMUNITY

Only a few reports have been published on the immune response to enteric adenoviruses. In a study in London on the acquisition of enteric adenovirus antibodies by children, it was shown that half of the children aged 4 years or more showed evidence of past infection.[78] In another study it was observed that the proportion of positive sera increased with increasing age and the positivity rate of antibody was more than double in children 2 to 4 years old than in children less than 2 years old.[47] In a comprehensive study from Japan, Shinozake et al.[79] also found that the incidence of antibodies gradually rose through childhood. Neutralizing antibodies to Ad 40 and Ad 41 were observed in 20% of the children between 1 and 6 months old and 50% of those 37 to 48 months old. Of serum samples from young adults, 48% had antibodies, whereas only 10% of people over 70 years of age were positive. The low seropositivity of sera from old people might indicate a decrease in immunity to infection among the elderly. In the Japanese investigation 20% of serum samples from pregnant women and their cord samples had antibodies to Ad 40 and Ad 41. Fifteen percent of sera from newborns was positive. These observations demonstrate the existence of passively transferred neutralizing antibodies to enteric adenovirus. Whether these transplacentally acquired antibodies can confer protection during infancy remains to be investigated. Reinfections with enteric adenoviruses have not been described to date.

There have been no reports on clinical immunity to Ad 40 and Ad 41 infections. It is not known whether circulating antibodies or local immune factors in the intestine are the major mediators of protection. The role of cell-mediated immunity in either protection or recovery from enteric adenovirus infection also needs to be investigated.

XII. PREVENTION AND CONTROL

It is not known how enteric adenovirus infection is transmitted, although studies involving outbreaks indicate person-to-person transmission by contact.[48,49] There is evidence of nosocomial spread of infection.[16] Enteric adenoviruses have seen to be restricted to the gastrointestinal tract and have not been identified in respiratory secretions. These data suggest that the viruses may be transmitted via the fecal-oral route.

Fluid replacement is the key to treatment of all infectious diarrheal illnesses regardless of causative enteropathogens. Children with enteric adenovirus infection usually have a clinically moderate disease and can be successfully rehydrated with oral glucose-electrolyte solutions. Patients with severe dehydration may require parenteral fluid replacement. Avoidance of lactose-containing formulas or food may be recommended in cases with prolonged diarrheal illness. At present, there is no specific treatment against enteric adenoviruses available.

Experimental vaccines against enteric adenovirus infections have not yet been developed. Major problems are the fastidious growth of Ad 40 and Ad 41 in cell cultures. Graham 293 cells which has proved to be the most susceptible cells to EAd infection, are a transformed cell line and hardly appropriate as a vaccine substrate. In addition, Ad 40 and Ad 41 have been shown to transform rat embryo cells,[80] although they are not oncogenic in newborn hamsters, as are some of the established adenoviruses. Live oral vaccines against respiratory adenoviruses, Ad 4 and Ad 7, have been used successfully in the U.S. to protect military recruits from the disease.[81]

However, there are still fundamental questions that have to be answered before vaccines against Ad 40 and Ad 41 can be developed. The impact of enteric adenovirus gastroenteritis in developing countries needs be to be further investigated. More knowledge has to be gained about the nature of enteric adenovirus immunity in humans, i.e., what type of immune response needs to be induced and to which viral antigens should these immune responses be directed in order to achieve protection.

REFERENCES

1. **Rowe, W. P., Heubner, R. T., Gilmore, L. K., Parott, R. H., and Ward, T. G.,** Isolation of a cytopathic agent from human adenoids undergoing spontaneous degeneration in tissue culture, *Proc. Soc. Exp. Biol. Med.*, 84, 570, 1953.
2. **Wigand, R., Bartha, A., Dreizin, R. S., Eschle, H., Ginsberg, H. S., Green, M., Heirholcer, J. C., Kalter, S. S., Mafferan, J. B., Pettersson, U., Russell, W. C., and Wadell, G.,** Adenoviridae; second report, *Intervirology,* 18, 169, 1982.
3. **Heirholzer, J. C., Wigand, R., Anderson, L. J., Adrian, T., and Gold, J. W. M.,** Adenoviruses from patients with AIDS; a plethora of serotypes and a description of five new serotypes of subgenus D (types 43 — 47), *J. Infect. Dis.*, 158, 804, 1988.
4. **Duncan, I. B. R. and Hutchison, J. G. P.,** Type-3 adenovirus infection with gastrointestinal symptoms, *Lancet,* 1, 530, 1961.
5. **Flewett, T. H., Bryden, A. S., Davies, H., and Morris, C. A.,** Epidemic viral enteritis in a long-stay childrens ward, *Lancet,* 1, 4, 1975.
6. **Ramos-Alvarez, M. and Sabin, A. B.,** Enteropathogenic viruses and bacteria. Role in summer diarrheal diseases of infancy and early childhood, *JAMA,* 167, 147, 1958.
7. **Brandt, C. D., Kim, H. W., Yolken, R. H., Kapikian, A. Z., Arrobio, J. O., Rodriguez, W. J., Wyatt, R. G., Chanock, R. M., and Parrott, R. H.,** Comparative epidemiology of two rotavirus serotypes and other viral agents associated with pediatric gastroenteritis, *Am. J. Epidemiol.*, 110, 243, 1979.
8. **Jacobsson, P. O., Johansson, M. E., and Wadell, G.,** Identification of an enteric adenovirus by immunoelectroosmophoresis (IEOP) technique, *J. Med. Virol.*, 3, 307, 1979.
9. **Retter, M., Middleton, P. J., Tam, S. J., and Petric, M.,** Enteric adenovirus: detection, replication and significance, *J. Clin. Microbiol.*, 10, 574, 1979.
10. **Kidd, A. H. and Madeley, C. R.,** *In vitro* growth of some fastidious adenoviruses from stool specimens, *J. Clin. Pathol.*, 34, 213, 1981.
11. **Uhnoo, I., Wadell, G., Svensson, L., and Johansson, M.,** Two new serotypes of enteric adenovirus causing infantile diarrhoea, *Dev. Biol. Stand.*, 53 311, 1983.
12. **Uhnoo, I., Wadell, G., Svensson, L., and Johansson, M.,** Importance of enteric adenoviruses 40 and 41 in acute gastroenteritis in infants and young children, *J. Clin. Microbiol.*, 20, 365, 1984.
13. **Madeley, C. R., Cosgrave, B. P., Bell, E. J., and Fallon, R. J.,** Stool viruses in babies in Glasgow, *J. Hyg. Camb.*, 78, 261, 1977.
14. **Appleton, H., Buckley, M., Robertson, M. H., and Thom, B. T.,** A search for fecal viruses in newborn and other infants, *J. Hyg. Camb.*, 81, 279, 1978.
15. **Konno, T., Suzuki, H., Imai, A., Kutsuzava, T., Ishida, N., Katsushima, N., Sakamoto, M., Kitaoka, S., Tsuboi, R., and Adachi, M.,** A long-term survey of rotavirus infection in Japanese children with acute gastroenteritis, *J. Infect. Dis.*, 138, 569, 1978.
16. **Middleton, P. J., Szymanski, H. T., and Petric, M.,** Viruses associated with acute gastroenteritis in young children, *Am. J. Dis. Child.*, 131, 733, 1977.
17. **Vesikari, T., Maki, M., Sarkkinen, H. K., Arstila, P. P., and Halonen, P. E.,** Rotavirus, adenovirus and non-viral enteropathogens in diarrhea, *Arch. Dis. Child.*, 56, 264, 1981.

18. **de Jong, J. C., Wigand, R., Kidd, A. H., Wadell, G., Kapsenberg, J. G., and Muzeric, C. J., Wermenbol, A. G., and Firtzlaff, R. G.,** Candidate adenoviruses 40 and 41: fastidious adenoviruses from human infant stool, *J. Med. Virol.*, 11, 215, 1983.
19. **Wadell, G., Allard, A., Johansson, M., Svensson, L., and Uhnoo, I.,** Enteric adenoviruses. Novel diarrhea viruses, in *CIBA Foundation Symp. 128*, John Wiley & Sons, Chichester, U.K., 1987, 63.
20. **Rosen, L.,** Hemagglutination inhibition technique for typing adenoviruses, *Am. J. Hyg.*, 71, 120, 1960.
21. **Huebner, R. J.,** Adenovirus-directed tumor and T antigens, in *Perspectives in Virology*, Pollard, M., Ed., Academic Press, New York, 1967, 147.
22. **Wadell, G., Hammarskjold, M. L., Winberg, G., Varsani, T. M., and Sundell, G.,** Genetic variability of adenoviruses, *Ann. N.Y. Acad. Sci.*, 354, 15, 1980.
23. **Wadell, G.,** Molecular epidemiology of human adenoviruses, *Curr. Top. Microbiol. Immunol.*, 110, 191, 1984.
24. **Wadell, G.,** Classification of human adenoviruses by SDS-polyacrylamide gel electrophoresis of structural polypeptides, *Intervirology*, 11, 47, 1979.
25. **Green, M., Mackey, J. K., Wold, W. S. M., and Rigden, P.,** Thirty-one human adenovirus serotypes (Ad 1 — Ad 31) form five groups (a — e) based upon DNA genome homologies, *Virology*, 93, 481, 1979.
26. **Li, O. and Wadell, G.,** The degree of genetic variability among adenovirus type 4 strains isolated from man and chimpanzee, *Arch Virol.*, 101, 65, 1988.
27. **van Loon, A. E., Rozijn, T. H., de Jong, J. C., and Sussenbach, J. S.,** Physicochemical properties of the DNAs of the fastidious adenovirus species 40 and 41, *Virology*, 140, 197, 1985.
28. **Takiff, H. E. and Strauss, S. E.,** Early relocative block prevents the efficient growth of fastidious diarrhoea-associated adenovirus cell culture, *J. Med. Virol*, 9, 93, 1982.
29. **Takiff, H. E. and Strauss, S. E., and Garon, C. F.,** Propagation and *in vitro* studies of previously noncultivable enteral adenoviruses in 293 cells, *Lancet*, 2, 832, 1981.
30. **Brown, M., Petric, M., and Middleton, P.,** Diagnosis of infectious enteric adenoviruses Ad 40 ad Ad 41 in stool specimens, *J. Clin. Microbiol.*, 20, 334, 1984.
31. **Perron-Henry, D. M., Herrman, J. E., and Blacklow, N. R.,** Isolation and propagation of enteric adenoviruses in Hep-2 cells, *J. Clin. Microbiol.*, 26, 1445, 1988.
32. **Shinozaki, T., Araki, K., Ushijima, M., Fuji, R., and Eshita, Y.,** Use of Graham 293 cells in suspension for isolating enteric adenoviruses from the stools of patients with acute gastroenteritis, *J. Infect. Dis.*, 156, 146, 1987.
33. **Robinson, A. J., Younghusband, H. B., and Bellett, A. J. D.,** A circular DNA-protein complex from adenoviruses, *Virology*, 56, 54, 1973.
34. **Rekosh, D. M. K., Russell, W. C., Bellett, A., and Robinson, A. J.,** Identification of a protein linked to the ends of adenovirus DNA, *Cell*, 11, 238, 1977.
35. **Arrand, J. R. and Roberts, R. J.,** The nucleotide sequences of the termini of adenovirus-2 DNA, *J. Mol. Biol*, 128, 577, 1979.
36. **Kidd, A. H., Berkowitz, F. E., Blaskovic, P. J., and Schoub, B. D,** Genome variants of human adenovirus 40 (subgroup F), *J. Med. Virol.*, 14, 235, 1984.
37. **Kidd, A. H.,** Genome variants of adenovirus 41 (subgroup G) from children with diarrhea in South Africa, *J. Med. Virol.*, 14, 49, 1984.
38. **Allard, A. K., Wadell, G., Evander, K. M., and Lindman, G. K. K.,** Specific properties of two enteric adenovirus 41 clones mapped with early region 1A, *J. Virol.*, 54, 145, 1985.
39. **van der Avoort, H., Wermenbol, A., Zomerdijk-Kleijne, J. A. F. W., van Asten, J. A. M. M., Jensma, P., Osterhaus, A. D. M. E., Kidd, A. H, and de Jong, J. C.,** Characterization of fastidious adenovirus type 40 and 41. DNA restriction enzyme analysis and neutralizing monoclonal antibodies, *Virus Res.*, in press.
40. **Allard, A. and Wadell, G.,** Physical organisation of the enteric adenovirus type 41 early region 1A., *Virology*, 164, 220, 1988.
41. **Ishino, M., Ohashi, Y., Emoto, T., Sawada, Y., and Fujinaga, K.,** Characterization of adenovirus type 41 El region, *Virology*, 165, 95, 1988.
42. **van Loon, A. E., Lightenberg, M., Reemst, A. M. C. B., Sussenbach, J. S, and Rozijn, T. H.,** Structure and organization of the left-terminal DNA regions of fastidious adenovirus types 40 and 41, *Gene*, 58, 109, 1987.
42a. **Toogood, C. E. A. and Hay, R. T.,** DNA sequence of the adenovirus type 41 hexon gene and predicted structure of the protein., *J. Gen. Virol.*, 69, 2291, 1988.
43. **Norrby, E. and Wadell, G.,** Immunological relationship between hexons of certain human adenoviruses, *J. Virol.*, 4, 663, 1967.
44. **Wadell, G. and Norrby, R.,** Immunological and other biological characteristics of pentons of human adenoviruses, *J. Virol.*, 4, 671, 1967.
45. **Philipson, L., Pettersson, U., and Linbgerg, U.,** Molecular biology of adenoviruses, *Virol. Monogr.*, 14, 1, 1975.
46. **Valentine, R. C. and Pereira, H. G.,** Antigens and structure of the adenovirus, *J. Mol. Biol.*, 13, 13, 1965.

47. **Kidd, A. H., Banatvala, J. E., and de Jong, J. C.,** Antibodies to fastidious adenoviruses (species 40 and 41) in sera from children, *J. Med. Virol.*, 11, 333, 1983.
48. **Richmond, S. J., Caul, E. O., Dunn, S. M., Ashley, C. R., and Clarke, S. K. R.,** An outbreak of gastroenteritis in young children caused by adenoviruses, *Lancet*, 1, 1178, 1979.
49. **Chiba, S., Nakata, S., Nakamura, I., Taniguchi, K., Urasawa, K., Fujinaga, K., and Nakao, T.,** Outbreak of infantile gastroenteritis due to type 40 adenovirus, *Lancet*, 2, 954, 1983.
50. **Yolken, R. H., Lawrence, F., Leister, F., Takiff, H. K., and Strauss, S. E.,** Gastroenteritis associated with enteric type adenovirus in hospitalized infants, *J. Paediatr.*, 101, 21, 1982.
51. **Riepenhoff-Talty, M., Saif, L. J., Barrett, H. J., Suzuki, H., and Ogra, P. L.,** Potential spectrum of etiological agents of viral enteritis in hospitalized infants, *J. Clin. Microbiol.*, 17, 352, 1983.
52. **Ellis, M. E., Watson, B., Mandahl, B. K., Dunbar, E. M., Craske, J., Curry, A., Roberts, J., and Lomaz, J.,** Micro-organisms in gastroenteritis, *Arch. Dis. Child.*, 59, 848, 1984.
53. **de Jong, J. C.,** unpublished data, 1986.
54. **Brandt, C. D., Kim, H. W., Rodriguez, W. J., Arrobio, J. O., Jeffries, B. C., Stallings, E. P., Lewis, C., Miles, A. J., Garner, M. K., and Parrott, R. H.,** Adenoviruses and pediatric gastroenteritis, *J. Infect. Dis.*, 151, 437, 1985.
55. **Leite, J. P. G., Pereira, H. G., Azeredo, R. S., and Schatzmayr, H. G.,** Adenoviruses in faeces of children with acute gastroenteritis in Rio de Janerio, Brazil, *J. Med., Virol.*, 15, 203, 1985.
56. **Kidd, A., H., Rosenblatt, A., Besselaar, T. G., Erasmus, M. J., Tiemesson, C. T., Berkowitz, F. E., and Shoub, B. D.,** Characterization of rotaviruses and subgroup F adenoviruses from acute summer gastroenteritis in South Africa., *J. Med. Virol.*, 18, 159, 1986.
57. **Herrman, J. E., Blacklow, N. R., Peron-Henry, D. M., Clements, E., Taylor, D. N., and Echeverria, P.,** Incidence of enteric adenoviruses among children in Thailand and the significance of these viruses in gastroenteritis, *J. Clin. Microbiol.*, 26, 1783, 1988.
58. **Bhan, M. K., Raj, P., Bhandari, N., Svensson, L., Stintzing, G., Prasad, A. K., Jayashree, S., and Srivastava, R.,** Role of enteric adenoviruses and rotaviruses in mild and severe acute enteritis, *Pediatr. Infect. Dis. J.*, 7, 320, 1988.
59. **Whitelaw, A., Davies, H., and Parry, J.,** Electron microscopy of fatal adenovirus gastroenteritis, *Lancet*, 1, 361, 1977.
60. **Mavromichalis, J., Evans, N., McNeish, A. S., Bryden, A. S., Davies, H. A., and Flewett, T. H.,** Intestinal damage in rotavirus and adenovirus gastroenteritis assessed by D-xylose malabsorption, *Arch. Dis. Child.*, 52, 589, 1977.
61. **Kagnoff, M. F., Austin, R. K., Hubert, J. J., Bernadin, J. E., and Kasarda, D. D.,** Possible role for a human adenovirus in the pathogenesis of celiac disease, *J. Exp. Med.*, 160, 1544, 1984.
62. **Zissis, G., Lambert, J. P., and Fonteyne, L.,** Enteric adenoviruses and diarrhoea, Abst. No. 27, presented at Int. Symp. Recent Adv. Enteric Infect., Brigge, Belgium, August, 1981.
63. **Svensson, L., Wadell, G., Uhnoo, I., Johansson, M., and von Bonsdorff, C. H.,** Cross-reactivity between enteric adenoviruses and adenovirus type 4: analysis of epitopes by solid-phase immune electron microscopy, *J. Gen. Virol.*, 64, 2517, 1983.
64. **Wood, D. J. and Bailey, A. S.,** Detection of adenovirus types 40 and 41 in stool specimens by immune electron microscopy, *J. Med. Virol.*, 21, 191, 1987.
65. **Johansson, M. E., Uhnoo, I., Kidd, A. H., Madely, C. R., and Wadell, G.,** Direct identification of enteric adenovirus, a candidate new serotype, associated with infantile gastroenteritis, *J. Clin. Microbiol.*, 12, 95, 1980.
66. **Johansson, M. E., Uhnoo, I., Svensson, L., Pettersson, C. A., and Wadell, G.,** Enzyme-linked immunosorbent assay for detection of human enteric adenoviruses, *J. Med. Virol.*, 17, 19, 1985.
67. **Singh-Naz, J. and Naz, N.,** Development and application of monoclonal antibodies for specific detection of human enteric adenoviruses, *J. Clin. Microbiol.*, 23, 840, 1986.
68. **Herrman, J. E., Peron-Henry, D. M., and Blacklow, N. R.,** Antigen detection with monoclonal antibodies for diagnosis of adenovirus gastroenteritis, *J. Infect. Dis.*, 155, 1167, 1987.
69. **Stalhandske, P., Hypia, R., Allard, A., Halonen, P., and Pettersson, U.,** Detection of adenovirus in stool specimens by nucleic acid spot hybridization, *J. Med. Virol.*, 16, 213, 1985.
70. **Takiff, H. E., Seidlin, M., Krause, P., Rooney, J. Brandt, C., Rodriguez, W., Yolken, R., and Strauss, S. E.,** Detection of enteric adenoviruses by dot-blot hybridization using a molecularly cloned viral DNA probe, *J. Med. Virol.*, 16, 107, 1985.
71. **Kidd, A. H., Harley, E. H., and Erasmus, M. J.,** Specific detection and typing of adenovirus types 40 and 41 in stool specimens by dot-blot hybridization, *J. Clin. Microbiol.*, 22, 934, 1985.
72. **Hammond, G., Hannan, C., Yeh, T., Fischer, K., Mauthe, G., and Strauss, S. E.,** DNA hybridization for diagnosis of enteric adenovirus infection from directly spotted fecal specimens, *J. Clin. Microbiol.*, 25, 1881, 1987.

73. **Niel, C., Gomes, S. A., Leite, J. G., and Pereira, H. G.,** Direct detection of fastidious and nonfastidious adenoviruses in stools by using a specific nonradioactive probe, *J. Clin. Microbiol.*, 24, 785, 1986.
74. **Brown, M., Petric, M., and Middleton, P. J.,** Solver staining of DNA restriction fragments for the rapid identification of adenovirus isolates; application during nosocomial outbreaks, *J. Virol. Med.*, 9, 87, 1987.
75. **Buitenwert, J. Louwerens, J. J., and de Jong, J. C.,** A simple and rapid method for typing adenovirus 40 and 41 without cultivation, *J. Virol. Meth.*, 10, 39, 1985.
76. **Willcocks, M. M., Carter, M. J., Laidler, F. R., and Madeley, C. R.,** Restriction enzyme analysis of faecal adenoviruses in Newcastle upon Tyne, *Epidem. Infect.*, 101, 445, 1988.
77. **Adrian, T., Wadell, G., Heirholzer, J. C., and Wigand, R.,** DNA restriction analysis of adenovirus prototypes 1 to 41, *Arch. Virol.*, 91, 277, 1986.
78. **Kidd, A. H., Chrystie, I. L., and Banatvala, J. E.,** Infection by fastidious enteric adenoviruses in childhood, *Lancet*, 2, 371, 1981.
79. **Shinozake, T., Araki, K., Ushijima, H., and Fujii, R.,** Antibody response to enteric adenovirus types 40 and 41 in sera from people in various age groups, *J. Clin. Microbiol.*, 25, 1679, 1987.
80. **van Loon, A. E., Maas, R., Vaessen, R. T. M. J., Reemst, A. M. C. B., Sussenbach, J. S., and Rozijn, T. H.,** Cell transformation by the left terminal regions of the adenovirus 40 and 41 genomes, *Virology*, 147, 227, 1985.
81. **Top, F. H., Jr., Buescher, E. L., Bancroft, W. H., and Russell, P. K.,** Immunization with live types 7 and 4 adenovirus vaccines. II. antibody response and protective effect against acute respiratory disease due to adenovirus type 7, *J. Infect. Dis.*, 124, 155, 1971.
82. **Wadell, G., Allard, A., Evander, M., and Li, Q.-G.,** Genetic variability and evolution of adenoviruses, *Chem. Scripta*, 26B, 325, 1986.
83. **Uhnoo, I., Olding-Stenkvist, E., and Kreuger, A.,** Clinical features of acute gastroenteritis associated with rotavirus, enteric adenovirus, and bacteria, *Arch. Dis. Child.*, 61, 732, 1986.

Chapter 6

ENTERIC ADENOVIRUS INFECTIONS OF ANIMALS

David A. Benfield

TABLE OF CONTENTS

I. INTRODUCTION

The nonhuman adenoviruses are currently divided into two genera, Mastadenovirus (mammalian adenoviruses) and Aviadenovirus. There are also candidate viruses for the family Adenoviridae among poikilothermic animals.[1] Hemagglutination-inhibition and viral neutralization tests indicate multiple adenovirus serotypes exist in most animal species.[1-3] Currently, there are 2 equine, 9 bovine, 4 porcine, 6 ovine, 2 caprine, 2 canine, 11 chicken, 4 turkey, 3 goose, 2 mice, and 25 simian serotypes of adenovirus.[1] Adenoviruses have also been isolated from quail, pheasants, opossum, rabbits, tree shrews (*Tupaia*), guinea pigs, rats, and sea lions.[1] Adenoviruses are highly host specific and are usually isolated from the upper respiratory tract, conjunctiva, gastrointestinal tract, or feces.[1-3] Most of these adenoviruses produce subclinical infections, but pneumonia, diarrhea, conjunctivitis, and hepatitis are frequently observed in animals infected with adenoviruses.[1-3] The etiological significance of the nonhuman enteric adenoviruses is not known, since in many species adenoviruses are isolated with equal frequency from the feces of clinically healthy animals and animals with diarrhea. Adenoviruses are also frequently isolated from both the respiratory and gastrointestinal tract of the same animal. Thus, the presence of adenovirus in feces may represent enteric viral replication or respiratory infection with swallowed virus shed in the feces. Except for hemorrhagic enteritis in turkeys, most experimental inoculations of animals with adenovirus have focused on reproduction of respiratory rather than enteric disease, despite the fact that diarrhea was frequently observed in these experimental animals after adenovirus inoculation.[4-14] However, the clinical signs of respiratory disease or diarrhea were usually so mild that many investigators concluded that adenoviruses were not primary pathogens.

However, our laboratory[122] and others[15-17,123] have observed the presence of adenovirus particles in the feces from pigs with diarrhea and intranuclear adenovirus inclusions in villous enterocytes in the intestinal tract of these pigs. Recently, enteritis has been experimentally produced in hysterectomy-derived, colostrum-deprived pigs infected with a porcine adenovirus type 3.[16,18,19] Circumstantial evidence from field cases of bovine enteritis,[20-24] foal enteritis,[25,26] and diarrhea in puppies[27,28] also implicates adenoviruses as causative agents of diarrhea. Hemorrhagic enteritis virus (HEV) of turkeys is an adenovirus, which causes a bloody diarrhea in naturally and experimentally infected birds.[29-33] Unlike the human enteric adenovirus types 40 and 41, the nonhuman enteric adenoviruses are not well characterized, principally because their role in the etiology of diarrheal diseases (with the exception of HEV in turkeys) has only recently become apparent.

II. MORPHOLOGY AND MORPHOGENESIS

The morphology of the nonhuman enteric adenoviruses is similar to other members of the family Adenoviridae. Virions observed by direct electron microscopy of fecal samples from pigs,[16,17,122] calves,[20-24] foals,[26] puppies,[27,28] and turkeys,[29,34,35] reveal nonenveloped icosahedra with a diameter of 70 to 80 nm (Figure 1).

The capsid is composed of 252 capsomers, 240 hexamers, and 12 pentamers which occupy the 12 corners of the virion. From each pentamer projects a fiber of lengths varying from 12 to 50 nm depending on the species of origin.[1] Some avian adenoviruses have two fibers attached to a penton base, but this is not a general morphologic characteristic of all Aviadenoviruses.[1]

The most notable morphologic feature of adenovirus infection and replication is the production of viral intranuclear inclusions. These Cowdry type A intranuclear inclusions have been observed in cell cultures inoculated with adenoviruses[36-41] and in the intestinal epithelial cells of naturally infected pigs and pigs experimentally inoculated with porcine adenovirus type 3;[16,18,19] mice inoculated with mouse adenovirus type 2;[42] a foal with prolonged diarrhea;[25] and

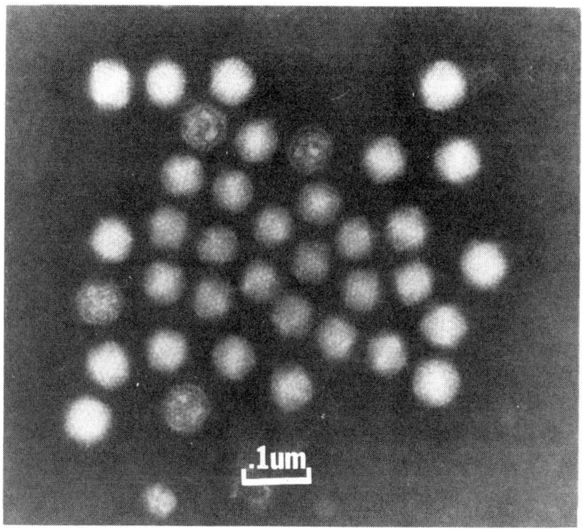

FIGURE 1. Immune electron microscopy of an enteric adenovirus in intestinal contents of a gnotobiotic pig challenged with a viral filtrate containing adenovirus from conventional diarrheic pig. Viral particles are aggregated in presence of antiserum from a gnotobiotic pig recovered from enteric adenovirus challenge. (Photo and results reported courtesy of Dr. Linda J. Saif.)

turkeys infected with HEV.[29-33,35,43,44] The sequential appearance of adenovirus intranuclear inclusions has been correlated with stages in adenoviral replication. These intracellular morphologic changes have been described in cell cultures infected with nonenteric adenoviruses isolated from man and animals.[37-39,41,45] Similar studies have also been done *in vivo* with pigs[18] and mice[42] inoculated with enteric adenoviruses and turkeys inoculated with HEV.[29,35,43] Light microscopic observations with the nonenteric human,[45,46] bovine,[36] equine,[41] ovine,[38] and avian adenoviruses[37] have indicated two distinct subgroups based on morphology of inclusion bodies. Subgroup I adenoviruses induced the appearance of irregular, intranuclear eosinophilic bodies, which increased in number to form a large basophilic inclusion in the center of the nucleus. Morphologic changes observed with subgroup II adenoviruses included several refractile "pearl-like" intranuclear inclusions which stained lightly eosinophilic. These inclusions were often surrounded by a halo and were more distinct than similar structures seen in the subgroup I adenoviruses. As the infection progressed, the nucleus of cells infected with subgroup II adenoviruses was filled with an irregular shaped, basophilic inclusion body, which was surrounded by a clear halo. The division of adenoviruses into two subgroups on the basis of cytopathology at the light microscopic level was confirmed by immunofluorescence and electron microscopy. Adenoviral antigens in the subgroup I and II inclusion bodies have been detected by immunofluorescence in cell cultures infected with nonenteric bovine,[36] ovine,[38] and avian[37] adenoviruses.

Ultrastructurally, four Types (I to IV) of inclusions have been observed in KB (epidermoid carcinoma) cells infected with human adenovirus type 12[45] and a fifth type of inclusion has been described in fetal equine kidney cells infected with equine adenovirus.[41] Within 24 h after infection of KB cells with human adenovirus type 2, Types I and II inclusions are most prominently observed by electron microscopy. Type I inclusions are extremely dense granular material forming small round patches in the nucleus. Type II inclusions are large irregular patches of fine fibrils 4 to 6 nm thick and are less dense than type I inclusions. Later (36 to 48 hours postinfection), the Type II inclusions were noted next to Type IV inclusions and paracrystalline arrays of adenovirus particles. Type III inclusions were observed at 24 to 36 h after infection and were characterized as large round foci of poorly stained loose fibrils. Type IV inclusions appeared at 36 h after inoculation and were described as a dense reticulated substance. Viral particles typical of adenoviruses were initially observed at 24 h after inoculation and tended to be located in the vicinity of Type II inclusions. As the infection progressed, paracrystalline arrays of virus particles were commonly observed in the nucleus of cells. The nuclear architecture was completely disorganized in most cells by 36 h postinoculation and the nucleus was occupied by numerous viral particles completely dispersed or aggregated into crystals. Nuclear chromatin was condensed and marginated along the nuclear membrane. At 72 h postinoculation the nuclear membrane disrupted and viral particles were seen in the cytoplasm. Ultrastructural cytochemistry and autoradiography revealed that the Types I and II inclusions contain newly synthesized viral DNA and protein, Type III inclusions are nucleoprotein, and type IV inclusions predominantly viral protein.[41,45]

The replication of an enteropathogenic porcine adenovirus,[18] nonenteropathogenic mouse adenovirus,[42] and the hemorrhagic enteritis virus of turkeys[29,35,43] have been studied by transmission electron microscopy in naturally or experimentally infected animals. In pigs experimentally infected with porcine adenovirus type 3, early ultrastructural changes in the small intestinal epithelial cells included an enlarged nucleus containing Type II intranuclear inclusions.[18] The Type II inclusions were often observed between paracrystalline arrays of adenoviral particles in the nucleus. Fine granular and electron dense aggregates (Type I inclusions) were located at the boundary of the central nuclear region. Type IV inclusion bodies were also seen in close proximity to the paracrystalline arrays of virus particles. Viral particles observed in the nucleus of villous epithelial cells in the small intestine were of two types: 65-nm particles with a 50-nm electron dense core surrounded by an outer capsid and 65-nm particles with an electron dense core bounded by an inner electron dense shell, which was separated from the capsid by a narrow clear space. The remainder of cellular changes in the intestinal epithelium were similar to those described in cell cultures infected with human,[45,46] equine,[41] ovine,[38] and bovine[36] nonenteric adenoviruses. Ultrastructural changes observed in intestinal epithelial cells in mice infected with mouse adenovirus Type 2, which does not produce clinical disease, included enlarged nuclei in the villous epithelial and crypt cells; nuclei of epithelial, crypt and goblet cells containing large masses of virus particles; margination of nuclear chromatin; and virions with an election dense or translucent center.[42] These two particle types are similar to those described in porcine intestinal epithelial cells infected with adenovirus type 3.[18] While Types I, II, III, or IV inclusions, were not described in intestinal epithelial cells infected with mouse adenovirus, the electron micrographs appeared to show evidence of such inclusions. Similar inclusions have also been observed in the nucleus of splenic cells and unidentified cells in the lamina propria of the intestine from turkeys naturally infected with hemorrhagic enteritis virus (HEV).[29,35,43] Cell cultures infected with nonenteric fowl,[37] bovine,[36] and ovine[38] adenoviruses also reveal intranuclear inclusions similar to those described for human adenoviruses.

III. *IN VITRO* CULTIVATION

Although the human enteric adenovirus types 40 and 41 are difficult to isolate in cell cultures,[48] adenoviruses have been readily cultivated from feces or gastrointestinal contents of calves,[7,23,36,49-51] dogs,[27,28] goats,[52] foals,[26] mice,[53,54] pigs,[55-58] and rabbits.[47] Primary or secondary kidney cell cultures prepared from the homologous species are the most common cell type for adenovirus isolation. Adenoviruses are species specific and the use of kidney cells derived from the same host species as the specimen, appears to be required to successfully isolate the virus. However, a porcine adenovirus serotype 4 has been cultivated in calf kidney, human kidney, and canine melanoma cells.[59,60]

Isolation of adenoviruses may require prolonged incubation times and several blind passages before cytopathic effects (CPE) are produced in cell culture.[1,3,55,61] After several passages CPE may be evident within 18 h or require 5 to 6 d. For example, Haig et al.[57] reported an 11-d incubation before adenoviral CPE was observed in swine kidney cells inoculated with a fecal suspension from a pig with diarrhea. Recently, Dea and ElAzhary[61] reported that secondary cultures of pig thyroid (PT) cells were highly susceptible to porcine adenovirus Type 4. A readily recognizable CPE could be observed on PT cells on the first passage and titers of porcine adenovirus Type 4 were approximately 100-fold higher on PT compared to similar pools of the virus propagated and titrated on an established pig kidney cell line (PK-15). However, these investigators did not attempt isolation of porcine adenovirus from clinical specimens. Adenoviral CPE is characterized by the rounding and often swelling of infected cells and aggregation into "grape-like" clusters.[2,55] Histologic examination of stained cell preparations will reveal intranuclear inclusions, nuclear hypertrophy, and chromatin margination.[1-3,55] Adenovirus specific antigen can be demonstrated by immunofluorescence within the inclusions and viral particles can be observed in the nucleus by electron microscopy.[1-3,36-38,55]

Most aviadenoviruses, except for HEV of turkeys, infect standard monolayer cultures of chick kidney or fibroblasts and can be readily propagated *in vitro*.[30,31,37] Early attempts failed to propagate HEV in chicken and turkey embryo fibroblasts; chicken, pheasant, or turkey kidney cells;[44,62] and turkey spleen cells.[63] Fasina and Fabricant[63] were able to demonstrate HEV antigens in turkey spleen cells, but infectious virus was not produced because attempts to serially passage HEV in these cells failed. Successful serial passage of virulent and avirulent isolates of turkey HEV was recently accomplished using lymphoblastoid B cells derived from turkeys infected with Marek's disease virus.[64] Spleen cells from HEV infected turkeys were cocultivated with the lymphoblastoid B cells with cell enlargement and lysis evident within 48 h after cocultivation. Adenoviral nuclear and cytoplasmic antigens were demonstrated by immunofluorescence 4 d after cocultivation. Cell extracts and supernatants from HEV infected cultures were positive for HEV specific antigens by the immunoprecipitation test and adenoviral particles were observed by electron microscopy within intranuclear inclusion bodies.[64] However, HEV did not replicate in lymphoblastoid T cells, indicating an affinity of HEV for B cells. The ability of turkey HEV to replicate in herpesvirus-transformed B lymphocytes parallels the replication of human enteric adenoviruses in Graham 293 cells, which is a human embryonic kidney cell line transformed by a type 5 adenovirus.[65] The Graham 293 cells may provide the human enteric adenoviruses with helper functions that aid replication.[48] It has not been determined why transformed B cells are permissive to HEV replication. However, B cells and an intact bursa are required for HEV inoculated turkey poults to develop clinical disease.[66]

IV. GENOME AND PHYSICOCHEMICAL PROPERTIES

Since many of the nonhuman Mastadenoviruses isolated from normal or diarrheic feces have not proven pathogenic and HEV of turkeys has only recently been propagated in cell culture, the properties of the viral genome and proteins of the nonhuman enteric adenoviruses have not been directly determined. However, it seems reasonable to assume that the genomic and physicochemical properties of nonhuman enteric adenoviruses are similar to other members of the family Adenoviridae.

A. GENOME

The adenoviral genome is a single linear DNA with a molecular weight of 17 to 28×10^6 Da.[1] The DNA of fowl adenovirus Type 1, the only virus among 11 avian serotypes for which the size of the DNA has been determined, is estimated to be $28 \times 10^6 \times$ Da.[1] This suggests that the average size of the aviadenovirus genome may be larger than those of the mastadenoviruses. Adenoviral DNA has two unique features: inverted terminal repetition (ITR) and a 55-kilodalton (kDa) terminal protein (TP) covalently linked to the 5′ end of each strand of DNA.[67,68] The functional significance of the ITR has not been fully determined, but nucleotide base sequences conserved within the ITR may bind an 80 kDa preterminal protein (pTP) to initiate DNA replication.[69]

The pTP is a viral coded protein, which initiates viral DNA replication by covalently binding to a recognized sequence of the ITR at the 5′ terminus.[69] The biological significance of the 55 kDa TP, which is a cleavage product of the 80 kDa pTP, has not been determined. However, the presence of the 55 kDa TP covalently linked to the 5′ terminus of adenoviral DNA enhances the ability of the viral DNA to transfect cells.[70] There are numerous published reports and reviews on the structure of the adenoviral genome[69-71] and the reader is referred to these for a detailed explanation of the properties of adenoviral DNA.

B. REPLICATION STRATEGY

Following adsorption, penetration, and uncoating of the virus, certain defined regions or "early transcription units" are transcribed by a cellular DNA dependent RNA polymerase to produce viral mRNA coding for nonvirion polypeptides. At least three early proteins have been identified and shown to function in preparing the genome for replication.[51] Adenoviral DNA replication proceeds from the 5′ end of both strands in a 5′ to 3′ direction using a virus coded DNA polymerase. Replication along both DNA strands is semiconservative, asynchronous, and does not require the synthesis of Okazaki fragments.[2,71] Most of the late mRNAs, which code for the virion structural proteins, is transcribed from newly synthesized progeny DNA. Several subgenomic polycistronic mRNAs are spliced to produce monocistronic mRNAs, which bind to ribosomes and are translated into viral proteins. These proteins are then shuttled back into the nucleus for viral assembly and maturation.[2,71]

The adenoviral genome contains sufficient genetic information to code for 10 polypeptides, although 10 to 15 distinct polypeptides have been observed on polyacrylamide gel electrophoresis of disrupted adenovirions (see review, Reference 71). The most important polypeptides are a hexon (120 kDa), penton base (85 kDa) and fiber protein (62 kDa) in the viral capsid. The inner surface of the hexon is a soluble protein, which is the genus-specific adenoviral antigen detected by complement fixation and immunodiffusion. An outer hexon protein and certain antigenic determinants on the fiber are type-specific antigens that give rise to neutralizing antibodies. The fiber also functions as the viral hemagglutinin (see review, Reference 71). Other important structural and nonstructural adenovirus polypeptides include three core proteins (48, 18, and 4 kDa); a 100 kDa scaffolding protein (to assemble the hexon trimer); a 72 kDa DNA binding protein; a 140 kDa virion DNA polymerase; and the 80 kDa pTp protein.

Human adenoviruses are currently classified into six subgenera (A to F) on the basis on hemagglutination patterns, oncogenic potential, size of viral polypeptides, degree of DNA homology, and GC content of DNA.[48] The human enteric adenoviruses are sufficiently divergent from the other human adenoviruses to be classified in their own subgenus F. A similar classification scheme for the animal adenoviruses has not been developed. Animal adenoviruses have been classified according to serotype except for the simian adenoviruses, which have been subgrouped according to hemagglutinin patterns (see review, Reference 1). Most nonhuman mastadenoviruses and the aviadenoviruses hemagglutinate red cells from various species.[1-3,55]

C. DENSITY

The nonhuman mastadenoviruses have a buoyant density of 1.32 to 1.34 g/cm^3 in cesium chloride or sucrose gradients.[1,71] Hemorrhagic enteritis virus of turkeys has a density of 1.34 g/cm^3 on sucrose gradients.[29]

D. RESISTANCE TO CHEMICAL AND PHYSICAL AGENTS

Studies on the relative resistance of nonhuman enteric adenoviruses to chemical and physical agents have been reported for porcine adenoviruses and hemorrhagic enteritis virus of turkeys.

The infectivity of porcine adenoviruses was not destroyed by ether or chloroform or at pH 4.0.[55] Porcine adenoviruses lose infectivity at 56°C after 10 min and are susceptible to chemical inactivation by sodium hypochorite, formaldehyde, phenolic compounds, ethyl alcohol, and sodium hydroxide.[72] Infectivity of HEV is destroyed by heating at 70°C for 1 h and exposure to sodium hypocholrite, sodium lauryl sulfate, chlorocide, Phenocide, Wescodyne, Lysol, or drying at 37 or 25°C for 1 week.[31,73,74] Hemorrhagic enteritis virus of turkeys is resistant to heat inactivation at 65°C for 1 h; storage for 1 month at 37°C; 6 months at 4°C; and 4 years at –40°C; the virus is also resistant to ether, chloroform, and maintenance at pH 3.0 at 25°C for 30 min.[73]

V. ANTIGENIC RELATIONSHIPS

A. *IN VITRO*

There is no antigen common to all members of the family Adenoviridae. This virus family is divided into two genera, Mastadenovirus and Aviadenovirus. Each of these genera have a distinct genus-specific soluble hexon antigen, which is detected by complement fixation or immunodiffusion.[1-3,71] Furthermore, HEV of turkeys does not cross react in immunodiffusion tests with other members of the genus Aviadenovirus.[30,31,75] These results indicate that HEV and related viruses (marble spleen disease in pheasants and avian adenovirus group II splenomegaly) constitute an immunologically distinct group of adenoviruses designated avian adenovirus group II.[30,31,75]

The number of serotypes of adenovirus isolated from each species was previously mentioned in the Introduction. Serotype specificity is a function of adenoviral surface proteins located on the hexon and fiber of the virion, which induce neutralizing antibodies *in vivo*.[71] Viral neutralization tests have been used to distinguish adenovirus serotypes both within and between species.[1-3] The fiber proteins also induce type specific antibodies which inhibit hemagglutination and hemagglutination-inhibition tests with type-specific antiserum have also been used for serological classification of the adenoviruses. Two large scale studies, which used reciprocal viral-neutralization (VN) tests to determine the degree of cross reaction between adenoviruses from different species, have been reported.[76,77] Adair and McFerran[76] compared the nine bovine, five ovine, and four porcine serotypes by reciprocal VN tests and found no cross-reaction. These investigators also failed to neutralize any of the 18 serotypes of animal adenoviruses with antisera to 28 different serotypes of human adenoviruses. Willimzik et al.[77] tested antisera prepared against 25 serotypes of simian adenovirus, 5 serotypes of ovine, 8 serotypes of avian, porcine adenovirus type 4, equine adenovirus type 1, rabbit adenovirus, canine adenovirus type 1, mouse adenovirus, and tree shrew adenovirus for their ability to neutralize 35 human adenovirus serotypes. Antisera to the 35 human adenovirus types were also tested against 25 simian adenovirus serotypes. The data showed that only two chimpanzee adenovirus serotypes had a limited cross-reaction with some human adenovirus serotypes. Viral neutralization tests have also failed to demonstrate antigenic relationships between porcine adenoviruses and those of other species.[55] As a general rule, neutralization tests have not detected cross-neutralization among different adenoviral serotypes within and between species.[1,76,77]

B. *IN VIVO*

Adenoviruses are species specific making it difficult to perform two way cross protection studies in animals. Pigs can be experimentally infected with human adenovirus serotypes.[58,78] Also, bovine adenovirus serotypes 2 and 7 have been isolated from lambs with respiratory disease.[79] Antisera collected from calves, each inoculated with bovine adenovirus serotypes 1, 2, and 3, neutralized only the homologous virus, no heterotypic neutralization was detected.[6] Convalescent antisera from turkeys infected with HEV will protect pheasants against infection with the serologically related marble spleen disease virus.[30,31] Thus, *in vitro* and *in vivo* studies indicate that adenoviruses are widely antigenic divergent both within and between species.

VI. EPIDEMIOLOGY

A. NONHUMAN MASTADENOVIRUSES

Little information is available on the epidemiology of the nonhuman enteric mastadenoviruses. Field isolates of enteric adenoviruses need to be characterized, serotyped, and used to conduct seroepidemiological surveys to determine the prevalence of enteric adenovirus infections. In general, adenoviruses in pigs,[55] calves,[3] sheep,[3] horses,[26,80] and presumably other species are prevalent worldwide. Serological surveys indicate that most adult animals have antibodies to adenoviruses, but the incidence of disease is low suggesting that infections are subclinical. Serological surveys done in California and Oregon indicate 10 to 70% of the adult cattle with antibodies to bovine adenovirus Type 3 and 51% with antibodies to bovine adenovirus Type 4.[81,82] In swine, surveys indicate that 26[83] to 52%[84] of the swine population in southeast England had group specific antibodies to adenoviruses. Among specific serotypes of porcine adenoviruses, Darbyshire[84] detected antibodies to porcine adenovirus Type 1 in less than 1% of the pigs, whereas antibodies to porcine adenovirus Type 4 have been reported in 66% of serum samples in a survey in England.[87] However, antibodies to porcine adenovirus Type 4 were detected in only 15% of serum samples collected from adult swine with respiratory disease.[88] In Bavaria, Bibrack (Reference 85 as quoted in Reference 86) demonstrated an increasing frequency and level of serum antibody titer against porcine adenoviruses with increasing age. Approximately 50% of weaned pigs and 80% of adult pigs contained antibodies to adenovirus. In horses, Studdert and Blackney[26] detected neutralizing antibodies to equine adenovirus Type 2, isolated from foals with diarrhea, in 77% of 339 serum samples.

Naturally acquired enteric adenoviral infections occur principally in young animals. In pigs, most enzootics of adenoviral diarrhea occur in pigs 1 to 4 weeks of age.[15,16] Adenoviruses are most frequently isolated from rectal swabs of weaned pigs and rarely from adults.[56,89] Adenoviral diarrhea is also most frequently observed in 1- to 4-week-old calves,[20-23,51] although there are reports describing an adenovirus necrotizing enterocolitis in a 10-month-old bullock[24] and a steer.[51] Other adenoviruses have been circumstantially linked to diarrhea in foals[26] and puppies,[27,28] suggesting that suckling and weaned animals are most susceptible to enteric adenoviral infections.

Enteric adenoviruses are horizontally transmitted from animal to animal either by the fecal-oral route or inhalation of aerosols.[1-3,55] Vectors are not involved, but the virus is relatively stable and could be transmitted on contaminated boots, clothing, bedding, and feeding utensils. Adenoviruses are also known to produce latent infections, and porcine and bovine adenoviruses have been spontaneously recovered in primary kidney and testicle cultures.[3,55] Porcine adenoviruses are shed in the feces for several weeks after weaning,[55,56] and ovine adenovirus Type 4 was isolated from the rectal swab of a lamb 80 d following experimental inoculation.[13] These results suggest adenoviruses may persist in certain tissues and can be shed into the environment through nasal secretions, urine, or feces. While further studies may define the role of latent-carrier animals in the transmission of adenovirus, present evidence suggests that most nonhuman mastadenoviruses are transmitted to susceptible animals from acutely or subclinically infected animals.[1,3,55] Until pathogenesis studies confirm adenoviruses as causative agents of diarrhea in animals, little epidemiological information will be forthcoming.

B. AVIADENOVIRUS (HEMORRHAGIC ENTERITIS VIRUS)

Like most members of the family Adenoviridae the epidemiology of hemorrhagic enteritis virus of turkeys is not fully understood. Hemorrhagic enteritis is an economically important disease of turkeys in the U.S., Australia, Canada, England, Germany, India, Israel, and Japan.[30,31] Serological surveys indicate that almost all flocks of adult turkeys have been infected by HEV.[30,31,75]

HEV typically occurs in 10- to 12- week old turkey poults but has been reported in poults as young as 2 1/2 weeks.[90] The disease is more prevalent in the summer and in turkeys on range. The virus can be experimentally transmitted orally or cloacally by inoculation of infectious feces.[33] The virus probably enters flocks on contaminated litter, feeding and watering equipment, and fecal contaminated shoes, boots, and coveralls worn by caretakers. The virus remains viable for several weeks in carcasses and feces protected from drying and in litter from HEV infected flocks.[30] The disease often recurs in flocks probably because the presence of dirt floors in many turkey operations prohibit complete disinfection of facilities, and most turkey operations introduce new susceptible poults into the flock each year. There is no evidence that the virus is transmitted by vectors, carrier animals, or through eggs.[30,31] Once the virus enters a flock the virus spreads horizontally through ingestion of infectious feces and the disease will usually run its course in about 2 weeks. Mortality in naturally infected flocks averages 10 to 15% (range 1 to 60%).[30,31] The mortality rate is probably influenced by the age of birds when infected, virulence of the infecting HEV strain, and husbandry practices. The infectivity rate in most natural infections is 100%.[30] HEV has been shown to replicate in the spleen of 3-day-old turkey poults[66] without producing clinical disease, and avirulent isolates of HEV also replicate in the spleen of poults and can be shed in feces.[91]

VII. PATHOGENESIS, PATHOLOGY, AND PATHOGENICITY

Adenovirus have been isolated from the feces of clinically healthy animals and from animals with diarrhea. Experimental attempts to reproduce enteric adenoviral enteritis in animals has not been consistently successful, except for experimental inoculation of pigs with porcine adenovirus type 3 and turkeys with HEV. Many of the early pathogenesis studies were also done in conventional and colostrum-deprived animals, which are susceptible to colonization and infection by indigenous or invading microorganisms, including viruses. In addition, the role of rotaviruses and coronaviruses as causes of diarrhea had not been defined when many of the early adenoviral pathogenesis studies were done, and there were no established protocols available to screen for these agents in fecal material, intestinal sections, or contents. Only recently have experimental studies to investigate the role of the nonhuman adenoviruses in enteric disease used techniques such as electron microscopy and immunofluorescence or immunoperoxidase staining of frozen intestinal sections or cell culture monolayers to exclude the presence of adventitious viruses and confirm adenovirus as an etiological agent of diarrhea. This section will review, by species, reports of naturally occurring diarrhea in animals associated with adenoviruses as well as experimental attempts to reproduce the disease. The reader should bear in mind that with the exception of a few studies in the pig and turkey there is only circumstantial evidence to denote adenovirus as a cause of diarrhea in animals.

A. BOVINE

Of the 9 serotypes of bovine adenovirus, serotypes 1 and 2 were isolated from feces of a normal calf.[50,92] Bovine adenovirus serotypes 5 and 6 were isolated from the lungs of naturally infected calves with clinical signs of rhinorrhea and diarrhea (pneumoenteritis).[49] Bovine adenovirus serotype 7 (Fukuroi strain) has been isolated from week-old calves with hemorrhagic diarrhea and pneumonia.[23,93] Experimental infections of colostrum-deprived calves with bovine adenovirus (BAV) serotypes 1, 2, 3, and 5 produced mild respiratory signs and diarrhea.[5-7,10,14,94] Virus was isolated from several organs and from nasal, conjunctival and fecal swabs of calves inoculated with BAV-1, -2, and -3.[4,6,7] Bovine adenovirus type 3 was shed in feces up to 10 d after infection, but gross and histologic lesions were confined to the respiratory tract.[6,7] However, immunofluorescence techniques were not used in any of these studies to localize the primary sites of replication of BAV-1, -2, and -3, making it difficult to determine if virus actually replicated in tissues from which it was isolated. In one experiment, calves inoculated intravenously within 2 hours of birth with BAV-5 had hemorrhagic lesions in the small intestine and colon; the intesinal epithelium was sloughed and intranuclear inclusions were observed in interstitial cells in the duodenum of one calf.[5] Immunofluorescence staining of frozen sections of several organs from this calf revealed viral antigen in most lymphoid tissue, liver, kidney, adrenal glands, and lamina propria of the gastrointestinal tract. The results suggest that BAV-1, -2, -3, and -5 are relatively apathogenic, and replicate primarily in the respiratory and intestinal tracts.

Previous field studies described the isolation of adenoviruses from 1-week to 4-month-old calves with clinical signs of pneumonia and enteritis (pneumoenteritis).[6,20,21,95] Recently, there have been three case reports describing the presence of adenoviral inclusion bodies in vascular endothelial cells in the intestine of calves and older cattle with hemorrhagic enteritis.[22,24,51] These animals have lesions of a necrotizing enteritis or enterocolitis, and in all three case reports, electron microscopic examination revealed the presence of adenoviral particles in intestinal contents or in the intranuclear inclusions. These lesions are similar to the focal hemorrhagic necrosis seen in calves infected with BAV-5.[5] An adenovirus isolated from the small intestine of calves in one study was determined to be serotype 7[23] while another isolate was uncharacterized.[51] The lesions in the calves infected with adenovirus are quite different than lesions observed in pigs and foals naturally infected with adenoviruses. In these latter species, inclusions are seen in the intestinal epithelium, whereas in the calf, inclusions are most frequently observed in the vascular endothelium. It has been suggested that damage to blood vessels may be the initial lesion leading to ischemic necrosis of the intestine of cattle infected with adenovirus.[22,24,51,96] Further studies are needed on the isolation and characterization of these "hemorrhagic enteritis viruses" in cattle and attempts to experimentally produce this disease in gnotobiotic calves. Unfortunately, experimental attempts to reproduce enteric disease with these isolates were not done.

B. OVINE

There are are least five and possibly six serotypes of ovine adenoviruses (OAV).[79] Serotypes 1, 2, and 3 were isolated from feces of normal sheep,[40] while serotypes 4, 5, and 6 were isolated from sheep with respiratory disease.[13,79,97,98] The pathogenicity of OAV-1, -4, and -5 have been described. Specific pathogen free lambs inoculated intranasally with OAV-4 did not develop clinical disease, but shed adenovirus in nasal secretions and feces for 8 and 9 d postinoculation, respectively.[13] In this same study, respiratory shedding preceded fecal shedding by 2 d, but OAV-4 attained higher titers in the alimentary tract. Adenovirus was also isolated from the rectal swab of one lamb 80 d postinoculation, which indicated OAV-4 tended to persist in the alimentary tract. Pulmonary edema was the only gross and histologic lesion in lambs inoculated with OAV-4.[13]

Colostrum-deprived lambs, inoculated with each of two different strains (one isolated from feces and one from nasal discharge) of OAV-5, developed serious nasal discharge and mild diarrhea.[4] Gross and histologic lesions were principally in the respiratory tract, but lesions were also seen in the intestinal tract and kidneys. Similar results have been obtained by Payla et al. (Reference 11 as quoted in Reference 1) in lambs inoculated with OAV-1. Thus, experimental evidence has established a role for adenovirus in ovine respiratory disease, but there is insufficient evidence to determine whether ovine adenoviruses replicated in the in the gastrointestinal tract.

C. EQUINE

Equine adenovirus (EAV) Type 1 has been isolated from respiratory secretions of clinically normal horses with respiratory disease.[99-101] Powell et al.[100] noted that clinical signs of diarrhea and respiratory disease were frequently observed in thoroughbred horses, which seroconverted from adenovirus negative to positive. Arabian foals with combined immunodeficiency disease develop a severe and fatal EAV-1 infection.[99,102] Clinical signs of diarrhea were observed in 50% of 31 adenovirus-infected Arabian foals and adenoviral intranuclear inclusions were observed in epithelial cells in the duodenum.[102] Villous atrophy was reported in the duodenum of a foal inoculated with EAV-1 at 44 d of age, but adenoviral antigens could not be demonstrated in duodenal enterocytes by fluorescent antibody and inclusion bodies were not observed at 6 d postinoculation.[9] Studdert et al.[26] isolated an equine adenovirus antigenically distinct from EAV-1 from diarrheic feces of foals. This EAV-2 was isolated in equine fetal kidney cell cultures from the feces of two foals, which also had rotavirus particles in their feces. This isolate appeared to be widespread in the equine population as 77% of 339 equine serum samples had neutralizing antibody to EAV-2. Adenoviral intranuclear inclusions were also found only in the enterocytes and crypt cells of the doudenum, jejunum, and ileum of an Arabian foal with a history of chronic diarrhea.[25] Focal to diffuse areas of villous atrophy were present throughout the small intestine and electron microscopy revealed 80 nm particles in the intranuclear inclusions. Experimental inoculation of gnotobiotic foals with EAV-2 or field isolates of EAV from diarrheic feces will be necessary to define the role of EAV, especially EAV-2, in enteric disease.

D. MURINE

Two serotypes of mouse adenovirus (MAV) have been identified. Serotype 1 induces a systemic and fatal disease in suckling, but not weaned and adult mice.[53] In contrast, athymic nude mice inoculated with MAV-1 develop a lethal wasting disease with characteristic lesions in the duodenum.[103] These mice do not develop diarrhea but hemorrhage into the duodenal lumen shortly before death. Light microscopic examination of the small intestine revealed adenoviral intranuclear inclusions in numerous villous and crypt cells compared to only a few endothelial cells. This suggests that the duodenal lesions are induced by MAV-1. The intestinal hemorrhage observed in nude mice infected with MAV-1 is similar to that described for HEV of turkeys.[32,33,104,105]

Adult and suckling mice inoculated with MAV-2 do not develop clinical disease, but the virus localized and replicated in the small intestine.[42,54] Although MAV-2 is apathogenic, the study by Takeuchi and Hashimoto[42] indicated that this virus is predominantly "enterotropic" (replicates in the ileum of the small intestine) and is one of the few studies where fluorescent antibody techniques (FA), light microscopy (LM), and electron microscopy (EM) were used to determine the site of virus replication. Briefly, in 4-week-old mice orally inoculated with MAV-2 and euthanized at 5 and 14 d postinoculation, viral antigens were demonstrated by FA in epithelial cells of the villi and crypts of the ileum and absent in cells of the lamina propria. Decreased villous crypt ratio, increased granularity in the lamina propria, and intranuclear inclusions in villous and crypt cells were observed by LM at 14 d postinoculation. Ultrastructural studies confirmed the LM observations and virions were most numerous in nuclei of the columnar epithelium and goblet cells on the villi and Paneth cells in the crypts. Thus, MAV-2 demonstrates a specific-tissue tropism (columnar epithelium, goblet, and Paneth cells of the small intestine), but does not produce a clinical disease.

E. PORCINE

The porcine adenoviruses (PAV) represent the one group of mastadenovirus, excluding human enteric adenoviruses, with sufficient experimental evidence to document enteric infections. Serotype 1 PAV was isolated form the rectal swab of a pig with diarrhea,[57] while PAV-2 and -3 were isolated from pigs with normal feces.[58] Serotype 4 PAV was isolated from the brain of a pigs with neurological signs and enteritis.[59] Early attempts to verify the pathogenicity of PAV-1, -2, -3, and -4 in conventional, colostrum-deprived, and gnotobiotic pigs were unsuccessful.[8,55,58] Porcine adenovirus adenovirus type 2 and 3 failed to produce clinical disease or lesions in conventional or colostrum-deprived pigs, respectively.[58] However, these two viruses could be isolated from tonsil, ileum, colon, and intestinal contents and there was little evidence that PAV-2 or -3 replicated in the respiratory tract.

Shadduck et al.[12] inoculated gnotobiotic and specific pathogen-free pigs with PAV-4, using five different routes of inoculation. The only consistent clinical sign was a mild diarrhea observed in all orally inoculated pigs, but no gross or microscopic lesions were evident in the intestinal tract. In a similar study, gnotobiotic pigs orally or intranasally inoculated with PAV-4 developed lesions characteristic of meningoencephalitis.[106]

Approximately 14 years after the original experiments of Sharpe and Jessett[58] failed to produce clinical disease in pigs inoculated with PAV-3, this same strain of PAV-3 was consistently shown to produce diarrhea in 2-d-old hysterectomy-derived colostrum-deprived pigs.[16,18,19] Piglets inoculated oronasally with PAV-4 developed a watery diarrhea 3 to 4 d postinoculation (PI). The diarrhea usually lasted 3 to 6 d and the pigs did not die. Adenoviral particles were demonstrated in intestinal contents of the experimentally infected pigs up to 9 d PI. Viral antigen was detected in the nucleus of villous epithelial cells as early as 24 h PI by immunoperoxidase staining. Occasionally, cells in the crypts were also positive. Usually only 10 to 20% of the villi contain viral antigen. Viral antigen was detected in the small intestinal epithelium, up to 15 d after infection and persisted up to 45 d PI in one pig. Grossly, the small intestine was thin walled and filled with fluid contents. Many inclusion bodies were observed by light microscopy in enterocytes covering the villi in the terminal jejunum and ileum. While the villi in the jejunum and ileum were shortened in the adenoviral infected pigs, the intestinal lesion is unremarkable compared to the villous atrophy and the massive destruction of the villous epithelium observed with rotavirus and coronavirus infections. Ultrastructurally, the lesions observed in the pigs inoculated with PAV-3, resembled those seen in MAV-2 infection in mice,[42] except that degenerative and cytolytic changes in the villous epithelial cells were more severe in pigs.[18,19] Thus, the consistent ability to produce diarrhea and the demonstration of adenoviral infected enterocytes by light microscopy, immunoperoxidase staining, and electron microscopy indicates that PAV-3 causes enteric disease in experimentally inoculated pigs. It is interesting that Coussement et al.[16] also reported on a natural adenoviral enteritis infection in a pig, but the field strain could not be isolated in cell culture for subsequent use in pathogenesis studies.

In our laboratory, we isolated an adenovirus of undetermined serotype from the diarrheic feces of a naturally infected pig and were able to reproduce diarrhea in 8-d old gnotobiotic pigs.[107] Diarrhea was observed in these pigs within 48 hours after inoculation, and adenoviral intranuclear inclusions and antigens were demonstrated by light microscopy and immunofluorescence in the short, blunt villi overlying lymphoid aggregates or Peyer's patches in the ileum. We also observed adenoviral particles by electron microscopy in the feces of the experimentally infected pigs, but no other viral particles were seen and intestinal sections did not contain antigen to rotavirus or transmissible gastroenteritis virus. Similar lesions have also been observed in scouring pigs naturally infected with adenovirus.[15-17] Thus, experimental evidence has established that porcine adenovirus produces diarrhea in 2- to 8-d-old pigs. These studies show that under experimental conditions certain serotypes of porcine adenoviruses are moderately pathogenic for pigs. Coussement et al.[16] suggested that adenoviruses produce diarrhea by destruction of villous enterocytes, which results in a loss of digestive enzymes producing a malabsorptive diarrhea. However, these results need further confirmation to determine the mechanism via adenovirus induces diarrhea in pigs.

F. OTHER NONHUMAN MASTADENOVIRUSES

Adenoviruses have also been isolated from rabbits with diarrhea;[47] baboons with pneumonia and diarrhea;[108] mucosal scrapings from the large intestine of goats;[52] and the intestine of puppies with hemorragic enteritis.[27,28] Both of the reports describing the isolation of adenovirus from the intestine of puppies identified the isolate as a variant of canine adenovirus type 2 (CAV-2). Canine adenovirus type 2 is usually only associated with respiratory disease in dogs.[109,110] Although the variant CAV-2 was serologically identical to the prototype CAV-2 (Toronto A26/61), restriction endonuclease "electropherotypes" of the DNA of the intestinal CAV-2 isolates were different from the DNA of the prototype CAV-2. The significance of the isolation of CAV-2 in canine feces is unclear. Dogs experimentally infected with CAV-2 shed virus in feces, but only after virus can be isolated from the respiratory tract and before viremia.[109,110] However, fecal virus could be recovered later in infection when virus was not isolated from lung. In the recent studies in dogs with enteric disease, virus was not isolated from the lungs[27] and no respiratory signs of disease were noted in dogs with diarrhea.[28] At present the experimental data is not sufficient to determine if CAV-2 or a variant CAV-2 is a primary agent of enteric disease in dogs. These "enteric" CAV-2 isolates will require further study in gnotobiotic or specific pathogen free puppies to determine their importance as an enteric pathogen.

G. AVIADENOVIRUS (HEMORRHAGIC ENTERITIS VIRUS)

Hemorrhagic enteritis is an acute disease of turkeys characterized by depression, bloody droppings, and sudden death.[105] All naturally occurring outbreaks of the disease have been reported in flocks 4 weeks of age or older, except for one outbreak in a flock of 2 1/2-week-old poults.[90] Younger birds appear to be refractory to the disease, but viral replication does occur in the spleen of these birds.[66] The reason why birds less then 3 weeks of age are resistant to clinical disease is not known. Recent evidence indicates that 2- to 5-week-old poults, chemically bursectomized with cyclophosphamide, were also resistant to clinical disease. These results imply that an intact bursa is necessary to induce clinical disease.[66]

Poults become depressed, excrete bloody droppings, and die within 3 to 4 d after intravenous inoculation or 6 d after oral or cloacal inoculation with homogenates of infectious spleen.[30,31] Antigens to HEV can be detected in the spleens of experimentally infected birds from 2 to 6 d PI.[111] Viral antigens were not detected in peripheral blood lymphocytes or the intestine until at least 4 d PI, suggesting that the spleen is the primary target organ with subsequent spread via viremia to other visceral organs including the liver, intestine, kidney, bone marrow, and thymus.[112] The HEV antigens were generally observed within reticuloendothelial (RE) cells in these organs.[112]

The distribution of gross and histologic lesions parallel the sequential progression of the virus. Enlarged and marbled spleens were detected in experimentally infected birds at 3 to 6 d PI, when HEV antigens were also found in this organ.[35,112] Hemorrhage was observed in the duodenum and cecum on days 5 and 6 PI. The gross lesions and histopathology of HEV infections has been described[32,33,35,104,105] and reviewed.[30] Briefly, the most remarkable gross lesions are found in the spleen and intestine. Spleens are enlarged and marbled or mottled. Dead poults are often pale and anemic due to massive hemorrhage into the intestine. The intestines are distended, dark in color, and filled with red to brownish blood. Lungs may also be congested. Histologic lesions included hyperplasia of the white pulp, necrosis of lymphoid cells, proliferation of enlarged RE cells, and presence of intranuclear inclusions in the RE cells in the spleen. Light microscopic lesions in the small intestine include severe congestion of the mucosa, degeneration and sloughing of epithelial cells from tips of the villi, followed by hemorrhage from broken capillaries into the lumen. The lesions are most pronounced in the jejunum and intranuclear inclusions are present in RE cells in the lamina propria. Similar but attenuated lesions are also observed in other portions of the digestive tract and the bursa. Intranuclear inclusions are also observed in the RE cells in most of these organs[29,32,33,104] and adenoviral particles can be demonstrated by electron microscopy within nuclei of RE cells in the spleen[29,35,104] and intestines.[29] In birds which survived, all major signs of infection dissipated by 10 days PI and intestinal villi appear to regain their normal length. Death is caused by the extensive hemorrhaging into the small intestinal lumen. The mechanism by which adenovirus induces hemorrhaging has not been determined, but HEV could destroy endothelial cells in the small intestine causing an ischemic necrosis of the intestine as described for adenoviral infections in cattle.[22,24,51,96]

VIII. DIAGNOSIS

A. NONHUMAN MASTADENOVIRUSES

Standard methods for the detection of enteric adenoviruses are not well defined, because many adenoviruses appear to be apathogenic and even pathogenic enteric adenoviruses may be of limited economic importance. However, diagnostic methods to detect adenoviruses in the intestinal cells or contents are needed to differentiate adenoviral diarrhea from the number of other viral and bacterial agents, which cause diarrhea in suckling and weaned animals. The clinical signs of diarrhea observed in experimental infections in pigs,[15,16] calves,[6] and in natural infections in pigs,[15-17] calves,[21,22,51] foals,[25,26] and dogs[27,28] are similar to those produced by rotavirus and coronavirus. There are also no gross or histologic pathognomonic lesions typical of adenovirus, although the presence of intranuclear inclusion bodies in intestinal epithelial cells may be suggestive of adenovirus infection.

Histologic examination of the intestinal epithelium for intranuclear inclusion bodies has been used to tentatively diagnose adenovirus as a cause of diarrhea in pigs and calves.[17,22,24,51,113] However, inclusion bodies are not diagnostic unless viral antigen is demonstrated in these cells by immunofluorescence or immunoperoxidase staining. Also, post-mortem tissue is often not well preserved resulting in autolysis and subsequent loss of cells with inclusion bodies. The section of intestine submitted for examination is also important. Adenoviruses replicate principally in terminal jejunum or ileum of most animal species, and inclusion bodies (at least in pigs) are commonly found in the enterocytes of the dome epithelium, which covers the short, blunt villi located over Peyer's patches (Figures 2 and 3).[17,107]

Most nonhuman mastadenoviruses grow readily on primary kidney or testicle cells from the homologous host species (see reviews, References 1, 3, 55). The isolation of adenovirus from fecal material may require extended incubation times or several blind passages before cytopathogenic effects appear in cell culture.[1,4,8,10,36,38,57,59,95,107] The isolated virus can then be characterized as an adenovirus by histologic (inclusions), immunofluorescence or immunoperoxidase staining (viral antigen), or the demonstration of viral particles in infected cells or cell lysates by electron microscopy. These isolates can also be identified as adenoviruses by complement-fixation or immunodiffusion tests, using infected cell culture supernatants against a reference serum possessing adenovirus group antibody[114] (see reviews, References 1, 3, 55). The new isolate can then be assigned to a particular serotype by viral neutralization tests using type specific immune sera. Isolation of adenoviruses from diarrheic feces is not sufficient evidence to incriminate the virus as a cause of diarrhea, since these viruses can also be isolated from normal feces.

Immunofluorescence or immunoperoxidase staining of enterocytes in frozen or paraffin fixed sections have detected viral antigens in intestinal epithelium in experimentally infected pigs.[16,107] However, these techniques are not widely used in the diagnosis of adenoviral enteritis, because reference antisera is not universally available.

Electron microscopic examination of feces or intestinal contents is probably the best method to determine unknown causative agents of viral diarrhea. While electron microscopy is expensive and requires skilled technical help, this technique offers the advantage of detecting other viral agents present in the same fecal samples. Specific antisera can also be used in immune electron microscopy to increase the sensitivity of the technique and confirm identity of the virus morphology (Figure 1). Electron microscopy is routinely used in the author's laboratory to identify the presence of adenoviral particles in the diarrheic feces of animals. Often a combination of electron microscopy to detect viral particles in feces and the demonstration of

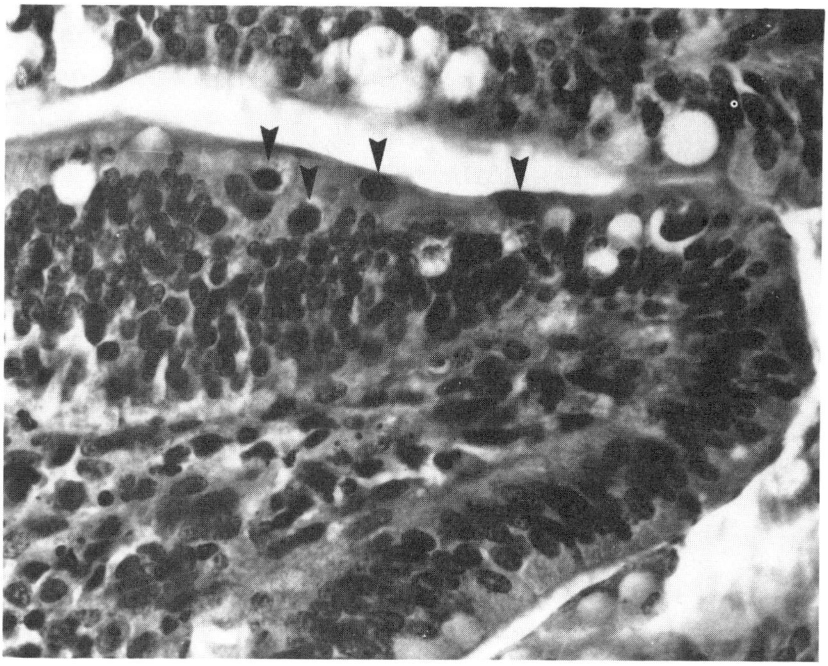

FIGURE 2. Enterocytes, containing adenoviral inclusion bodies (arrows), observed in the villus in the ileum of a pig with diarrhea. Photomicrograph courtesy of Dr. Bruce Janke, South Dakota Animal Disease Research and Diagnostic Laboratory. (Hematoxylin-eosin; magnification ×140.)

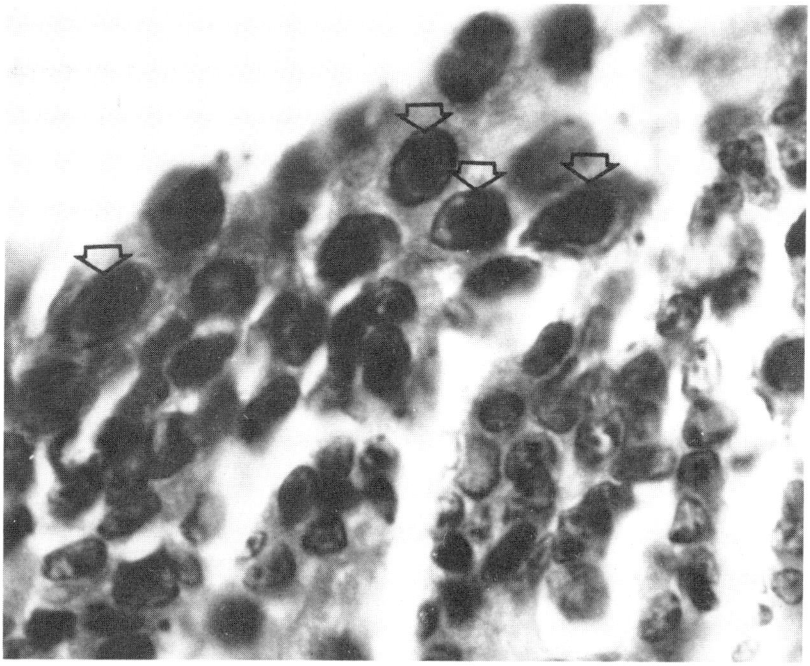

FIGURE 3. Adenoviral inclusion bodies (arrows) in enterocytes from villus in ileum of pig with diarrhea. Note inclusions often fill the entire nucleus and are surrounded by a clear halo. Photomicrograph courtesy of Dr. Bruce Janke, South Dakota Animal Disease Research and Diagnostic Laboratory. (Hematoxylin-eosin; magnification ×400.)

adenoviral inclusions in the small intestinal epithelium can give a tentative diagnosis of adenovirus diarrhea.[15,122] Serological diagnosis of suspected adenovirus infections in animals can be done by either demonstrating seroconversion or rising antibody titers in viral neutralization or immunodiffusion tests.[1,3,6,55]

B. AVIADENOVIRUSES (HEMORRHAGIC ENTERITIS VIRUS)

A tentative diagnosis of HEV can be obtained from the clinical history and gross lesions.[30] Virus can be isolated from intestinal contents and spleens by oral, intravenous, or cloacal inoculation of 6- to 10-week-old turkey poults.[30] Death should occur within 3 d of intravenous and 5 to 6 d of oral or cloacal inoculations.

Cocultivation of spleen cells from suspected HEV infected turkeys with a lymphoblastoid B cell line of turkey origin has successfully isolated HEV from experimentally infected turkeys.[102] The technique apparently requires viable spleen cells from infected birds. Therefore, this *in vitro* isolation procedure may not be practical on post-mortem tissues unless these tissues are from live birds submitted for necropsy.

Viral antigen has been detected in a variety of tissues by immunofluorescence[112] and immunoperoxidase staining[111] from experimental but not naturally infected birds. The most popular means for diagnosis of HEV is to demonstrate by immunodiffusion the presence of HEV antigens in splenic extracts from live or dead birds.[30,115] A blocking enzyme-linked immunosorbent assay (ELISA) to detect HEV antigens in spleens of infected poults has also been described.[116] Adenoviral particles can also be detected by direct electron microscopy in negatively stained extracts of spleens from turkey poults infected with HEV.[29,35]

Seroconversion can be demonstrated as early as 3 and 5 d PI using the immunodiffusion[30,115] and ELISA assays.[111,116]

IX. IMMUNITY, PREVENTION, AND CONTROL

Since the role of the nonhuman mastadenoviruses as etiological agents of diarrhea has not been established or only recently determined, there is little or no information on immunity, prevention, and control of enteric adenoviral infections. Most experimental studies with the nonhuman mastadenoviruses have addressed only questions of pathogenicity. Since many of these adenoviruses appear to be apathogenic, studies have not proceeded beyond this aspect of the infection.

In most species which have been experimentally inoculated with adenoviruses isolated from feces, serum neutralizing antibodies can be detected within 1 to 2 weeks after inoculation.[4,6,8,10,13,55] Precipitating antibodies are usually detected by immunodiffusion at 3 to 4 weeks PI and persist for at least 3 to 4 months.[6,7,13] Since adenoviruses have a predilection for mucous membranes of the oral and nasopharyngeal cavities, the presence or absence of local IgA-type antibodies will probably determine the degree of initial multiplication of the virus.[1] The appearance of neutralizing activity resembling IgA has been described in mice infected with the apathogenic MAV-2, but the significance of this antibody in preventing infection of mice by MAV-2 has not been determined.[117,118] Whether serum or local antibodies moderate or prevent the course of enteric adenovirus infections will probably depend on the pathogenesis of the adenovirus strain. In the pig, the intestine appears to be the primary site of replication for enteric adenoviruses, since there is no evidence of infection in other tissues.[16,66] Whereas, in the calf, circumstantial evidence suggests that adenovirus reaches the intestine via infected endothelial cells. Thus, prevention of the local intestinal infection in pigs can best be accomplished by stimulation of IgA antibodies as described in the mouse system.[117,118] Serum antibodies, which reduce or eliminate the viremic phase of adenovirus pathogenesis, may prevent infections in calves in a manner analogous to that described for HEV in turkeys.[30,119]

Since most adult animals are seropositive to adenovirus,[1-3,55] it is presumed that passive antibodies are passed from the dam to the fetus or neonate via placenta, colostrum, or milk. Colostrum deprived and newborn conventional pigs were free of serum antibodies before nursing. After nursing, the antibody titer rose to a peak within a few days and persisted for 8 to 9 weeks after weaning (reference 85 as quoted in reference 86). It is not known if these passively acquired antibodies are protective, but infection must be moderated because the frequency of adenovirus shedding increases after weaning.[56,89]

The nonhuman enteric mastadenoviruses have not been shown to be of sufficient pathogenicity or economic importance to justify the development of vaccines for prevention of adenoviral infections. Control of adenoviral infections has also been difficult because adenoviruses are known to establish persistent and latent infections in lymphoid tissue[1,3,55] and some animals have been shown to shed virus in feces up to several months after infection.[1,3,7,13,55] Thus, adequate control of enteric adenoviral infections will require the development of methods to identify and remove carrier animals. The best preventive approach is probably to follow good sanitation practices that reduce the level of environmental fecal contamination and thoroughly disinfect facilities before introducing new animals.

In contrast to the paucity of information on the immunity, prevention, and control of the nonhuman enteric mastadenoviruses, several studies have been done with HEV, an economically important disease of turkeys. Poults, which survive natural or experimental infection with HEV, are refractory to reinfection.[30,31] After natural or experimental infections serum antibody can be detected by the immunodiffusion in most poults by 7 to 10 d postinoculation (PI).[30,111,115] Antibodies apparently persist in most birds for at least 40 months PI.[30] Antibody has also been detected by an ELISA as early as 3 to 4 d PI with peak titers at 14 d PI.[111,116] Passive immunity can be conferred by injection of convalescent antisera obtained from recovered birds. Injection of as little as 1 to 0.5 ml of convalescent antisera prevented disease, whereas turkey poults given 0.25 to 0.1 ml volumes of antisera developed splenic, but not intestinal lesions.[30,119]

Prior to the demonstration that HEV could be propagated in a Marek's disease-induced lymphoblastoid B-cell line,[64] control of HEV in turkeys was achieved with a live, water administered, turkey spleen propagated vaccine.[120] The virus used in this vaccine originated from the spleen of a pheasant with marble spleen disease and the virus was propagated in turkey poults. The spleens of turkey poults were harvested and homogenized to produce the vaccine virus. Poults vaccinated with the turkey spleen propagated, pheasant origin live avirulent virus were protected from challenge with virulent HEV at 17 and 23 d post vaccination (PV). Protection against HEV can also be obtained by vaccination with a cell culture live virus vaccine (CC-HEV).[91,121] Poults inoculated with the CC-HEV develop precipitating antibody within 1 week PV and vaccinated poults were protected against challenge exposure as early as 1 to as late as 8 weeks PV. Antigens to HEV can be detected in spleens of vaccinated poults, and vaccine virus spreads horizontally to contact exposed controls as indicated by seroconversion and resistance of contact exposed controls to challenge. The CC-HEV offers the advantage of purity over the turkey spleen propagated vaccine, which can carry adventitious agents. The CC-HEV vaccine is commercially available, but reports on its efficacy in field trials are not available. HEV can also be treated by injection of all poults in a flock with convalescent antisera. This treatment will markedly reduce mortalities if convalescent antisera is administered as soon as possible after the initial diagnosis of HEV.[31,119] The mechanism by which antisera and the live cell culture vaccines prevent clinical disease has not been described. Presumably, the serum antibodies neutralize HEV in the bloodstream, thereby preventing the localization in reticuloendothelial cells in the intestine and other organs.

REFERENCES

1. **Ishibashi, M. and Yasue, H.,** Adenoviruses of animals, in *The Adenoviruses*, Ginsberg, H. S., Ed., Plenum Press, New York, 1984, 497.
2. **Fenner, F., Bachmann, P. A., Gibbs, E. P. J., Murphy, F., Studdert, M. J., and White, D. O.,** *Veterinary Virology*, Academic Press, Orlando, FL, 1987, 55, 329.
3. **Gillespie, J. H. and Timoney, J. F.,** Eds., The Adenoviridae, in *Hagan and Bruner's Infectious Diseases of Domestic Animals*, 7th ed., Cornell University Press, Ithaca, NY, 1981, 506.
4. **Belak, S., Vetesi, F., Palfi, V., and Papp, L.,** Isolation of a pathogenic strain of ovine adenovirus type 5 and a comparison of its pathogenicity with that of another strain of the same serotype, *J. Comp. Pathol.*, 90, 169, 1980.
5. **Cutlip, R. C. and McClurkin, A. W.,** Lesions and pathogenesis of disease in young calves experimentally induced by a bovine adenovirus type 5 isolated from a calf with weak calf syndrome, *Am. J. Vet. Res.*, 36, 1095, 1975.
6. **Darbyshire, J. H.,** Bovine adenoviruses, *J. Am. Vet. Med. Assoc.*, 152, 786, 1968.
7. **Darbyshire, J. H., Jennings, A. R., Dawson, P. S., Lamont, P. H., and Omar, A. R.,** The pathogenesis and pathology of infection in calves with a strain of bovine adenovirus type 3, *Res. Vet. Sci.*, 7, 81, 1966.
8. **Derbyshire, J. B., Clarke, M. C., and Collins, A. P.,** Serological and pathogenicity studies with some unclassified porcine adenoviruses, *J. Comp. Pathol.*, 85, 437, 1975.
9. **Gleeson, L. J., Studdert, M. J., and Sullivan, N. D.,** Pathogenicity and immunologic studies of equine adenoviruses in specific pathogen-free foals, *Am. J. Vet. Res.*, 39, 1636, 1978.
10. **Lehmkuhl, H. D., Smith, M. H., and Dierks, R. E.,** A bovine adenovirus type 3: isolation, characterization and experimental infection in calves, *Arch. Virol.*, 48, 39, 1975.
11. **Payla, V., Belak, S., and Palifi, V.,** Adenovirus infection in lambs. II. Experimental infection of lambs, *Zentralbl. Veterinaermed. Reihe B*, 24, 529, 1977.
12. **Shadduck, J. A., Koestner, A., and Kasza, L.,** The lesions of porcine adenoviral infection in germ-free and pathogen-free pigs, *Pathol. Vet.*, 4, 537, 552, 1967.
13. **Sharp, J. M., Rushton, B., and Rimer, R. D.,** Experimental infection of specific pathogen-free lambs with ovine adenovirus type 4, *J. Comp. Pathol.*, 86, 621, 1976.
14. **Darbyshire, J. H., Kinch, D. A., and Jennings, A. R.,** Experimental infection of calves with bovine adenovirus types 1 and 2, *Res. Vet. Sci.*, 10, 39, 1969.
15. **Abid, H. N., Holscher, M. A., and Byerly, C. S.,** An outbreak of adenovirus enteritis in piglets, *Vet. Med.*, 79, 105, 1984.
16. **Coussement, W., Ducatelle, R., Charlier, G., and Hoorens, J.,** Adenovirus enteritis in pigs, *Am. J. Vet. Res.*, 42, 1905, 1981.
17. **Sanford, S. E. and Hoover, D. M.,** Enteric adenovirus infection in pigs, *Can. J. Comp. Med.*, 47, 396, 1983.
18. **Ducatelle, R., Coussement, W., and Hoorens, J.,** Replication of porcine enteric adenoviruses *in vivo*, *Arch. Virol.*, 69, 219, 1981.
19. **Ducatelle, R., Coussement, W., and Hoorens, J.,** Sequential pathological study of experimental porcine adenovirus enteritis, *Vet. Pathol.*, 19, 179, 1982.
20. **Aldasy, P., Csontos, L., and Bartha, A.,** Pneumo-enteritis in calves caused by adenoviruses, *Acta Vet. Hung.*, 15, 167, 1964.
21. **Bulmer, W. S., Tsae, K. S., and Little, P. B.,** Adenovirus infection in two calves, *J. Am. Vet. Med. Assoc.*, 166, 233, 1975.
22. **Orr, J. P.,** Necrotizing enteritis in a calf infected with adenovirus, *Can. Vet. J.*, 25, 72, 1984.
23. **Reed, D. E., Wheeler, J. G., and Lupton, H. W.,** Isolation of bovine adenovirus type 7 from calves with pneumonia and enteritis, *Am. J. Vet. Res.*, 39, 1968, 1978.
24. **Smyth, J. A., Cush, P. F., Adair, B. M., and Young, J. A.,** Adenovirus associated enterocolitis in a bullock, *Vet. Rec.*, 119, 574, 1986.
25. **Corrier, D. E., Montgomery, D., and Scutchfield, W. L.,** Adenovirus in the intestinal epithelium of a foal with prolonged diarrhea, *Vet. Pathol.*, 19, 564, 1982.
26. **Studdert, M. J. and Blackney, M. H.,** Isolation of an adenovirus antigenically distinct from equine adenovirus type 1 from diarrheic foal feces, *Am. J. Vet. Res.*, 43, 543, 1982.
27. **Hamelin, C., Jouvenne, P., and Assaf, R.,** Genotypic characterization of type-2 variants of canine adenovirus, *Am. J. Vet. Res.*, 47, 625, 1986.
28. **Macartney, L., Cavanaugh, H. M. A., and Spibey, N.,** Isolation of canine adenovirus-2 from the faeces of dogs with enteric disease and its unambiguous typing by restriction endonuclease mapping, *Res. Vet. Sci.*, 44, 9, 1988.
29. **Carlson, H. C., Al-Sheikhly, F., Pettit, J. R., and Seawright, G. L.,** Virus particles in spleens and intestines of turkeys with hemorrhagic enteritis, *Avian Dis.*, 18, 67, 1974.

30. **Domermuth, C. H. and Gross, W. B.,** Hemorrhagic enteritis and related infections, in *Diseases of Poultry,* Hofstad, M. S., Barnes, H. J., Calnek, B. W., Reid, W. M., and Yoder, H. W., Eds., Iowa State University Press, Ames, IA, 1984, 511.
31. **Domermuth, C. H. and Gross, W. B.,** Hemorrhagic enteritis of turkeys, in *Animal Models for Intestinal Disease,* Pfeiffer, C. J., Ed., CRC Press, Boca Raton, FL, 1985, Chap. 22.
32. **Gross, W. B.,** Lesions of hemorrhagic enteritis, *Avian Dis.,* 11, 684, 1967.
33. **Gross, W. B. and Moore, W. E. C.,** Hemorrhagic enteritis of turkeys, *Avian Dis.,* 11, 296, 1967.
34. **Tolin, S. A. and Domermuth, C. H.,** Hemorrhagic enteritis of turkeys: electron microscopy of the causal virus, *Avian Dis.,* 19, 118, 1975.
35. **Gross, W. B. and Domermuth, C. H.,** Spleen lesions of hemorrhagic enteritis of turkeys, *Avian Dis.,* 20, 455, 1976.
36. **Adair, B. M. and Curran, W.,** Studies on the cytopathology of bovine adenoviruses in calf testis and calf kidney cell cultures, *Vet. Microbiol.,* 3, 241, 1978.
37. **Adair, B. M., Curran, W. L., and McFerran, J. B.,** Ultrastructural studies on the replication of fowl adenoviruses in primary cell cultures, *Avian Pathol.,* 8, 133, 1979.
38. **Adair, B. M., Curran, W. L., and McFerran, J. B.,** Light and electron microscope studies of the growth of ovine adenoviruses in primary lamb kidney cell cultures, *Vet. Microbiol.,* 4, 293, 1979.
39. **Derbyshire, J. B., Chandler, R. I., and Smith, K.,** Observations on inclusion bodies in pig kidney tissue culture cells infected with porcine adenoviruses, *Res. Vet. Sci.,* 9, 300, 1968.
40. **McFerran, J. B., Nelson, R., and Knox, E. R.,** Isolation and characterization of sheep adenoviruses, *Arch. Gasamte Virusforsch.,* 35, 232, 1971.
41. **Shahrabadi, M. S., Marusyk, R. G., and Crawford, T. B.,** Electron-microscopic study of the development of an equine adenovirus in cultured fetal equine kidney cells, *Can. J. Microbiol.,* 23, 497, 1977.
42. **Takeuchi, A. and Hoshimoto, K.,** Electron microscope study of experimental enteric adenovirus infection in mice, *Infect. Immun.,* 13, 569, 1976.
43. **Itakura, C. and Carlson, H. C.,** Electron microscopic findings of cells with inclusion bodies in experimental hemorrhagic enteritis of turkeys, *Can. J. Comp. Med.,* 39, 299, 1975.
44. **Wyand, D. S., Jakowski, R. M., and Berke, C. N.,** Marble spleen disease in ring-necked pheasant. Histology and ultrastructure, *Avian Dis.,* 16, 319, 1972.
45. **Martinez-Palome, A., Lebuis, J., and Bernhard, W.,** Electron microscopy of adenovirus 12 replication. I. Fine structural changes in the nucleus of infected KB cells, *J. Virol.,* 1, 817, 1967.
46. **Weber, J. and Stich, H. F.,** Electron microscopy of cells infected with adenovirus type 2, *J. Virol.,* 3, 198, 1969.
47. **Bodon, L., Prohaszka, L., Adam, E., and Nasz, I.,** Isolation of an adenovirus from rabbits, *Acta Vet. Acad. Sci. Hung.,* 27, 73, 1979.
48. **Wadell, G., Allard, A., Johansson, M., Svensson, L., and Uhnoo, I.,** Enteric adenoviruses, in *Novel Diarrhoea Viruses,* Beck, G. and Whelan, J., Ed., Ciba Foundation Symp. 128, John Wiley & Sons, New York, 1987, 63.
49. **Bartha, A. and Aldasy, P.,** Further two serotypes of bovine adenoviruses (serotype 4 and 5), *Acta Vet. Acad. Sci. Hung.,* 16, 107, 1966.
50. **Klein, M., Earley, E., and Zellat, J.,** Isolation from cattle of a virus related to human adenovirus, *Proc. Soc. Exp. Biol. Med.,* 102, 1, 1959.
51. **Thompson, K. G., Thomson, G. W., and Henry, J. N.,** Alimentary tract manifestations of bovine adenovirus infections, *Can. Vet.,* 22, 68, 1981.
52. **Gibbs, E. P. J., Taylor, W. P., and Lawman, M. J. P.,** The isolation of adenoviruses from goats affected with peste des petits ruminants in Nigeria, *Res. Vet. Sci.,* 23, 331, 1977.
53. **Hartley, J. W. and Rowe, W. P.,** A new mouse virus apparently related to the adenovirus group, *Virology,* 11, 645, 1960.
54. **Sugiyama, T., Hashimoto, K., and Sasaki, S.,** An adenovirus isolated from the feces of mice. II. Experimental infection, *Jpn. J. Microbiol.,* 11, 33, 1967.
55. **Derbyshire, J. B.,** Porcine adenovirus infection, in *Diseases of Swine,* Leman, A. D., Straw, B., Glock, R. D., Mengeling, W. L., Penny, R. H. C., and Schoell, E., Eds., Iowa State University Press, Ames, IA, 1986, 321.
56. **Derbyshire, J. B., Clarke, M. C., and Jessett, D. M.,** Observations on the fecal excretion of adenoviruses and enteroviruses in conventional and "minimal disease" pigs, *Vet. Rec.,* 79, 595, 1966.
57. **Haig, D. A. and Clarke, M. C.,** Isolation of an adenovirus from a pig, *J. Comp. Path.,* 74, 81, 1964.
58. **Sharpe, H. B. A. and Jessett, D. M.,** Experimental infection of pigs with 2 strains of porcine adenovirus, J. Comp. Pathol., 77, 45, 1967.
59. **Kasza, L.,** Isolation of an adenovirus from the brain of a pig, *Am. J. Vet. Res.,* 27, 751, 1966.
60. **Mayr, A., Bibrack, B., and Bachman, P. A.,** Propagation of porcine adenoviruses in calf kidney cell cultures, *Zentralbl Bakteriol. (Orig. B),* 203, 59, 1967.
61. **Dea, S. and ElAzhary, M.A.S.Y.,** Cultivation of a porcine adenovirus in porcine thyroid cell culture, *Cornell Vet.,* 74, 208, 1984.

62. **Iltis, J. P., Jakowski, R. M., and Wyand, D. S.,** Transmission of marble spleen disease in turkeys and pheasants, *Am. J. Vet. Res.*, 36, 97, 1975.
63. **Fasina, S. O. and Fabricant, J.,** *In vitro* studies of hemorrhagic enteritis virus with immunofluorescent antibody technique, *Avian Dis.*, 26, 150, 1982.
64. **Nazerian, K. and Fadly, A. M.,** Propagation of virulent and avirulent turkey hemorrhagic enteritis virus in cell culture, *Avian Dis.*, 26, 816, 1982.
65. **Graham, F. L., Simely, J., Russell, W. C., and Nairn, R.,** Characteristics of a human cell line transformed by DNA from human adenovirus type 5, *J. Gen. Virol.*, 36, 59, 1977.
66. **Fadly, A. M. and Nazerian, K. C.,** Evidence of bursal involvement in the pathogenesis of hemorrhagic enteritis virus of turkeys, *Avian Dis.*, 26, 525, 1982.
67. **Rekosh, D. M. K., Russel, W. C., Bellett, A. J. D., and Robinson, A. J.,** Identification of a protein linked to the ends of adenovirus DNA, *Cell*, 11, 283, 1977.
68. **Robinson, A. J. and Bellett, A. J. D.,** A circular DNA-protein complex from adenoviruses and its possible role in DNA replication, *Cold Spring Harbor Symp. Quant. Biol.*, 39, 523, 1974.
69. **Futterer, J. and Winnacker, E. L.,** Adenovirus DNA replication, in *Current Topics in Microbiology*, Volume III, Doerfler, W., Ed., Springer-Verlag, New York, 1984, 41.
70. **Sussenbach, J. S.,** The structure of the genome. in *The Adenoviruses*, Ginsberg, H. S., Ed., Plenum Press, New York, 1984, 44.
71. **Horwitz, M. S.,** Adenoviruses and their replication, in *Virology*, Fields, B. N., Ed., Raven Press, New York, 1985, 433.
72. **Derbyshire, J. B. and Arkell, S.,** The activity of some chemical disinfectants against talfan virus and porcine adenovirus type 2, *Br. Vet. J.*, 127, 137, 1971.
73. **Domermuth, C. H. and Gross, W. B.,** Effect of disinfectant and drying on the virus of hemorrhagic enteritis of turkeys, *Avian Dis.*, 16, 94, 1972.
74. **Domermuth, C. H. and Green, W. B.,** Effect of chlorine on the virus of hemorrhagic enteritis of turkeys, *Avian Dis.*, 16, 952, 1972.
75. **Domermuth, C. H., Weston, C. R., Cowen, B. S., Colwell, W. M., Gross, W. B., and DuBose, R. T.,** Influence and distribution of avian adenovirus group II splenomegaly of chickens, *Avian Dis.*, 24, 591, 1980.
76. **Adair, B. M. and McFerran, J. B.,** Comparative serological studies with mammalian adenoviruses, *Arch. Virol.*, 51, 319, 1976.
77. **Willimzik, H. F., Kalter, S. S., Lester, T., and Wigand, R.,** Immunological relationship among adenoviruses of humans, simians, and nonprimates, as determined by the neutralization test, *Intervirology*, 15, 28, 1981.
78. **Betts, A. O., Jennings, A. R., Lamont, P. H., and Page, Z.,** Inoculation of pigs with adenoviruses of man, *Nature (London)*, 193, 45, 1962.
79. **Adair, B. M., McFerran, J. B., and McKillop, E. R.,** A sixth species of ovine adenovirus isolated from lambs in New Zealand, *Arch. Virol.*, 74, 269, 1982.
80. **Studdert, M. J., Wilks, C. R., and Coggins, L.,** Antigenic comparisons and serologic survey of equine adenoviruses, *Am. J. Vet. Res.*, 35, 693, 1974.
81. **Bibrack, B. and McKercher, D. G.,** Serologic evidence for adenovirus infection in California cattle, *Am. J. Vet. Res.*, 32, 805, 1971.
82. **Mattson, D. E., Smith, P. P., and Schmitz, J. A.,** Isolation of bovine adenovirus type 4 from cattle in Oregon, *Am. J. Vet. Res.*, 38, 2029, 1977.
83. **Darbyshire, J. H. and Pereira, H. G.,** An adenovirus precipitating antibody present in some sera of different animal species and its association with bovine respiratory disease, *Nature (London)*, 201, 895, 1964.
84. **Darbyshire, J. H.,** Adenovirus antibodies in the sera of pigs, *Vet. Rec.*, 81, 118, 1967.
85. **Bibrack, B.,** Untersuchungen iiber das vorkommen von adenovirus-Antikorpern bei schweinen verschiedenen, *Alters. Berlin Muench. Tieraerztl. Wochschr.*, 81, 327, 1968.
86. **Koestner, A. and Kasza, L.,** Adenoviruses, in Diseases of Swine, Dunne, H. W. and Leman, A. D., Eds., Iowa State University Press, Ames, IA, 1975, 422.
87. **Kasza, L., Hodges, R. T., Betts, A. O., and Trexter, P. C.,** Pneumonia in gnotobiotic pigs produced by simultaneous inoculation of a swine adenovirus and *Mycloplasma hyopneumoniae*, *Vet. Rec.*, 84, 262, 1969.
88. **Dea, S. and ElAzhary, M.A.S.Y.,** Prevalence of antibodies to porcine adenovirus in swine by indirect fluorescent antibody test, *Am. J. Vet. Res.*, 45, 2109, 1984.
89. **Derbyshire, J. B. and Collins, A. P.,** Virological studies on an experimental minimal disease herd of pigs, *Br. Vet. J.*, 127, 436, 1971.
90. **Harris, J. R. and Domermuth, C. H.,** Hemorrhagic enteritis in two-and one-half-week-old turkey poults, *Avian Dis.*, 21, 120, 1977.
91. **Fadly, A. M. and Nazerian, K.,** Efficacy and safety of cell-culture live-virus vaccine for hemorrhagic enteritis of turkeys: laboratory studies, *Avian Dis.*, 28, 183, 1984.
92. **Klein, M., Earley, E., and Zellat, J.,** Isolation from cattle of a virus related to human adenovirus, *Proc. Soc. Exp. Biol. Med.*, 105, 340, 1959.

93. **Inaba, Y., Tanaka, Y., Sato, K., Ito, H., Ito, Y., Omori, T., and Matumoto, M.**, Bovine Adenovirus, II. A serotype Fukoroi, recovered from Japanese cattle, *Jpn. J. Microbiol.*, 12, 219, 1968.
94. **McClurkin, A. W. and Coria, M. F.**, Infectivity of bovine adenovirus type 5 recovered from a polyarthritic calf with weak calf syndrome, *J. Am. Vet. Med. Assoc.*, 167, 139, 1975.
95. **Mattson, D. E.**, Naturally occurring infection of calves with a bovine adenovirus, *Am. J. Vet. Res.*, 34, 623, 1973.
96. **Fujiwara, H. and Konno, S.**, Histopathology of an enzootic diarrheal disease in cattle, *Natl. Inst. Anim. Health Q.*, 8, 16, 1968.
97. **Davies, D. H. and Humphreys, S.**, Characterization of two strains of adenovirus isolated from New Zealand sheep, *Vet. Microbiol.*, 2, 97, 1977.
98. **Sharp, J. M., McFerran, J. P., and Rae, A.**, A new adenovirus from sheep, *Res. Vet. Sci.*, 17, 268, 1974.
99. **McChesney, A. E., England, J. J., and Rich, L. J.**, Adenoviral infection in foals, *J. Am. Vet. Med. Assoc.*, 162, 545, 1973.
100. **Powell, D. G., Burrows, R., and Goodridge, D.**, Respiratory viral infections among thoroughbred horses in training during 1972, *Equine Vet. J.*, 6, 19, 1974.
101. **Roberts, A. W., Whitenack, D. L., and Carter, G. R.**, Recovery of adenoviruses and slow herpesviruses from horses having respiratory tract infection, *Am. J. Vet. Res.*, 35, 1169, 1974.
102. **McChesney, A. E., England, J. J., Whiteman, C. E., Adcock, J. L., Rich, L. J., and Chow, T. L.**, Experimental transmission of equine adenovirus in arabian and non-arabian foals, Am. J. Vet. Res., 35, 1015, 1974.
103. **Winters, A. L. and Brown, H. K.**, Duodenal lesions associated with adenovirus infection in athymic "nude" mice (40862), *Proc. Soc. Exp. Biol. Med.*, 164, 280, 1980.
104. **Fujiwara, H., Tanaami, S., Yamaguchi, M., and Yoshino, T.**, Histopathology of hemorrhagic enteritis in turkeys, *Natl. Inst. Anim. Health Q.*, 15, 68, 1975.
105. **Pomeroy, B. S. and Fenstermacher, R.**, Hemorrhagic enteritis in turkeys, *Poult. Sci.*, 16, 378, 1937.
106. **Edington, N., Kasza, L., and Christofinis, G. J.**, Meningo-encephalitis in gnotobiotic pigs inoculated intranasally and orally with porcine adenovirus 4, *Res. Vet. Sci.*, 13, 289, 1972.
107. **McAdaragh, J. P., Eustis, S., and Benfield, D. A.**, Adenovirus associated with diarrhea in pigs, Abstr., Conf. Res. Workers in Anim. Dis., Chicago, IL, November 10–11, 1980.
108. **Eugster, A. K., Kalter, S. S., Kim, C. S., and Pinkerton, M. E.**, Isolation of adenoviruses from baboons (*Papio* sp.) with respiratory and enteric infections, *Arch. Gesamte. Virusforsch.*, 26, 260, 1969.
109. **Emery, J. B., House, S. A., and Brown, W. R.**, Cross protective immunity to canine adenovirus type 2 by canine adenovirus type 1 vaccination, *Am. J. Vet. Res.*, 39, 1778, 1978.
110. **Swango, L. J., Wooding, W. L., and Binn, L. N.**, A comparison of the pathogenesis and pathogenicity of infectious canine hepatitis virus and the A26/61 virus strain (Toronto), *J. Am. Vet. Med. Assoc.*, 156, 1687, 1970.
111. **Silim, A. and Thorsen, J.**, Hemorrhagic enteritis: virus distribution and sequential development of antiobdy in turkeys, *Avian Dis.*, 25, 444, 1982.
112. **Fasina, S. O. and Fabricant, J.**, Immunofluorescent studies on the early pathogenesis of hemorrhagic enteritis virus infection in turkeys and chickens, *Avian Dis.*, 26, 158, 1982.
113. **Fujiwara, H., Minamimoto, S., and Namioka, S.**, Enteric lesion with intranuclear inclusion bodies in piglets, *Natl. Inst. Anim. Health. Q.*, 8, 53, 1968.
114. **Hafez, S. M. and Liess, B.**, Immunodiffusion and neutralization studies on porcine adenoviruses, *Vet. Microbiol.*, 5, 101, 1980.
115. **Domermuth, C. H., Gross, W. B., DuBose, R. T., Douglass, C. S., and Reubush, C. B., Jr.**, Agar gel diffusion precipition test for hemorrhagic enteritis of turkeys, *Avian Dis.*, 16, 852, 1972.
116. **Ianconescu, A., Smith, E. J., Fadly, A. M., and Nazerian, K.**, An enzyme-linked immunosorbent assay for detection of hemorrhagic enteritis virus and associated antibodies, *Avian Dis.*, 28, 677, 1984.
117. **Hashimoto, K., Sugiyama, T., Yoshikawa, M., and Sasaki, S.**, Intestinal resistance in the experimental enteric infection of mice with a mouse adenovirus. I. Growth of the virus and the appearance of a neutralizing substance in the intestinal tract, *Jpn. J. Microbiol.*, 14, 381, 1970.
118. **Hashimoto, K., Yoshikawa, M., Sugiyama, Y., and Sasaki, S.**, Intestinal resistance to experimental enteric infection of mice with a mouse adenovirus. II. Determination of the neutralizing substance in the intestinal tract as an IgA antibody, *Jpn. J. Microbiol.*, 15, 499, 1971.
119. **Domermuth, C. H. and Gross, W. B.**, Hemorrhagic enteritis of turkeys. Antiserum-efficacy, preparation and use, *Avian Dis.*, 19, 657, 1975.
120. **Domermuth, C. H., Gross, W. B., Douglass, C. S., DuBose, R. T., Harris, J. R., and Davis, R. B.**, Vaccination for hemmorhagic enteritis of turkeys, *Avian Dis.*, 21, 557, 1977.
121. **Fadly, A. M., Nazerian, K., Nagaraja, K., and Below, G.**, Field vaccination against hemorrhagic enteritis of turkeys by a cell-culture live-virus vaccine, *Avian Dis.*, 29, 768, 1985.
122. **Benfield, D. A.**, unpublished observations, 1981.
123. **Saif, L. J.**, personal communication, 1987.

Chapter 7

THE 22 TO 30 NM GASTROENTERITIS AGENTS OF MAN

Harry B. Greenberg, Marna Skaar, and Stephan S. Monroe

TABLE OF CONTENTS

I. INTRODUCTION

In the 1940s and 1950s several investigators demonstrated that an acute gastroenteritis syndrome could be experimentally transmitted to volunteers by oral inoculation of bacteria-free fecal filtrates derived from individuals who had acute gastroenteritis.[1-9] Despite considerable effort, pathogenic viruses could not be cultivated from such infectious filtrates and studies of nonbacterial gastroenteritis fell out of favor. In 1972, new life was brought to the field when workers at the National Institutes of Health using the technique of immune electron microscopy (IEM) demonstrated that virus-like particles could be visualized in an infectious fecal filtrate obtained from an outbreak of diarrheal disease in Norwalk, Ohio.[10,11]

Since this seminal discovery 15 years ago, a multitude of small 22- to 30-nm nonenveloped particles have been identified in the stool specimens of patients with acute nonbacterial diarrhea.[12-40] Particles have been observed in a variety of epidemiologic settings, but most frequently they have been identified in fecal specimens obtained during the investigation of acute common source epidemics of diarrhea and vomiting. Despite the frequency with which new gastroenteritis agents have been visualized, the universal failure to propagate these fastidious agents efficiently in cell culture or in an animal model has severely restricted progress in this field. In addition, the level of scientific rigor employed in establishing an etiologic relationship between a newly visualized gastroenteritis virus and disease has varied enormously from report to report. Hence, it should not be surprising that frequent visualizations of putative new gastroenteritis agents coupled with the inability of investigators to fully characterize these new agents on a biochemical or serologic basis has resulted in a confusing and disorganized series of publications.

Currently, ultrastructural analysis is the only analytic method available to compare and contrast all the 22- to 30-nm particles which have been visualized. Comparisons based on serologic relationships, protein or nucleic acid content, or epidemiologic behavior are not yet possible because of the limited amount of published data for many agents. Recently, a useful interim classification scheme for the 22- to 30-nm gastroenteritis has been presented.[41] This scheme, based entirely on ultrastructural analysis, described five distinct groups of small round gastroenteric viruses (SRGVs). This interim scheme can be simplified to include only three separate groups of agents if one considers only viruses that have been associated with gastroenteritis and takes into account recent biochemical data.

The first group of viruses includes those viral strains which have an ultrastructure similar or identical to caliciviruses and those agents which have a less distinct, but not featureless ultrastructure, such as Norwalk virus and Snow Mountain virus.[41] Recent biochemical studies indicate that all these viruses share some of the features of caliciviruses. These agents are clearly pathogenic for man. Indeed, most of the epidemiologic data concerning the pathogenicity of the 22- to 30-nm gastroenteritis viruses comes from studies of viruses in the calicivirus and calici-like virus group.

The second category of viruses can be grouped under the heading of astroviruses. The astrovirus group shares a common ultrastructure and appears to have a protein structure distinct from the calicivirus group.[42,43] There are some good epidemiologic and experimental data demonstrating that the astrovirus group of enteric agents is associated with acute gastroenteritis.[27,45-47] This association is not as strong, however, as is the association with the calicivirus group.

Finally, there exists a group of poorly characterized 22- to 28-nm agents without any detectable ultrastructure. These particles most closely resemble parvoviruses, although some data indicates that they are actually picornaviruses.[34-36,48] As will be seen, the association of these small particles with disease has yet to be well documented.

In this chapter, we will use the modified, ultrastructure-based, classification scheme of three groups of small round gastroenteric viruses to organize and present the available data concerning

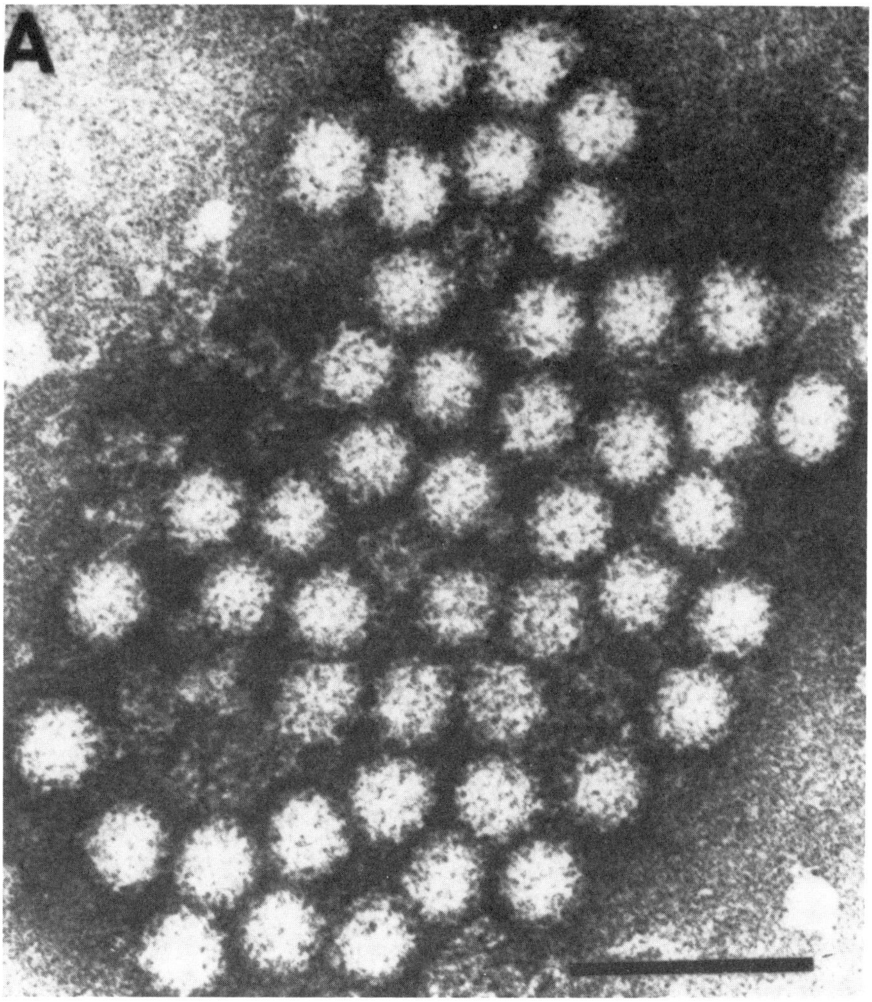

FIGURE 1. An aggregate of Norwalk virus after incubation of a stool filtrate with a 1:5 dilution of pre-challenge volunteer antiserum. Bar = 100 nm. (From Adler, I. and Zickl, R., *J. Infect. Dis.*, 119, 668, 1969. With permission.)

these agents. At present, there is insufficient data to determine if this tentative classification scheme is valid. It does represent, however, a logical and useful way of presenting the SRGVs in an organized manner.

II. INFECTIOUS AGENTS

A. MORPHOLOGY

Four morphologically distinct groups of round, nonenveloped, 22- to 30-nm gastroenteritis viruses have been described.[41] Norwalk virus, Hawaii virus, Snow Mountain virus, and the Taunton virus are all 27- to 32-nm spherical viruses characterized by a definite, yet amorphous, surface structure with ragged edges and a hint of surface indentations (Figures 1 and 2).[19,40,41,49,50] The indistinct surface ultrastructure seen in this morphologic group is most reminiscent of the calicivirus families (Figure 1). Of note, recent biochemical data (see Section B below) supports the hypothesis that these viruses belong to the calicivirus family.

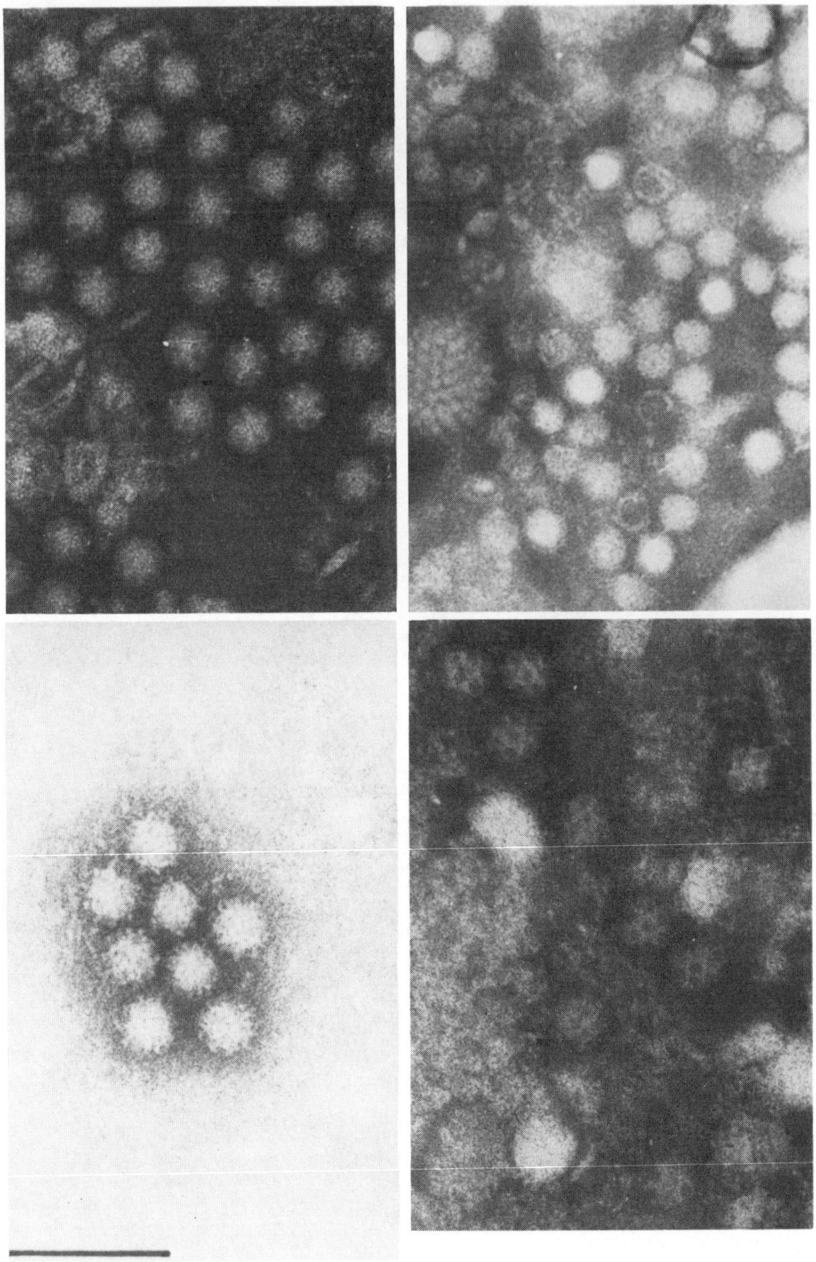

FIGURE 2. Electron micrographs of astrovirus (upper left), adenoassociated parvovirus and an adenovirus (upper right) Norwalk-like virus (lower left) and human calicivirus (lower right). Bar = 100 nm. (Electron micrographs kindly provided by Dr. C. R. Madeley.)

A second morphological group of gastroenteritis agents consists of viruses with classical calicivirus ultrastructure, including a feathery edge, a six-pointed surface star with a dark central indentation, and oval cuplike surface hollows (Figure 2). These viruses are morphologically identical to the cultivatable animal caliciviruses.[42,44] The calici-like viruses (Norwalk, Hawaii, etc.) and the typical caliciviruses will be grouped together under the calicivirus heading for the purposes of this discussion.

The third morphologic group of gastroenteritis agents have been labeled astroviruses.[41,42] These viruses have an unbroken outer edge, either a five- or six-pointed surface star with a white center, and triangular surface hollows (Figure 2). As will be seen, these viruses have a distinct protein structure which clearly distinguishes them from the calici and calici-like strains.

The final group of virus particles are characterized by their small size (22 to 28 nm) and their featureless ultrastructure (Figure 2). These agents include the Wollan, Ditchling, Cockle, and Paramatta agents.[12,13,17,41,48] Because of their size, they appear to resemble parvoviruses, but their featureless ultrastructure is characteristic of the picornavirus family as well.

B. BIOPHYSICAL CHARACTERISTICS

Because the 27-nm gastroenteritis agents have not been adapted to tissue cultures, there is very little information available concerning their biophysical characteristics or genomic structure.

All of the 27- to 30-nm viruses thus far described (with the exception of the Snow Mountain agent) have buoyant densities in cesium chloride of between 1.36 and 1.41 g/cm^3.[26,51,52,53] Snow Mountain agent has a somewhat lower density of 1.34 g/cm^3.[54] Unfortunately, this range of buoyant densities does not distinguish between parvoviruses and picornaviruses. All of the SRGVs lack a lipid envelope. As would be expected from their lack of a lipid coat, the Norwalk virus and the Wollan agent have been shown to retain infectivity after exposure to 20% ether.[55]

The nucleic acid content and structure have not been fully determined for any of the SRGVs isolated from humans. In a preliminary report, workers in England have adapted one human calicivirus strain to grow in primary human embryonic kidney cells.[126] This virus was demonstrated to incorporate ^{3}H uridine in the presence of actinomycin D, indicating that it had an RNA genome. This finding is consistent with the ultrastructure-based classification of these agents as caliciviruses. Caliciviruses contain a plus-stranded RNA genome with a 3′poly-A tail.[44] Caliciviruses also have an unusual protein structure consisting of a single structural protein of approximately 60 kDa.[44] A calici-like virus particle isolated from an infant with diarrhea in Japan did indeed have a single structural protein of approximately 60 kDa.[57]

The genomic composition of the Norwalk family of viruses (ultrastructure reminiscent of, but not identical to, caliciviruses) has not been determined. The protein structure of two viruses in this family has been studied, however. Both the Norwalk virus and the Snow Mountain virus have been purified using standard ultracentrifugation techniques.[53,54] These two purified viruses were iodinated and specifically immunoprecipitated with paired infection sera. In both cases, the virions appeared to be composed of only one structural protein with an apparent molecular weight of approximately 60 kDa. This data provides strong supportive evidence for the case that these viruses (Norwalk, SMA, etc.) with uncharacteristic ultrastructures are truly related to the calicivirus family.

Astroviruses have been isolated from the feces of humans, lambs, mice, birds, cats, dogs, and cattle.[27,55-59b] The availability of an animal model and the finding that astroviruses appear to be shed in the stool in greater quantity than the calici-like group of viruses have led to some progress in the characterization of this group of gastroenteritis agents. The protein structure of an ovine astrovirus has been described.[43] Density-gradient purified ovine astrovirus contained two structural proteins of similar size (approximately 33 kDa when analyzed by slab gel electrophoresis. Further investigation of the ovine astrovirus disclosed that the virion contained a 35S single-stranded RNA with a poly A tail.[43] Recently, a similarly sized RNA has been isolated from a human astrovirus.[147] The Marin County virion, which resembles an astrovirus on electron microscopic examination, has also been shown to contain one or two structural proteins of about 30 kDa[148] and to be serologically related to other astroviruses.[140] It would seem, therefore, that astroviruses are biochemically distinct from caliciviruses since they have a different protein structure. However, both of these agents appear to be plus-stranded RNA viruses.

The final group of SRGVs, the picorna-parvovirus-like group, has not been extensively examined. Although these agents have a featureless ultrastructure and buoyant density similar to the parvovirus,[41] there aren't any biochemical data that would indicate that they have a DNA genome or a protein structure similar to parvoviruses. In fact, in very preliminary studies, some of these viruses appear to have an RNA genome.[48] At present, there is insufficient biochemical or biophysical information available to make even a tentative classification of this group.

C. SEROLOGIC STUDIES

The antigenic nature of the SRGVs has been studied primarily by IEM. Some agents have also been investigated using radioimmunoassay (RIA) or enzyme-linked immunosorbent assay (ELISA) or by immunofluorescence.[60-67] Unfortunately, serologic studies of these agents have been done piecemeal in a variety of different laboratories using a number of different serum reagents. Since the assays have not been done with the same reagents at the same time, all conclusions regarding relationships between agents are tentative and await confirmation when the viruses are adapted to cell culture.

Human enteric caliciviruses have been described in England, Japan, Australia, Norway and Canada.[15,16,24,37,68] There is very little information available concerning the serologic relationship between these strains. The strain isolated in Japan does not appear to be antigenically related to Norwalk virus.[69] Workers in England have presented evidence for at least four distinct strains of human caliciviruses.[70] One of these four serotypes was a Japanese isolate.[70] Several recent reports indicate that Norwalk virus might be serologically related to one of the English calicivirus strains.[38,141]

The Norwalk virus has been demonstrated to be serologically related to the Montgomery Co. isolate and a large number of other strains (see below). The Norwalk strain does not appear to be serologically related to the Hawaii isolate, the SMA isolate, or a variety of cultivatable caliciviruses including feline calicivirus and San Miguel sea lion virus.[149] The Norwalk virus is not serologically related to the featureless 22- to 28-nm particles including the W agent, the Paramatta agent, or the cockle agent, and is not related to several astrovirus strains.[150] The Norwalk strain is also not related to a variety of cultivatable parvovirus strains.[60] The SMA strain of calici-like virus has been shown to be antigenically distinct from the Norwalk virus, and the Marin Co. agent.[19] Calici-like viruses isolated in Japan were not serologically related to Norwalk virus, typical enteric calicivirus, or W agent when tested by IEM.[69]

At least five serologically distinct strains of human astrovirus have been described.[71] In addition, a variety of animal astroviruses have been isolated.[38,55-59b] A bovine astrovirus strain was not serologically related to a human isolate.[59] The relationship of the other animal strains to human isolates is unknown. Astrovirus strain 624 is serologically unrelated to Norwalk virus.[151] The JW strain of human astrovirus is not related to the W agent.[46] An astrovirus strain isolated in Japan was shown to be serologically distinct from Norwalk virus, a human calicivirus, the Sapporo/Otofuke agent and several cultivatable enteroviruses.[45] Recently, the Marin Co. agent has been shown to be serologically related to astrovirus type 5.[140]

The picorna-parvovirus group of enteric viruses has had the least extensive serologic study. The cockle agent did not appear to be related to the W agent, the Norwalk agent, or the Hawaii agent.[13,41] The Ditchling and W agents appear to be serologically related.[12] The Paramatta agent is distinct from the Norwalk virus by IEM.

It is clear from the above discussion that a moderate amount of serologic diversity seems to exist in each of the three groups of pathogenic enteric viruses. Until these agents can be easily propagated, clear conclusions concerning their antigenic relationship will remain speculative at best. Although the initial serologic data implies great diversity among these viruses, it is encouraging to realize that a single serotype, the Norwalk virus, appears to account for over one third of all cases of epidemic gastroenteritis in the U.S.[72] To account for this apparent contradiction, it is possible that a large number of the existing isolates are actually rather infrequent causes of gastroenteritis. Alternatively, many of the existing strains may be demonstrated to be serologically related when hyperimmune antisera and standardized assays are employed to study them.

III. EPIDEMIOLOGY

A. ANTIBODY PREVALENCE

The lack of widely available viral detection assays or serologic assays to identify the SRGVs directly or to detect antibody to the SRGVs has interfered with our gaining an understanding of the epidemiology of these agents. However, within the past decade, a variety of solid phase immunoassays has been developed for several of these viruses[60-66,142] and these assays have led to some understanding of the epidemiologic niche the SRGVs occupy. The Norwalk virus has been most extensively studied and this agent will be used as a model for the type of epidemiologic investigations which are lacking for most of the other agents. A sensitive and specific solid phase radioimmunoassay for Norwalk antigen and a blocking type solid phase radioimmunoassay for antibody to Norwalk virus were developed in 1978.[60] These assays enabled investigators to carry out large antibody prevalence surveys in a variety of different geographic locations.[73-76] Greater than 50% of adults (over 18 years of age) in the U.S., Ecuador, Belgium, Switzerland, Yugoslavia, Bangladesh, Taiwan, the Philippines, and Nepal had serum antibody to the Norwalk virus.[73,76] A very isolated and small Indian tribal group in Ecuador represents the only population thus far identified which does not show serologic evidence of Norwalk infection.[73] There was no difference in antibody prevalence to Norwalk virus between homosexual men or women and random blood bank donors in New York City. A striking difference was observed when the rate of acquisition of antibody to Norwalk virus was studied in developed and less developed countries. In the U.S. and Yugoslavia, antibody prevalence gradually increased over the first two decades of life.[73] In contradistinction, in the less developed countries of Ecuador, Bangladesh, and the Philippines, virtually the entire population had evidence of Norwalk infection by the age of 5 (Figure 3). This antibody prevalence data were considered consistent with the assumption that Norwalk virus is transmitted by the fecal-oral route and exposure risks decrease as sanitation improves.

None of the other SRGVs have been studied as extensively as the Norwalk virus. Antibody prevalence data are not presently available for other members of the calici-like virus group, the Snow Mountain, Taunton, and Hawaii agents. In Japan, the prevalence of antibody to a typical enteric calicivirus strain has been studied by IEM.[77] Antibody to the Japanese calicivirus strain was acquired rapidly so that over half the children studied had antibody by the age of 5. Unfortunately, this same group of children were not studied for evidence of Norwalk antibody. It would be very interesting to know if the pattern of antibody acquisition for these two related viruses is similar or different in a single population. Antibody to the UK1 strain of human calicivirus has a worldwide distribution.[77a] In England, the peak incidence for acquisition of antibody to the UK1 calicivirus occurred between 3 months and 6 years of age.[77a] In a recent study from Houston, Texas, antibody acquisition to human caliciviruses appeared to occur rapidly in early childhood.[142]

The prevalence of antibody to one strain of human astrovirus was studied by indirect immunofluorescence.[78] In England, 75% of a group of 3- to 4-year-olds had detectable antibody to this strain of astrovirus. Again, as with the Japanese calicivirus strain, it appears that the rate of antibody acquisition for astroviruses in a developed country might be different than for Norwalk virus. Unfortunately, since this English population was not simultaneously studied for Norwalk virus antibody, such inference must remain speculative at present. Antibody prevalence data for the featureless 22- to 28-nm particles is not yet available.

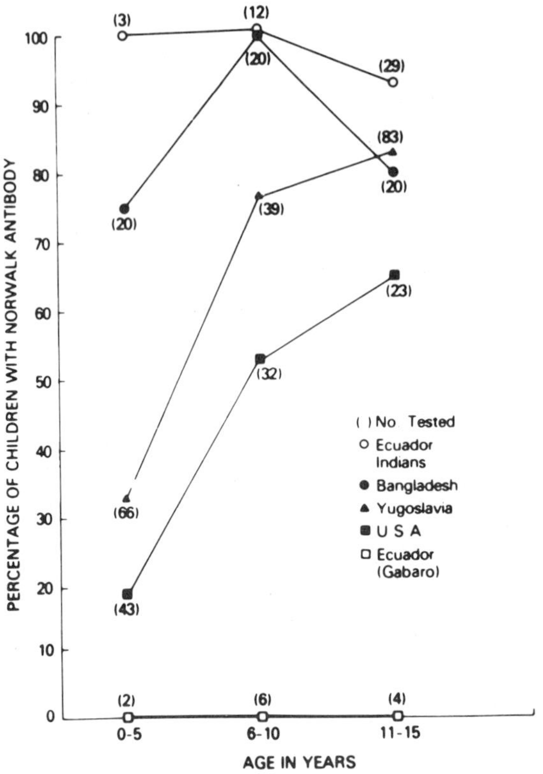

FIGURE 3. Age-related prevalence of serum antibody to Norwalk virus in children from various countries. (From Greenberg, H. B., Valdesuso, J., Kapikian, A. Chanock, R. M., Wyatt, R. G., Szmuness, W., Larrick, J., Kaplan, J., Gilman, R. H., and Sak, D. A., *Infect. Immun.*, 26, 270, 1979. With permission.)

B. THE ROLE OF SRGVS IN EPIDEMIC AND ENDEMIC GASTROENTERITIS

Attention was originally focused on the SRGVs because they were found in apparent association with large common-source epidemics of gastroenteritis. The calici-like group of viruses has been strongly associated with epidemic diarrhea. The Norwalk virus was originally isolated from a community-wide epidemic in Ohio.[10] The Hawaii virus came from a small family epidemic in Hawaii[40] while the Snow Mountain agent was found in association with an epidemic at a resort camp in Colorado. A second epidemic of food-borne gastroenteritis caused by an SMA-like virus has been described.[32,79] The Taunton agent caused an epidemic of gastroenteritis on an orthopedic and geriatric ward of a hospital in England.[50]

Interestingly, the enteric caliciviruses with typical ultrastructure have been infrequently associated with clear cut community-wide epidemics of diarrhea. In most cases (see below) the typical calicivirus isolates have been identified with epidemics of diarrhea in infants.[16,68] An epidemic of calicivirus associated diarrhea among young school children and in an elderly nursing home population has been described, but large community-wide epidemics have not yet been reported.[80,81] It is not clear if this apparent difference in epidemiologic characteristics between the "calici-like" and typical caliciviruses is real or not. The most convincing proof of the association of the Norwalk virus with epidemic diarrhea comes from serologic analysis of paired sera derived from epidemics of gastroenteritis. Similar studies have not yet been carried out using specific antibody assays for the typical caliciviruses.

The astroviruses have been associated with a variety of epidemic settings including kindergarteners and their teachers, nursing homes, and a children's ward in a hospital.[14,27,45] In each of these epidemics, adults were symptomatically infected. No single serotype of the astroviruses has been associated with a large number of separate epidemics. The Marin Co. agent, a probable astrovirus, caused a prolonged epidemic of gastroenteritis in a nursing home.[33]

The featureless "picorna-parvo" group of viruses (Ditchling, W, cockle, and Paramatta) were all isolated from epidemics of gastroenteritis. The cockle agent was isolated from a large community-wide epidemic caused by contaminated seafood[13] while the other three epidemics occurred in primary or secondary schools.[12,17,18]

It seems clear that all the SRGVs, with the possible exception of those viruses with a typical calicivirus ultrastructure, are capable of causing epidemic gastroenteritis in both adults and children. This feature clearly distinguishes these agents from the rotaviruses and the enteric adenoviruses. While the rotavirus and the enteric adenoviruses can symptomatically infect adults, they are, with rare exception, not the cause of epidemic diarrhea in adults.[82] It should be mentioned, however, that the epidemic potential for the SRGVs has only been noted in developed countries. The role these agents play in less developed areas is unclear. It is certain, however, that exposure to the SRGVs occurs at an earlier age in the less developed world.[73]

The role of the Norwalk strain of SRGVs in epidemic gastroenteritis has been extensively studied. Twenty-four of 70 outbreaks of gastroenteritis analyzed by workers at the NIH were characterized as being caused by the Norwalk virus, because over half of the paired serum specimens from affected people in these outbreaks demonstrated an immune response to Norwalk virus.[72,74] The Centers for Disease Control (CDC) published a detailed review of all the outbreaks of nonbacterial gastroenteritis in which Norwalk serology was studied between 1976 and 1980.[83] Many of these outbreaks were also included in the study from the NIH. Of the 74 epidemics described 42% appeared to have been caused by a Norwalk-like virus. These outbreaks occurred in a variety of specific settings including camps and recreational areas, schools, nursing homes, cruise ships, families, and communities. Outbreaks occurred in every month of the year and included people of all ages with a relative sparing of infants and very young children.

Most of the outbreaks had evidence of a common source origin and a likely vehicle of transmission was identified in 17 of 38 episodes.[83] Water was the most frequent vehicle. Contaminated municipal water,[86,88] semi-public water,[84,91] stored water on cruise ships,[87] and recreational swimming water (lake and pool)[85,89,90] were all implicated in Norwalk transmission. Food-borne outbreaks were less frequent than water-borne episodes. Contaminated salad ingredients and oysters have both been implicated as transmission vehicles.[92] In other investigations not reported by the CDC, cake icing and chicken salad have been implicated as vehicles for Norwalk transmission[93,94] and additional epidemics of oyster-associated disease have been recently noted.[95,98]

In most of the outbreaks reported by the CDC, secondary person-to-person transmission was observed. The attack rate of secondary cases varied from 4 to 32%. Outbreaks had a median duration of 1 week, but some reported epidemics were rather prolonged.[84,99] The prolonged outbreaks were frequently observed in the setting of a recreation camp or cruise ships where new "susceptible" populations continued to be introduced into a contaminated setting.

The size of reported epidemics has varied considerably. In the CDC study, epidemics ranged from small family outbreaks involving two to three people to massive country-wide episodes caused by contaminated oysters involving several thousand people. A recently reported Norwalk epidemic caused by eating contaminated cake icing caused over 2,000 cases of diarrhea.[93]

The symptoms of disease reported with the outbreaks of Norwalk disease are extremely similar to signs and symptoms observed during volunteer studies.[83,100-102] Nausea, vomiting, diarrhea, abdominal pain, and cramps are all very common. Less common, but still frequent symptoms include headache and low grade fever. It is interesting to note that analysis of several epidemics has demonstrated that when children and adults are exposed to the same common source of Norwalk virus, vomiting is more frequently encountered in the children and diarrhea in the adults.[83] Norwalk epidemics have only rarely been associated with severe illness requiring hospitalization. In most cases, the disease is self-limited and lasts 1 to 2 d. The signs and symptoms specifically ascribed to Norwalk-associated gastroenteritis appear to be similar or identical to the illness caused by the other SRGVs.

The incubation period in the CDC review ranged from 4 to 77 h with a mean of between 24 and 48 h. This is similar to the incubation period during volunteer studies.[103,104]

Using the criterion of serum immune response in over 50% of affected individuals, approximately one third of all outbreaks of nonbacterial gastroenteritis appear to be caused by the Norwalk virus. In an additional 20% of outbreaks studied, some, but less than 50%, of the affected individuals have evidence of serum responses to the Norwalk particle.[23,74] The exact cause of the low frequency of serum response to Norwalk virus in this group of epidemics is unknown. One interpretation of these data is that these epidemics were caused by a virus which is antigenically related to the Norwalk virus and, hence, induced a seroresponse to the Norwalk agent in only a minority of affected individuals. Similar cross-reactivity has been observed during enterovirus infection.

By analyzing the epidemiologic characteristics of the documented epidemics of Norwalk virus, certain criteria can be developed which characterize these episodes: (1) stool cultures must be negative for bacterial pathogens, (2) the mean duration of illness is between 12 and 60 h, (3) vomiting is common and occurs in at least 50% of cases, and (4) the mean incubation period prior to illness is 24 to 48 h. Using these criteria, epidemiologists at the CDC reviewed 642 outbreaks of acute gastroenteritis.[100] Fifty-four outbreaks met the criteria of a Norwalk-like pattern. These 642 included 460 food-borne outbreaks, 18 cruise-ship epidemics, 14 nursing-home episodes, 142 water-borne outbreaks, and 8 summer-camp epidemics.

The CDC workers estimated, using the above set of criteria, that 23% of all water-borne epidemics, approximately 60% of summer-camp and nursing-home outbreaks, 28% of cruise-ship epidemics, and 4% of food-borne outbreaks were caused by Norwalk or Norwalk-like viruses. When paired sera from 14 of these 54 outbreaks were actually studied, the predictive value of the "Norwalk" criteria was quite good. Ten of 14 epidemics predicted to be caused by Norwalk virus showed seroresponses to Norwalk virus.[100] From the above data, it is clear that the Norwalk virus is a very frequent cause of epidemic gastroenteritis, accounting for at least one third of all episodes. This finding appears to be in some conflict with the large number of serologically distinct SRGV particles thus far identified. Perhaps many of the other particles, if studied by solid phase immunoassay, would prove to be antigenically related. Alternatively, a number of the antigenically distinct SRGVs described in the literature are not very commonly encountered, at least in the setting of epidemic diarrhea.

The role of the SRGVs in infantile diarrhea has been studied by several workers but much more information will be needed before a clear picture emerges regarding the role of these agents in young children.

Norwalk virus is an infrequent cause of severe diarrhea in young children. In a serologic study of paired sera obtained from 51 rotavirus-negative children admitted for severe diarrhea to the Children's Hospital, Washington, D.C., none had evidence of an immune response to Norwalk virus.[105] Black and co-workers collected 31 paired serum samples from children with severe diarrhea in whom no enteric pathogens could be identified. Only one of these serum pairs showed evidence of a seroresponse to the Norwalk virus.[106] The paucity of cases of severe infantile diarrhea caused by the Norwalk virus is very consistent with the epidemiologic data obtained during epidemics of Norwalk virus which indicate that Norwalk infection is brief, relatively mild, and rarely necessitates hospitalization.

In developed countries, there has not been an extensive investigation of the role of Norwalk virus in mild gastroenteritis in young children. The antibody prevalence surveys imply that exposure to Norwalk virus is limited.[73] Unfortunately, we do not know how many times a child must be infected with Norwalk virus before a measurable and enduring immune response is produced. As mentioned earlier, preliminary observations indicate that exposure in developed countries to astroviruses and typical caliciviruses may occur at an earlier age than exposure to Norwalk virus. In a small prospective study of diarrheal disease in families with infants, 2 of 14 outbreaks of diarrhea appeared to be caused by the Norwalk virus.[107] In one of these outbreaks a 10-month-old infant was affected.

Black and colleagues have tried to estimate the importance of Norwalk virus in young children in rural Bangladesh using a serologic analysis of serial serum samples.[106] In a prospective, longitudinal study of village children over a 1-year period, they showed an annual incidence of Norwalk infection of 29% for all children (ages 2 to 49 months). Infections were most frequent in December through March. In prospective serologic studies of young children in Panama, a similar annual incidence rate of Norwalk infection was observed.[108] In Bangladesh, infection incidence was highest for 14- to 18-month-old children and declined to low levels by 38 months of age. In order to assess the pathogenic potential of Norwalk infection, Black compared the incidences of diarrheal episodes without an identified cause during periods with and without fourfold increases in antibody titer to Norwalk virus. Children with diarrhea and no identifiable fecal pathogen had an excess incidence of seroresponses to Norwalk virus. Extrapolating from these data, Black estimated Norwalk infection could account for between 1 and 2% of the episodes of mild diarrhea in children in rural Bangladesh. Since Norwalk virus represents only a single serotype of SRGVs, the aggregate pathogenic potential of these agents in mild infantile diarrhea in less developed countries might be considerable. Prospective, longitudinal serologic studies of infection rates for the other calici or calici-like viruses, the astroviruses, or the picorna-parvoviruses are not available.

Several investigators have prospectively studied fecal specimens from young children with diarrhea in order to determine the etiologic role of various viral agents as causes of diarrhea.[24,37,70,109-111] These investigators have used EM or IEM as a method for detecting enteric viruses. These studies suffer from several generic defects which limit their reliability as surveys to assess the importance of SRGVs. First of all, EM and IEM are extremely insensitive means of detecting SRGVs such as Norwalk virus. EM also underestimates the number of SMA- and calicivirus-positive fecal specimens. In other words, antigen detection studies have probably substantially underestimated the prevalence of SRGVs. In addition, most of the published antigen detection studies have been surveys based on patients brought to or in hospitals. These surveys would tend to study more severely ill children and, hence, enrich for rotavirus-positive specimens and select against the SRGVs which seem to cause milder disease. In light of these deficiencies, it is not surprising that the IEM surveys have not identified the SRGVs as common enteric pathogens in young children. In an 8-year prospective study of diarrheal disease in hospitalized children in Washington, D.C., Norwalk-like particles were identified in only 25 of 1,537 fecal specimens.[109] In a 2-year study in Buffalo, New York of hospitalized children with diarrhea, workers identified SRGVs in 14% of 314 virus-positive fecal specimens.[112] Most positive specimens contained rotaviruses. Of the 45 specimens containing SRGVs, 20 appeared to be astroviruses, 5 caliciviruses, and 20 picorna-parvo-like viruses.[112] None of these specimens contained Norwalk virus when tested by RIA. Many of these SRGV infections appeared to have had a nosocomial origin.

Kogasaka et al. used IEM to evaluate stool specimens from 233 pediatric cases of acute diarrhea seen at the Sapporo Medical College Hospital over a 4$^1/_2$-year period. Viruses were identified in 178 cases. SRGVs were detected in only nine cases. These nine isolates resembled Norwalk virus morphologically and biophysically.[26] They did not, however, react with paired sera to Norwalk virus, calicivirus, or W agent, but did react with convalescent sera from an infected child.

Spratt et al. studied hospital-acquired diarrheal disease in infants in Montreal over a 2-month period. Caliciviruses were associated with 18% of diarrheal episodes and picorna-parvoviruses with 9%.[37] In Australia, Grohmann conducted a 5-year study during which he examined a total of 7482 stool samples, 3576 from children under 5. In the latter group, he identified 20 caliciviruses (0.6%), 155 astroviruses (4.3%), 144 small round viruses (4.0%), and only 4 Norwalk-like viruses.[143]

Over a 2¹/2-year period, 592 cases of gastroenteritis in North West London were studied by IEM. Thirty-nine patients (6.6%) were found to be excreting caliciviruses. Most of these patients were hospitalized children under the age of 3. Infection occurred year round, but peaked in the winter. These enteric calicivirus isolated were serologically distinct from two epidemic strains, but might have been serologically related to a third strain.[70]

Cameron and colleagues described the shedding of an SRGV similar to Norwalk virus and/or astrovirus by 7 of 10 newborns in a newborn nursery.[113] Some of these infants were also infected with rotavirus. Unfortunately, other investigators have not studied SRGV infections in newborns.

In a 3-year prospective study of 647 hospitalized infants and young children with acute gastroenteritis in Japan, 8 patients were shown to shed calicivirus. Seven other patients shed SRGVs of an unknown nature.[110] In a second survey from Japan, Oishi et al. prospectively examined fecal specimens from outpatient children with diarrhea over a 3-year period. Of 395 fecal specimens studied by EM, 1% contained caliciviruses, 2.3% contained astroviruses, and 1.8% contained picorna-parvo-like agents. Viruses resembling Norwalk virus (calici-like) were not identified.[111] The SRGVs in this study were found primarily in children under 18 months of age and occurred primarily, but not exclusively, in the summer months.

In a 6-month IEM survey of fecal specimens from 143 patients with acute nonbacterial gastroenteritis sent to a public health laboratory in England for study, three specimens were shown to contain astrovirus and three contained uncharacterized 20- to 30-nm viruses.[114] Most of the patients in this study were 2 years of age or younger.

Finally, in a 6-month study of 238 fecal specimens from gastroenteritis patients in Norway, Kjeldsberg detected 18 samples which contained SRGVs. Four specimens had particles resembling Norwalk virus, ten specimens had particles with an ultrastructure similar to astroviruses, three resembled caliciviruses, and one picorna-parvovirus. Most of the specimens were obtained from children. In one case of astrovirus infection, immune electron microscopy was used to demonstrate a seroresponse to the virus.[24]

From the preceeding discussion, it is clear that SRGVs of all groups can be identified in low numbers in fecal specimens obtained from children with gastroenteritis. Unfortunately, in most reported studies, normal controls were not studied, and IEM demonstration of seroresponses to the isolated SRGVs was not attempted. Given the insensitivity of EM/IEM examination and the emphasis on relatively ill children in reported series, it is hard to obtain a clear picture of the actual role these small viruses play in mild endemic infantile diarrhea in either developed or less developed countries. A few findings seem clear. The SRGVs are not frequent causes of severe diarrhea in young children. The caliciviruses appear to be particularly involved in epidemic episodes of diarrhea in young children. There is a good deal of evidence that the SRGVs do cause some cases of mild infantile diarrhea (approximately 15%) in several countries and this percentage might be underestimated. Until sensitive detection assays are developed, it is unlikely that the role of these enteric agents in infantile gastroenteritis can be adequately evaluated.

Norwalk virus appears to be a minor cause of travelers' diarrhea. If the other SRGVs also cause travelers' diarrhea with a similar frequency, then the aggregate effect may be substantial.[137-139]

The role of SRV as a cause of gastroenteritis in AIDS patients is unknown. A recent study from Australia indicates that these viruses may be a common pathogen in patients with AIDs.[146]

IV. PATHOLOGY AND PATHOGENESIS

There is a paucity of information available concerning the histopathologic lesions produced by the SRGVs and even less concerning the mechanism by which these agents produce diarrhea. Gastric, duodeno-jejunal, and rectal histology has been studied in adult volunteers experimentally infected with either Norwalk virus or Hawaii virus. These are the only histopathologic studies of SRGV infection in man. Both Norwalk virus and Hawaii virus produce a characteristic, but not pathognomonic, lesion in the small bowel. This lesion is characterized by lamina propria and mucosal inflammation with both polymorphonuclear cells and monocytic cells, villous shortening and crypt cell hyperplasia with increased mitosis. Also seen is absorptive cell vacuolization, loss of normal nuclear polarity, and decreased absorptive cell height.[115-118] However, the mucosa remained intact throughout the illness. Electron microscopic changes in the small intestinal epithelial cells at the time of Norwalk illness included dilated endoplasmic reticulum, the presence of a large number of multivesiculate bodies, widening of intracellular spaces and shortening of microvilli.[115] The small bowel lesions produced by Norwalk agent and Hawaii virus were similar but not identical.[116] Virtually all ill volunteers appeared to develop histologic abnormalities.[117] Some, but not all, volunteers who remained well after challenge also developed pathologic changes in the small bowel. Histopathologic alterations persisted for at least 5 to 6 d after illness,[117] but were normal 2 weeks and 6 weeks after recovery. In 15 volunteers inoculated with Norwalk virus, gastric biopsies (both fundus and antrum) did not demonstrate any changes during illness.[119] In two cases of Norwalk disease, rectal biopsies obtained at the height of illness were normal.[115]

Although there is not any information available concerning histopathologic changes with the other SRGVs in man, there is some information from animal studies that might be relevant. Infection of young pigs with a calicivirus-like agent led to villous atrophy in the duodenum.[120] Astrovirus infection in young calves caused mild villous atrophy.[121] Infection of calves with the Newbury agent, a virus with an ultrastructure similar to Norwalk and other calici-like viruses, led to villous shortening, thickening, and increased cellularity in the lamina propria, most prominently in the middle portion of the small bowel.[121] Lambs infected with astrovirus show villous shortening, increased lamina propria cellularity, and a variety of electron microscopic changes limited exclusively to the small intestine.[122]

The mechanism by which the SRGVs produce diarrhea has not been demonstrated. Presumably, diarrhea is a consequence of the loss of mature absorptive epithelium as is the case with rotavirus and TGE infection. It is clear that the histologic changes that occur during Norwalk and Hawaii virus infection are accompanied by abnormalities in absorptive function, brush border enzyme content, and gastric motor activity.

Volunteers infected with Norwalk virus had reduced levels of D-xylose absorption after oral[102] and intraduodenal administration of D-xylose.[117] D-xylose malabsorption was still present 5 days after inoculation. Lactose intolerance was evaluated in two volunteers who became ill after inoculation with Norwalk virus. In both volunteers, lactose malabsorption was exacerbated during illness.[102] Fat absorption was evaluated in five Norwalk volunteers, two of whom were ill, the other three were well. All five transiently excreted increased amounts of fat (between 6 and 15 g/d). Absorption returned to normal by day 9 postinoculation.[102]

Brush border enzymes have been studied in Norwalk volunteers. Intestinal alkaline phosphatase and trehalase content were significantly depressed 24 to 48 h after inoculation in ill volunteers, but sucrase and lactase content were unchanged.[102,123] Jejunal adenylate cyclase activity was evaluated in volunteers inoculated with either Norwalk or Hawaii virus. Cyclase activity was not increased during illness.[123]

Because nausea and vomiting are prominent components of the clinical syndrome produced by the SRGVs, Meeroff and colleagues evaluated gastric motor function in ten volunteers inoculated with either Norwalk or Hawaii virus.[124] Ill volunteers demonstrated a marked delay in gastric emptying, but did not evidence changes in acid, pepsin, or intrinsic factor secretion.[124]

In man, abnormalities in intestinal function, or enzyme content have not been studied during infection with the other SRGVs. Calves infected with the Newbury agent also demonstrated an acute fall in D-xylose absorption during infection.[121]

V. INFECTION OF TISSUE CULTURE AN EXPERIMENTAL ANIMALS

The SRGVs have proven extremely difficult to cultivate *in vitro*. Norwalk virus, Hawaii virus, Ditchling virus, W virus, cockle virus, Tauton virus, Marin County virus, Paramatta virus, and Snow Mountain virus have all failed to grow in a variety of cell lines and organ cultures. Over the years, fecal specimens derived from a large number of unnamed epidemics of nonbacterial gastroenteritis have been inoculated into a variety of cell culture lines without identifying a cultivatable SRGV.

Recently, however, some success has been achieved with attempts to cultivate human astroviruses and caliciviruses. Kurtz and colleagues serially passed an astrovirus isolate 13 times in primary human embryonic kidney cells.[125] They also adapted their virus to growth in a continuous monkey kidney cell line, $LLCMK_2$. However, the virus could not be adapted to VERO, Hep II, MRC-5, BHK, and HRT 18 cells. Viral growth was absolutely dependent on the addition of trypsin to the growth media. Kurtz has used the astrovirus infected cells as a substrate in a fluorescent antibody screening assay. These encouraging results have recently been repeated by other.[140]

Several investigators have failed in attempts to propagate various calicivirus strains. Cubbitt and colleagues, however, have been able to propagate one strain of human enteric calicivirus in HEK cells.[126] These investigators also used trypsin (100 μg/ml) to enhance the growth of their calicivirus strain. They serially passed their viruses an unspecified number of times, and demonstrated that it had an RNA genome by showing that the virion incorporated ^3H uridine. Other workers have not yet reproduced these exciting preliminary findings but there is recent encouraging success passaging a porcine enteric calicivirus in pig kidney cells.[144] The porcine virus could only be propagated in primary cells, and only with the addition of an intestinal contents extract to the growth medium. Successful propagation of human caliciviruses may require similar additions to the culture medium. Other SRGVs have not been successfully cultivated.

Attempts to passage human SRGVs in animal hosts have been rather disappointing. With the exception of chimpanzees, other nonhuman primates including rhesus, baboons, owl, patas, marmosets, and wooley monkeys have not proven susceptible to Norwalk infection.[127] In chimpanzees, Norwalk inoculation is followed by a low level immune response and the shedding of soluble Norwalk protein, but not by disease or shedding of detectable virion. Attempts to passage Norwalk virus and Hawaii virus in a variety of small rodents were not successful. Serologic surveys of a wide variety of animal species did not demonstrate antibody to Norwalk virus.[149]

There is very little published data concerning attempts to passage other SRGVs in animal hosts. In several instances, SRGV strains have been inoculated into suckling mice without apparent effect. Two of three rabbits fed a strain of human calicivirus developed a detectable immune response to the virus, but did not develop disease.[128] Despite the fact that animals have not proven to be a very successful host for human SRGVs, it is becoming increasingly clear that many animals are naturally infected with these enteric strains. These natural animal models of SRGV infection have not yet been studied extensively and represent a potential useful resource for further investigations.

VI. CLINICAL FEATURES

The clinical syndrome of Norwalk virus infection has been thoroughly studied in volunteers and in a large series of Norwalk epidemics.[83,101,102] In one volunteer study, 34 of 59 inoculated individuals developed illness.[104] The mean incubation period of illness was 26 h (range 21 to 37) and the mean duration 21.6 (range 4 to 48). All the ill volunteers complained of nausea, 91% vomited and 56% developed diarrhea. The mean number of vomiting episodes was 3.3, and the mean number of episodes of diarrhea was 1.9. All volunteers recovered within 72 h.

In a review of 38 outbreaks of Norwalk infection, 79% of ill individuals had nausea, 69% vomiting, 66% diarrhea, 70% abdominal cramps, 50% headaches, and 37% fever. Chills, myalgias and sore throats occurred less frequently.[83] In this review, most illnesses lasted between 12 and 60 h and had incubation periods of between 24 and 48 h. Severe Norwalk illness requiring IV hydration or hospitalization has been seen very infrequently in either the volunteer or field setting. The Hawaii, W virus, and Snow Mountain virus have also been studied in a volunteer setting. Illness with these agents is clinically indistinguishable from Norwalk infection.

One strain of astrovirus in England and the Marin County virus, a putative astrovirus from the U.S., have also been given to adult volunteers. Seventeen volunteers received a bacteria-free fecal filtrate containing astrovirus strain J.W.[46] Of note, only one volunteer developed illness. This volunteer vomited once on days 3 and 4, following inoculation. On days 4, 5, and 6, the patient had some or all of the following symptoms: nausea, anorexia, headache, malaise, abdominal discomfort, and fever. On day 6, the patient developed watery diarrhea which lasted 1d. None of the other 16 adult volunteers developed vomiting or diarrhea. The Marin County agent was administered to 19 volunteers.[152] Only one developed illness and this illness (diarrhea) occurred on day 4. From these limited volunteer studies, it appears that the astroviruses are substantially less virulent in adults than Norwalk, Hawaii, SMA, or W virus.

The clinical syndrome associated with the typical caliciviruses, other astroviruses and the picorna-parvo-featureless-viruses has not been studied in the volunteer setting. From the brief descriptions of the clinical syndrome seen in patients, it does not appear that these viruses produce an illness that is recognizably different from the Norwalk agent.

From volunteer studies and epidemiologic data, it appears that the SRGVs are transmitted by the fecal oral route. Norwalk has been identified in vomitus and this may provide an alternative vehicle for transmission.[136]

A. IMMUNITY

Information is not presently available concerning the nature of immunity to enteric astrovirus or picorna-parvoviruses. While astrovirus epidemics have affected adults[45] the relative inability to transmit disease to adult volunteers makes it likely that long-term immunity exists following infection. The relationship of astrovirus serotypic diversity to immunity has not been studied.

The nature of humoral immunity to calicivirus has recently been examined by Nakata and co-workers.[69,130] These investigators have prospectively followed diarrheal disease in an orphanage for young children in Japan. They used a specific RIA to analyze paired sera from children exposed to a calicivirus epidemic at the orphanage. This study demonstrated that the presence of pre-existing serum antibody in children was associated with a significantly decreased risk of becoming ill. A similar study of children with Norwalk virus does not exist. In adult volunteers however, pre-existing serum antibody titer is not correlated with protection (see below).

Clinical immunity to Norwalk virus has been studied more extensively than immunity to the other SRGVs. Numerous volunteer studies with Norwalk, Hawaii, and SMA have demonstrated that at least half of all volunteers are susceptible to illness after challenge. In outbreaks of Norwalk virus, attack rates of over 50% are common.[83] The reason some volunteers are resistant

TABLE 1
Level of Antibody to Norwalk Virus Prior to Inoculation of Volunteers

Response of volunteers to oral administration of Norwalk virus	Number of volunteers	Geometric mean pre-inoculation antibody titer[a] (reciprocal)	
		Serum	Jejunal fluid
Diarrheal illness	11	291	53[b]
No illness	12	178	9[c]

[a] Measured by radioimmunoassay blocking technique.
[b] Only 7 of 11 volunteers tested successfully.
[c] Only 8 of 12 volunteers tested successfully.

From Greenberg, H. B., Wyatt, R. G., Kalica, A. R., Yolken, R. H., Black, R., Kapikian, A. A., and Chanock, R. M., *Perspect. Virol.*, 11, 163, 1981. With permission.

and some susceptible to Norwalk infection is poorly understood. In studies of short term immunity, it was demonstrated that infection with Norwalk virus protected against challenge for at least 2 months.[131,132] Norwalk infected volunteers were not protected against challenge with Hawaii virus however.[116]

In studies aimed at trying to understand the basis of long-term immunity to Norwalk infection, Parrino and colleagues administered Norwalk virus to a series of volunteers and then rechallenged these volunteers with the same inoculum 2 years later.[132] In this study, it was observed that the same volunteers who became ill on initial challenge, redeveloped illness on challenge 2 years later. Analysis of the prechallenge serum samples from these volunteers disclosed that the titer of antibody to Norwalk virus was not related to susceptibility to illness. In another study, the prechallenge serum and jejunal RIA blocking antibody titer to Norwalk virus was not directly correlated with susceptibility to illness (Table 1). Instead, there was an inverse correlation demonstrating that those volunteers with high levels of both circulating and/or local antibody were more likely to develop Norwalk illness when challenged.[74] The explanation for this inverse relationship is not clear. It does seem likely, however, that humoral immune mechanisms (local or systemic) are not the primary determinants of protection from Norwalk illness, at least in a volunteer setting. In a study of the immune response of young teenagers to a natural Norwalk epidemic, the pre-existing Norwalk antibody titer did not correlate with susceptibility. However, those teenagers who became ill did develop higher antibody titer than those children who were asymptomatically infected.[84]

Studies of jejunal biopsies from volunteers acutely infected with either Norwalk virus or Hawaii virus failed to demonstrate detectable levels of interferon.[133]

The development of *in vitro* neutralization tests for the SRGVs will be a great aid to further studies on clinical immunity to these agents.

B. DIAGNOSIS

Widely available rapid diagnostic assays are not yet available for detection of any of the small round enteric pathogens. Because a highly permissive cell line has not been identified, detection strategies have relied on immunologic assays and serologic surveys. Several sensitive and specific solid-phase immunoassays have been developed for Norwalk virus, Snow Mountain virus, Marin County agent, and enteric calicivirus, but the gold standard for viral detection is still IEM. This technique has lead to the discovery of virtually all the known 27-nm agents and has been widely used as a serologic assay to study immune responses during infection.[49,134,135]

The technique of IEM is based on the actual direct observation of specific antigen-antibody complexes in the electron microscope. Unfortunately, the small round gastroenteritis viruses are shed in small amounts and their morphology is not sufficiently distinct to differentiate them from other small objects in the feces. The addition of detecting antibody to the fecal sample enables the electron microscopist to specifically identify and highlight particular viruses. In addition, if the assay is carried out using pooled human gamma globulin, the test is relatively open ended, identifying a wide variety of serologically distinct agents. Specificity can be obtained by using well-characterized, acute and convalescent infection sera from specific outbreaks. The technique of immune electron microscopy has been well described in the past.[49] IEM has several major drawbacks, however. Expensive and complex equipment is required and the microscopist must be well trained and very patient. The technique is not readily ammendable to testing large numbers of samples and requires considerable amounts of reagents.

In order to circumvent many of these drawbacks, a variety of solid phase immunoassays have recently been developed for SRGV detection.[60-66,140,142] These assays have been demonstrated to be useful for virus detection as well as measuring serum responses in infected individuals. In most cases, but not all, it would appear that the detection of serum responses in affected individuals is the most sensitive method for documenting SRGV infection.

The solid phase immunoassays that have been described to date have utilized human infection sera as immunologic reagents. The specificity of these assays has been based on the differential binding of virus or viral proteins to preinfection and convalescent sera. Both RIAs and ELISAs have been developed and both appear to be at least as sensitive and specific as IEM. The ELISAs have the advantage of not requiring radioactive material and having reagents with a longer shelf life. Because all the assays developed to date utilize human infection sera, standardization from laboratory to laboratory has been difficult. The recent isolations of monoclonal antibodies directed against astroviruses,[140] the Snow Mountain agent,[145] and the Marin County virus[148] offer promise that standardized detection reagents will be generally available in the near future. Until the genome of the 27- to 30-nm viruses is isolated and characterized, it is highly unlikely that nucleic acid probe technology will be utilized for SRGV detection.

C. TREATMENT AND PREVENTION

The illness caused by the SRGVs appears to be self-limited and relatively mild. In most instances, no specific therapeutic intervention is indicated. In rare instances, short courses of oral rehydration or intravenous rehydration are required. Bismuth subsalicylate has been shown to have a small but statistically significant effect on the duration of abdominal cramps, but no effect on the amount of diarrhea.[104]

In light of the difficulty in demonstrating long-term immunity to Norwalk infection in a volunteer setting, it seems unlikely that an effective vaccine strategy will be developed in the near future. Since Norwalk and presently most of the other SRGVs are transmitted by the fecal-oral route, classic public health and hygiene interventions are likely to reduce the spread of these agents.

REFERENCES

1. **Reiman, H. A., Price, A. H., and Hodges, J. A.,** The cause of epidemic diarrhea, nausea and vomiting (viral dysentery?), *Proc. Soc. Exp. Biol. Med.*, 59, 8, 1945.
2. **Gordon, I., Ingraham, H. S., and Korns, R. F.,** Transmission of epidemic gastroenteritis to human volunteers by oral administration of fecal filtrates, *J. Exp. Med.*, 86, 409, 1947.
3. **Gordon, I., Ingraham, H. S., Korns, R. F., and Trussell, R. E.,** Gastroenteritis in man due to a filtrable agent, *N.Y. State J. Med.*, 49, 1918, 1949.
4. **Gordon, I., Meneely, J. K., Jr., Currie, G. D., and Chicoine, A.,** Clinical laboratory studies in experimentally-induced epidemic nonbacterial gastroenteritis, *J. Lab. Clin. Med.*, 41, 133, 1953.

5. **Gordon, I., Patterson, P. R., and Whitney, E.,** Immunity in volunteers recovered from nonbacterial gastroenteritis, *J. Clin. Invest.,* 35, 200, 1956.
6. **Kojima, S., Fukumi, H., Kusama, H., Yamamoto, S., Suzuki, S., Uchida, T., Ishimarau, T., Oka, T., Kuretani, K., Ohmur, K., Nishikawa, F., Fujimoto, J., Fujita, K., Nakano, A., and Sunakawa, S.,** Studies on the causative agent of the infectious diarrhea: records of the experiments on human volunteers, *Jpn. Med. J.,* 1, 467, 1948.
7. **Yamamoto, A., Zennyogi, H., Yanagita, K., and Kato, S.,** Research into the causative agent of epidemic gastroenteritis which prevailed in Japan in 1948, *Jpn. Med. J.,* 1, 379, 1948.
8. **Fukumi, H., Nakaya, R., Hatta, S., Noriki, H., Yunoki, H., Akagi, K., Saito, T., Uchiyama, K., Kobari, K., and Nakanishi, R.,** An indication as to identity between the infectious diarrhea in Japan and the afebrile infectious nonbacterial gastroenteritis by human volunteer experiments, *Jpn. J. Med. Sci. Biol.,* 10, 1, 1957.
9. **Jordan, W. S., Gordon, I., and Dorrance, W. R.,** A study of illness in a group of Cleveland families. VII. Transmission of acute nonbacterial gastroenteritis to volunteers: evidence for two different etiologic agents, *J. Exp. Med.,* 98, 461, 1973.
10. **Adler, I. and Zickl, R.,** Winter vomiting disease, *J. Infect. Dis.,* 119, 668, 1969.
11. **Kapikian, A. Z., Wyatt, R. G., Dolin, R., Thornhill, T. S., Kalica, A. R., and Chanock, R. M.,** Visualization by immune electron microscopy of a 27 nm particle associated with acute infectious nonbacterial gastroenteritis, *J. Virol.,* 10, 1075, 1972.
12. **Appleton, H., Buckley, M., Thom, B. T., Cotton, J. L., and Henderson, S.,** Virus-like particles in winter vomiting disease, *Lancet,* 1, 409, 1977.
13. **Appleton, H. and Pereira, M. S.,** A possible virus etiology in outbreaks of food-poisoning from cockles, *Lancet,* 1, 780, 1977.
14. **Ashley, C. R., Caul, E. O., and Paver, W. K.,** Astrovirus-associated gastroenteritis in children, *J. Clin. Pathol.,* 31, 939, 1978.
15. **Cameron, D. J. S., Bishop, R. F., Veenstra, A. A., Barnes, G. L., Holmes, I. H., and Ruck, B. J.,** Pattern of shedding of two noncultivatable viruses in stools of newborn babies, *J. Med. Virol.,* 2, 7, 1978.
16. **Chiba, S., Sakuma Y., Kogasaka, R., Akihara, M., Horino, K., Nakao, T., and Fukui,** S., An outbreak of gastroenteritis associated with calicivirus in an infant home, *J. Med. Virol.,* 4, 249, 1979.
17. **Christopher, P. J., Grohmann, G. S., Millsom, R. H., and Murphy, A. M.,** Parvovirus gastroenteritis—a new entity for Australia, *Med. J. Aust.,* 1, 121, 1978.
18. **Clarke, S. K. R., Cook, G. T. Egglestone, S. I., Hall, T. S., Miller, D. L., Reed,** S. E., **Rubenstein, D., Smith, A. J., and Tyrrell, D. A. J.,** A virus from epidemic vomiting disease, *Br. Med. J.,* 3, 86, 1972.
19. **Dolin, R., Reichman, R. C., Roessner, K. D., Tralka, T. S., Schooley, R. T., Gary, W., and Morens, D.,** Detection immune electron microscopy of the Snow Mountain agent of acute viral gastroenteritis, *J. Infect. Dis.,* 146, 184, 1982.
20. **Eggleston, S. I.,** Norwalk-like particles in epidemic gastroenteritis in the UK, *Lancet,* 1, 147, 1980.
21. **Flewet, T. H. and Davies, H.,** Caliciviruses in man, *Lancet,* 1, 311, 1976.
22. **Gill, O. N., Cubitt, W. D., McSwiggan, D. A., Watney, B. M., and Bartlett, C. L. R.,** Epidemic of gastroenteritis caused by oysters contaminated with small round structured viruses, *Br. Med. J.,* 287, 1532, 1983.
23. **Kapikian, A. Z. and Chanock, R. M.,** Norwalk group of viruses, in *Virology,* Fields, R. N., et al., Raven Press, New York, 1985, 1495.
24. **Kjeldsberg, E.,** Small spherical viruses in faeces from gastroenteritis patients, *Acta Pathol. Microbiol. Scand. B,* 85, 351, 1977.
25. **Kogasaka, R., Nakamura, S., Chiba, S., Sakuma, Y., Terashima, H., Yokahama, T., and Nakao, T.,** The 33 to 39 nm virus-like particles, tentatively designated as Sapporo agent, associated with an outbreak of acute gastroenteritis, *J. Med. Virol.,* 8, 187, 1981.
26. **Kogasaka, R., Sakuma, Y., Chiba, S., Akihara, M., Horino, K., and Nakao, T.,** Small round virus-like particles associated with acute gastroenteritis in Japanese children, *J. Med. Virol.,* 5, 151, 1980.
27. **Kurtz, J. B., Lee, T. W., and Pickering, D.,** Astrovirus associated gastroenteritis in a children's ward, *Clin. Pathol.,* 30, 948, 1977.
28. **Madeley, C. R. and Cosgrove, B. P.,** Viruses in infantile gastroenteritis, *Lancet,* 2, 124, 1975.
29. **Madeley, C. R. and Cosgrove, B. P.,** Caliciviruses in man, *Lancet,* 1, 199, 1976.
30. **McSwiggan, D. A., Cubitt, D., and Moore, W.,** Calicivirus associated with winter vomiting disease, *Lancet,* 1, 1215, 1978.
31. **Middleton, P. J. and Szymanski, M. T.,** Viruses associated with acute gastroenteritis in young children, *Am. J. Dis. Child.,* 131, 733, 1977.
32. **Morens, D. M., Zweighaft, R. M., Vernon, T. M., Gary, G. W., Eslien, J. J., Wood, B. T., Holman, R. C., and Dolin, R.,** A waterborne outbreak of gastroenteritis with secondary person-to-person spread: association with a virus agent, *Lancet, 1,* 964, 1979.

33. **Oshiro, L. S., Haley, C. E., Roberto, R. R., Riggs, J. L., Croughan, M., Greenberg, H. B., and Kapikian, A. Z.,** A 27 virus from an outbreak of acute infectious nonbacterial gastroenteritis in a convalescent home: a possible new serotype, *J. Infect. Dis.*, 143, 791, 1981.
34. **Paver, W. K., Caul, E. O., Ashley, C. R., and Clarke, S. K. R.,** A small virus in human faeces, *Lancet*, 1, 237, 1973.
35. **Paver, W. K., Caul, E. O., and Clarke, S. K. R.,** Parvovirus-like particles in human faeces, *Lancet*, 1, 691, 1975.
36. **Paver, W. K. and Clarke, S. K. R.,** Comparison of human fecal and serum parvo-like viruses, *J. Clin. Microbiol.*, 4, 67, 1976.
37. **Spratt, H. C., Marks, M. I., Gomersall, M., Gill, P., and Pai, C. H.,** Nosocomial infantile gastroenteritis associated with minirotavirus and calicivirus, *J. Pediatr.*, 93, 922, 1978.
38. **Bock, G. and Whelan, J., Eds.,** Novel Diarrhoea Viruses, Ciba Symp. 128, John Wiley & Sons, New York, 1987, 92.
39. **Taniguchi, K., Urasawa, S., and Urasawa, T.,** Virus-like particles 35 to 40 nm, associated with an institutional outbreak of acute gastroenteritis in adults, *J. Clin. Microbiol.*, 10, 730, 1979.
40. **Thornhill, T. S. Wyatt, R. G., Kalica, A. R., Dolin, R., Chanock, R. M., and Kapikian, A. Z.,** Detection by immune electron microscopy of 26—27 nm virus-like particles associated with two family outbreaks of gastroenteritis, *J. Infect. Dis.*, 135, 20, 1977.
41. **Caul, E. O. and Appleton, H.,** The electron microscopical and physical characteristics of small round human fecal viruses: an interim scheme for classification, *J. Med. Virol.*, 9, 257, 1982.
42. **Madeley, C. R.,** Comparison of the features of astroviruses and caliciviruses seen in samples of feces by electron microscopy, *J. Infect. Dis.*, 139, 519, 1979.
43. **Herring, A. J., Gray, E. W., and Snodgrass, D. R.,** Purification and characterization of ovine astrovirus, *J. Gen. Virol.*, 53, 47, 1981.
44. **Schaffer, F. L.,** Caliciviruses, in *Comprehensive Virology*, Vol. 14, Fraenkel-Conrat, and Wagner, R. R., Plenum Press, New York, 249.
45. **Konno, T., Suzuki, H., Ishida, N., Chiba, R., Mochizuki, K., and Tsunoda, A.,** Astrovirus-associated epidemic gastroenteritis in Japan, *J. Med. Virol.*, 9, 11, 1982.
46. **Kurtz, J. B., Lee, T. W., Craig, J. W., and Reed, S. E.,** Astrovirus infection in volunteers, *J. Med. Virol.*, 3, 221, 1979.
47. **Midthun, K., Walsh, M., Kapikian, A., Wyatt, R., and Greenberg, H. B.,** Transmission and initial characterization of the Marin County agent, in Abstr. of the Annu. Meet. Am. Soc. Microbiology, Abstr. C47, 1983, 319.
48. **Grohmann, G. S.,** Role of non-rotaviral agents in viral diarrhoeas, Virology Dept., Inst. of Clin. Pathol. and Med. Res., Westmead Hospital, Sydney, Australia, 1984.
49. **Kapikian, A. Z., Feinstone, S. M., Purcell, R. H., Wyatt, R. G., Thornhill, T. S., Kalica, A. R., and Chanock, R. M.,** Detection and identification by immune electronmicroscopy of fastidious agents associated with respiratory illness, acute nonbacterial gastroenteritis, and hepatitis A, *Perspect. Virol.*, 9, 9-47, 1975.
50. **Caul, E. O., Ashley, C., and Pether, J. V. S.,** "Norwalk" like particles in epidemic gastroenteritis in the UK, *Lancet*, 2, 1292, 1979.
51. **Kapikian, A. Z., Gerin, J. L., Wyatt, R. G., Thornhill, T. S., and Chanock, R. M.,** Density of cesium chloride of the 27 nm "8Flla" particle associated with acute infectious nonbacterial gastroenteritis: determination by ultracentrifugation and immune electron microscopy, *Proc. Soc. Exp. Biol. Med.*, 142, 874, 1974.
52. **Paver, W. K., Caul, E. O., and Clarke, S. K. R.,** Comparison of a 22 nm virus from human faeces with animal parvoviruses, *J. Gen. Virol.*, 22, 447, 1974.
53. **Greenberg, H. B., Valdesuso, J., Kalica, A. R., Wyatt, R. G., McAuliffe, V. J., Kapikian, A. Z., and Chanock, R. M.,** Proteins of Norwalk virus, *J. Virol.*, 37, 994, 1981.
54. **Madore, H. P., Treanor, J. J., and Dolin, R.,** Characterization of the Snow Mountain agent of viral gastroenteritis, *J. Virol.*, 58, 487, 1986.
55. **Snodgrass, D. R., Angus, K. W., Gray, E. W., Menzeis, J. D., and Paul, G.,** Pathogenesis of diarrhea caused by astrovirus infections in lambs, *Arch. Virol.*, 60, 217, 1979.
56. **Snodgrass, D. R. and Gray, E. W.,** Detection and transmission of 30 nm virus particles (astroviruses in faeces of lambs with diarrhea, *Arch. Virol.*, 55, 287, 1977.
57. **Terashima, H., Chiba, S., Sakuma, Y., Kogasaka, R., Nakata, S., Minami, R., Horino, K., and Nakao, T.,** The polypeptide of a human calcivirus, *Arch. Virol.*, 78, 1, 1985.
58. **Hoshino, Y., Zimmer, J. F., Moise, N. S., and Scott, F. W.,** Detection of astroviruses in feces of a cat with diarrhea, *Arch. Virol.*, 70, 373, 1981.
59. **Woode, G. N. and Bridger, J. C.,** Isolation of small viruses resembling astroviruses and caliciviruses from acute enteritis of calves, *J. Med. Microbiol.*, 11, 441, 1978.
59a. **Williams, F. P., Jr.,** Astrovirus-like, coronavirus-like, and parvovirus-like particles detected in the diarrheal stools of beagle pups, *Arch. Virol.*, 66, 215, 1980.

59b. **Marshall, J. A., Healey, D. S., Studdert, M. J., Scott, P. C., Kennett, M. L., Ward, B. K., and Gust, I. D.,** Viruses and virus-like particles in the faeces of dogs with and without diarrhoea, *Aust. Vet. J.*, 61, 1984.
60. **Greenberg, H. B., Wyatt, R. G., Valdesuso, J., Kalica, A. R., London, W. T., Chanock, R. M., and Kapikian, A. Z.,** Solid-phase microtiter radioimmunoassay for detecion of the Norwalk strain of acute nonbacterial epidemic gastroenteritis and its antibodies, *J. Med. Virol.*, 2, 97, 1978.
61. **Dolin, R., Roessner, K. D., Treanor, J. J., Reichman, R. C., Phillips, M., and Madore, H. P.,** Radioimmunoassay for detection of the Snow Mountain agent of viral gastroenteritis, *J. Med. Virol.*, 19, 11, 1986.
62. **Madore, H. P., Treanor, J. J., Pray, K. A., and Dolin, R.,** Enzyme-linked immunosorbent assays for Snow Mountain and Norwalk agents of viral gastroenteritis, *J. Clin. Microbiol.*, 24, 456, 1986.
63. **Gary, G. W., Jr., Kaplan, J. E., Stine, S. E., and Anderson, L. J.,** Detection of Norwalk virus antibodies and antigen with a biotin-avidin immunoassay, *J. Clin. Microbiol.*, 22, 274, 1985.
64. **Herrmann, J. E., Nowak, N. A., Blacklow, N. R.,** Detection of Norwalk virus in stools by enzyme immunoassay, *J. Med. Virol.*, 17, 127, 1985.
65. **Nakata, S., Chiba, S., Terashima, H., Sakuma, Y., Kogasaka, R., and Nakao, T.,** Microtiter solid-phase radioimmunoassay for detection of human calicivirus in stools, *J. Clin. Microbiol.*, 17, 198, 1983.
66. **Greenberg, H. B. and Kapikian, A. Z.,** Detection of Norwalk agent antibody and antigen by solid-phase radioimmunoassay and immune adherence hemagglutination assay, *J. Am. Vet. Med. Assoc.*, 173, 620, 1978.
67. **Lee, T. W. and Kurtz, J. B.,** Astrovirus detected by immunofluorescence, *Lancet*, 2, 406, 1977.
68. **Madeley, C. R., Cosgrove, B. P., Bell, E. J., and Fallon, R. J.,** Stool viruses in babies in Glasgow. I. Hospital administrations with diarrhea, *J. Hyg. (Camb.)*, 78, 261, 1977.
69. **Nakata, S., Chiba, S., Terashima, H., Yokoyama, T., and Nakao, T.,** Humoral immunity in infants with gastroenteritis caused by human calicivirus, *J. Infect. Dis.*, 152, 274, 1985.
70. **Cubitt, W. D. and McSwiggan, D. A.,** Calicivirus gastroenteritis in North West London, *Lancet*, 2, 975, 1981.
71. **Kurtz, J. B. and Lee, T. W.,** Human astrovirus serotypes, *Lancet*, 2, 1405, 1984.
72. **Greenberg, H. B., Valdesuso, J., Yolken, R. H., Gangarosa, E., Gary, W., Wyatt, R., Konno, T., Suzuki, H., Chanock, R. M., and Kapikian, A. Z.,** Role of Norwalk virus in outbreaks of nonbacterial gastroenteritis, *J. Infect. Dis.*, 139, 564, 1979.
73. **Greenberg, H. B., Valdesuso, J., Kapikian, A. Z., Chanock, R. M., Wyatt, R. G., Szmuness, W., Larrick, J., Kaplan, J., Gilm, R. H., and Sack, D. A.,** prevalence of antibody to the Norwalk virus in various countries, *Infect. Immun.*, 26, 270, 1979.
74. **Greenberg, H. B., Wyatt, R. G., Kalica, A. R., Yolken, R. H., Black, R., Kapikian, A. Z., and Chanock, R. M.,** New insights in viral gastroenteritis, *Perspect. Virol.*, 11, 163, 1981.
75. **Cukor, G., Blacklow, N. R., Echeverria, P., Bedigian, M. K., Puruggan, H., and Basaca-Sevilla, V.,** Comparative study of the acquisition of antibody to Norwalk virus in pediatric populations, *Infect. Immun.*, 29, 822, 1980.
76. **Blacklow, N. R., Cukor, G., Bedigian, M. K., Echeverria, P., Greenberg, H. B., Schreiber, D. S., and Trier, J. S.,** Immune response and prevalence of antibody of Norwalk enteritis virus as determined by radioimmunoassay, *J. Clin. Microbiol.*, 10, 903, 1979.
77. **Sakuma, Y., Chiba, S., Kogasaka, R., Terashima, H., Nakamura, S., Horino, K., and Nakao, T.,** Prevalence antibody to human calicivirus in general population of northern Japan, *J. Med. Virol.*, 7, 221, 1981.
77a. **Cubitt, W. D. and McSwiggen, D. A.,** Seroepidemiological survey of the prevalence of antibodies to a strain of human calicivirus, *J. Med. Virol.*, 21, 361, 1987.
78. **Kurtz, J. B. and Lee, T. W.,** Astrovirus gastroenteritis: age distribution of antibody, *Med. Microbiol. Immunol.*, 166, 227, 1978.
79. **Brondum, J., Spitalny, K. C., Vogt, R. L., Godlewski, K., Madore, H. P., and Dolin, R.,** An outbreak of gastroenteritis in Vermont attributed to Snow Mountain agent, *J. Infect. Dis.*, 152, 834, 1985.
80. **Cubitt, W. D., Pead, P. J., and Saeed, A. A.,** A new serotype of calicivirus associated with an outbreak of gastroenteritis in a residential home for the elderly, *J. Clin. Pathol.*, 34, 924, 1981.
81. **Cubitt, W. D., McSwiggan, D. A., and Moore, W.,** Winter vomiting disease caused by calicivirus, *J. Clin. Pathol.*, 32, 786, 1979.
82. **Kapikian, A. Z., Greenberg, H. B., Wyatt, R. G., Kalica, A. R., Kim, H. W., Brandt, C. D., Rodriguez, W. J., Parrott, R. H., and Chanock, R. M.,** Viral gastroenteritis, in *Viral Infections of Humans. Epidemiology and Control*, 2n, Evans, A. S., Ed., Plenum Press, New York, 282.
83. **Kaplan, J. E., Gary, G. W., Baron, R. C., Singh, N., Schonberger, L. B., Feldman, R., and Greenberg, H. B.,** Epidemiology of Norwalk gastroenteritis and the role of Norwalk virus in outbreaks of acute nonbacterial gastroenteritis, *Ann. Intern. Med.*, 96, 756, 1982.
84. **Baron, R. C., Greenberg, H. B., Cukor, G., and Blacklow, N. R.,** Serological responses among teenagers following natural exposure to Norwalk virus, *J. Infect. Dis.*, 150, 531, 1984.
85. **Baron, R. C., Murphy, F. D., Greenberg, H. B., Davis, C. E., Bregman, D. J., Gary, G. W., Hughes, J. M., and Schonberg, L. B.,** Norwalk gastrointestinal illness: an outbreak associated with swimming in a recreational lake and secondary person-to-person transmission, *Am. J. Epidemiol.*, 115, 163, 1982.

86. **Goodman, R. A., Buehler, J. W., Greenberg, H. B., McKinley, T. W., and Smith, J. D.,** Norwalk gastroenteritis associated with a water system in a rural Georgia community, *Arch. Environ. Health,* 37, 358, 1982.
87. **Gunn, R. A., Terranova, L. A., Greenberg, H. B., Yashuk, J., Gary, G. W., Wells, J. G., Taylor, P. R., and Feldman, R. A.,** Norwalk virus gastroenteritis aboard a cruise ship: an outbreak on five consecutive cruises, *Am. J. Epidemiol.,* 112, 820, 1980.
88. **Kaplan, J. E., Goodman, R. A., Schonberger, L. B., Lippy, E. C., and Gary, G. W.,** Gastroenteritis due to Norwalk virus: an outbreak associated with a municipal water system, *J. Infect. Dis.,* 146, 190, 1982.
89. **Kappus, K. D., Marks, J. S., Holman, R. C., Bryant, J. K., Baker, C., Gary, G. W., and Greenberg, H. B.,** An outbreak of Norwalk gastroenteritis associated with swimming in a pool and secondary person-to-person transmission, *Am. J. Epidemiol.,* 116, 834, 1982.
90. **Koopman, J. S., Eckert, E. A., Greenberg, H. B., Strohm, B. C., Isaacson, R. E., and Monto, A. S.,** Norwalk virus enteric illness acquired by swimming exposure, *Am. J. Epidemiol.,* 15, 173, 1982.
91. **Taylor, J. W., Gary, G. W., and Greenberg, H. B.,** Norwalk related viral gastroenteritis due to contaminated drinking water, *Am. J. Epidemiol.,* 114, 584, 981.
92. **Griffin, M. R., Surowiec, J. J., McCloskey, D. I., Capuano, B., Pierzynski, B., Quinn, M., Wojnarski, R., Parkin, W. E., Greenberg, H. B., and Gary, G. W.,** Foodborne Norwalk virus, *Am. J. Epidemiol.,* 15, 178, 1982.
93. **Kuritsky, J. N., Osterholm, M. T., Greenberg, H. B., Korlath, J. A., Godes, J. R., Hedberg, C. W., Forfang, J. C. Kapikian, A. Z., McCullough, J. C., and White, K. E.,** Norwalk gastroenteritis: a community outbreak associated with bakery product consumption, *Ann. Intern. Med.,* 100, 519, 1984.
94. **White, K. E., Osterholm, M. T., Mariotti, J. A., Korlath, J. A., Lawrence, D. H., Ristinen, T. L., and Greenberg, H. B.,** A foodborne outbreak of Norwalk virus gastroenteritis: evidence for post-recovery transmission, *Am. Epidemiol.,* 124, 120, 1985.
95. **Grohmann, G. S., Murphy, A. M., Christopher, P. J., Auty, E., and Greenberg, H. B.,** Norwalk virus gastroenteritis in volunteers consuming depurated oysters, *Aust. J. Exp. Biol., Med. Sci.,* 59, 219, 1981.
96. **Gunn, R. A., Janowski, H. T., Lieb, S., Prather, E. C., and Greenberg, H. B.,** Norwalk virus gastroenteritis following raw oyster consumption, *Am. J. Epidemiol.,* 115, 348, 1982.
97. **Murphy, A. M., Grohmann, G. S., Christopher, P. J., Lopez, W. A., Davey, G. R., and Millson, R. H.,** Norwalk virus gastroenteritis following raw oyster consumption, *Am. J. Epidemiol.,* 115, 348, 1982.
98. **Morse, D. L., Guzewich, J. J., Hanrahan, J. P., Stricof, R., Shayegani, M., Deibel, R., Grabau, J. C., Nowak, N. A., Herrman, J. E., Cukor, G., and Blacklow, N. R.,** Widespread outbreaks of clam- and oyster-associated gastroenteritis: role of Norwalk virus, *N. Engl. J. Med.,* 314, 678, 1986.
99. **Oliver, B., Ng, S., Marshall, J., Greenberg, H., Gust, I. D., Creswell, V., Ward, B., Kennett, M., and Birch, C.,** Prolonged outbreak of Norwalk gastroenteritis in an isolated guest house, *Med. J. Aust.,* 142, 391, 1985.
100. **Kaplan, J. E., Feldman, R., Campbell, D. S., Lookabaugh, C., and Gary, G. W.,** The frequency of a Norwalk-like pattern of illness in outbreaks of acute gastroenteritis, *Am. J. Public Health,* 72, 1329, 1982.
101. **Dolin, R., Blacklow, N. R., DuPont, H., Formal, S., Buscho, R. J., Wyatt, R. G., Kasal, J. A., Chames, R. P., Hornick, R., and Chanock, R. M.,** Transmission of acute infectious nonbacterial gastroenteritis to volunteers by oral administration of stool filtrates, *J. Infect. Dis.,* 123, 307, 1971.
102. **Blacklow, N. R., Dolin, R., Federson, D. S., DuPont, H., Northrup, R. S., Hornick, R. B., and Chanock, R. M.,** Acute infectious nonbacterial gastroenteritis: etiology and pathogenesis: a combined clinical staff conference at the Clinical Center of the National Institutes of Health, *Ann. Intern. Med.,* 76, 993, 1972.
103. **Dolin, R., Blacklow, N. R., DuPont, H., Buscho, R. J., Wyatt, R. G. Kasal, J. A., Hornick, R., and Chanock, R. M.,** Biological properties of Norwalk agent of acute infectious nonbacterial gastroenteritis, *Proc. Soc. Exp. Biol. Med.,* 140, 578, 1972.
104. **Steinhoff, M. C., Douglas, R. G., Jr., Greenberg, H. B., and Callahan, D. R.,** Bismuth subsalicylate therapy of viral gastroenteritis, *Gastroenterology,* 78, 1495, 1980.
105. **Kapikian, A. Z., Greenberg, H. B., Cline, W. L., Kalica, A. R., Wyatt, R. G., James, H. D., Jr., Lloyd, N. L., Chanock, R. M., Ryde, R. W., and Kim, H. W.,** Prevalence of antibody to the Norwalk agent by a newly developed immune adherence hemagglutination assay, *J. Med. Virol.,* 2, 281, 1978.
106. **Black, R. E., Greenberg, H. B., Kapikian, A. Z., Brown, K. H., and Becker, S.,** Acquisition of serum antibody to Norwalk virus and rotavirus and relation to diarrhea in a longitudinal study of young children in rural Bangladesh, *J. Infect. Dis.,* 145, 483, 1982.
107. **Pickering, L. K., DuPont, H. L., Blacklow, N. R., and Cukor, G.,** Diarrhea due to Norwalk virus in families, *Infect. Dis.,* 146, 116, 1982.
108. **Ryer, R., Singh, N., Reeves, W., Kapikian, A., Greenberg, H., and Sack, B.,** Evidence of immunity induced by naturally acquired rotavirus and Norwalk virus infection on two remote Panamanian islands, *J. Infect. Dis.,* 151, 99, 1985.

109. **Brandt, C. D., Kim, H. W., Rodriguez, W. J., Arrobio, J. D., Jeffries, B. C., Stallings, E. P., Lewis, C., Miles, A. J., Chanock, R. M., Kapikian, A. Z., and Parrott, R. H.,** Pediatric viral gastroenteritis during eight years of study, *J. Clin. Microbiol.,* 18, 71, 1983.
110. **Suzuki, H., Konno, T., Katsuzawa, T., Imai, A., Tazawa, F., Ishida, N., Katsushima, N., and Sakamoto, M.,** The occurrence of calicivirus in infants with acute gastroenteritis, *J. Med. Virol.,* 4, 321, 1979.
111. **Oishi, I., Yamazaki, K., Minekawa, Y., Nishimura, H., and Kitaura, T.,**Three-year survey of the epidemiology of rotavirus, enteric adenovirus, and some small spherical viruses including "Osaka-agent" associated with infantile diarrhea, *Biken J.,* 28, 9, 1985.
112. **Riepenhoff-Talty, M., Saif, L. J., Barrett, H. J., Suzuki, H., and Ogra, P. L.,** Potential spectrum of etiological agents of viral enteritis in hospitalized infants, *J. Clin. Microbiol.,* 17, 352, 1983.
113. **Cameron, D. J. S., Bishop, R. F., Veenstra, A. A., and Barnes, G. L.,** Noncultivatable viruses and neonatal diarrhea: fifteen-month survey in a newborn special care nursery, *J. Clin. Microbiol.,* 8, 93, 1978.
114. **Narang, H. K. and Codd, A. A.,** Frequency of preclumped virus in routine fecal specimens from patients with acute nonbacterial gastroenteritis, *J. Clin. Microbiol.,* 13, 982, 1981.
115. **Agus, S. G., Dolin, R., Wyatt, R. G., Tousimus, A. J., and Northrup, R. S,** Acute infectious nonbacterial gastroenteritis: intestinal histopathology; histologic and enzymatic alterations during illness produced by the Norwalk agent in man, *Ann. Intern. Med.,* 79, 18, 1973.
116. **Dolin, R., Levy, A. G., Wyatt, R. G., Thornhill, T. S., and Gardner, J. D.,** Viral gastroenteritis induced by the Hawaii agent: jejunal histopathology and seroresponse, *Am. J. Med.,* 59, 761, 975.
117. **Schreiber, D. S., Blacklow, N. R., and Trier, J. S.,** The mucosal lesion of the proximal small intestine in acute infectious nonbacterial gastroenteritis, *N. Engl. J. Med.,* 228, 1318, 1973.
118. **Schreiber, D. S., Blacklow, N. R., and Trier, J. S.,** The small intestinal lesion induced by the Hawaii agent in infectious nonbacterial gastroenteritis, *J. Infect. Dis.,* 124, 705, 1974.
119. **Widerlite, L., Trier, J. S., Blacklow, N. R., and Schreiber, D. S.,** Structure of the gastric mucosa in acute infectious nonbacterial gastroenteritis, *Gastroenterology,* 68, 425, 1975.
120. **Saif, L. J., Bohl, E. H., Theil, K. W., Cross, R. F., and House, J. A.,** Rotavirus-like, calicivirus-like, and 23-nm virus-like particles associated with diarrhea in young pigs, *J. Clin. Microbiol.,* 12, 105, 1980.
121. **Woode, G. N. and Bridger, J. C.,** Isolation of small viruses resembling astroviruses and caliciviruses from acute enteritis of calves, *J. Med. Microbiol.,* 11, 441, 1978.
122. **Gray, E. W., Angus, K. W., and Snodgrass, D. R.,** Ultrastructure of the small intestine in astro-virus infected lambs, *J. Gen. Virol.,* 49, 71, 1980.
123. **Levy, A. G., Widerlite, L., Schwartz, P. R., Dolin, R., Blacklow, N. R., Garnder, J., Kimberg, D. V., and Trier, J. S.,** Jejunal adenylate cyclase activity in human subjects during viral gastroenteritis, *Gastroenterology,* 70, 321, 1976.
124. **Meeroff, J. C., Schreiber, D. S., Trier, J. S., and Blacklow, N. R.,** Abnormal gastric motor functions in viral gastroenteritis, *Ann. Intern. Med.,* 92, 370, 1980.
125. **Lee, T. W. and Kurtz, J. B.,** Serial propagation of astrovirus in tissue culture with the aid of tryspin, *J. Gen. Virol.,* 57, 421, 1981.
126. **Cubitt, W. D. and Barrett, A. D. T.,** Propagation of human candidate calicivirus in cell culture, *J. Gen. Virol.,* 65, 123, 1984.
127. **Wyatt, R. G., Greenberg, H. B., Dalgard, D. W., Allen, W. P., Sly, D. L., Thornhill, T. S., Chanock, R. M., and Kapikian, A. Z.,** Experimental infection of chimpanzees with the Norwalk agent of epidemic viral gastroenteritis, *J. Med. Virol.,* 2, 89, 1978.
128. **Kjeldsberg, E. and Mortensson-Egnund, K.,** Antibody response in rabbits following oral administraiton of human rota-, calici- and adenovirus, *Arch. Virol.,* 78, 97, 1983.
129. **Dolin, R., Reichman, R. C., and Fauci, A. S.,** Lymphocyte populations in acute viral gastroenteritis, *Infect. Immun.,* 14, 422, 1976.
130. **Nakata, S., Chiba, S., Terashima, H., and Nakao, T.,** Prevalence of antibody to human calicivirus in Japan and Southeast Asia determined by radioimmunoassay, *J. Clin. Microbiol.,* 22, 519, 1985.
131. **Wyatt, R. G., Dolin, R., Blacklow, N. R., DuPont, H. L., Buscho, R. F., Thornhill, T. S., Kapikian, A. Z., and Chanock, R. M.,** Comparison of three agents of acute infectious nonbacterial gastroenteritis by cross-challenge volunteers, *J. Infect. Dis.,* 129, 709, 1974.
132. **Parrino, T. A., Schreiber, D. S., Trier, J. S., Kapikian, A. Z., and Blacklow, N. R.,** Clinical immunity in acute gastroenteritis caused by the Norwalk agent, *N. Engl. J. Med.,* 297, 86, 1977.
133. **Dolin, R. and Baron, S.,** Absence of detectable interferon in jejunal biopsies, jejunal aspirates and sera in experimentally induced viral gastroenteritis in man, *Proc. Soc. Exp. Biol. Med.,* 150, 337, 1975.
134. **Kapikian, A. Z., Dienstag, J. L., and Purcell, R. H.,** Immune electron microscopy as a method for the detection, identification and characterization of agents not cultivatable in an in vitro system, in *Manual of Clinical Immunology,* 2nd ed., Rose, N. R. and Friedman, H., Eds., American Society for Microbiology, Washington, D. C., 1980, 70.

135. **Kapikian, A. Z., Yolken, R. H., Greenberg, H. B., Wyatt, R. G., Kalica, A. R., Chanock, R. M., and Kim, H. W.,** Gastroenteritis viruses, in *Diagnostic Procedures for Viral, Rickettsial, and Chlamydial Infections*, 5th ed., Lennette, E. H. and Schmidt, N. J., Eds., American Public Health Association, Washington, D. C., 1979, 927.
136. **Greenberg, H. B., Wyatt, R. G., and Kapikian, A. Z.,** Norwalk virus in vomitus, *Lancet*, 2, 55, 1979.
137. **Ryder, R. W., Oquist, C. a., Greenberg, H., Taylor, D. N., Orskov, F., orskov, I., Kapikian, A. Z., and Sack, R. B.,** Travelers' diarrhea in Panamanian tourists in Mexico, *J. Infect. Dis.*, 144, 442, 1981.
138. **Santosham, M., Sack, R. B., Froehlich, J., Greenberg, H., Yolken, R., Kapikian, A., Javier, C., Medina, C., Orskov, F., and Orskov, I.,** Biweekly prophylactic doxycyline for travelers' diarrhea, *J. Infect. Dis.*, 143, 598,
139. **Sack, R. B., Froelich, J. L., Zulich, A. W., Hidi, D. S., Kapikian, A. Z., Orskov, F., Orskov, I., and Greenberg, H. B.,** Prophylactic doxycycline for travelers' diarrhea, *Gastroenterology*, 76, 1368, 1979.
140. **Hermann, J. E., Hudson, R. W., Perron-Henry, D. M., Kurtz, J. B., and Blacklow, N. R.,** Antigenic characterization of cell-cultivated astrovirus serotypes and development of astrovirus-specific monoclonal antibodies, *J. Infect. Dis.*, 158, 182, 1988.
141. **Cubitt, W. D., Blacklow, N. R., Herrmann, J. E., Nowak, N. A., Nakata, S., and Chiba, S.,** Antigenic relationships between human calicivirus and Norwalk virus, *J. Infect. Dis.*, 156, 806, 1987.
142. **Nakata, S., Estes, M. K., and Chiba, S.,** Detection of human calicivirus antigen and antibody by enzyme-linked immunosorbent assays, *J. Clin. Microbiol.*, 26, 2001, 1988.
143. **Grohmann, G.,** Viral diarrhoea, in *Australian Infectious Diarrhoea in the Young*, Tzipori, S., E, Elsevier, Amsterdam, 1985, 25.
144. **Flynn, W. T. and Saif, L. J.,** Serial propagation of porcine enteric calicivirus-like virus in primary porcine kidney cell culture, *J. Clin. Microbiol.*, 26, 206, 1988.
145. **Treanor, J., Dolin, R., and Madore, H. P.,** Production of a monoclonal antibody against the snow mountain agent of gastroenteritis by in vitro immunization of murine spleen cells, *Proc. Natl. Acad. Sci. U.S.A.*, 85, 3613, 1988.
146. **Cunningham, A. L., Grohman, G. S., Harkness, J., Law, C., Marriott, D., Tindall, B., Cooper, D. A.,** Gastrointestinal viral infections in homosexual men who were symptomatic and seropositive for human immunodeficiency virus, *J. Infect. Dis.*, 158, 386, 1988.
147. **Grohmann, G.,** personal communication, 1987.
148. **Midthun, K.,** personal communication, 1987.
149. **Greenberg, H.,** unpublished results, 1987.
150. **Greenberg, H. and Kapikian,A.,** unpublished data, 1983.
151. **Kurtz, J. B. and Kapikian, A.,** unpublished data, 1987.
152. **Greenberg, H. and Midthun, K.,** unpublished observation, 1987.

Chapter 8

SMALL VIRUSES ASSOCIATED WITH GASTROENTERITIS IN ANIMALS

J. C. Bridger

TABLE OF CONTENTS

I. INTRODUCTION

Since the early 1970s, fecal samples from diarrheic animals have been examined by electron microscopy in many laboratories. This has revealed viruses with morphologies which were unknown previously. Rotaviruses are the best known example (and are reviewed in Chapters 3 and 4) but astroviruses are another which come into the remit of the present chapter. Astroviruses were named after identification in human feces in 1975,[1] but they have since been identified in the feces or enteric tracts of lambs,[2] calves,[3,4] pigs,[5,5a] cats,[6-8] dogs,[9-11] deer,[12] mice,[13] ducks,[14] and turkeys,[15,16] often in association with enteritis (Table 1.)

Electron microscopy also revealed viruses, such as caliciviruses, whose morphology was known, but which had not been identified previously in the enteric tract in association with diarrhea. Caliciviruses had been described in association with respiratory disease of cats (feline calicivirus) and vesicular lesions in pigs (vesicular exanthema virus) and sea lions (San Miguel sea lion virus). Electron microscopy revealed them in the feces of pigs,[5,22] chickens,[23] navel orangeworms,[24] and dogs,[19,25] often in association with diarrhea. Electron microscopy also revealed small viruses which were similar to caliciviruses but were without the characteristic morphology. These viruses have an amorphous surface structure and have been called calici-like viruses. They have been described in cattle[3,26-29,95] and humans (see Chapter 7).

Some members of a third group of small viruses, the parvoviruses, were known to cause enteritis in animals well before the application of electron microscopy to the examination of feces. Parvoviruses differ in their pathogenesis from the known pathogenesis of the enteric caliciviruses, calici-like viruses, and astroviruses, as they cause generalized, often severe, infection in addition to infection of the enteric tract. Feline parvovirus was shown to be the causal agent of feline enteritis and panleucopenia in 1965,[30] but more recently, in 1978, canine parvovirus emerged as a new enteric pathogen of dogs. Parvoviruses with an enteric involvement are known also in mink[31] and cattle.[32] They have been incriminated recently in the infectious stunting syndrome of chickens[33,34] and turkeys,[35] a syndrome which has an enteric involvement but which is manifested clinically also by poor growth and feathering.[36] They have also been identified recently in association with diarrhea in unweaned piglets.[37]

In addition to astroviruses, caliciviruses, calici-like viruses, and parvoviruses, there are a few reports of other small viruses associated with enteric disease. Entero-like viruses were described in association with infectious stunting of chickens.[38,39] Entero-like viruses were reported also in association with diarrhea and the runting and stunting syndrome of turkeys in Northern Ireland, France, and the U.S.[40-44]

Thus, this chapter covers viruses of several morphological types with diameters ranging from 20 to 40 nm (Table 1). At least two virus families are represented, the Parvoviridae and the Caliciviridae. The Parvoviridae have been extensively studied and thoroughly reviewed previously.[45-47] Further information on the enteric feline, canine, and bovine parvoviruses can be found in reviews by Povey,[21] McCandlish,[48] Afshar,[49] Pollock,[50] and Storz and Leary.[32] The nonenteric caliciviruses, which include the feline respiratory calicivirus, vesicular exanthema virus of swine, and San Miguel sea lion virus, have also been reviewed previously.[51-53] Classification of the astroviruses, calici-like viruses and entero-like viruses remains to be established. Kurtz and Lee[54] suggested that astroviruses were members of the Picornaviridae while Herring et al.[55] suggested that they were intermediate between the Picornaviridae and the Caliciviridae.

Some properties of parvoviruses, caliciviruses, and astroviruses are summarized in Table 2 Candidate enteric viruses will need to meet these requirements to be classified as one of these three groups. At present, the most useful property which allows them to be distinguished from one another is their morphology, but the high buoyant density of parvoviruses sets them apart from the other viruses. Caul and Appleton[58] proposed an interim classification scheme for the

TABLE 1
Small Viruses Associated with Gastroenteritis in Animals Included in the Present Chapter

Virus type	Species found in	Countries where recognized	Date of first report
Astroviruses[a]	Sheep	Britain	1977
	Cattle	Britain, U.S.	1978
	Pig	Britain, U.S.	1980
	Dog	U.S., Australia	1980
	Turkey	N.I., U.S.	1980
	Cat	U.S., Britain, Australia	1981
	Deer	Britain	1981
	Mice	Norway	1985
Caliciviruses[b]	Pig	Britain, U.S.	1980
	Chicken	Britain	1981
	Navel orange-worm (insect)	U.S.	1981
	Dog	U.S.	1985
Calici-like viruses	Cattle	Britain, U.S. Germany	1978
Parvoviruses[c]	Cattle	Widespread	1961
	Cat	Widespread	1965
	Mink	Widespread	1974
	Dog	Widespread	1978
	Turkeys	U.S.	1983
	Chickens	Hungary	1984
Entero-like viruses	Chickens	N.I., Belgium	1984
	Turkeys	France, N.I., U.S.	1979

Note: N.I. = Northern Ireland; U.S. = United States

[a] Astroviruses which cause hepatitis are known also in ducks.[17] Particles are found in the feces which may be loose and white.

[b] Caliciviruses have been identified also in feces of walruses[18] and the enteric tract of a coyote.[19]

[c] Parvoviruses with some enteric involvement are known also in geese.[20] The feline parvovirus infects nondomesticated members of the families Mustelidae, Procyonidae, and Viverridae.[21]

small enteric viruses of humans based on morphology, particle size, and buoyant density. This scheme can be applied equally well to the small enteric viruses of animals. Other properties, such as type of nucleic acid, site of replication, and polypeptide composition, should be useful for differentiation as the difficulties in obtaining adequate quantities of purified virus are solved.

II. MORPHOLOGY

The four morphological types of viruses associated with enteritis in man (see previous chapter) have been associated with enteritis in animals. They are caliciviruses, calici-like

TABLE 2
Properties of Parvoviruses, Astroviruses and Caliciviruses which Enable Them to Be Distinguished from One Another

	Parvoviruses[a]	Astroviruses[b]	Caliciviruses[c]
Virion			
Size (mm)	18—26	28	35—40
Morphology	Featureless	Characteristic	Characteristic
Density in CsCl (g/ml)	1.39—1.42	1.34—1.40[d]	1.36—1.39
S value	110—125	?	170—183
Genome			
Nucleic acid	DNA	RNA	RNA
Strandedness	Single	Single	Single
Molecular weight	$1.5—2.0 \times 10^6$	2.7×10^6	$2.6—2.8 \times 10^6$
Infectivity	Infectious	?	Infectious
Polypeptides			
Number	3—4	2—4	1
Size	50,000—100,000	33,000—37,000	60,000—71,000
Lipid	None	None	None
Replication	Nuclear	Cytoplasmic	Cytoplasmic

[a] Based on data from References 47 and 56.
[b] Based on data for lamb, dog, and human astroviruses.
[c] Based on data for nonenteric caliciviruses.[57]
[d] Precise determinations were hampered by virus aggregation: see text.

viruses, astroviruses, and parvoviruses. In addition, particles with the featureless morphology of enteroviruses have been associated with enteric disease in chickens and turkeys.[38-42] In contrast to the enteric viruses of man, animal parvoviruses have been established longest and most firmly as enteric pathogens.

Of the four morphological types, parvoviruses are the simplest visually. Virions are isometric, nonenveloped particles, 18 to 26 nm in diameter with an icosahedral symmetry.[47] They have no surface morphology (Figure 1a). Astroviruses are larger, approximately 28 nm diameter, and their periphery may appear smooth or slightly crenated (Figure 1b). A proportion of particles show a surface configuration of a five- or six-pointed star with a white center,[1,54] but, with ammonium molybdate as the negative stain, their surface morphology can be improved.[59] Bridging structures have been seen between adjacent particles.[2,6,13]

Enteric caliciviruses and calici-like viruses are more complex visually and particles are larger. Diameters fall within the range of 30 to 40 nm.[60] The surface of classical caliciviruses has dark spherical spots, about 10 nm in diameter, which are formed by deposits of negative stain filling cup-shaped surface depressions (hence the name calicivirus from the Latin "calyx" meaning cup). Particles can exhibit three distinctive surface patterns depending on their orientation. Four dark areas may form a cross, a central dark area may be surrounded by six dark areas, or a central dark area may be surrounded by five dark areas and the edges of the virion may appear to have ten evenly spaced projections[61] (Figure 2). The calicivirus from navel orangeworms was shown to lose its characteristic calicivirus morphology after treatment with alpha chymotrypsin.[24] Smaller, smooth-edged particles were produced similar in morphology to picornaviruses. The authors note that this could lead to the misdiagnosis of caliciviruses in feces.

The calici-like viruses, represented in animals by the Newbury agents of cattle,[3,26] fail to show the clear distinctive surface patterns of the caliciviruses. They have an indefinite pattern on their surface and a ragged outer edge (Figure 3). However, Bridger et al.[26] noted that some particles of one strain, SRV-1 showed ten spikes at their edges, the pattern produced by classical caliciviruses when viewed along their fivefold axis of symmetry. Also, the dark surface hollows of caliciviruses were visible, although the characteristic surface patterns of caliciviruses were

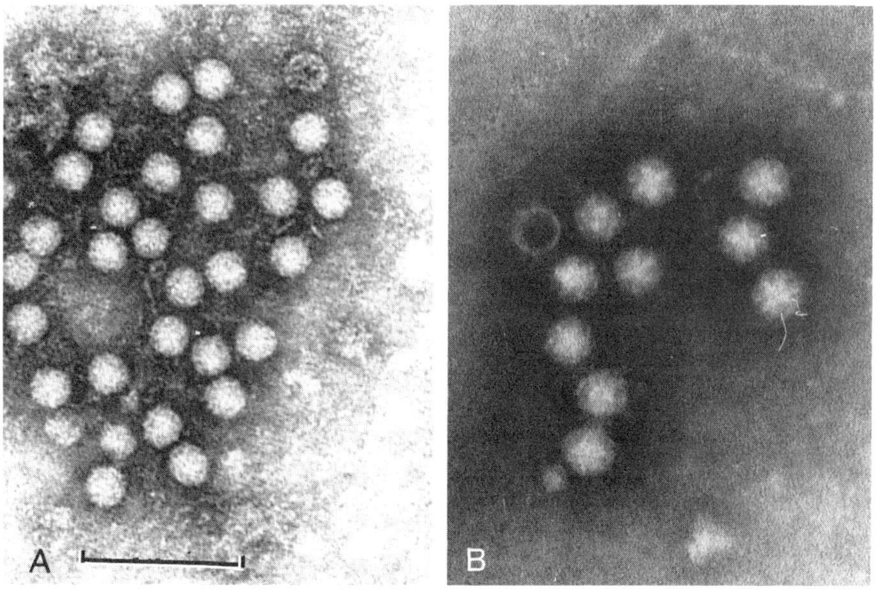

A B

FIGURE 1. (A) Particles of the feline parvovirus, which causes enteritis and panleucopenia, negatively stained with potassium phosphotungstate. (Courtesy of Dr. E. O. Caul.) (B) Particles of porcine astrovirus negatively stained with potassium phosphotungstate. Some particles show the characteristic star surface pattern. Bar indicates 100 nm. (From Bridger, J. C., *Vet. Rec.*, 107, 532, 1980. With permission.)

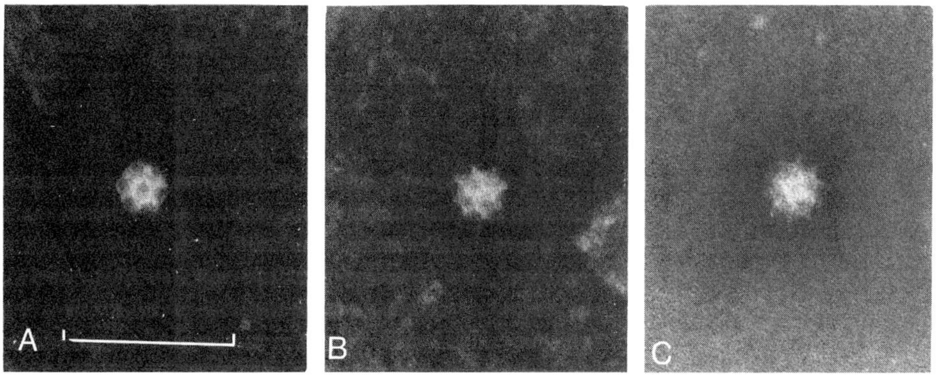

A B C

FIGURE 2. Particles of porcine enteric calicivirus in 3 orientations stained with potassium phosphotungstate. (A) The one-plus-six arrangement of dark surface hollows; (B) four surface hollows; (C) a particle with ten spikes at the periphery. Bar indicates 100 nm. (From Bridger, J.C., *Vet. Rec.*, 107, 532, 1980. With permission.)

not seen (Figure 3b). Whether the animal calici-like viruses belong to the Caliciviridae remains to be proven, but several facts suggest that this could be so. All strains of the established feline respiratory calicivirus and a human enteric calicivirus do not show the typical calicivirus morphology by negative stain electron microscopy and can appear indistinct or fuzzy.[60,62,63] Secondly, the surface patterns of a human calici-like virus were enhanced by the Markham rotation technique to reveal the distinctive patterns of classical caliciviruses.[60] Third, the

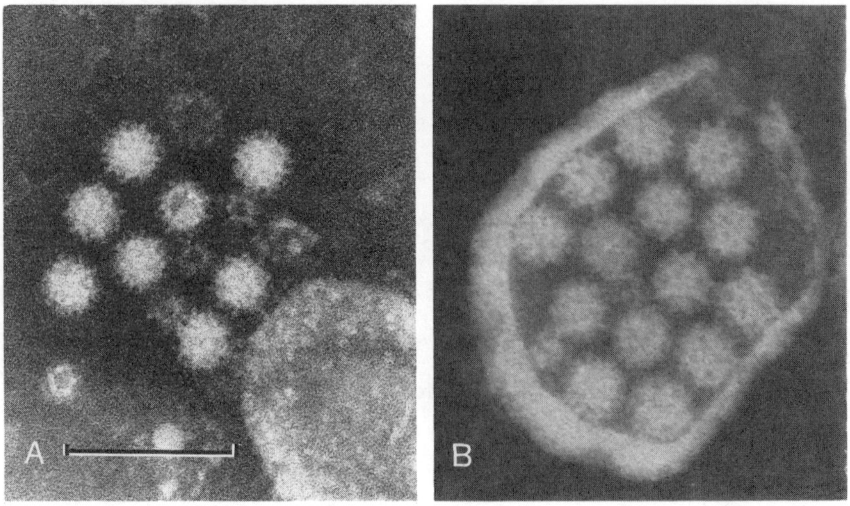

A B

FIGURE 3. Particles of bovine calici-like virus stained with potassium phosphotungstate. (A) Newbury agent SRV-2 with an indefinite surface pattern; (B) Newbury agent SRV-1 with a surface pattern suggestive of caliciviruses. Bar indicates 100 nm. (From Bridger, J. C. and Hall, G. A., *Les Colloques de l'INSERM*, 90, 233, 1979. With permission.)

polypeptide composition of two human calici-like viruses, the Norwalk and Snow Mountain viruses,[64,65] was similar to that of classical caliciviruses: only one major structural polypeptide with a molecular weight of about 60 kDa was demonstrated.

III. NUCLEIC ACID AND POLYPEPTIDE COMPOSITION

Both DNA and RNA genomes are represented by the small enteric viruses. Parvoviruses contain one piece of linear, single-stranded, 15 to 18S DNA with a molecular weight between 1.5 to 2.0×10^6 Da (Table 2).[47] Much information has been accrued on the structure and coding strategy of the genome and the viral life cycle of various members of the parvovirus genus, and the reader wishing more information is referred to Hauswirth[66] and Cotmore and Tattersall.[56]

Of the animal astroviruses, only the lamb astrovirus genome has been studied. The virus contains a 34S single-stranded RNA with an apparent molecular weight of 2.7×10^6 Da and a poly A tail.[55] The sedimentation behavior of the RNA is similar to that of caliciviruses and picornaviruses.

The chicken and canine caliciviruses have RNA genomes. The chicken calicivirus multiplied in the presence of actinomycin D and tritiated uridine was incorporated into the nucleic acid.[67] The nucleic acid of the canine calicivirus sedimented with an S value of 34 to 36.[25] The nucleic acid from the calicivirus of navel orangeworms was sensitive to RNAse, had a molecular weight of 2.5×10^6 Da and an S value of 36S.[24] These properties compare favorably with those of the nonenteric caliciviruses which are composed of one molecule of infectious, positive-sense, single-stranded RNA with a covalently linked protein. The RNA has a sedimentation coefficient of 36 to 38S and a molecular weight of 2.6 to 2.8×10^6 Da.[57]

Restriction endonuclease mapping of the genomes of the feline, mink, and canine parvovirus have helped to establish the origins of these parvoviruses which are closely related antigenically. The mink and feline parvoviruses are virtually identical. The canine and feline viruses differ in about 15% of their restriction sites, but one feline vaccine strain showed a closer resemblance to the canine parvovirus.

Available evidence indicates that the polypeptide composition of the small enteric viruses will differentiate them from one another (Table 2). Parvoviruses contain three or four polypeptides with high molecular weights, ranging from 50,000 to 100,000.[47,56] The ovine astrovirus contained two polypeptides of approximately 33,000 mol wt,[55] but a human astrovirus contained four polypeptides, although the molecular weights were similar to that of the ovine virus, 32,000 to 36,500,[54] Herring et al.[55] concluded that lamb astrovirus had a polypeptide structure intermediate between that of picornaviruses and calicicivurses, while Kurtz and Lee[54] concluded the polypeptide pattern found with the human virus suggested that they are members of the Picornaviridae.

The nonenteric caliciviruses have a characteristic polypeptide composition. They are composed of one major polypeptide of 60,000 to 71,000 mol wt.[57] Of the animal enteric caliciviruses and calici-like viruses, those from dogs and navel orangeworms have this characteristic structure. The others have not been studied. A single major polypeptide of 58,000 was detected in the dog virus[25] and a single polypeptide of 70,000 in the calicivirus from navel orangeworms.[24] Analyses of two human calici-like viruses, the Norwalk and Snow Mountain viruses, revealed that they had one major polypeptide of about 60 kDa.[64,65] When it becomes possible practically, the polypeptide composition of the candidate enteric caliciviruses and calici-like viruses should enable their classification to be resolved.

IV. PHYSICOCHEMICAL PROPERTIES

The physicochemical properties of the parvoviruses and the nonenteric caliciviruses have been well studied,[47,57] but there is little information available on the other viruses. The buoyant density of parvoviruses is high, 1.39 to 1.42 g/ml in caesium chloride (Table 2). Sedimentation constants range from 100 to 125S. They are a resistant group of viruses; infectivity is not lost after incubation at 56°C for 30 min. The Haden strain of bovine parvovirus was resistant to 56°C for 8 h[68] and the canine virus survived in feces at room temperature for 6 months. They resist ether and chloroform treatment.[46,47]

Nonenteric caliciviruses have buoyant densities of between 1.36 and 1.39 g/ml and sedimentation coefficients of 170 to 183S. They are not disrupted by ether, chloroform, or mild detergents. The buoyant density in caesium chloride of the enteric chicken calicivirus was determined as 1.38 to 1.39 g/ml[67] while that of the enteric canine calicivirus was outside the range for nonenteric caliciviruses, 1.335 to 1.35 g/ml.[25] The sedimentation value of the chicken virus was estimated to be 205S while that for the feline respiratory calicivirus, run in a parallel gradient, was 204S.[67] The morphological appearance of the chicken virus was unaffected by exposure to ether or chloroform.

The buoyant density of ovine astrovirus in caesium chloride was between 1.36 and 1.40 g/ml,[55] while that of a canine astrovirus was 1.34 g/ml.[9] Precise determinations for the ovine virus were hampered by strong aggregation of the virus particles. A human astrovirus survived treatment at pH 3 and 60°C for 5 but not 10 min.[54] It was resistant to chloroform for 10 min.

The ability to hemagglutinate red blood cells has not been reported for the enteric caliciviruses, calici-like viruses or astroviruses, but parvoviruses hemagglutinate red blood cells from at least one animal species. Appropriate conditions for hemagglutination are summarized by Siegl.[46]

V. *IN VITRO* CULTIVATION

One property that the astroviruses, enteric calici-, and calici-like viruses have in common is that they have specific requirements for serial passage *in vitro*. These requirements are only just being recognized. The inability to identify them by serial passage in cell culture by commonly used methods is the reason why they were not identified until electron microscopy was applied to the examination of fecal samples. Passage in cell culture gives a readily available source of antigen for immunological and biochemical studies. It also allows viruses to be purified by cloning before their pathogenicity is assessed. Because it has not been possible to passage the small enteric viruses readily, such studies have been hampered. Hopefully, as the requirements for their cultivation are satisfied, they will be isolated in cell culture more often.

Most animal astroviruses have not been passed serially in cell culture, but, at the first passage, bovine and deer astroviruses infected bovine kidney and testicular cells to give single immunofluorescent cells.[3,4,12,26] There was no cytopathic effect and serial passage was not achieved. Fluorescent cells were not produced when fetal lamb kidney cells were inoculated with ovine astrovirus.[2] A human astrovirus gave similar nonproductive infections until trypsin was incorporated into the maintenance medium when it was serially passed in primary human embryo kidney, primary baboon kidney, and a continuous line of rhesus monkey kidney.[69] More recently, a feline astrovirus was serially passaged six times in feline embryo cells when trypsin was included in the maintenance medium.[7] Virus was detected by immunofluorescence which was absent in cultures maintained without trypsin. The entero-like virus associated with infectious stunting of chickens[38] produced single cells with cytoplasmic fluorescence after inoculation of chick embryo liver and chick kidney cells but attempts to passage the virus were unsuccessful.

More success has been achieved with the enteric caliciviruses, but often with considerable effort. The chicken calicivirus, identified in association with the infectious stunting syndrome, was propagated in chick embryo fibroblasts when trypsin was incorporated into the maintenance medium.[67] Canine enteric caliciviruses have been propagated in cultures of canine, dolphin, and feline cells.[19,25] Schaffer et al.[25] found the pH of the medium was critical for success whereas inclusion of trypsin appeared beneficial but not essential. Cytopathic effects were produced and enough virus for biochemical analyses. However, neither inclusion of trypsin or pancreatin into the medium nor alteration in the pH of the medium enabled the porcine enteric calicivirus to be propagated serially.[70] The virus was propagated in primary porcine kidney cultures with production of cytopathic effects when intestinal fluid from uninfected gnotobiotic piglets was added to the maintenance medium.

Parvoviruses have been cultured for many years, but they also have a specific requirement for growth in cell culture. They require functions of the replicating host cell which are expressed only transiently in late S- or early G-2 phase of mitosis.[46] For efficient parvovirus replication, cultures should be infected at the time of seeding the cells or when monolayers are not completely confluent. Parvoviruses can be propagated serially in primary, secondary, and permanent cell cultures of homologous or closely related hosts.[20] The canine parvovirus multiplies in a wider variety of host cells.[49] Infected cells develop cytopathic changes which include intranuclear Cowdry type A inclusion bodies.[46] However, cytopathic changes may be transient as further pyknotic cells will not be produced unless cells are dividing. Infectivity can be determined by reading unstained cytopathic effects, examination for intranuclear inclusions after histologic staining, staining with fluorescent antibody, or by plaque assay.

VI. ANTIGENIC RELATIONSHIPS

There are no known antigenic relationships between astroviruses, parvoviruses, and the enteric calici/calici-like viruses. Parvoviruses which do not cause enteric disease occur in many animal species and they can be separated into distinct serotypes which do not share common antigens.[46] However, of the parvoviruses under discussion, the feline, canine, and mink parvoviruses are considered to be variants of one serotype. These three viruses are closely related antigenically by hemagglutination inhibition, immunofluorescence, and neutralization, but they can be differentiated from one another (see section on diagnosis).

Current knowledge indicates that astroviruses from each animal species do not share common antigens, but those from the same species do. Bovine astroviruses did not cross-react by immunofluorescence with human,[3,71] ovine,[26,72] porcine,[5] and deer astroviruses;[112] ovine astrovirus did not cross-react with human and deer astroviruses;[12,71,72] and the feline astrovirus did not cross-react with a human astrovirus.[7] However, two British and two American isolates were strongly related by immunofluorescence,[4,26,73] but antigenic differences between the american bovine astroviruses were demonstrated by cross-neutralization tests.[73] Differences were sufficient to divide them into two serotypes. Other bovine astroviruses were not neutralized by antisera to the two serotypes, indicating the existence of more than two serotypes of bovine astrovirus.

There is some information about the antigenic relatedness and diversity of the enteric calicivirus and calici-like viruses to each other and to the established nonenteric caliciviruses. Cross-relationships between members of the nonenteric caliciviruses exist. Vesicular exanthema virus of pigs and the San Miguel sea lion viruses have broad cross-reactivity, but the feline respiratory calicivirus is not closely related to either.[53] The enteric canine caliciviruses identified by Evermann et al.[19] were related to the feline respiratory virus, but the enteric canine calicivirus isolated by Schaffer et al.[25] was not closely related to either of the three nonenteric caliciviruses by immunodiffusion, immunofluorescence, immune electron microscopy, or radioimmune precipitation. Nor did the canine virus of Schaffer et al. react with several enteric caliciviruses and calici-like viruses, but it was strongly related to the chicken enteric calicivirus.[25,67] The chicken enteric calicivirus was unrelated to the three nonenteric caliciviruses and five enteric caliciviruses and calici-like viruses.[67] Porcine enteric caliciviruses were unrelated to the porcine vesicular exanthema virus and the feline respiratory calicivirus.[5,22]

Sufficient antigenic diversity occurs within the nonenteric caliciviruses for them to be divided into many serotypes.[61,74] The only report of antigenic differences within the enteric caliciviruses and calici-like viruses is between two enteric bovine calici-like viruses which did not cross-protect.[26]

VII. PATHOGENICITY AND PATHOGENESIS

The small enteric viruses can be divided into two groups with respect to their pathogenesis: those which cause enteric damage only and those which cause enteric and systemic damage. The astroviruses, caliciviruses, and calici-like viruses come into the first group, the parvoviruses and virus(es) associated with infectious stunting of chickens and turkeys into the second.

The pathogenicity of many of the viruses has been investigated by animal inoculation. However, the pathogenicity of all but the parvoviruses has yet to be confirmed with cloned viruses because of the difficulties encountered with serial passage in cell culture. It is important that this is done, when it becomes possible, as mixed viral infections are common in the enteric tract. For instance, bovine astroviruses have been present in fecal filtrates containing bovine rotavirus, bovine coronavirus, and bovine Breda agent.[3,4,26]

A. CALICIVIRUSES AND CALICI-LIKE VIRUSES

The pathogenicity of several caliciviruses and calici-like viruses has been examined. Two uncultured bovine calici-like viruses (Newbury agents), caused disease in gnotobiotic calves aged 3 to 60 day.[3,26] Inocula were free of bovine rotaviruses, bovine enteric coronaviruses, bovine astroviruses, and bovine parvoviruses. Change in fecal color, quantity and consistency, anorexia, and D-xylose malabsorption were produced and virus was detected in feces. Villus:crypt ratios were reduced throughout the small intestine with one strain.[3] With a second strain, SRV-1, damage was restricted to the anterior half of the small intestine.[75] The severity of this strain was similar to that of virulent bovine rotaviruses in calves of a similar age.[76]

The bovine enteric calici-like viruses are similar to rotaviruses in their pathogenesis as they infect the mature differentiated, absorptive enterocytes of the villus epitheleum. The first

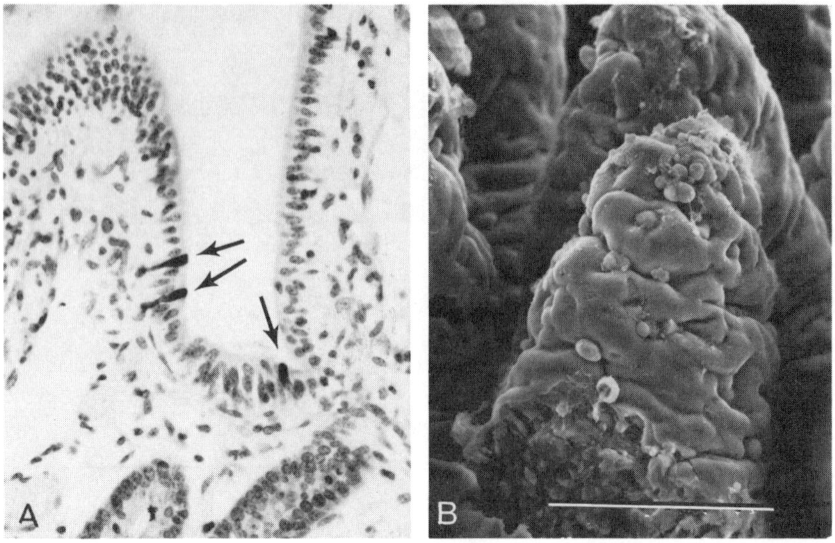

A B

FIGURE 4. (A) Arrows indicate enterocytes at the base of a villus of the mid-ileum from a calf infected with the bovine calici-like virus, Newbury agent SRV-1. Original magnification × 500. (From Parsons, K. R., Wilson, A. M., Hall, G. A., Bridger, J. C., Chanter, N., and Reynolds, D. J., *J. Clin. Pathol.*, 37, 645, 1984. With permission.) (B) Scanning electron micrograph of a villus in the anterior small intestine of a gnotobiotic calf infected with Newbury agent SRV-1. Bar indicates 0.1 mm. (From Hall, G. A., in *Novel Diarrhea Viruses*, Ciba Foundation Symp. 28, Bock, G. and Whelan, J., Eds., John Wiley & Sons, Chichester, U.K., 1987, 192. Reprinted by permission of John Wiley & Sons, Ltd.)

infected enterocytes were detected by immunoperoxidase staining at the base of villi 12 h after infection (Figure 4) and, later, scattered over the entire surface.[75,77] Damaged enterocytes became swollen, their microvilli were abnormal or absent, and they exfoliated. Villi became stunted and fused, mucosal B-galactosidase activity fell and D-xylose malabsorption occurred. No viruses were seen by transmission electron microscopy in infected enterocytes, although calici-like viruses were detected in intestinal contents. Virus particles were not seen in duodenal biopsies from volunteers infected with the Norwalk agent, a calici-like virus of man.[78]

Diarrhea was produced after oral inoculation of pigs, aged 4 days, with an uncultured porcine enteric calicivirus.[79] Specific immunofluorescence was confined to villus epithelial cells primarily in the duodenum and ileum. The effects of infection included separation and loss of microvilli, swollen enterocytes, and shortened and fused villi. Typical calicivirus particles were detected in gut contents and feces. Loose droppings, failure to thrive, and death occurred in day-old chickens after feeding the chicken calicivirus associated with infectious stunting.[67] The virus was demonstrated in feces and the intestinal contents of infected specific pathogen free chicks. It was not shown conclusively that the calicivirus was responsible for the clinical signs as reovirus was also present.

In contrast to the successful reproduction of diarrhea with the bovine, porcine, and chicken caliciviruses and calici-like viruses, disease was not reproduced with a canine calicivirus, identified in association with bloody diarrhea, even though seronegative puppies were inoculated.[25]

B. ENTERO-LIKE VIRUSES

An entero-like virus from infectious stunting of chickens produced fecal changes, depressed weight gains, and slow feathering in broilers inoculated at 1-day old.[38] Virus antigens and enterovirus-like particles were demonstrated in villus epithelial cells but again the presence of reovirus prevented an etiological association being made. However, Meulmans et al.[39] partially reproduced the syndrome in chickens with an entero-like virus alone and not with reovirus or parvovirus. In turkeys, oral inoculation of crude gut contents, in which only entero-like viruses were detected, produced diarrhea, reduced appetite, and poor feathering.[44]

C. ASTROVIRUSES

The ability of astroviruses to cause disease appears to differ between isolates from different species. Two uncultured British bovine astroviruses did not produce disease in gnotobiotic calves aged 3 to 49 d, although mild villus atrophy was found in the mid small intestine with the one isolate examined.[3,26] Gnotobiotic calves, aged 0 to 3 d, did not develop clinical disease when inoculated with two American strains, although feces became yellow and slightly soft at the time of astrovirus excretion.[4,73] With the American strains, infection was found in the epithelium of the dome villi of the jejunal and ileal Peyer's patches. Astrovirus particles were seen in degenerate cells. Retrospective studies with one of the U.K. astroviruses showed that it also multiplied in the epithelium of the dome villi.[80] Woode et al.[73] postulated that infection by astroviruses may have a deleterious effect on gut immune response which might exacerbate infection with the other pathogens which are commonly found in the enteric tract.

In contrast to bovine astroviruses, ovine, turkey, and feline astroviruses have produced diarrhea experimentally. Ovine astrovirus, first identified in association with diarrhea in 4-to-6-week-old lambs, caused diarrhea after oral inoculation of 2- and 3-day-old lambs with an uncultured inoculum.[2] Virus was detected in feces by electron microscopy and in villus enterocytes by immunofluorescence. Crystalline arrays of virus were present in the cytoplasm of mature columnar epithelial cells of the apical two thirds of villi[2,81] (Figure 5). Villus atrophy was found in the mid-gut and ileum but not the jejunum. Levels of lactase were low in the mid-gut of infected lambs. The ovine virus was not reported to infect the M cells of the domes. Inoculation of turkey poults aged 1 d, with fecal preparations shown to be free of other known pathogens, produced diarrhea, decrease in weight gain, and d-xylose malabsorption.[82] Inoculation of specific pathogen free cats with an uncultured British feline astrovirus produced less severe signs than seen in the original clinical case: mild diarrhea with pyrexia.[7]

D. PARVOVIRUSES

In contrast to the other viruses, only mitotic cells will support replication of parvoviruses. Parvoviruses exert their cytocidal effect in differentiating and differentiated tissues with a high rate of cellular proliferation. Intestinal damage occurs after infection of the mitotically active cells of the crypts of Leberkühn and not the mature, differentiated absorptive cells of the villi. Cells which are lost normally from the tips of villi are not replaced from the crypts leading to shortened, nonabsorptive villi and hence to diarrhea. Crypts become dilated and distended with mucus and cell debris. Changes can be most severe in areas of epithelium associated with Peyer's patches.

There are additional aspects to parvovirus pathogenesis. Leukopenia occurs after infection with the feline, mink, canine, and bovine viruses, cerebellar hypoplasia occurs with the feline virus and myocarditis can be produced with the canine virus.[21,31,83,84] Calves inoculated with the bovine virus also developed nasal and ocular discharge and coughing.[85] Now that infection with canine parvovirus is endemic (see epizootiology), canine parvovirus infection occurs as a panleucopenia-enteritis syndrome in weaned pups between the ages of 7 and 14 weeks. Maternal antibody protects pups beyond the age at which myocarditis develops. McCandlish[48] summarizes the relationship between the age of the infected dog and occurrence of the various disease syndromes.

The pathological changes in the gut associated with feline and canine parvovirus infections

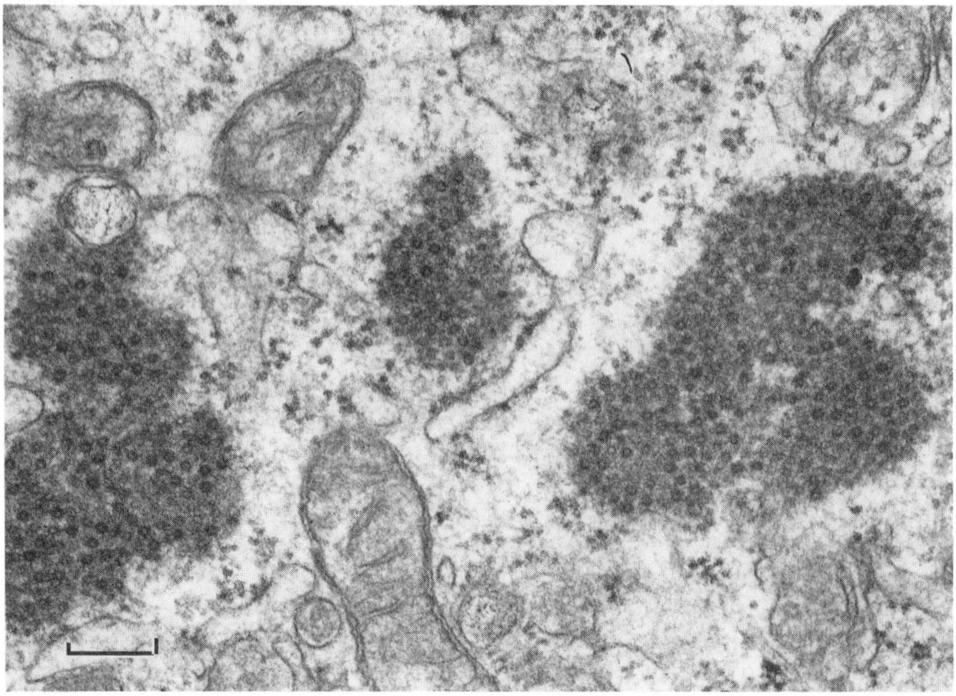

FIGURE 5. Aggregates of astrovirus virus particles in an enterocyte from the mid-gut of an infected lamb. Bar indicates 0.2 μm. (From Gray, E. W., Angus, K.W., and Snodgrass, D. R., *J. Gen. Virol.*, 49, 71, 1980. With permission.)

are similar.[86] After infection by the oral or nasal route, the incubation period is 2 to 10 d, but may be shorter if the virus is administered intravenously. Virus replicates in the lymphoid tissues of the oropharynx, followed by a brief viremia 3 to 5 d after infection. Viremia is not thought to occur with the other small enteric viruses. Leukopenia and fever develop soon after infection. Virus antigens can be detected by immunofluorescent staining in crypt epithelial cells in the small intestine, thymus, spleen, lymph nodes, and bone marrow. In contrast to the other viruses, parvoviruses are thought to reach the intestine hematogenously. Virus shedding in the feces can commence 3 to 4 d after infection and concentrations of up to 10^9 $TCID_{50}$ (tissue culture infectious dose) per gram of feces can be found. Necrosis of the crypt epithelial cells and lymphoid tissues, including Peyer's patches, lymph nodes, spleen, and thymus, occur. The diarrhea may be hemorrhagic and watery and accompanied by pyrexia, anorexia, depression, and dehydration. There can be a high mortality. However, it has often been difficult to reproduce the full spectrum of the enteric disease experimentally and, it is thought that other microorganisms play a role in producing the severe natural disease.[46] Overt enteritis was more frequently reproduced after infection of animals by the respiratory or intravenous route than after oral exposure.

There are conflicting reports on the pathogenicity of the enteric avian and bovine parvoviruses. Kisary[34] reproduced the infectious stunting syndrome in broiler chickens with an uncultured avian parvovirus, but Meulemans et al.[39] failed to do so with the same strain. Lucas and Westcott[87] failed to produce disease in the U.K. with serologically negative calves aged 4 days and 3 months and inoculated with bovine parvovirus. Other workers have produced mild to moderate diarrhea with viremia and leukopenia experimentally.[83,88-90] Both antibody-free newborn and 3- to 7-month-old calves were susceptible after intranasal, intragastric, and oral inoculation. Calves given the virus intravenously were more severely affected. The virus infected duodenum, jejunum, ileum, caecum, and colon and several other tissues including lymphoid tissues. Moderate villus atrophy and fusion occurred in the small intestine. Nuclear immunofluorescence was seen in crypt epithelial cells and particles were found in nuclei and cytoplasm of infected cells. Disease was severe when animals had concurrent subclinical coccidia infection.[91]

VIII. DIAGNOSIS

Electron microscopy identified the astroviruses, entero-like viruses, caliciviruses, and calici-like viruses and is the most useful general method. However, the numbers of bovine calici-like viruses in feces are low, even after concentration and purification of fecal samples collected soon after the onset of clinical signs.[26] This may be why so few laboratories have reported the identification of these viruses. The Norwalk virus, a calici-like virus of humans, was also shed in low amounts and briefly.[92] Bovine astroviruses required concentration by ultracentrifugation before they were readily detected in feces.[4] However, concentration was not necessary for the detection of ovine astrovirus.[2,59] Immune electron microscopy has not been used widely. Electron microscopy excels as a diagnostic method in the research laboratory but is time consuming for routine diagnosis.

Serial passage in cell cultures has not been a useful method for the diagnosis of astroviruses, entero-like viruses, enteric caliciviruses, and calici-like viruses. Astroviruses from cattle were identified using cell cultures when monolayers were stained by immunofluorescence 24 to 48 h after inoculation.[3,4,26] Single immunofluorescent cells were produced and the virus could be assayed to give titers of up to 10^5/ml. A feline astrovirus was also detected by immunofluorescence of feline embryo cells inoculated with the feces from three experimentally infected cats inoculated with the same strain of feline astrovirus.[7] This method is more sensitive than electron microscopy.[7,93] One porcine enteric calicivirus has been serially passed recently in primary porcine kidney cells,[70] but it is not known how many strains will be successfully cultured. Feline, canine, and bovine parvoviruses have been routinely isolated when actively dividing cells have been used (see section on *in vitro* cultivation).

Unlike some of the small enteric viruses of humans (see previous chapter), solid phase immunoassays (radioimmune assays and enzyme-linked immunoassays) are still to be developed for the animal astroviruses, entero-like viruses, caliciviruses, and calici-like viruses. This will become more feasible as the viruses are propagated in cell cultures and potent sera are produced. The importance of these viruses as enteric pathogens can then be assessed more readily.

While the methods available for diagnosis are limited for most of the small viruses, there are a variety of methods suitable for parvoviruses.They include serial passage in cell cultures (when virus may be demonstrated by the presence of intranuclear inclusions, immunofluorescence or cytopathic effect), electron microscopy, hemagglutination by fecal samples, and enzyme-linked immunoassays. Other methods are demonstration of intranuclear inclusion bodies in small intestinal tissues and demonstration of leukopenia for the feline virus. Immunofluorescent staining has revealed parvovirus antigens in tissues of infected cats, cattle, and dogs.[89,94,96]

Differences between the feline, mink, and canine parvoviruses, which are considered to be variants of one serotype, have been detected by *in vivo* host range, *in vitro* host cell range, hemagglutination patterns, and restriction enzyme digestion of their nucleic acids.[46,56,84] There are also antigenic differences which can be detected by hemagglutination inhibition and neutralization, immunodiffusion, and monoclonal antibodies.

IX. EPIZOOTIOLOGY

A. PREVALENCE

Feline, mink, canine, and bovine parvoviruses are geographically widespread but the other small enteric viruses have been identified in only a limited number of countries so far (Table 1). Where studies have been conducted, infection with astroviruses, caliciviruses, and calici-like viruses has been found quite commonly, but not all animals have evidence of infection. Eleven of 22 herds examined in England and Scotland had antibody to bovine astroviruses and 30% of cattle sera tested in the U.S. were positive.[3,4,73] On three farms in southern England, 60 to 100% of calves excreted astroviruses in the feces in the first 5 weeks of life,[26] although in the U.S. only 2.6% of 1060 field samples of diarrhea from calves aged 3 d to 3 weeks were astrovirus positive.[73] Turkey astroviruses were common in the U.S. They were geographically widespread and were identified in poults aged 3 d to 4 weeks in all of four flocks examined, often in association with other enteric viruses, loose droppings, listlessness, and stunting.[43,82]

Of dog sera collected in Tennessee during the years 1976 to 1981, 76% contained antibody to the canine calicivirus, although antibody was uncommon in dog sera from the U.K.[25] Five of five sera from adult swine in the U.S. were positive for antibody to the porcine enteric calicivirus.[22] The bovine Newbury agents (calici-like viruses) were found commonly in association with calf enteritis in the U.K.[97] The age of infected calves ranged from 2 to 21 d.

Today, antibodies to feline, mink, canine, and bovine parvoviruses are common, but mink and canine parvoviruses have not always been ubiquitous. Disease caused by these viruses appeared suddenly in 1947 and 1978, respectively, and quickly spread through their respective populations. Pandemics of fatal myocarditis and enteritis in puppies, associated with canine parvovirus, occurred virtually simultaneously throughout the world beginning in 1978.[84] Dog sera collected before 1976 were free of antibody to canine parvovirus which is thought to be a host range mutant of the feline virus which spread rapidly through the susceptible canine population. This newly acquired virus of dogs rapidly established the epidemiological nature of an endemic pathogen. All ages of cats and dogs are susceptible but the disease now occurs as maternal antibody falls.[45] The number of fatal cases of canine parvovirus infection is much reduced and the course of illness rarely exceeds a week.[84] Antibodies to bovine parvovirus were found when they were first sought in 1961,[68] but all herds tested have not been positive and between 14% and 100% of animals in infected herds had antibody.[88,89,98]

B. ASSOCIATION WITH DISEASE

Feline, mink, and canine parvoviruses have an important role in enteric disease, and vaccination is widely practiced in domestic cats, dogs, and farm mink. The importance of the other small viruses in enteric disease remains to be elucidated. Isolations of bovine parvovirus have been made from diarrheic calves aged 1 week to 12 months,[32] but the role of bovine parvovirus in bovine enteric disease is unclear at present. In three Australian cattle herds, diarrhea was observed in only one of three herds at the time of parvovirus excretion at weaning.[99] Lucas and Westcott[87] also concluded that the role of bovine parvovirus in calf diarrhea was unclear in the U.K. Bovine calici-like viruses (Newbury agents) were associated with 26% of outbreaks of calf diarrhea investigated in southern Britain in 1981 to 1983, but whether this was a causal association was not established as healthy animals were not studied.[97]

Many of the small enteric viruses have been found in association with enteric disease in the presence of other micro-organisms. Bovine calici-like viruses and astroviruses were found together in diarrheic feces in association with rotavirus and coronavirus;[3,100] turkey astroviruses were observed in samples from diarrheic turkey poults in association with rotaviruses;[15] porcine enteric caliciviruses were found with astroviruses, rotaviruses, and 23-nm diameter particles in association with diarrhea;[5,22] and canine caliciviruses were identified in association with fatal enteritis in association with *Salmonella* sp., canine parvovirus, canine coronavirus, and canine rotavirus.[19] In a survey of enteric viruses in turkey poults in the U.S., astroviruses were never detected alone but in combination with adenoviruses, enteroviruses, rotaviruses, and reoviruses.[16] Parvoviruses have also been found as mixed infections. Bovine parvoviruses were detected with coronaviruses, adenoviruses, and enteroviruses as naturally occurring mixed viral infections.[90]

Some of the viruses have been demonstrated in feces of healthy animals as well as diseased animals. Astroviruses were identified first in both loose and normal stools of children[1] and have since been demonstrated in gut contents from normal and diarrheic mice[13] and cats.[8] In the U.S., turkey astrovirus was found in 86% of 43 flocks with diarrhea and 41% of 22 healthy flocks when samples from poults aged 6 to 35 days were examined.[82] Of 157 feces from healthy dogs, 2% contained astroviruses.[11] Apparently healthy cattle, mink, and cats can shed parvoviruses in feces.[21,68,101]

C. TRANSMISSION

Transmission of astroviruses, enteric caliciviruses, and calici-like viruses is assumed to be by the fecal-oral route. Feline parvovirus is shed in vomit, saliva, and urine as well as feces. High levels of the feline virus are produced and its resistance means that it persists in the environment. The virus may be acquired by direct contact with other cats or fomites. Vertical transmission by transplacental infection has been demonstrated for feline and bovine parvoviruses and carriers can occur.[46] After *in utero* infection, kittens shed feline parvovirus in feces for at least 43 d.[46] High titers of antibody are characteristically found in animals with persistent, clinically inapparent infection.

D. CROSS-SPECIES INFECTIVITY

Excluding the parvoviruses, there is little information on the infectivity of the small enteric animal viruses for heterologous species. As far as is known, none of the animal viruses is infectious to man. Human and bovine astroviruses failed to infect gnotobiotic piglets.[26] However, some rabbits seroconverted after oral administration of a human calicivirus but not human astrovirus;[102] antibody to a human enteric calicivirus was found in dog sera,[103] and antibody to feline astrovirus was detected in a human serum.[8] Antibody to bovine parvoviruses has been reported in heterologous species, in monkeys, guinea pigs, goats, dogs, and horses, but not in sheep.[104] The presence of antibody in these species may indicate cross-species infectivity, as parvoviruses from different animal species are said to be serologically distinct, with the exception of the feline, canine, and mink parvoviruses.

Infection of heterologous species has occurred with serious consequences with parvoviruses and nonenteric caliciviruses. The feline parvovirus infected mink and dogs and the nonenteric calicivirus, San Miguel sea lion virus, infected pigs. Feline enteritis was well known before the turn of the century and caused huge epidemics of enteritis in cats in Europe, North, and South America but neither circumstantial nor experimental evidence pointed to naturally susceptible species other than domestic and wild members of the Felidae. The sudden appearances of severe enteritis in ranch mink (a member of the Mustelidae) in 1947 in Canada and of pandemics of enteritis and myocarditis in 1978 in dogs were associated with viruses closely related to the feline virus. All attempts to transmit feline parvovirus disease to young or adult dogs before 1978 had failed.[45] There is good evidence that the virus involved in the canine disease is a host-range mutant of the feline virus which did not exist previously. Siegl[84] raises and discusses several questions about the origins of the canine virus and the possibility of transference to humans.

The nonenteric calicivirus, vesicular exanthema virus of swine, is thought to have originated from the sea lion caliciviruses.[105] When inoculated into swine, San Miguel sea lion virus produced lesions similar to those produced by the calicivirus of vesicular exanthema, which is troublesome, because the lesions can be confused with those produced by foot and mouth disease virus.[110,111] Furthermore, a monkey was infected experimentally with the San Miguel sea lion virus, producing vesicular lesions.[106] Feline respiratory calicivirus has been isolated occasionally from feces[61] and a newly described bovine calicivirus, which was not associated with gastrointestinal symptoms, was isolated from the rectums of infected calves.[107] Thus, it is possible that caliciviruses may have mutated or may mutate in the future to cause damage in the enteric tract of homologous or heterologous species.

X. IMMUNITY

There have been few studies on active or passive immunity to astroviruses, caliciviruses, and calici-like viruses. Protective immune responses were induced after oral inoculation of two bovine calici-like viruses (Newbury agents.). Calves resisted homologous challenge up to 22 d after initial exposure.[26]

Active immune responses to feline parvovirus were not suppressed in cats suffering from feline panleukopenia.[46] Cytotoxic T cells and extremely high levels of hemagglutinating and neutralizing antibody are found after infection. Immunity was established by 48 to 72 h after vaccination with live modified virus.[22] A neutralizing antibody titer of 1:4 is protective and immunity can last up to 4 years. Immunity after infection appears life-long and is probably maintained by subclinical infections. Virus and antibody may coexist and an apparently normal animal can be a reservoir for infectious virus. Hence, the chance of natural boosting is high.

Neonatal disease caused by endemic agents occurs when levels of passive maternal immunity fall below protective levels. With viruses whose pathogenesis is confined to the enteric tract, disease occurs as local gut antibody falls below protective levels. Thus, diarrhea associated with the bovine calici-like viruses occurred in calves with a mean age of 10 d, the point at which passive maternal antibody waned.[97] Systemic and local immunity play a role in resistance to feline and canine parvovirus whose pathogenesis is not confined to the enteric tract.[108] Depending on the level of immunity of the mother, and hence the level of passive immunity transferred to the offspring, offspring become susceptible at different ages, from a few weeks of age to as long as 22 weeks of age.[74]

XI. PREVENTION AND TREATMENT

Vaccines are available for feline, mink, and canine parvoviruses, but not for any of the other small viruses. The role of astroviruses and some caliciviruses in enteric disease needs to be defined more clearly before the need for vaccination is established. A vaccine for the calici-like viruses of cattle is likely to be needed as these viruses were associated with 11 of 43 (26%) outbreaks of calf diarrhea in southern Britain.[97] A vaccine for dams to elevate levels of passive maternal antibody will only be possible if adequate amounts of virus can be produced *in vitro*. Antigenically different bovine calici-like viruses have been demonstrated,[26] but it may be unnecessary to incorporate them in a potential vaccine if they share some common antigens. Incorporation of numerous serotypes in vaccines of feline respiratory calicivirus has not been necessary even though the feline virus showed antigenic heterogenicity *in vitro*.[53]

Both live attenuated and inactivated vaccines are available from several commercial companies to stimulate active immunity to feline, canine, and mink parvoviruses. Systemic passive antibodies are important in protection of the young against parvovirus disease, but they also interfere with active vaccination.[21] There have been considerable problems in devising vaccination schedules because of the variable levels of maternal antibodies in offspring. However, successful active vaccination of the young can be achieved. Vaccines are administered on one to three occasions as maternal antibody falls. In mink colonies, vaccination is advised at the first sign of an outbreak, because the onset of protection occurs at about 6 d after vaccination.[31] With the bovine parvovirus, vaccination of dams to elevate passive maternal antibodies has been described but increases over unvaccinated dams were low.[109]

Treatment by replacement of food with supportive oral electrolytes is advised for all the viruses. Broad-spectrum antibiotics are also advised for parvovirus infection to control secondary bacterial invaders.[21,31] Reduction of environmental contamination is advised, as far as is possible, but parvoviruses are a resistant group of viruses. Sodium hypochlorite is an effective disinfectant.[21] There is no information on suitable chemical disinfectants for the animal astroviruses, caliciviruses, and calici-like viruses.

XII. SUMMARY

Astroviruses, caliciviruses, calici-like viruses, parvoviruses, and entero-like viruses are small viruses, with diameters ranging from 20 to 40 nm, which have been associated with gastroenteritis in animals. Application of electron microscopy to the examination of fecal samples was instrumental in identification of all but the parvoviruses. Caliciviruses were known to cause nonenteric infections before they were found in feces, but astroviruses were unknown before they were identified by electron microscopy in feces. At least two virus families are represented, the Caliciviridae and Parvoviridae. Classification of astroviruses, calici-like viruses, and entero-like viruses remains to be established.

Properties which distinguish the small enteric viruses of animals from one another are size and morphology, nucleic acid and polypeptide composition, buoyant density, and site of replication. Morphology has been the most useful property. Particles can be featureless (the parvoviruses and entero-like viruses) or with a surface structure which differentiates caliciviruses, calici-like viruses, and astroviruses. The genome of parvoviruses is DNA, whereas that of the other viruses is RNA. Parvoviruses, astroviruses, and caliciviruses differ in the number and molecular weight of their polypeptides. In addition, parvoviruses have high buoyant densities and replicate in the nucleus.

Serial passage in cell culture has not been successful until specific requirements have been satisfied. Parvoviruses require functions expressed by replicating cells. Cytopathic effects with intranuclear Cowdry type A inclusions are produced when replicating cells are infected. Enteric caliciviruses have been propagated in cell cultures on individual occasions in specific conditions. Many astroviruses have not been propagated serially, but bovine and deer astroviruses infect single cells at the first passage.

Antigenic relationships are not known between astroviruses, parvoviruses, and the enteric calici-/calici-like viruses. Parvoviruses from different animals belong to distinct serotypes except for the cat, dog, and mink enteric parvoviruses. These are considered to be variants of one serotype. Current knowledge indicates that astroviruses from different animals do not share common antigens but those from the same species do. Cross-relationships between caliciviruses have been reported, but more studies are required to determine the relationships of the enteric caliciviruses to one another and to the nonenteric caliciviruses. Antigenic differences have been demonstrated between bovine astroviruses by neutralization tests, and two bovine calici-like viruses failed to cross-protect.

Damage is confined to the enteric tract, with caliciviruses, calici-like viruses, and astroviruses, or extends beyond the enteric tract with parvoviruses and entero-like viruses. Parvoviruses differ as damage occurs only in mitotically active cells. Cells lining the intestinal crypts, not mature enterocytes, are attacked. The pathogenicity of parvoviruses has been studied with cloned viruses. Feline, mink, and canine parvoviruses are proven pathogens; there are conflicting reports on the pathogenicity of the avian and bovine parvoviruses. Of the viruses which are confined to the enteric tract, bovine calici-like viruses, a porcine and a chicken calicivirus, ovine, turkey, and feline astroviruses were pathogenic to their homologous species given uncloned inocula. Bovine astroviruses and a canine calicivirus were not pathogenic.

For diagnosis, electron microscopy is a good general method. For all but the parvoviruses, serial passage in cell cultures has not been useful although some astroviruses can be identified in cell cultures stained 24 to 48 h after infection. Solid phase immunoassays are still to be developed. A variety of methods are available for the enteric parvoviruses, and the feline, mink, and canine viruses can be distinguished from one another *in vitro*.

Feline, mink, canine, and bovine parvoviruses are geographically widespread, but enteric caliciviruses, calici-like viruses, astroviruses, and entero-like viruses have been identified in only a limited number of countries so far. Where studied, infection with astroviruses, calici-, and calici-like viruses has often been common. Mink and canine parvoviruses have not always been common in their respective populations however. Disease caused by these viruses appeared suddenly in 1947 and 1978, respectively. The canine parvovirus is thought to be a host range mutant of the feline virus, which spread rapidly through the susceptible canine population to become an endemic pathogen.

Feline, mink, and canine parvoviruses have an important role in enteric disease. The other small viruses have been associated with enteric disease, but their importance remains to be established. They are often found in association with one another or with rotaviruses, coronaviruses, adenoviruses, or reoviruses. They have also been found in the feces of healthy animals. Apart from feline parvovirus, there is little information on the infectivity of the small enteric viruses for heterologous species.

Immunity to those viruses whose pathogenesis is confined to the enteric tract relies on local gut immunity, whereas systemic immunity is also important with the parvoviruses. Vaccines are available for feline, mink, and canine parvoviruses but not for any of the other small viruses. The necessity for vaccines will become apparent as the role of these viruses in enteric disease is established. Treatment of the enteric disease caused by all the viruses should include replacement of food with supportive electrolytes.

REFERENCES

1. **Madeley, C. R. and Cosgrove, B. P.,** Viruses in infantile gastroenteritis, *Lancet,* 2, 124, 1975.
2. **Snodgrass, D. R. and Gray, E. W.,** Detection and transmission of 30 nm virus particles (astroviruses) in faeces of lambs with diarrhoea, *Arch. Virol.,* 55, 287, 1977.
3. **Woode, G. N. and Bridger, J. C.,** Isolation of small viruses resembling astroviruses and caliciviruses from acute enteritis of calves, *J. Med. Microbiol.,* 11, 441, 1978.
4. **Woode, G. N., Pohlenz, J. F., Kelso Gourley, N. E., and Fagerland, J. A.,** Astrovirus and Breda virus infections of dome cell epithelium of bovine ileum. *J. Clin. Microbiol.,* 19, 623, 1984.
5. **Bridger, J. C.,** Detection by electron microscopy of caliciviruses, astroviruses and rotavirus-like particles in the faeces of piglets with diarrhoea, *Vet. Rec.,* 107, 532, 1980.

5a. **Saif, L. J., Bohl, E. H., Theil, K. W., Kohler, E. M., and Cross, R. F.,** 30 nm virus-like particles resembling astrovirus in intestinal contents of a diarrheic pig, Abstr. 149, *Proc. Conf. Res. Workers in Animal Diseases,* Chicago, IL, 1980.

6. **Hoshino, Y., Zimmer, J. F., Moise, N. S., and Scott, F. W.,** Detection of astroviruses in feces of a cat with diarrhoea, *Arch. Virol.,* 70, 373, 1981.
7. **Harbour, D. A., Ashley, C. R., Williams, P. D., and Gruffydd-Jones, T. G.,** Natural and experimental astrovirus infection of cats, *Vet. Rec.,* 120, 555, 1987.
8. **Marshall, J. A., Kennett, M. L., Rodger, S. M., Studdert, M. J., Thompson, W. L., and Gust, I. D.,** Virus and virus-like particles in the faeces of cats with and without diarrhoea, *Aust. Vet. J.,* 64, 100, 1987.
9. **Williams, F. P.,** Astrovirus-like, coronavirus-like and parvovirus-like particles detected in the diarrhoeal stools of Beagle pups, *Arch. Virol.,* 66, 215, 1980.
10. **Hammond, M. M. and Timoney, P. J.,** An electron microscopic study of viruses associated with canine gastroenteritis, *Cornell Vet.,* 73, 82, 1983.
11. **Marshall, J. A., Healey, D. S., Studdert, M. J., Scott, P. C., Kennett, M. L., Ward, B. K., and Gust, I. D.,** Viruses and virus-like particles in the faeces of dogs with and without diarrhoea, *Aust. Vet. J.,* 61, 33, 1984.
12. **Tzipori, S., Menzies, J. D., and Gray, E. W.,** Detection of astroviruses in the faeces of red deer, *Vet. Rec.,* 108, 286, 1981.

13. **Kjeldsberg, E. and Hem, A.,** Detection of astroviruses in gut contents of nude and normal mice, *Arch. Virol.,* 84, 135, 1985.
14. **Gough, R. E.,** Duck hepatitis type 2 associated with an astrovirus, in *Acute Virus Infections of Poultry,* Mcferran, J. B. and McNulty, M. S., Eds., Martinus Nijhoff, Dordrecht, 1986, 223.
15. **McNulty, M. S., Curran, W. L., and McFerran, J. B.,** Detection of astroviruses in turkey faeces by direct electron microscopy, *Vet. Rec.,* 106, 561, 1980.
16. **Saif, L. J., Saif, Y. M., and Theil, K. W.,** Enteric viruses in diarrheic turkey poults, *Avian Dis.,* 29, 798, 1985.
17. **Gough, R. E., Collins, M. S., Borland, E., and Keymer, I. F.,** Astrovirus-like particles associated with hepatitis in ducklings, *Vet. Rec.,* 114, 279, 1984.
18. **Smith, A. W., Ritter, D. G., Ray, G. C., Skilling, D. E., and Wartzok, D.,** New calicivirus isolates from feces of walrus (*Odobenus rosmarus*), *J. Wildlife Dis.,* 19, 86, 1983.
19. **Evermann, J. F., McKeirnan, A. J., Smith, A. W., Skilling, D. E. and Ott, R. L.,** Isolation and identification of caliciviruses from dogs with enteric infections, *Am. J. Vet. Res.,* 46, 218, 1985.
20. **Kurstak, E. and Tijssen, P.,** Animal Parvoviruses: comparative aspects and diagnosis, in *Comparative Diagnosis of Viral Diseases, Vol. III,* Academic Press, New York, 1981, 4.
21. **Povey, R. C.,** *Infectious Diseases of Cats. A Clinical Handbook,* Centaur Press, Guelph, Canada, 1985.
22. **Saif, L. J., Bohl, E. H., Theil, K. W., Cross, R. F., and House, J. A.,** Rotavirus-like, calicivirus-like and 23 nm virus-like particles associated with diarrhoea in young pigs, *J. Clin. Microbiol.,* 26, 206, 1980.
23. **Wyeth, P. J., Chettle, N. J., and Labram, J.,** Avian calicivirus, *Vet. Rec.,* 109, 477, 1981.
24. **Hillman, B., Morris, T. J., Kellen, W. R., Hoffman, D. and Schlegel, D. E.,** An invertebrate calici-like virus: evidence for partial virion disintegration in host excreta, *J. Gen. Virol.,* 60, 115, 1982.
25. **Schaffer, F. L., Soergel, M. E., Black, J. W., Skilling, D. E., Smith, A. W., and Cubitt, W. D.,** Characterization of a new calicivirus isolated from feces of a dog, *Arch. Virol.,* 84, 181, 1985.
26. **Bridger, J. C., Hall, G. A., and Brown, J. F.,** Characterisation of a calici-like virus (Newbury agent) found in association with astrovirus in bovine diarrhea, *Infect. Immun.,* 43, 133, 1984.
27. **Gunther, H., Otto, P., and Heilmann, P.,** Untersuchungen zum Durchfall junger Kalber. 6. Mitteilung: Nachweis der Pathogenität eines bovinen Coronavirus und eines unbestimmten ikosaedrischen Virus, *Arch. Exp. Vet. Med. Leipzig,* 38, 781, 1984.
28. **Granzow, H. and Schirrmeier, H.,** Elektronenmikroskopischer direktnachweis von 32 nm virus Partikeln im kot durchfallkranker Kalber (Kurzmitteilung), *Monatsh. Veterinarmed.,* 40, 228, 1985.
29. **Herbst, W., Lange, H., and Krauss, H.,** Elektronenmikroskopischer Nachweis von calicivirus-ahnlichen Partikeln im kot durchfallkranker Kalber. *Dtsch. Tieraerztl Wochenschr.,* 94, 406, 1987.
30. **Johnson, R. H.,** Feline panleucopaenia. I. Identification of a virus associated with the syndrome, *Res. Vet. Sci.,* 6, 466, 1965.
31. **Tomson, F. N.,** Mink, in *The Veterinary Clinics of North America. Small Animal Practice,* Vol. 17, Harkness, J. E. Ed., W. B. Saunders, Philadelphia, 1987, 1145.
32. **Storz, J. and Leary, J. J.,** Bovine parvoviruses: epidemiology and host-virus relationships, *Les Colloques de l'INSERM,* 90, 63, 1979.
33. **Kisary, J., Nagy, B. and Bitay, Z.,** Presence of parvoviruses in the intestine of chickens showing stunting syndrome, *Avian Pathol.,* 13, 339, 1984.
34. **Kisary, J.,** Experimental infection of chicken embryos and day-old chickens with parvovirus of chicken origin, *Avian Pathol.,* 14, 1, 1985.
35. **Trampel, D. W., Kinden, D. A., Solorzano, R. F., and Stogsdill, P. L.,** Parvovirus-like enteropathy in Missouri turkeys, *Avian Diseases,* 27, 49, 1983.
36. **Bracewell, C. D. and Randall, C. J.,** The infectious stunting syndrome, *World's Poultry Sci. J.,* 40, 31, 1984.
37. **Dea, S., Elazhary, M. A. S. Y., Martineau, G. P., and Vaillancourt, J.,** Parvovirus-like particles associated with diarrhea in unweaned piglets, *Can. J. Comp. Med.,* 49, 343, 1985.
38. **McNulty, M. S., Allan, G. M., Connor, T. J., McFerran, J. B., and McCracken, R. M.,** An entero-like virus associated with the runting syndrome in broiler chickens, *Avian Pathol.,* 13, 429, 1984.
39. **Meulemans, G., Decaesstecker, M., and Charlier, G.,** Runting syndrome in broiler chickens. Experimental reproduction studies, in *Acute Virus Infections of Poultry,* McFerran, J. B. and McNulty, M. S., Eds., Martinus Nijhoff, Dordrecht, 1986, 1979.
40. **McNulty, M. S., Curran, W. L., Todd, D., and McFerran, J. B.** Detection of viruses in avian faeces by direct electron microscopy, *Avian Pathol.,* 8, 239, 1979.
41. **Andral, B. and Toquin, D.,** Observations et isolements de pseudopicornavirus a partir de dindonneux malades, *Avian Pathol.,* 13, 377, 1984.
42. **Andral, B., Toquin, D., L'Haridon, R., Jestin, A., Metz, M. H., and Rose, R.,** Les diarrhees du dindonneau: un bilan des recherches virales effectuees (rotavirus, reovirus, adenovirus, pseudopicornavirus), *Avian Pathol.,* 14, 147, 1985.
43. **Reynolds, D. L., Saif, Y. M., and Theil, K. W.,** A survey of enteric viruses of turkey poults, *Avian Dis.,* 31, 89, 1987.

44. **McLoughlin, M. F., McLoone, D. A., and Connor, T. J.,** Runting and stunting syndrome in turkeys, *Vet. Rec.*, 121, 583, 1987.
45. **Siegl, G.,** The Parvoviruses, in *Virology Monographs, Vol. 15*, Gard, S. and Hallauer, C., Eds., Springer Verlag, Berlin, 1976.
46. **Siegl, G.,** Biology and pathogenicity of autonomous parvoviruses, in *The Parvoviruses*, Berns, K. I., Ed., Plenum Press, New York, 1984, chap. 8.
47. **Siegl, G., Bates, R. C., Berns, K. I., Carter, B. J., Kelly, D. C., Kurstak, E., and Tattersall, P.,** Characteristics and taxonomy of Parvoviridae, *Intervirology*, 23, 61, 1985.
48. **McCandlish, I. A. P.,** Canine parvovirus infection, in *Veterinary Annual*, Vol. 21, Grunsell, C. S. G. and Hill, F. W. G., Eds., John Wright, Bristol, 1981, 259.
49. **Afshar, A.,** Canine parvovirus infections — a review, *Vet. Bull.*, 51, 605, 1981.
50. **Pollock, R. V. H.,** Canine viral enteritis, in *Current Veterinary Therapy, Vol. VIII, Small Animal Practice*, Kirk, R. W., Ed., W. B. Saunders, Philadelphia, 1983, 1164.
51. **Studdert, M. J., Oda, C., Riegl, C. A., and Roston, R. P.,** Aspects of the diagnosis, pathogenesis and epidemiology of canine parvoviruses, *Aust. Vet. J.*, 60, 197, 1983.
52. **Schaffer, F. L.,** Caliciviruses, in *Comprehensive Virology, Vol. 14, Newly Characterized Vertebrate Viruses*, Fraenkel-Conrat, H. and Wagner, R. R., Eds., Plenum Press, New York, 1979, 249.
53. **Povey, R. C.,** The pathobiology of feline caliciviruses, in *Comparative Pathology of Viral Diseases, Vol. 2*, Olsen, Krakowka, and Blackeslee, Eds., CRC Press, Boca Raton, FL, 1985, ch. 5.
54. **Kurtz, J. B. and Lee, T. W.,** Astroviruses: human and animal, in *Novel Diarrhoea Viruses*, Ciba Foundation Symp., Bock, G. and Whelan, J. Eds., John Wiley & Sons, Chichester, U.K., 1987, 92.
55. **Herring, A. J., Gray, E. W., and Snodgrass, D. R.** Purification and characterisation of ovine astrovirus, *J. Gen. Virol.*, 53, 47, 1981.
56. **Cotmore, S. F. and Tattersall, P.,** The autonomously replicating parvoviruses of vertebrates, *Adv. Virus Res.*, 33, 91, 1987.
57. **Matthews, R. E. F.,** Classification and nomenclature of viruses, *Intervirology*, 17, 1, 1982.
58. **Caul, E. O. and Appleton, H.,** The electron microscopical and physical characteristics of small round human fecal viruses: an interim scheme for classification, *J. Med. Virol.*, 9, 257, 1982.
59. **Snodgrass, D.,** Astroviruses in diarrhoea of young animals and children, in *Comparative Diagnosis of Viral Diseases, Vol. IV, Part B*, Kurstak and Kurstak, Eds., Academic Press, New York, 1981, 659.
60. **Cubitt, W. D.,** The candidate caliciviruses, in *Novel Diarrhoea Viruses*, Ciba Foundation Symposium, 128, G. Bock and J. Whelan, Eds., John Wiley & Sons, Chichester, U.K., 1987, 126.
61. **Studdert, M. J.,** Caliciviruses, *Arch. Virol.*, 58, 157, 1978.
62. **Zwillenberg, L. O. and Burki, F.,** On the capsid structure of some feline and bovine RNA viruses, *Arch. gesamte Virusforcsh.*, 19, 373, 1966.
63. **Caul, E. O., Ashley, C., and Pether, J. V. S.,** "Norwalk"-like particles in epidemic gastroenteritis in the UK, *Lancet*, 2, 1292, 1979.
64. **Greenberg, H. B., Valdesuso, J., Kalica, A. R., Wyatt, R. G., McAuliffe, V. J., Kapikian, H. Z., and Chanock, P. M.,** Proteins of Norwalk virus, *J. Virol.*, 37, 994, 1981.
65. **Madore, H. P., Treanor, J. J., and Dolin, R.,** Characterization of the Snow Mountain agent of viral gastroenteritis, *J. Virol.*, 58, 487, 1986.
66. **Hauswirth, W. W.,** Autonomous parvovirus DNA structure and replication, in *The Parvoviruses*, Berns, K. I. Ed., Plenum Press, New York, 1984, chap. 4.
67. **Cubitt, W. D. and Barrett, D. T.,** Propagation and preliminary characterization of a chicken candidate calicivirus, *J. Gen. Virol.*, 66, 1431, 1985.
68. **Abinanti, F. R. and Warfield, M. S.,** Recovery of a hemadsorbing virus (HADEN) from the gastrointestinal tract of calves, *Virology*, 14, 288, 1961.
69. **Lee, T. W. and Kurtz, J. B.,** Serial propagation of astrovirus in tissue culture with the aid of trypsin, *J. Gen. Virol.*, 57, 421, 1981.
70. **Flynn, W. T. and Saif, L. J.,** Serial propagation of porcine enteric calicivirus-like virus in primary porcine kidney cell cultures, *J. Clin. Microbiol.*, 26, 206, 1988.
71. **Lee, T. W. and Kurtz, J. B.,** Human astrovirus serotypes, *J. Hyg., Camb.*, 89, 539, 1982.
72. **Snodgrass, D. R., Angus, K. W., Gray, E. W., Menzies, J. D. and Paul, G.,** Pathogenesis of diarrhoea caused by astrovirus infections in lambs, *Arch. Virol.*, 60, 217, 1979.
73. **Woode, G. N., Kelso Gourley, N. E., Pohlenz, J. F., Liebler, E. M., Mathews, S. L., and Hutchinson, M. P.,** Serotypes of bovine astrovirus, *J. Clin. Microbiol.*, 22, 668, 1985.
74. **Fenner, F., Bachmann, P. A., Gibbs, E. P. J., Murphy, F. A., Studdert, M. J., and White, D. O., Eds.,** Parvoviridae, in *Veterinary Virology*, Academic Press, New York, 1987, 407.
75. **Hall, G. A., Bridger, J. C., Brooker, B. E., Parsons, K. R. and Ormerod, E.,** Lesions of gnotobiotic calves experimentally infected with a calicivirus-like (Newbury) agent, *Vet. Pathol.*, 21, 208, 1984.
76. **Bridger, J. C. and Pocock, D. H.,** Variation in virulence of bovine rotaviruses, *J. Hyg. Camb.*, 96, 257, 1986.

77. **Parsons, K. R., Wilson, A. M., Hall, G. A., Bridger, J. C., Chanter, N., and Reynolds, D. J.**, Localisation of enteropathogens in paraffin embedded tissue by immunoperoxidase, *J. Clin. Pathol.*, 37, 645, 1984.
78. **Dolin, R.**, Norwalk agent-like particles associated with gastroenteritis in human beings, *J. Am. Vet. Med. Assoc.*, 173, 615, 1978.
79. **Flynn, W. T., Saif, L. J., and Moorhead, P. D.**, Pathogenesis of porcine enteric calicivirus-like virus in four-day-old gnotobiotic pigs, *Am. J. Vet. Res.*, 49, 819, 1988.
80. **Hall, G. A.**, Comparative pathology of infection by novel diarrhoea viruses, in *Novel Diarrhoea Viruses*, Ciba Foundation Symp. 128, Bock, G. and Whelan, J., Eds., John Wiley & Sons, Chichester, U.K., 1987, 192.
81. **Gray, E. W., Angus, K. W., and Snodgrass, D. R.**, Ultrastructure of the small intestine in astrovirus-infected lambs, *J. Gen. Virol.*, 49, 71, 1980.
82. **Reynolds, D. L. and Saif, Y. M.**, Astrovirus: a cause of an enteric disease in turkey poults, *Avian Dis.*, 30, 728, 1986.
83. **Durham, P. J. K., Lax, A., and Johnson, R. M.**, Pathological and virological studies of experimental parvoviral enteritis in calves, *Res. Vet. Sci.*, 38, 209, 1985.
84. **Siegl, G.**, Canine parvovirus. Origin and significance of a "new" pathogen, in *The Parvoviruses*, Berns, K. I., Ed., Plenum Press, New York, 1984, chap. 9.
85. **Weiblen, R., Mock, R. E., Woods, G. T., Hoffmann, W. E., and Shivaprasa, D.**, Bovine parvovirus infection, *Proc. 3rd Int. Symp. World Assoc. of Vet. Lab. Diagnosticians*, 1983, 369.
86. **Kahn, D. E.**, Pathogenesis of feline panleukopenia, *J. Am. Vet. Med. Assoc.*, 173, 628, 1978.
87. **Lucas, M. H. and Westcott, D. G. F.**, Bovine parvovirus, *Vet. Rec.*, 116, 698, 1985.
88. **Spahn, G. J., Mohanty, S. B., and Hetrick, F. M.**, Experimental infection of calves with hemadsorbing enteric (Haden) virus, *Cornell Vet.*, 56, 377, 1966.
89. **Storz, J. and Bates, R. C.**, Parvovirus infections in calves, *J. Am. Vet. Med. Assoc.*, 163, 884, 1973.
90. **Storz, J., Leary, J. J., Carlson, J. H. and Bates, R. C.**, Parvoviruses associated with diarrhea in calves, *J. Am. Vet. Med. Assoc.*, 173, 624, 1978.
91. **Durham, P. J. K., Johnson, R. H., and Parker, R. J.**, Exacerbation of experimental parvoviral enteritis in calves by coccidia and weaning stress, *Res. Vet. Sci.*, 39, 16, 1985.
92. **Thornhill, T. S., Kalica, A. R., Wyatt, R. G., Kapikian A. Z., and Chanock, R. M.**, Pattern of shedding of the Norwalk particle in stools curing experimentally induced gastroenteritis in volunteers determined by immune electron microscopy, *J. Infect. Dis.*, 132, 28, 1975.
93. **Woode, G. N.**, In *Novel Diarrhoea Viruses*, Ciba Foundation Symp. 128, Bock, G. and Whelan, J., Eds., John Wiley & Sons, Chichester, U.K., 1987, 104.
94. **Csiza, C. K., de Lahunta, A., Scott, F. W., and Gillespie, J. H.**, Pathogenesis of feline panleukopenia virus in susceptible newborn kittens. II. Pathology and immunofluorescence, *Infect. Immun.*, 3, 838, 1971.
95. **Saif, L. J.**, Discussion, in *Novel Diarrhoea Viruses*, Ciba Foundation Symp. 128, Bock, G. and Whelan, J., Eds., John Wiley & Sons, Chichester, U.K., 1987, 124.
96. **Carmichael, L. E. and Binn, L. N.**, New enteric viruses in the dog, in *Advances in Veterinary Science and Comparative Medicine*, Vol. 25, Cornelius, C. E. and Simpson, C. F., Eds., Academic Press, New York, 1981, 1.
97. **Reynolds, D. J., Morgan, J. H., Chanter, N., Jones, P. W., Bridger, J. C., Debney, T. G., and Bunch, K. J.**, Microbiology of calf diarrhoea in southern Britain, *Vet. Rec.*, 119, 34, 1986.
98. **Huck, R. A. and Woods, D. W.**, Isolation of a bovine parvovirus in the United Kingdom, *Vet. Rec.*, 96, 155, 1975.
99. **Durham, P. J. K., Johnson, R. H., Isles, H., Parker, R. J., Holroyd, R. G., and Goodchild, I.**, Epidemiological studies of parvovirus infections in calves on endemically infected properties, *Res. Vet. Sci.*, 38, 234, 1985.
100. **Almeida, J. D., Craig, C. R., and Hall, T. E.**, Multiple viruses present in the faeces of a scouring calf, *Vet. Rec.*, 102, 170, 1978.
101. **Bouillant, A. and Hanson, R. P.**, Epizootiology of mink enteritis. III. Carrier state in mink, *Can. J. Comp. Med. Vet. Sci.*, 29, 183, 1985.
102. **Kjeldsberg, E. and Moetensson-Egnund, K.**, Antibody response in rabbits following oral administration of human rota-, calici-, and adenovirus, *Arch. Virol.*, 78, 97, 1983.
103. **Humphrey, T. J., Cruickshank, J. G., and Cubitt, W. D.**, An outbreak of calicivirus associated gastroenteritis in an elderly persons home. A possible zoonosis? *J. Hyg. Camb.*, 92, 293, 1984.
104. **Storz, J., Bates, R. C., Waren, G. S., and Howard, T. H.**, Distribution of antibodies against bovine parvovirus 1 in cattle and other animal species, *Am. J. Vet. Res.*, 33, 269, 1972.
105. **Fenner, F., Bachmann, P. A., Gibbs, E. P. J., Murphy, F. A., Studdert, M. J., and White, D. O., Eds.**, Caliciviridae, in *Veterinary Virology*, Academic Press, New York, 1987, 445.
106. **Smith, A. W., Prato, C., and Skilling, D. E.**, Caliciviruses infecting monkeys and possibly man, *Am. J. Vet. Res.*, 39, 287, 1978.

107. **Smith, A. W., Mattson, D. E., Skilling, D. E., and Schmitz, J. A.,** Isolation and partial characterization of a calicivirus from calves, *Am. J. Vet. Res.*, 44, 851, 1983.
108. **Rice, J. B., Winters, K. A., Krakowka, S., and Olsen, R. G.,** Comparison of systemic and local immunity in dogs with canine parvovirus gastroenteritis, *Infect. Immun.*, 38, 1003, 1982.
109. **Wieda, J., Bengelsdorff, H. J., Bernhardt, D., and Hungerer, K. D.,** Antibody levels in milk of vaccinated and unvaccinated cows against organisms of neonatal diarrhoea, *J. Vet. Med. B.*, 34, 495, 1987.
110. **Smith, A. W., Akers, T. G., Madin, S. H., and Vedros, N. A.,** San Miguel sea lion virus isolation, preliminary characterization and relationship to vesicular exanthema of swine virus, *Nature (London)*, 244, 108, 1973.
111. **Breese, S. S. and Dardiri, A. H.,** Electron microscope observations on a virus transmissible from pinnipeds to swine, *J. Gen. Virol.*, 36, 221, 1977.
112. **Bridger, J. C.,** personal observation.

IV. Enveloped Enteropathogenic Viruses

Chapter 9

ENTEROPATHOGENIC CORONAVIRUSES

Linda J. Saif and Robert A. Heckert

TABLE OF CONTENTS

PART 1. INTRODUCTION

The Coronaviridae is a monogeneric family comprising at least 11 coronavirus species which infect vertebrates.[1-3] Under natural conditions, the respiratory and intestinal tracts may be the primary site of replication for all coronaviruses, with localization of infection at these sites for most coronaviruses or subsequent spread by others to different target organs, such as the central nervous system, liver, kidney, and parotid gland.[1-3] Coronaviruses infect a wide variety of animals including man, swine, cattle, mice, rats, chickens, turkeys, dogs, and cats.

Coronaviruses are characterized on the basis of their distinctive morphology as enveloped, pleomorphic particles with diameters of 60 to 220 nm, surrounded by a "corona" of club-shaped peplomers, 15 to 20 nm in length.[1-3] In 1968, the name coronavirus was applied to avian infectious bronchitis virus (IBV) and morphologically similar viruses.[4] By electron microscopy, only one type of peplomer has been observed on most coronaviruses including transmissible gastroenteritis virus (TGEV) as illustrated in Figure 1a. However as illustrated in Figure 1b, bovine coronavirus (BCV) consistently shows 2 types of peplomers, a shorter and longer layer. It is possible that the second type may represent the additional 140 kDa glycoprotein described for BCV as the hemagglutinin protein.[5]

More recently biochemical and molecular criteria have been applied for coronavirus classification. The virions are composed of three major proteins including the nucleocapsid (N) protein (50 to 60 kDa) and two glycoproteins, the peplomer (E2) protein (90 to 180 kDa) and the transmembrane matrix (E1) protein (20 to 35 kDa).[1-3] Some coronaviruses (BCV) contain a third glycoprotein (140 kDa) believed to be the viral hemagglutinin (E3).[5] The genomic RNA is a single-stranded, infectious linear molecule which is capped and polyadenylated.[1-3] During

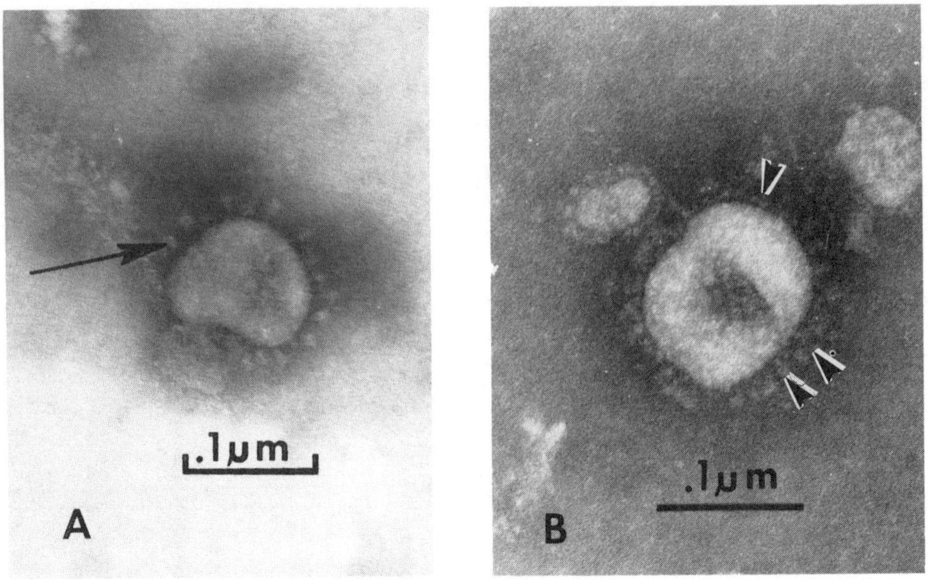

A B

FIGURE 1. Electron micrograph of negatively stained coronavirus particles. (A) Typical transmissible gastroenteritis virus particle showing single layer of peplomers of uniform length (arrow). (B) Typical bovine coronavirus particle showing shorter (single arrowhead) and longer (double arrowheads) layers of peplomers. (From Saif L. J., Bohl, E. H., Kohler, E. M., Hughes, J. H., *Am. J. Vet. Res.*, 38, 13, 1977. With permission.)

maturation in the cytoplasm, virions assemble by budding into the cisternae of rough endoplasmic reticulum, acquiring their lipid envelopes from the host cell.

Based on immunologic studies, four antigenic groups of coronaviruses have been defined, with several coronaviruses still unclassified within these groups.[1,2] The enteropathogenic coronaviruses, and their antigenic relationships to one another and to other coronaviruses, are listed in Table 1.

Preliminary data is available on the nucleic acid sequence homology between a few coronaviruses. No homology between murine hepatitis coronavirus (MHV) and human coronavirus (HCV) 229E was evident using an MHV-specific cDNA probe.[6] Similarly, there was little genome sequence homology (37% for the N protein gene) between BCV and TGEV or BCV and IBV (43% for the N protein gene).[7] Genome sequence homology (98%) was reported between BCV and HCV OC43 and 72% genome sequence homology was observed for the N protein genes of MHV and BCV.[5,7] Others noted genome sequence homology between the peplomer protein genes of TGEV and feline infectious peritonitis virus (FIPV) consisting of 39% homology in one domain (coding for amino acids 1 to 294), but 93% homology in the second domain (coding for amino acids 295 to 1147).[8] Thus, the genomic analysis of the coronaviruses studied to date correlates closely with the antigenic relationships established by serology (Table 1).

Although the enteropathogenic coronaviruses do not share a common antigen and comprise at least three distinct antigenic groups, they possess a number of common characteristics in addition to those described previously. Field strains of enteropathogenic coronaviruses remain difficult to adapt to growth in cell culture on primary isolation. This property has hindered analysis of the degree of temporal or geographic variation which may occur among coronavirus strains. However, at the present time, only one serotype has been reported for any of the enteric

TABLE 1
Antigenic Relationships among Enteropathogenic Coronaviruses and Other Coronaviruses

Mammalian

Enteropathogenic Coronaviruses

Group 1	**Group 2**
—Transmissible gastroenteritis virus of swine	—Bovine coronavirus
—Canine coronavirus	—Diarrhea virus of infant mice
—Feline enteric coronavirus	

Other Coronaviruses

—Human coronavirus 229E	—Human coronavirus — OC43
—Feline infectious peritonitis virus	—Mouse hepatitis virus
	—Rat coronavirus
	—Rat sialodacryoadenitis
	—Hemaglutinating encephalomyelitis virus of swine

Avian

Group 3	**Group 4**
—Infectious bronchitis virus	—Bluecomb disease of turkeys

Unclassified

—Human enteric coronavirus
—Porcine epidemic diarrhea virus
—Rabbit enteritis coronavirus
—Equine coronavirus

coronaviruses. Inclusion of trypsin or pancreatin in the growth medium enhanced the *in vitro* replication of some enteric coronaviruses[9,10] and was a necessity for the recent serial propagation of porcine epidemic diarrhea virus in Vero cells.[11] The mechanism of proteolytic enhancement of coronavirus infectivity is unknown, but may relate to proteolytic cleavage of the E2 protein.[3]

Whereas adult seronegative animals may be susceptible to infection, resulting in inapparent or mild diarrhea, enteropathogenic coronaviruses cause the most severe lesions and sometimes fatal diarrhea in young animals which appear to be more highly susceptible to infection and disease. Enteropathogenic coronaviruses selectively infect the absorptive villous epithelial cells of the small intestine, resulting in villous atrophy and consequently a malabsorptive diarrhea.

None of the enteropathogenic coronaviruses have been shown to induce viremia or generalized infections. However several of the enteropathogenic coronaviruses such as BCV and TGEV are capable of causing pneumoenteric infections.[12] Preliminary data suggests BCV may persist in individual animals in a herd by causing repeated respiratory infections which are usually asymptomatic in older cattle.[13] The same strains of virus apparently cause both enteritis and respiratory infections in susceptible calves.[12] Another interesting aspect of the relationship between respiratory and enteric coronaviruses is the recent emergence of a respiratory strain of TGEV in swine.[14] This coronavirus replicates in the upper respiratory tract but not the intestine, causes no gastroenteritis, and is antigenically indistinguishable in conventional serologic tests from enteric strains of TGEV.[14] Infection of pigs with the respiratory TGEV strain does not confer protection upon challenge with an enteric TGEV strain.[14] The origin of this porcine coronavirus with altered tissue tropism for cells of the upper respiratory tract is unclear. However, it is of interest to note that live attenuated strains of TGEV have been described previously which replicate in the upper respiratory tract, but not the intestinal tract of newborn pigs.[15]

Likewise it is possible that the recently identified feline enteric coronavirus, which is antigenically indistinguishable from FIPV,[16] may represent a naturally occurring feline coronavirus variant with tissue tropism for the villous enterocytes of the intestinal tract and unable to cause systemic disease associated with FIPV infections. Thus, additional antigenic relationships, new species, and viral variants continue to be identified among members of the Coronaviridae family. Molecular analysis of these emerging species and strains of coronavirus may assist in their further detailed classification and provide possible information on their origin.

In this chapter, the properties of each of the currently known enteropathogenic coronaviruses are reviewed in detail. Described in separate sections are TGEV of swine, canine coronavirus, feline enteric coronavirus, bovine coronavirus, diarrhea virus of infant mice, bluecomb disease of turkeys, human enteric coronavirus, and porcine epidemic diarrhea virus. A lack of detailed information about rabbit enteric coronavirus or equine coronavirus precluded their review in this chapter.

PART 2. ANTIGENENIC GROUP 1 — MAMMALIAN

A. TRANSMISSIBLE GASTROENTERITIS VIRUS

I. INTRODUCTION

Transmissible gastroenteritis virus (TGEV) was the first specific viral diarrheal disease of swine to be identified.[1] The disease was first reported in the U.S. in 1945[1] and since then has been shown to have almost a worldwide distribution.[2] Transmissible gastroenteritis is characterized by vomiting, severe diarrhea, and a high mortality rate (often 100%) in piglets under 2 weeks of age. In densely swine-populated areas, such as the midwestern U.S., TGE is recognized as one of the major causes of morbidity and mortality in young pigs. Although swine of all ages are susceptible to TGEV infection, clinical signs in adult pigs are milder and the mortality rate is low. Serologic surveys indicated 19 to 54% of sampled swine herds in Europe and North America were seropositive for TGEV antibodies.[3,4] However, the prevalence of seropositive herds is increasing in some countries such as England and Belgium where a respiratory form of TGEV infection was reported recently.[5,6]

II. MORPHOLOGY AND MORPHOGENESIS

As a typical member of the Coronaviridae family, TGEV is enveloped, pleomorphic, and 60 to 160 nm in diameter.[7,8] The virus has a single layer of club-shaped surface projections or peplomers which are 12 to 25 nm in length (Figure 2). These projections are readily lost following repeated freezing and thawing of the virus or after digestion by the enzyme bromelain.[9]

As for other coronaviruses, TGEV maturation occurs in the cell cytoplasm by budding through the endoplasmic reticulum.[10,11,12] Viral particles are often observed within cytoplasmic vacuoles or lining the host cell membranes after exit from infected cells.[2]

III. CULTIVATION

Oral exposure of newborn piglets is probably the most sensitive method for isolating or

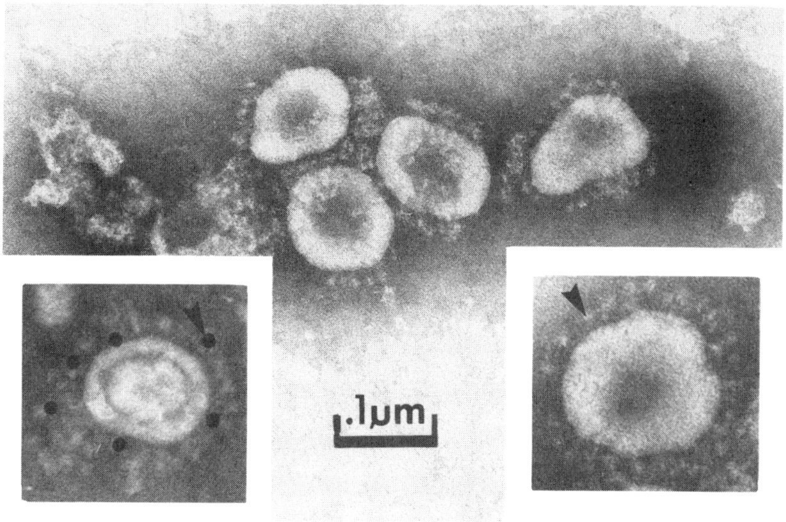

FIGURE 2. Electron micrograph of negatively stained coronavirus particles. A typical transmissible gastroenteritis virus (TGEV) particle with a single layer of peplomers of uniform length (arrow) is shown in the right inset. The left inset contains a TGEV particle whose peplomers are coated with TGEV antiserum and then reacted with protein A gold (arrow), insets × 195,000. The other TGEV particles in the photo are aggregated by reaction with TGEV antiserum (immune electron microscopy).

detecting TGEV.[13] However, because of the expense of this procedure, pig cell cultures are used for the routine isolation of TGEV. Cells commonly used include primary and secondary pig kidney or thyroid cells[14,15] and the McClurkin swine testes (ST) cell line.[16] Distinct cytopathogenic effects (CPE) may not be observed upon primary isolation of field strains, but may develop upon additional passages. Characteristic CPE seen in ST or porcine thyroid cells includes enlarged, rounded, or elongated cells which have a balloon-like appearance.[17] Other cells in which TGEV has been cultivated include: pig salivary gland cells[18] and organ cultures from pig esophagus, ileum, cecum, and colon.[19] TGEV also has been propagated in nonporcine tissues including canine kidney cells[20] and an established feline whole fetus cell line.[21] However neither of the two antigenically related coronaviruses, canine coronavirus (CCV) or feline infectious peritonitis virus (FIPV), replicate *in vitro* in pig cells including ST or secondary pig thyroid cells.[22] Enhanced titers of TGEV from cell cultures have been achieved by addition of proteolytic enzymes to the culture medium,[21,23] use of older cells (5 to 6 d after reaching confluency), and maintenance of a slightly acidic culture medium.[24]

IV. PHYSICOCHEMICAL PROPERTIES

TGEV has a buoyant density of 1.19 to 1.21 g/ml in sucrose and a sedimentation coefficient of 495S.[9,25] The virus contains glycolipid and phospholipid in the viral envelope, derived from the endoplasmic reticulum of the host cell during virus maturation.[26] Examination of the polypeptides of TGEV revealed at least three major and two minor polypeptides (Figure 3).[9,27-29] The three structural polypeptides include: (1) a 160 to 200 kDa glycoprotein which is associated with the surface peplomers (E2) and elicits neutralizing antibodies;[30-33] (2) a 25 to 33 kDa matrix glycoprotein located in the viral envelope (E1), which also elicits neutralizing antibodies but of low titer;[31,33] and (3) a 47 to 56 kDa nucleoprotein (N) which is phosphorylated and associated with the RNA genome. The amino acid sequences for both TGEV glycoproteins, deduced from nucleotide sequencing have been reported.[34] The two minor polypeptides, whose functions are

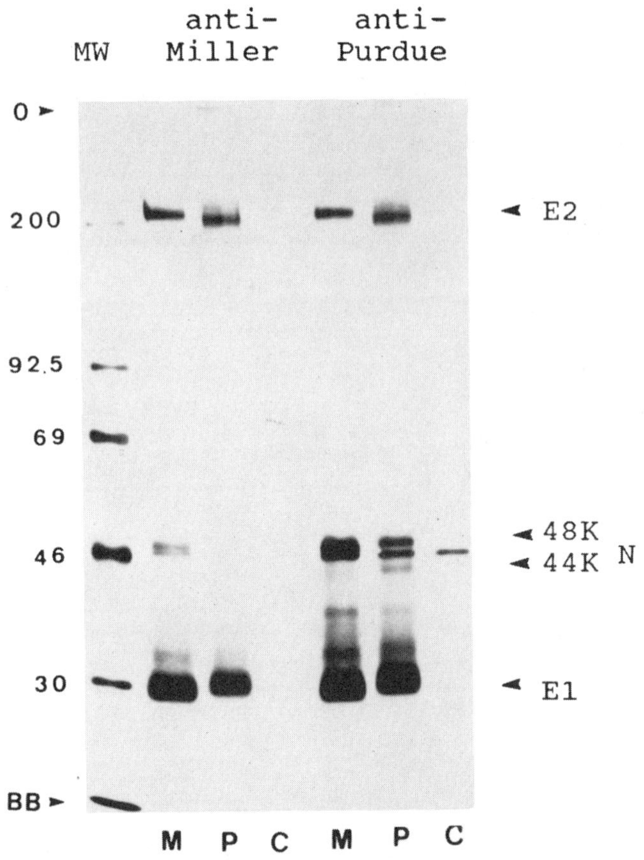

FIGURE 3. Radioimmunoprecipitaion of TGE viral proteins using hyperimmune serum to Miller (M) virulent and Purdue (P) attenuated strains of TGE virus. M = Miller TGEV infected swine testicle (ST) cell lysate; P = Purdue attenuated TGEV infected ST cell lysate; C = Mock infected ST cell lysate. O = origin; BB = bromophenol blue dye marker; E1, membrane protein; E2, peplomer protein; N, nucleocasid protein. Note 44K N protein evident in Purdue ST cell lysate reacted with anti-Purdue serum.

unknown, have molecular weights of 105 kDa and 80.5 kDa.[9] Recently, 17-kDa and 14-kDa intracellular proteins were identified in cells infected with the Purdue or Miller strains of TGEV, respectively; their functions are unknown.[29,33,35] Horzinek et al.[27] also reported a minor 21 kDa polypeptide associated with sucrose-purified TGEV.

Monoclonal antibodies (MAbs) have been produced recently against Purdue[31,32,34,36,37] and Miller strains[33,37] of TGEV and used to further characterize TGEV proteins. All major neutralization determinants were associated with the virus peplomer (E2)[31,33] and these epitopes were highly conserved on most TGEV strains.[32] Anti-E1 MAbs had little or no neutralizing activity,[31,33] but the neutralizing activities of some anti-E1, but no anti-E2 MAbs were enhanced by complement from guinea pigs, rabbits, or swine.[32,34,37] Furthermore certain anti-E1 MAbs were shown to block interferon induction *in vitro*, suggesting E1 may play a role in the induction of interferon *in vivo*.[34] Use of neutralizing MAbs to map epitopes on the E2 glycoprotein, revealed four to six different antigenic sites,[32,34,36] with sites A and B containing the highly conserved epitopes recognized by strongly neutralizing MAbs. Results of studies in mice suggested that all the critical epitopes in virus neutralization are conformational, but this has not been evaluated in swine.[32]

V. GENOME

The TGEV genome consists of single-stranded RNA of positive polarity which is polyadenylated and infectious.[25,38] It is slightly greater than 20 kilobases (kb) in length. Positive-stranded mRNAs are produced via a negative strand replicative intermediate. Five to ten subgenomic species of mRNAs have been identified for TGEV.[29,39,41] These mRNAs are arranged in a 3′ coterminal nested set. The sequence relationships between the genome, the mRNA species, and the protein products have been partially defined.[40,41]

The structural protein genes of the Purdue attenuated strain of TGEV (P115) have been cloned by several research groups,[34,41-44] and nucleotide sequence data on approximately 2.6 kb from the genome 3′ end has been published.[42,43] The E2 protein of the attenuated Purdue and virulent Miller strains of TGEV was also cloned, but complete nucleotide sequence data was not published.[29,34,44] Rasschaert et al.[34] showed that the structural protein genes of P115 TGEV are located in the first 8.22 kb from the 3′ end, with the N protein gene closest to the 3′ end, flanked by the E1 gene and followed by the E2 gene (but with intervening sequences present). The P115 TGEV N, E1, and E2 protein genes are 1,146; 786; and 4,341 nucleotides in length and code for proteins with relative molecular mass (mr) of 43, 29, and 160 kDa, respectively.[34,41-44] The N protein gene of a virulent British strain of TGEV was also cloned and sequenced recently.[45] It encoded a polypeptide with an mr of 43 kDa which is almost identical to the mr for the N protein from attenuated TGEV.[42] Results of Hu[29] indicated that the E2 protein gene of the virulent Miller TGEV was smaller (3.9 kb) than the P115 E2 protein gene. These same investigators reported that the Miller E2 protein gene encoded an unglycosylated protein of mr 145 kDa which is smaller than the mr, as noted above, for the P115 E2 unglycosylated protein.[34] The expression of the cloned E2 protein gene in *Escherichia coli* and vaccinia was reported.[46] The E2 protein fragments expressed in *E. coli* were not glycosylated, were difficult to isolate due to aggregation and insolubility, and did not induce TGEV neutralizing antibodies in mice.[46] The E2 protein expressed in vaccinia had an mr of 185 kDa, was glycosylated, and induced low neutralizing antibody titers in mice.[46] Studies of the immune response of swine to this TGE-vaccinia rDNA virus have not been reported.

VI. ANTIGENIC RELATIONSHIPS

TGEV is not antigenically related to the two other porcine coronaviruses, hemagglutinating encephalomyelitis virus[48] and porcine epidemic diarrhea virus.[49,50] The latter virus causes a similar disease syndrome to TGE, but has been documented only in swine in Europe and Taiwan.

At present, only one serotype of TGEV is known.[47] However, recently a respiratory strain of TGEV (porcine respiratory coronavirus), antigenically indistinguishable (using polyclonal antisera in two-way virus neutralization tests) from the enteric strain, has been described in some countries.[5,6] Cross-reactivity between the porcine respiratory coronavirus and TGEV at the level of the E1, E2, and N proteins was also shown using polyclonal antisera in immunoblotting.[51] Use of MAbs specific for TGEV in virus neutralization (VN) and radioimmunoassays (RIA) demonstrated the presence of similar epitopes on E1 and N proteins as well as the neutralization-mediating epitopes on the E2 protein.[51] However, E2-specific MAbs without VN activity (antigenic sites designated B, C, and D) did not react with the porcine respiratory coronavirus, indicating that these sites were modified or absent on the equivalent polypeptide of the respiratory coronavirus. In another study, a MAb which recognized TGEV, but not porcine respiratory coronavirus, was used in a competitive ELISA to discriminate between sera from pigs experimentally infected with TGEV (blocking observed) or porcine respiratory coronavirus (no blocking observed).[52] However, the applicability of this test for the differentiation of the two coronaviruses using antisera from naturally infected swine has not been reported.

TGEV, canine coronavirus (CV), and a coronavirus of cats (feline infectious peritonitis virus,

FIPV) are antigenically related[21,22,48] and form one of the two major subtypes of mammalian coronaviruses. It has been suggested that these 3 viruses may represent host range mutants of a single virus strain.[27] Antigenic relationships among TGEV, CCV, and FIPV, reported from *in vitro* studies using VN and immunofluorescence (IF) tests include the following observations: (1) antisera to CCV and TGEV cross-reacted with both viruses in IF and two-way cross-neutralization tests;[21,22,48] and (2) a two-way relationship between TGEV and FIPV was shown by IF and cross-neutralization studies, but only a one-way relationship between FIPV and CCV was evident by cross-neutralization.[21,48] In all cases, VN titers were consistently higher in homologous than heterologous reactions, suggesting these three viruses could be distinguished serologically (using sera from naturally infected animals) by using two-way cross-neutralization tests.[22] That antigenic cross-reactivity exists at the level of TGEV structural proteins, was reported by Horzinek et al.[27] who showed cross-reactivity between analogous E1, E2, and N proteins of TGEV, FIPV, and CCV using radioimmunoprecipitation, electroblotting, and ELISA. Investigators from this laboratory further examined the nucleic acid sequence homology between the peplomer protein genes of TGEV and FIPV.[44] They found only 39% homology in one domain (amino acids 1 to 274), but 93% homology in the second domain (amino acids 275 to 1447). Other *in vitro* biologic differences also have been reported; whereas both TGEV and CCV grow in either canine kidney cells[20,22] or an established feline cell line,[21] neither CCV nor FIPV grow in porcine thyroid or swine testes cells, both of which support the growth of TGEV strains.[22]

In vivo biologic differences have been noted among these three coronaviruses in their pathogenicity for neonatal pigs.[53,54] Virulent FIPV was as pathogenic for newborn pigs as virulent TGEV, producing diarrhea, extensive intestinal lesions, and infection of predominately villous epithelial cells. On the other hand, CCV caused no clinical signs and only slight villous atrophy in infected pigs, infecting mainly epithelial cells of the crypts. Although there are no reports of the natural infection of pigs with FIPV or CCV, in reference to the above findings, this possibility cannot be dismissed. However, pigs infected with FIPV[54] or CCV[53,54] failed to seroconvert to TGEV.

Dogs and cats also have been proposed as possible carriers of TGEV. Fecal shedding of infectious TGEV and seroconversion to TGEV, in the absence of clinical signs, has been shown in dogs and cats fed TGEV.[55-58] However, cats infected with TGEV did not seroconvert to FIPV and were not protected from FIPV challenge.[57,58] At present, it is unclear if TGE VN and IF antibodies frequently detected in dogs and cats are due to infection with TGEV or with the antigenically related CCV and FIPV. The latter possibility is favored by the widespread occurrence of these antibodies in dogs and cats, particularly in those with no previous contact with swine.

VII. PATHOGENESIS AND PATHOGENICITY

A. INTESTINAL REPLICATION

The first clinical sign of TGE in young piglets is usually vomiting, rapidly followed by profuse watery yellow or greenish diarrhea. This severe and prolonged diarrhea results in rapid loss of weight, dehydration, and high morbidity and mortality in piglets under 2 weeks of age. Death occurs after 2 to 7 d in most piglets infected at this age. Severity of clinical signs, duration of the disease, and mortality are inversely related to age. Most suckling pigs over 3 weeks of age will survive, but may remain unthrifty or succumb to secondary infections. Mixed infections with *E. coli*[59] and rotavirus[60] may also complicate recovery from TGEV leading to increased mortality or prolonged unthriftiness among pigs. Clinical signs in finishing pigs and sows are usually limited to inappetence and diarrhea for 1 or a few days, with vomiting occasionally observed. The few deaths observed are probably due to complicating factors such as stress or concurrent infections. Some lactating animals become very sick, with elevated temperatures, agalactia, vomiting, inappetence, and diarrhea. Such severe signs may be due to the high level of viral exposure of sows in close contact with their affected litters or hormonal factors which may influence susceptibility or immunity.[62]

Details of the pathogenesis of TGE have been described.[63,64] Whether initial infection occurs by the oral or nasal route, the virus is swallowed and, because of resistance to low pH and proteolytic enzymes, remains viable until it comes in contact with the susceptible villous epithelial cells of the small intestine. The virus invades these cells which function in digestion and absorption of nutrients from ingested food. Loss of the infected cells leads to marked shortening of the villi and replacement by immature cells originating from the crypts. Villous atrophy is most marked in the jejunum and ileum and often absent in the proximal portion of the duodenum (Figure 4).[63] Virus particles have also been observed in M cells and lymphocytes in the dome regions of Peyer's patches. The cell destruction and rapid cell turnover results in: (1) malabsorption caused by loss of cell function and an increase in osmotic pressure due to undigested milk lactose;[63] and (2) electrolyte imbalance due to altered sodium transport by immature epithelial cells.[66] Both factors lead to diarrhea, dehydration, acidosis, and death in young pigs.[67]

Young piglets may be more highly susceptible to infection by TGEV than older animals, thus accounting for the high mortality rates observed in young pigs. Both virus production and villous atrophy were greater in newborn piglets than in 3-week-old pigs.[68,69] Several mechanisms were proposed to account for this age-dependent resistance to clinical disease. First, intestinal epithelial cells are replaced more rapidly in older pigs and the newly replaced cells are resistant to TGEV infection.[70,71] Resistance may be due to the onset of the immune response including maturation of NK cells,[62] stimulated by presence of intestinal interferon which is also virocidal,[72] or inability of the regenerating cells to support virus growth. Second, TGEV replicates and accumulates in the apical tubulovascular system of villous absorptive cells, a system limited to pigs less than 3 weeks old.[73] Third, virus dose may be important since Witte and Walther,[74] demonstrated that the infectious dose needed to infect a 6 month-old pig was 10^4 times that required to infect a 2-d-old piglet.

Cell culture attenuated strains of TGEV failed to infect epithelial cells in the cranial portion of the piglet small intestine, probably accounting for the mild or lack of diarrhea reported for such strains.[75,76] In addition, there was a reverse correlation between the degree of cell culture attenuation of TGEV and the extent of intestinal infection.[76]

B. EXTRAINTESTINAL REPLICATION

Although the fecal-oral route may be the most common means of TGEV transmission, airborne infection may be important also. Lung lesions without accompanying pneumonia were observed in gnotobiotic pigs inoculated intranasally (IN) and orally with TGEV and in market weight swine from which TGEV was isolated from the lung.[59] TGEV was present also in alveolar macrophages of infected neonatal pigs, suggesting a possible role for such cells in lung infection. However, *in vitro*, only the cell culture adapted but not virulent TGEV replicated in cultures of alveolar macrophages.[77] Attenuated strains of TGEV that replicate in the upper respiratory tract and lung, but not in the intestine of newborn pigs, have been reported.[78] More recently, a respiratory strain of TGEV has been identified in Denmark, Belgium, and England.[5,6] This porcine respiratory coronavirus replicates almost exclusively in the upper respiratory tract of swine, causing mild respiratory disease (coughing and nasal discharge), but no gastroenteritis.[5,51] It is spread by the aerogenic route.[51] This respiratory TGEV strain is serologically indistinguishable from the enteric TGEV strains in conventional tests (infected animals produce neutralizing antibodies to enteric TGEV strains) and does not provide cross-protection against experimental challenge with an enteric TGEV strain.[5] The origin of this respiratory strain of TGEV is uncertain. It may represent a mutant showing altered tissue tropism which originated from enteric TGEV, the live vaccine strain of TGEV, or the antigenically-related canine or feline coronaviruses.

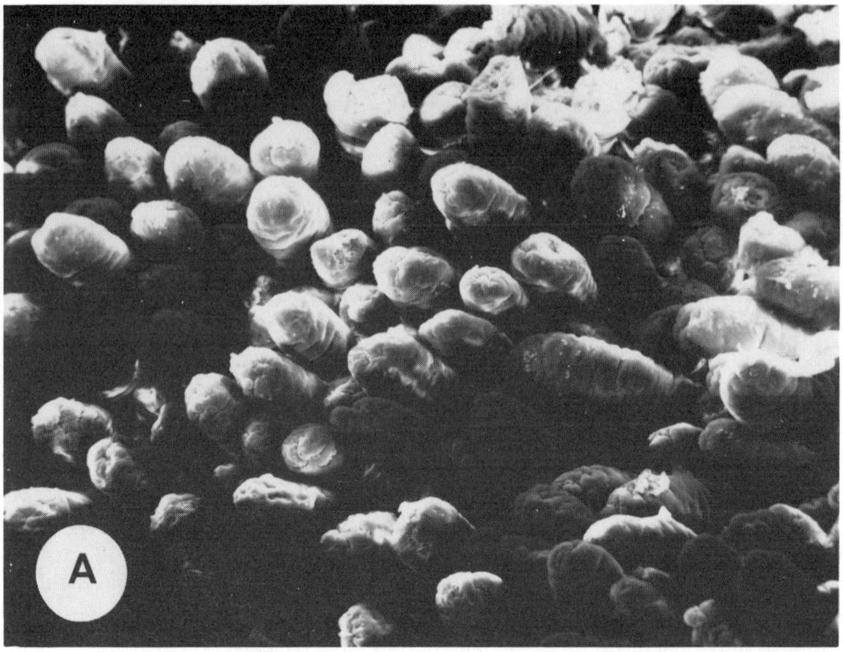

A

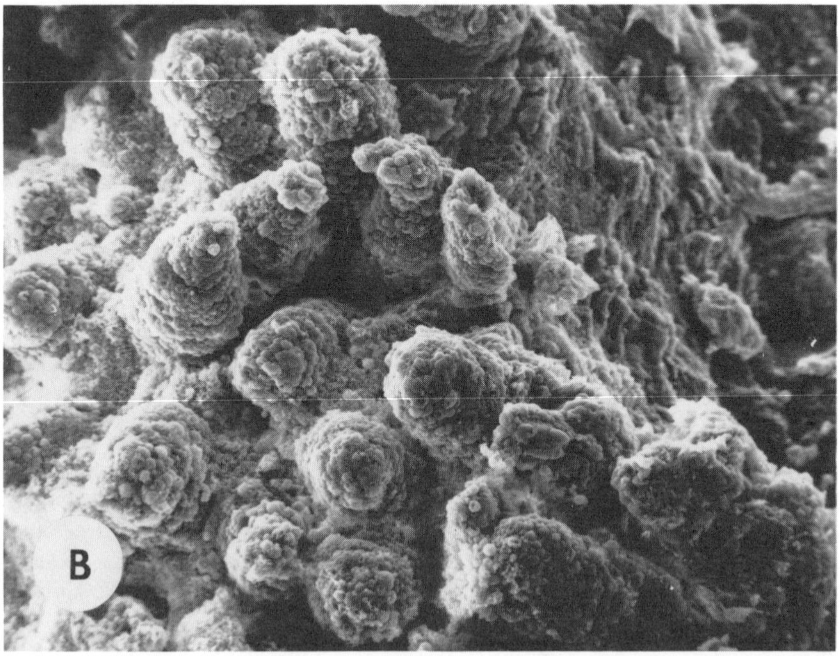

B

FIGURE 4. Scanning electron micrograph of small intestinal villi from a gnotobiotic pig challenged 28 h previously with virulent TGE virus. (A) Mainly intact villi in the duodenum of the infected pig. (B) Severely shortened villi containing swollen enterocytes in the ileum of the same pig. (Magnification × 100.)

Additional studies have shown that TGEV replicates in mammary tissues of lactating sows[79] and that infected animals shed virus in milk.[79,80] The significance of the mammary-gland route of infection or transmission of virus is unclear, but it may play a role in the agalactia often noted in TGEV-infected sows, or rapid spread of infection among piglets.

VIII. DIAGNOSIS

The diagnosis of TGE may be accomplished by one or more of the following procedures: detection of viral antigen, electron microscopic detection of virus, isolation and identification of virus, or detection of rising antibody titers in serum. However the routine diagnosis of TGEV in certain herds has recently been complicated by the finding that in some countries (Denmark, Belgium, England, and Switzerland), a respiratory strain of TGEV occurs which cannot be antigenically distinguished from the enteric strains.[5,6,51] In these countries, respiratory TGEV infections have become endemic, thereby precluding any serologic diagnosis of enteric TGEV or diagnosis based on use of polyclonal antiserum in conventional immunoassays. However, the recent generation of MAbs, which can discriminate between the respiratory and enteric strains of TGEV, may provide the basis for development of future competitive ELISA tests for the serologic differentiation of the two virus strains.[51,52] Diagnosis based on clinical signs or histologic lesions alone is not accurate since similar signs and lesions may frequently occur following other enteric infections. However, the routine evaluation of clinical signs, histologic lesions and patterns of virus shedding in feces vs. respiratory secretions may be of great importance in distinguishing possible respiratory and enteric forms of TGEV infection. Presently, wholly respiratory strains of TGEV have not been reported to occur in countries other than those noted earlier.

A. DETECTION OF TGEV OR VIRAL ANTIGEN

The most widely used technique for diagnosis of TGE in young pigs is probably immunofluorescent (IF) staining of small intestinal sections or smears from infected pigs.[12,81] Problems which may be encountered in IF tests include a lack of sensitivity or specificity of antibody reagents and loss of infected epithelial cells in pigs euthanatized at a later stage of infection. The problem of reagent specificity may be overcome in the future by use of MAbs for direct or indirect IF staining of TGEV antigens (Figure 5). Two studies reported generation of MAbs which had high IF titers to TGEV in infected cell cultures.[31,33] Some of these MAbs showed different degrees of reactivity with heterologous TGEV strains suggesting their possible application for virus strain differentiation. Preliminary data, employing strain-specific MAbs in competitive ELISA to differentiate the recently identified respiratory strain of TGEV from the enteric strain have been reported.[51,52] Other frequently used methods for detection of TGEV include: (1) isolation of virus in cell culture and identification of infected cells by IF;[23] and (2) detection in feces, intestinal contents, or infected cell culture fluids using negative contrast electron microscopy or immune electron microscopy.[82] Other less commonly used procedures include: agar gel precipitation,[83] immunoelectrophoresis and counterimmunoelectrophoresis,[84] indirect hemagglutination,[85] immunoperoxidase staining of infected tissues,[86,87] and ELISA.[88]

B. SEROLOGIC DIAGNOSIS

Demonstration of a rise in serum antibody titers between acute and convalescent samples is a useful diagnostic method, especially for epidemic TGEV. Serologic tests also are used to monitor the TGE infection status of a herd. However, in some countries as noted previously, use of these tests may be complicated by the occurrence of a respiratory form of TGEV, causing swine of all ages to seroconvert to TGEV in the absence of any confirmed enteric infections with TGEV.[5] Neutralizing antibodies to enteric TGEV infections can be detected as early as 7 to 8 d after infection and may persist for at least 18 months;[89,90] little is known about onset or duration of neutralizing antibodies generated against a respiratory strain of TGEV.

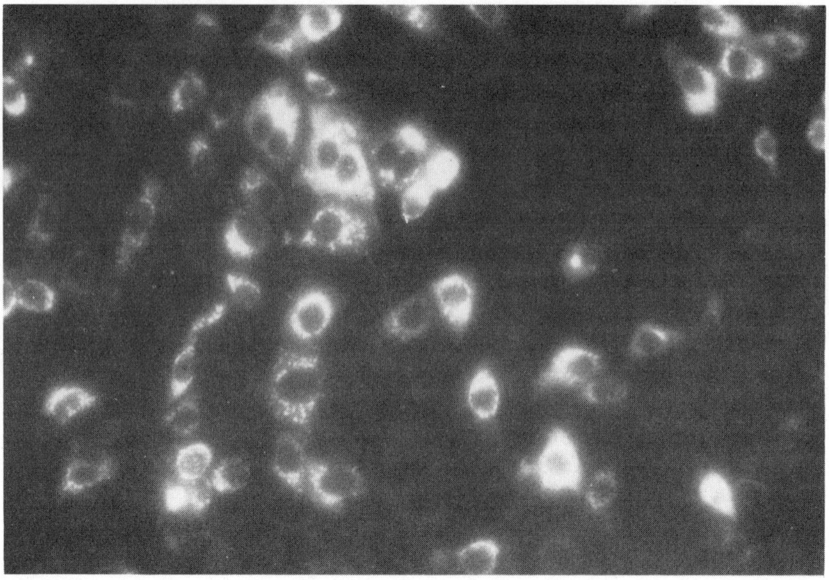

FIGURE 5. Immunofluorescence observed after indirect staining of Miller TGEV-infected ST cells with TGEV monoclonal antibody directed against the nucleocapsid protein. Notice the bright cytoplasmic particulate fluorescence. No fluorescence was observed in uninfected ST cells stained in a similar manner. (Magnification × 200.)

The VN test has been the most common antibody assay, using cell culture-adapted viruses in a variety of procedures and cell culture systems. These include inhibition of CPE[3,14] and plaque reduction.[91] An indirect IF test was rapid,[92] but was less sensitive and reliable than the VN test.[93] An agar gel precipitation test was also relatively insensitive.[94,95] More sensitive passive hemagglutination tests[96,97] and ELISA tests[98-100] have been described, but both required concentrated purified virus. Both ELISA and indirect immunoperoxidase tests have been adapted to detect immunoglobulin isotypes.[99,101] Other recently described tests include radioimmunoprecipitation[102] and an autoradiographic test.[103]

IX. EPIDEMIOLOGY

On a herd basis, two epidemiologic forms of TGEV are observed: epidemic and endemic. The major characteristics of each will be noted below.

A. EPIDEMIC TGE

Epidemic TGE follows introduction of TGEV into a herd in which most, if not all, of the animals are susceptible (seronegative). TGEV spreads very rapidly within such a herd, with swine of all ages infected and showing some degree of inappetence, vomition or diarrhea. Virus is spread to uninfected swine in feces (with as little as 2×10^{-7} ml of intestinal contents infectious for 2-d-old piglets), and possibly also as aerosols and in milk.[79,80]

Suckling pigs become very sick, dehydrating rapidly, and mortality may approach 100% in piglets under 3 weeks of age. Lactating sows often develop anorexia and agalactia, contributing further to piglet mortality. Occurrence of such severe clinical signs aids in diagnosis of epidemic TGE in the U.S., where other similar diseases have not been reported, but might be difficult to distinguish from PED in Europe.[49]

B. ENDEMIC TGE

Endemic TGE refers to persistence of TGEV and disease in partially immune herds which have frequent or continuous farrowings or other frequent additions of susceptible swine.[2,61,104,105] Seropositive breeding females will transfer to their suckling piglets a variable degree of passive immunity via their colostrum and milk. When the level of viral exposure exceeds the piglets' passive immunity, mild but typical signs of TGEV occur frequently in piglets from 6 to 21 d of age. Mortality is usually less than 10 to 20%, being influenced by the piglets' age when infected and the variable degree of immunity from the sow. Sows usually do not become sick. In some herds, the disease occurs primarily during the postweaning period, or is first established in older swine, eventually spreading to the young piglets. Endemic TGE in suckling or recently weaned pigs can be difficult to diagnose and could be confused with other types of enzootic diarrheal problems, such as rotaviral diarrhea and *E. coli*, unless an etiologic diagnosis is done.[2,61] Endemic TGE may persist in herds as long as susceptible or partially immune swine are exposed to TGEV. In England, endemic TGE was found in ~50% of herds surveyed between 1981 to 1983, and recurred much more frequently in larger herds (>100 sows).[61]

C. TRANSMISSION AND RESERVOIRS

A major epidemiologic feature of TGEV is its seasonal occurrence, with the peak incidence during the winter, usually from November to April. Explanations given for this seasonal pattern include: (1) increased survival of the virus in the cold,[55] allowing direct transmission between herds through transport of contaminated feces on vehicles or animals or other objects; (2) the effect of reduced or fluctuating ambient temperatures in predisposing feeder pigs to clinical manifestations of TGE[100,107] or reactivation of quiescent virus causing increased transmission; and (3) possible increased contact in the winter between swine and nonporcine carriers of TGEV.

An important, but largely unresolved question, is what constitutes the reservoir of TGEV between seasonal epidemics? At least three reservoirs have been suggested: pig farms in which the virus spreads subclinically; carrier swine; and nonporcine hosts.[55] There is evidence for maintenance of the diseases in the endemic form in feeder pig operations,[108] and a similar situation probably prevails in continuous farrowing operations.[61] TGEV infections can spread slowly through growing swine during the summer months,[109] thus constituting a foci of infection for dissemination of the disease during the winter months.

The possible role of carrier swine in transmission of TGEV is poorly understood and difficult to assess. Under natural conditions, the fecal shedding period for TGEV was about 2 weeks,[12] with only one report suggesting a more prolonged shedding period of up to 8 weeks post-exposure.[110] Respiratory shedding, as detected by isolation of TGEV from nasal swabs, persisted for post-exposure periods up to 11 d.[111] TGEV was also recovered from milk of infected sows during the acute phase of the disease[80] or after intramammary infusion or injection of live TGEV into lactating sows.[79] Although TGEV has been isolated from intestinal contents or lung homogenates of infected gnotobiotic pigs for post-exposure periods of up to 104 d,[59] it is uncertain whether such virus can be eliminated from the body in a viable state which will result in new infections. Evidence to the contrary was reported in one study in which addition of susceptible pigs to a herd at 3, 4, or 5 months after an earlier TGE outbreak, resulted in no infections or seroconversions in the introduced pigs.[112]

Experimental evidence exists for replication of TGEV in nonporcine hosts which were fed the virus. Fecal excretion of TGEV in the absence of clinical signs has been reported in cats, dogs, and foxes, and these species may serve as potential carriers of TGEV from one herd to another.[55,56,58] TGEV was found to replicate in intestinal epithelial cells in dogs which were shown to excrete TGEV in feces that was infectious for pigs.[55,113,114] Other investigators demonstrated neutralizing antibodies to TGEV in the sera of puppies, kittens, skunks, and opossums, a month or longer after experimental exposure to TGE.[115] However, the detection of

TGEV neutralizing antibodies in sera of nonchallenged dogs and cats must be interpreted with caution since feline and canine coronaviruses, antigenically related to TGEV, occur in domestic animals. In comparison, no seroconversion was detected in TGEV-exposed laboratory rodents including mice, rats, gerbils, and guinea pigs.

Both starlings (*Sturnus vulgaris*) and house flies (*Musca domestica*) have been proposed as possible vectors for TGEV, the virus surviving passage through these hosts,[116,117] but probably without causing infection in them. Congregation of large flocks of starlings in feeding areas of swine in winter may permit mechanical transmission of TGEV from one farm to another. House flies may carry TGEV among swine within a herd in which TGEV is endemic.

X. IMMUNITY

A. ACTIVE IMMUNITY

Mechanisms and duration of active immunity to TGEV infections in swine are poorly characterized. This is due in part to the problems associated with sampling intestinal fluids and the assay of antibodies or mediators of cellular immunity in such fluids. Active immunity is essential in the sow to transmit effective passive (or lactogenic) immunity to pigs and to provide herd immunity for control of the disease. There is little information on the persistence of serum antibodies after TGEV infection of breeding age swine beyond about 6 months.[104] However, it is well documented that circulating antibodies (actively or passively acquired) provide little protection against TGEV infections,[118,119] although they do provide a means for serologic diagnosis of TGE. Swine that have recovered from TGEV infections develop immunity to subsequent challenge, presumably due to local immunity within the intestinal tract. The completeness and duration of active immunity may be influenced by the age of the animal at initial challenge and the severity of the challenge.

The mechanism of antibody-mediated active immunity in the gut is unclear in regard to TGEV infections, but by analogy with mucosal immunity in other species may relate to production of secretory (S) IgA antibodies by lymphoid cells within the lamina propria.[120,121] TGEV antibodies of the SIgA class are present in serum and intestinal fluids of orally, but not parenterally TGEV-inoculated swine.[102,122] Kodama et al.[102] proposed that the presence of such serum IgA antibodies may signal a prior gut infection and development of active immunity. Oral inoculation of gnotobiotic piglets with TGEV resulted in the sequential appearance of IgM and IgA immunocytes in the lamina propria of the gut[123] and the development of serum and intestinal neutralizing antibodies by 5 d post-exposure. These and other studies[124] indicate that the pig is immunocompetent at birth, but antibody production and antibody producing cell numbers in the intestine require additional maturation time to reach adult levels.

Cell-mediated immune (CMI) responses in the intestines of infected pigs may also be important in active immunity against TGEV infections. Cell-mediated immunity to TGEV has been detected using a number of tests including macrophage migration inhibition,[125] leukocyte migration inhibition,[126,127] direct lymphocyte cytotoxicity,[128] lymphocyte proliferative responses,[129,130] spontaneous cell-mediated cytotoxicity (SCMC), and antibody-dependent cell-mediated cytotoxicity (ADCMC).[131] At present, only indirect evidence exists concerning the role of CMI in resistance to or recovery from TGEV infections. CMI was demonstrated with lymphocytes from gut-associated lymphoid tissues of pigs orally infected with virulent TGEV; parenterally inoculated pigs, or pigs given attenuated TGEV orally developed CMI only in systemic sites (spleen or peripheral blood lymphocytes).[125,129,130] These latter observations may be especially important in the failure of parenterally administered TGEV vaccines to evoke lactogenic immunity[132,133] or in the susceptibility of pigs given attenuated TGEV to infection following challenge with virulent TGEV.[167] CMI persisted with gut associated lymphoid tissue, but not systemic lymphocytes for at least 110 d after oral infection of 6 month-old swine,[129] but only up to approximately 14 d after infection of younger (7-d-old) swine.[130]

Recent reports indicated lymphocyte cytotoxicity was greatly decreased in newborn piglets and in parturient sows.[62] This lack of killer (K) and natural killer (NK) cell activity against TGEV correlates with the increased susceptibility of newborn piglets and parturient sows to TGEV infection.[62] Thus SCMC, ADCMC, or sensitized T-lymphocyte-mediated cytotoxicity alone or in combination may play a role in either recovery from TGEV infection or resistance to reinfection via the rapid elimination of TGEV-infected epithelial cells.

B. PASSIVE IMMUNITY

The pigs at greatest risk from TGEV infections are the newborn which may not have time prior to viral exposure to develop active immunity. Hence, passive immunity is of primary importance in providing immediate protection. Piglets are born devoid of immunoglobulins (Igs) which they acquire after birth via colostrum. Colostral Igs consist primarily of IgG and represent a serum exudate transferred from the dam across the piglet's intestinal epithelium and into its circulation.[120] The neonate is thereby provided with the same complement of serum antibodies as in the dam. These circulating antibodies function mainly in protection against systemic infections and provide little protection against intestinal infections.[63] To be effective against enteric infections, antibodies must be present on a frequent or almost continual basis in the lumen of the gut.[134] This is accomplished naturally when pigs suckle immune sows, or it can be artificially accomplished by continuous feeding of TGEV antiserum to piglets.[134] The latter, although presently impractical, might eventually be feasible by use of MAbs in a type of slow-release delivery system.

Mechanisms of passive immunity to TGEV infection have been reviewed.[2,135-138] TGEV antibodies in the colostrum and milk of sows naturally infected or orally inoculated with virulent TGEV are primarily of the SIgA class. Parenteral or systemic inoculation with TGEV results in mainly IgG antibodies in colostrum and milk. Antibodies of the SIgA class in milk provide optimal passive immunity to TGEV,[135-141] probably because they are the predominant antibodies in milk[120] and they are more resistant to proteolytic degradation in the intestine. IgM and IgG antibodies may also be effective in passive immunity,[141,142] but generally these antibodies do not persist in high levels in milk. The occurrence of SIgA TGEV antibodies in milk following an intestinal infection, was explained by the proposed migration of TGEV-committed IgA immunocytes from the gut to the mammary gland. These cells then localize in the mammary gland and secrete TGEV antibodies of the IgA class into the colostrum and milk.[2,136,138] This "gut-mammary" immunologic axis is an important concept in designing effective immunization protocols to provide lactogenic immunity against enteric infections.

1. Vaccination of the Seronegative Dam

There have been numerous attempts to stimulate passive immunity to TGEV by vaccination of the pregnant dam using experimental or commercial vaccines. Vaccination has been attempted using virus in a variety of forms (virulent, attenuated, inactivated and subunits) and administration routes (oral, intranasal [IN], intramuscular [IM], and intramammary [IMm]).[132,135-145] However, there are a number of variables which complicate evaluation of TGEV vaccines, often producing conflicting data. These variables include: the immune status of the dam at the time of vaccination; the age of the pig at challenge; the dose and strain of TGEV used for challenge; environmental conditions during challenge, particularly temperatures; and the milking efficiency of the dam.

2. Intramuscular Vaccination

In general, IM inoculation of seronegative sows with virus, whether inactivated, attenuated or virulent, resulted in stimulation of only IgG TGEV antibodies in serum, colostrum, and milk and little protection of litters (38 to 56% mortality) against challenge.[132,133,135-143,145] Only two studies reported favorable protection results after IM inoculations using commercial TGEV vaccines.[146,147] The failure of IM vaccination of the dam to provide adequate passive immunity may relate first to a lack of stimulation of gut immunity in the dam; sows may get sick when exposed to TGEV during lactation, depriving their piglets of adequate milk. Secondly TGEV antibodies in milk are usually of low nonprotective titers of the IgG class.[132,135-143]

3. Intramammary Vaccination

Intramammary inoculation of seronegative pregnant swine with the live attenuated Purdue strain of TGEV resulted in high titers of primarily IgG antibodies in milk,[132] whereas similar inoculations in seronegative lactating sows resulted in IgA and IgM antibodies in milk.[79] Specific antibody occurred in milk from both injected and noninjected glands. Protection was enhanced (14 to 26% mortality) in litters of IMm-vaccinated dams compared to controls, presumably because exceptionally high levels of IgG antibodies persisted in milk at the time of challenge at 3 d post-farrowing.[141,148] Other investigators[144] also noted greatly enhanced IgG milk antibody titers and complete litter protection (0% mortality) after IM/IN inoculation of sows with a high titered (10^8 to $10^{9.3}$ $TCID_{50}$) attenuated TO163 strain of TGEV. These data suggest IgG TGEV antibodies are also protective when present in high persisting levels in milk.

4. Oral/Intranasal Vaccination

Natural infection or oral administration of virulent TGEV to pregnant swine led to significant quantities of TGEV antibodies of the SIgA class in milk which persisted throughout lactation and provided optimal passive protection (0 to 6% litter mortality).[2,136-138] This finding confirms early observations that sows which recover from TGE transmit protection to their piglets.[149] This procedure has been used to minimize losses in the face of a TGEV outbreak by the "planned" infection of pregnant swine at least 3 weeks prior to farrowing with minced guts from acutely infected piglets. Such methods, using virulent TGEV from pig intestines, have the disadvantage of the possible introduction of other pathogens into pregnant sows or dissemination of other pathogens throughout the premises.

Attempts to mimic the natural route of infection by oral inoculation of pregnant swine with attenuated strains of TGEV have been generally unsuccessful in either eliciting high levels of SIgA milk antibodies or providing passive protection (25 to 100% litter mortality).[2,136-138,145] Concerns that attenuated strains of TGEV are more acid labile and sensitive to degradation by proteolytic enzymes than virulent strains (which has been shown in some studies,[150-152] but not others),[153] prompted studies of the administration of virus in enteric-coated capsules or by direct inoculation into the gut lumen. Results varied depending on the attenuated strain of TGEV tested: use of the B1 or Reims strain of attenuated TGEV provided greater protection (10% litter mortality)[154,155] than did use of the Purdue strain of TGEV (44 to 62% mortality).[145] A variant of the attenuated Purdue strain of TGEV showing enhanced resistance *in vitro* to proteolytic enzymes[152] and a low cell culture passage level of the Miller virulent strain of TGEV,[143] was also evaluated as a possible vaccine in pregnant swine. Both types of vaccines produced inconsistent results in terms of passive protection of challenged piglets, with litter mortality ranging from 0 to 73%. Conflicting litter protection data were also reported for vaccination studies using a commercial vaccine administered twice orally (in feed) and once IM to pregnant sows. Whereas one study noted good protection (8% litter mortality),[156] other investigations by two independent groups of researchers reported high mortality, similar to that seen in piglets suckling unvaccinated sows.[133,157,158]

The generally poor results obtained with oral or IN vaccination of seronegative sows using attenuated strains of TGEV may be attributed to the superficial or limited replication of such strains in the adult intestine.[75,76] Limited antigenic stimulation of intestinal IgA lymphocytes would provide seeding of few IgA TGE-sensitized lymphocytes to the mammary gland and consequently little IgA antibody secretion in milk. The dilemma still exists of how to commercially engineer vaccine strains of TGEV capable of generating an IgA antibody response in the gut of sows, but sufficiently attenuated so as not to cause disease in newborn piglets.

5. Other Experimental Vaccines

Other recent approaches to development of TGEV vaccines include evaluation of (1) heterologous coronavirus vaccines, based on the antigenic relationships which exist between TGEV, FIPV and CCV;[54,159] (2) a live attenuated small plaque (SP) variant vaccine;[159,160] and (3) subunit vaccines.[161] Results of studies using FIPV or CCV as vaccines in pregnant swine[54,159] were similar to those reported using the live attenuated strain of TGEV.[136,143] A moderate level of protection was obtained (25 to 52% litter mortality). Better protection (14 to 24% litter mortality) was noted after challenge of suckling pigs from dams vaccinated oral/IN and/or IMm with the SP variant TGEV strain.[159,160] Further studies of the site of SP viral replication (this virus reportedly replicates in cells in the lamina propria and not gut epithelial cells)[159] and mechanisms of antibody induction are warranted.

Virus subunits purified from virulent TGEV and inoculated with adjuvants by parenteral routes were evaluated as vaccines in two studies.[30,161] In the first study, Garwes et al.[30] injected pregnant sows IMm with TGEV peplomers, and observed mainly IgG antibodies in colostrum. Poor protection resulted after challenge of the suckling pigs (100% mortality in 4 of 6 litters). In the second investigation, Gough et al.[161] used a low molecular weight (about 23,000) undefined subunit of TGEV to IM inject pregnant gilts. These researchers reported only 4% litter mortality among challenged piglets, but the Ig class of TGEV antibodies in the milk was not determined. Although rDNA technology might eventually provide the means to produce large quantities of TGEV subunits, the conflicting protection data generated in these two studies suggests further work is needed to more thoroughly characterize TGEV subunits and the immune reponses they elicit in swine.

6. Vaccination of Previously Infected Swine

Use of vaccines in pregnant swine that have been previously naturally infected with TGEV, compared with their use in seronegative swine, may result in major differences in immune responses and piglet protection. Such differences may account for some of the conflicting results reported in vaccine challenge studies, if previously infected swine were unknowingly used.[2] Use of very sensitive TGEV antibody assays (such as plaque reduction VN) and knowing the exposure history of the test animals are necessary to detect and avoid the use of previously infected swine.

Results of limited experimental studies have indicated that parenteral inoculation of previously infected pregnant swine with live attenuated TGEV boosts both IgA and IgG TGEV antibody titers in milk.[79,129,130] Others reported four- to sevenfold greater TGEV antibody titers in milk after IMm inoculation of previously infected swine with inactivated TGEV.[162] Thus, the currently available TGEV vaccines may be more efficacious for boosting immunity in previously infected swine than for initiating immunity in seronegative pregnant swine. Such vaccines may be particularly useful in herds troubled by endemic TGE.[104,163,164]

Although little information is presently available, preliminary data suggests that prior infection of sows with the porcine respiratory coronavirus does not prevent infection or disease associated with enteric TGEV, which results in a typical outbreak of diarrhea and high mortality in litters affected when under 2 weeks of age.[165] However, previous exposure of sows to respiratory coronavirus did appear to moderate the severity of TGEV infections among litters from sows which were infected with TGEV late in pregnancy. Lactogenic immunity was rapidly enhanced in such sows, presumably due to rapid booster responses, which occurred as a result of their prior exposure to porcine respiratory coronavirus.

XI. PREVENTION AND CONTROL

Because the commercially available vaccines are of limited effectiveness in providing the level of lactogenic immunity needed to prevent the epidemic form of TGEV,[132,133,157,158] management factors play an important role in the control of TGE. Care must be taken to eliminate or control the spread of TGEV within a herd or between herds via mechanical transmission of contaminated feces (carried on boots, shoes, clothing, feeds, vehicles, or by starlings or flies). Attention should be given to the possible role of the carrier animal in the introduction of TGEV into a herd. This includes isolation of newly acquired animals or previously infected swine for at least 4 weeks before adding them to the herd proper or to a TGEV free herd, respectively. An alternative is to introduce only TGEV seronegative animals into a TGE free herd. In certain situations, cats, dogs, foxes, or other wildlife may also play a role in the dissemination of TGEV between herds, so their movement or access to swine should be controlled whenever possible.

One approach to the control of endemic TGE is use of the commercial vaccines to boost passive immunity in suckling pigs, thereby delaying the onset of TGEV in exposed pigs and reducing mortality.[79,104,138] Second, breaking the cycle of infection by an "all in, all out" management system can be especially important in preventing the continual influx of susceptible animals into the herd or facility.

Finally, in the face of a TGEV outbreak, death losses can be minimized by providing pigs with a very warm (above 32°C), dry, draft-free environment and free access to water, electrolyte solutions or milk replacer. Antibacterial therapy might be useful in 2- to 5-week-old pigs if there are concurrent or secondary bacterial infections. Cross-suckling or putting infected or susceptible pigs onto TGE-immune sows was found useful in various field outbreaks.[104,166] When TGE has occurred on a farm and some animals are due to farrow more than 2 weeks hence, purposely exposing them to virulent virus, by feeding the guts of infected pigs, may render them immune at farrowing time.

B. CANINE CORONAVIRUS

I. INTRODUCTION

The study of canine coronavirus infections in dogs has been complicated by the finding that dogs are susceptible to infection with two coronaviruses: the canine coronavirus (CCV)[1] and transmissible gastroenteritis of swine.[2] Results of most studies suggest that only CCV causes disease in dogs and TGEV infections are asymptomatic.[2,3] However, the relationship between TGEV and CCV as a cause of gastroenteritis in dogs required further investigation, since in one study a TGE-like virus (based on virulence for pigs) was recovered from dogs with gastroenteritis.[4] Although TGEV and CCV are antigenically closely related, they can be differentiated serologically by reciprocal cross neutralization tests[5] and differences in infectivity for canine or porcine cell cultures.[1,5] In 1970 Norman and colleagues[6] first detected antibodies to TGEV in dogs, many of which had never been in contact with pigs. Cartwright and Lucas,[7] 2 years later, showed rising titers of antibodies which neutralized TGEV in a kennel of dogs following an outbreak of gastroenteritis. These early reports, documenting coronavirus infections in dogs not closely associated with pigs, may reflect the presence of CCV antibodies in dogs.

In 1971, a CCV designated I-71 was isolated in primary dog kidney cell cultures from military dogs during an outbreak of diarrhea.[1] The virus was antigenically related to TGEV, but could be differentiated from TGEV by its infectivity for puppies in which it caused diarrhea and dehydration, and its failure to infect piglets or grow in porcine cell cultures.[1] In 1978, widespread epidemics of diarrhea occurred among dogs in the U.S., and CCV was isolated repeatedly from dogs during the disease outbreaks.[8-10] CCV causes vomiting and diarrhea in dogs, with infections ranging in severity from inapparent to fatal, but with most dogs recovering after 1 to 2 weeks. Puppies are usually more severely affected.[9,10]

II. MORPHOLOGY AND MORPHOGENESIS

Morphologically, CCV is a typical member of the Coronaviridae family. It is enveloped and pleomorphic with a single layer of club-shaped surface projections or peplomers as revealed by negative stain electron microscopy[8,11,12] (Figure 6). Virion size ranges were not described in the literature except for atypical coronavirus-like particles seen in stools from clinically normal Australian dogs.[13] Since similar particles were seen in human stools from persons in contact with the dogs, these particles may have been of human origin and ingested by the dogs. In our laboratory, we have found CCV particles ranging in diameter from 50 to 180 nm, with peplomers which are 12 to 18 nm in length (Figure 6).

Typical of coronaviruses, CCV matures in the cell cytoplasm by budding through the endoplasmic reticulum.[1,14] Virions accumulate in cytoplasmic vesicles and appear oval or round, ranging from 59 to 90 nm in diameter with inner cores, 35 to 45 nm.

III. CULTIVATION

CCV propagates in a number of canine and feline cells. It was first isolated in primary dog kidney cells and a canine thymus cell line.[1] CCV replicates and produces CPE in canine embryo, synovium and fibroma (A72) cell lines, with CPE induced most consistently in the latter cell line.[9,10,15] The CPE is usually observed at 2 to 3 days postinfection, and consists of the appearance of enlarged, abnormal cells followed by focal detachment. Formation of large syncytia was seen with some isolates in certain cells when inoculated at a low multiplicity of infection.[10,16] Feline cells susceptible to CCV infection include the CRFK cat kidney cell line[9,10,16] and a feline embryo fibroblast or whole fetus cell line.[9,17] Plaque formation by CCV was reported in A72 and CRFK cells.[10,18] CCV (I71 isolate) failed to replicate in porcine cells including primary swine kidney, pig kidney PK15, swine testes and secondary pig thyroid.[1,5] This characteristic of CCV provides an important basis for differentiation of CCV from TGEV, which replicates in the aforementioned cells.

IV. PHYSICOCHEMICAL PROPERTIES

Physicochemical properties of CCV resemble those reported for other coronaviruses.[1,9,10] CCV has a buoyant density of 1.18 g/ml in sucrose gradients. The virus is sensitive to chloroform (indicative of a lipid-containing envelope) and also inactivated by formalin, phenol, hypochlorite solutions, and beta-propiolactone. CCV is acid stable at pH 3.0 (20 to 22°C) and growth in cell culture was not inhibited by iododeoxyuridine, consistent with an RNA genome.

The structural polypeptides of CCV have been examined by several investigators.[19-21] Three major structural proteins were identified: (1) a peplomer glycoprotein (E2) with a relative molecular mass (M_r) of 130 to 203.8 kDa; (2) a membrane glycoprotein (E1) with an M_r of 28 to 31.8 kDa; and (3) a nucleocapsid protein with an M_r of 45 to 49.8 kDa. A fourth major protein, a glycoprotein of M_r 21.6 kDa was described in one study[20] as being specific to CCV and not found in TGEV, but in a second study a minor polypeptide with a M_r of 21 kDa was associated with TGEV.[21] The three major proteins of CCV correspond closely with those of TGEV.[20,21]

VI. ANTIGENIC RELATIONSHIPS

In vitro — Only one serotype of CCV is presently known, but antigenic differences have been

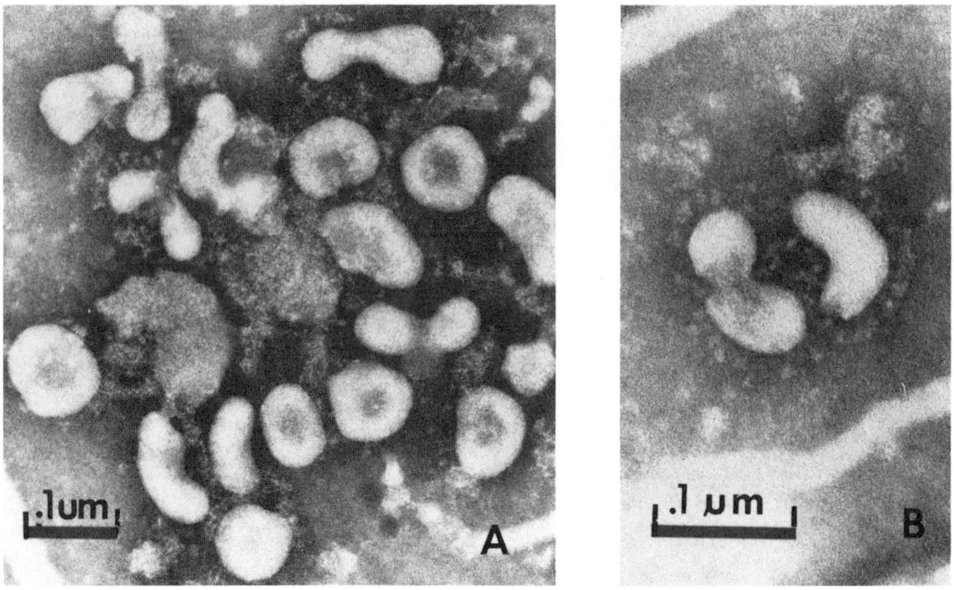

FIGURE 6. Electron micrograph of negatively stained canine coronavirus (CCV) grown in feline whole fetus cell. (A) The CCV was incubated with hyperimmune serum to Miller TGEV, resulting in specific aggregation of coronavirus particles. (B) Two CCV particles in the absence of specific antiserum. Note some particles have retained their peplomers (arrow) while others are missing peplomers.

reported between CCV isolates.[8,10] CCV is antigenically related to feline infectious peritonitis virus (FIPV) and TGEV of swine, and it has been proposed that these three viruses may represent host range mutants of a single virus strain.[21] Either one-way or two-way cross-reactions among the three viruses have been observed using immunofluorescence (IF), virus neutralization tests (VN) or immune electron microscopy (IEM, see Figure 6).[5,17,22,23] However, the viruses could be distinguished serologically in reciprocal VN tests, with homologous antibody titers consistently higher than heterologous titers.[5] Studies by Horzinek and colleagues[21] revealed common antigenic determinants associated with each of the major viral proteins (E1, E2, and N) of CCV, TGEV, and FIPV. Specific details of the various patterns of antigenic reactivities observed among the three viruses using a variety of assays are discussed in the TGEV chapter, as are more extensive comparisons of their *in vivo* infectivities for dogs, cats or pigs.

In vivo — Infection of neonatal pigs with CCV was demonstrated in one study,[24] but in agreement with an earlier report, the pigs did not develop disease or seroconvert to TGEV.[1] Likewise, dogs could be infected with TGEV as demonstrated by fecal shedding. However, no disease developed, dogs did not seroconvert to CCV, but did seroconvert to TGEV[2] and dogs were not protected upon subsequent challenge with CCV.[10] Dogs given FIPV orally were not protected against CCV challenge.[10] More recently, it was shown that cats are susceptible to infection with a field isolate of CCV, as evident by seroconversion to both CCV and TGEV (VN antibody responses).[25] However, there was no apparent clinical disease or detectable shedding of virus in feces. Furthermore, the inoculated cats were not protected upon challenge with FIPV. Therefore both *in vivo* and *in vitro* studies suggest differences exist among these three antigenically similar coronaviruses in their infectivities for the homologous versus heterologous hosts or cell cultures.

VI. PATHOGENESIS AND PATHOGENICITY

Clinical signs of CCV in dogs are variable and inversely related to age, with young dogs more severely affected.[8-10] After a short incubation period of 1 to 3 d, infected dogs develop vomiting and diarrhea, characterized by watery or mucoid orange-colored feces, occasionally streaked with blood. Affected dogs may show anorexia, depression, and dehydration in severe cases, requiring fluid therapy. Leukopenia or elevated temperatures have not been observed in dogs infected with CCV, in contrast to diarrhea caused by canine parvovirus in which these two features are common.[8] Diarrhea usually continues for several days, but may persist for 2 weeks or more in some dogs. Most dogs recover spontaneously after 7 to 10 d, and the mortality rate is usually low, but stress and the presence of multiple or secondary infections may complicate recovery and increase mortality rates.[1,8,10,26]

The intestinal pathogenesis of CCV resembles that of TGEV in pigs and bovine coronavirus in calves.[26] Virus is spread by the fecal-oral route. CCV was recovered from feces within 1 to 2 d after onset of clinical signs and shedding persisted for 6 to 9 d. Virus was rarely recovered from pharyngeal secretions. Electron microscopic studies of CCV infections indicated that the virus entered small intestinal enterocytes between the microvilli via pinocytosis.[14] Viral antigen was identified by immunofluorescence (IF) in the apical cytoplasm of the small intestinal enterocytes but not crypt cells.[26] Fluorescence was visualized in the duodenum at 2 d postinoculation (DPI) and spread throughout the small intestine and to the ileum at 4 DPI. Specific fluorescence was not observed in the colon, but coronavirus-like particles were seen in the colonic mucosa of 10-week-old dogs with fatal diarrhea.[27] Virus was recovered from the small intestine, colon, mesenteric lymph nodes, and occasionally the liver and spleen at 2 to 10 DPI, but no viremia or generalized infection occurred.[26]

Histologic changes were restricted to the small intestine and characterized by atrophy and fusion of villi, crypt hyperplasia and increased cellularity in the lamina propria.[26] Villous epithelial cells became cuboidal and vacuolated and the contents of goblet cells were discharged. Cellular digestive enzymes were generally depressed, contributing to the occurrence of the malabsorptive diarrhea.

VII. DIAGNOSIS

An etiologic diagnosis is necessary to distinguish contagious gastroenteritis caused by CCV from similar syndromes caused by canine parvovirus or other enteric viruses.[8-10] Preliminary diagnosis of CCV infections can be done by electron microscopy,[8-12,28] virus isolation,[1,8,15,16,26] IF,[1,26] and serologic tests.[1,9,10]

Electron microscopy (EM) of negatively stained fecal suspensions is a rapid and sensitive method for detecting CCV and aids in detection of fastidious CCV strains which fail to replicate in cell culture.[8,11,12,28] Aggregation of virus by specific immune serum (IEM) is helpful in visualization of coronavirus particles (Figure 6)[29] and differentiation of coronavirus from other pleomorphic membranous particles commonly present in feces.[30] Use of a protein-A gold-IEM procedure provides additional sensitivity and specificity for coronavirus detection.[31]

A-72 canine and CRFK feline cell lines have been used successfully for primary isolation of CCV from feces.[9,10,15,16] Specific identification of the virus can be done by neutralization of CPE by CCV antiserum, or detection of virus or viral antigen by IEM and IF, respectively. IF staining of intestinal tissues or smears from infected dogs is another means for diagnosis of CCV infections,[26] but requires tissues freshly obtained at necropsy before autolysis of the villi and from dogs in the early stages of the infection before infected villi are desquamated.

Serology is also useful for diagnosis of CCV infections, particularly if paired sera are available. However, neutralizing antibody titers are slow to reach their peak and are generally low.[9,32]

Since CCV and TGEV are antigenically related, confirmatory tests to differentiate the two viruses may be needed, although results of most studies suggest TGEV infections are asymptomatic in dogs.[2,3] Differential tests for the two viruses include infectivity of CCV for canine or feline cells, but a lack of infectivity for porcine cells or in newborn piglets.[1] Serologic differentiation is possible by testing convalescent sera in reciprocal cross-neutralization tests.[5]

VIII. EPIDEMIOLOGY

The mode of transmission of CCV is ingestion of virus from a fecally contaminated environment and fomites. CCV infections occur frequently in the winter months.[1,9] Dogs of all breeds and ages are susceptible. Although it is assumed that wild Canidae are also susceptible, the virus has been isolated only from coyotes with diarrhea.[33] However, neutralizing antibodies to TGEV were detected in foxes,[2] which raises the possibility that those antibodies may result from prior infections with CCV. Canine coronaviral infections in domestic dogs are widespread, with evidence of infection reported in Europe,[1,7,11,32,34-36] the U.S.,[8,9,37] and Thailand.[9] Serologic evidence of CCV infection ranged from a seroprevalence of about 20% in family dogs to 60 to 80% in kennel populations.[34,37] In surveys of coronaviral shedding from dogs with diarrhea, investigators in Germany reported a detection rate of 8.2%.[36] In a recent survey among family dogs in the U.S., coronavirus was the third most commonly detected enteric virus in dogs with diarrhea (10.5% detection rate from 1980 to 1982).[28]

IX. IMMUNITY, PREVENTION, AND CONTROL

Recovery from oral infection with CCV renders dogs immune to reinfection, but the duration of immunity is unknown. Immunity does not necessarily correlate with serum antibodies, since dogs given live CCV parenterally developed serum antibodies, but were still susceptible to oral challenge with CCV.[10] Thus, local intestinal immunity appears to be important for protection. Inoculation of dogs by the oral route with heterologous, antigenically related coronaviruses (TGEV and FIPV) resulted in intestinal infections, but failed to induce immunity to CCV.[10] Maternal antibodies to CCV were present in dogs from affected kennels, but titers were low.[10] The possible role of milk antibodies in passive protection of pups against CCV has not been defined.

A modified-live CCV vaccine was briefly marketed in the U.S. in 1983, but was withdrawn following neurological reactions in dogs when administered together with modified live canine distemper and canine parvovirus.[38] Recently, an inactivated CCV vaccine was developed for intramuscular or subcutaneous administration to dogs.[39] Preliminary trials showed efficacy against CCV infection in experimental challenge trials, but no information on vaccine efficacy under field conditions has been reported.

Presently, other measures which may aid in control of the disease include avoiding contact of dogs, particularly pups with infected dogs and their excretions. Disinfection of kennels with a 3% hypochlorite solution is effective in killing CCV. However, once CCV becomes established in a kennel, spread of infection is difficult to control.[10] After the disease is contracted, symptomatic treatment (such as fluid replacement therapy) should be instituted early in the course of the disease in severely affected dogs to control vomiting and diarrhea. Supportive care, such as avoiding stress to the dogs and providing extra warmth in the environment, may speed recovery.

C. FELINE ENTERIC CORONAVIRUS

I. INTRODUCTION

Initially, the only coronavirus recognized in cats was feline infectious peritonitis virus (FIPV) which causes a generalized infection characterized by inflammatory exudate in the body cavities (peritonitis, pleuritis) or disseminated granulomatous disease.[1] Since this coronavirus does not remain localized in the intestinal tract or consistently produce a clinically apparent enteritis, it will not be considered further except as related to a recently described feline enteric coronavirus.

Feline enteric coronavirus (FEC) infections of cats were first recognized in 1981.[2,3] The FEC causes inapparent to moderately severe enteritis in cats and is antigenically related to FIPV.[2] Because only limited information is available regarding this coronavirus, the following will provide a brief review of current knowledge of this disease.

II. MORPHOLOGY, MORPHOGENESIS, PHYSICOCHEMICAL PROPERTIES, AND GENOME

FEC is morphologically similar to FIPV or other coronaviruses. The virions are pleomorphic, 75 to 150 nm in diameter with peplomers (15 to 30 nm long) radiating from the envelope.[2,4]

Limited information is available describing the viral proteins and nucleic acid of FEC. Preliminary data suggests the viral proteins of FEC and FIPV have very similar relative molecular masses (M_r) consisting of 220 kDa for the E2 peplomer protein, 27 to 32 kDa for the E1 matrix protein, and 47 or 50 kDa for the nucleocapsid proteins of FEC and FIPV, respectively.[5,6] Unknown proteins with M_r of 58, 53, 28, and 21 kDa were also reported for both viruses.[6] FIPV proteins are recognized by antibodies to FEC in immunoblotting assays, indicating extensive cross-reactivity among these two viruses at the level of the individual proteins.[5]

III. CULTIVATION AND ANTIGENIC RELATIONSHIPS

Most strains of FEC do not grow in cell culture, including the FEC-UDC strain.[4] This strain has been propagated by fecal-oral passage in specific pathogen free (SPF) cats.[2] However, an FEC (strain FEC-79-1683) was isolated in Crandell feline kidney cells (CFK) from mesenteric lymph nodes and intestinal wash of an adult cat with fatal enteritis, but no symptoms of FIP.[3] This virus produced CPE in the CFK cells, similar to that observed with canine coronavirus (CCV).

To date, only the aforementioned two strains of FEC have been characterized to any degree. Both FEC-UDC and FECV-79-1683 are antigenically similar to FIPV strains UDC1, Black, and 79-1146.[5] Another coronavirus, antigenically related to FIPV was observed by Hayashi and colleagues[7] in the intestine of a cat with enteritis. However whether this virus was FIPV or FEC is uncertain. Other investigators[8,9] also detected coronavirus-like particles in the feces of cats with or without diarrhea. Dea and colleagues[9] reported that the coronavirus they detected from diarrheic cats was antigenically related to bovine coronavirus which is antigenically unrelated to FIPV.

FEC is antigenically related to FIPV and two other cross-reactive coronaviruses, transmissible gastroenteritis of swine (TGEV), and CCV, all members of the same coronavirus antigenic group.[2,4,10-12] A more detailed discussion of these antigenic relationships is presented in the TGEV section in this book. Cats infected with FEC showed immunofluorescence (IF) in intestinal epithelial cells stained with FITC-conjugated anti-TGEV serum or homologous FEC antiserum. Similarly, serum antibodies to FEC cross-reacted with FIPV, CCV, and TGEV antigens in IF tests.[2,12] Thus it is proposed that these latter coronaviruses may represent host-range mutants of a single virus strain,[11] whereas FEC and FIPV may represent variants of one another with altered tissue tropism. At present, no tests have been devised that can distinguish FIPV and FEC serologically.[11,12] Antibodies to FEC cross-react with FIPV in IF, virus neutralization (VN), ELISA, and immunoblotting.[2,4,5,12] Although titers were often higher against the homologous virus, particularly in VN tests, this response was not consistent for all sera tested.[12] Absorption of antiserum to FIPV with FEC, failed to reveal any FIPV-specific antibodies, demonstrating a lack of antigens unique to FIPV by this methodology.[6]

However, FIPV and FEC differ in their pathogenicities for cats.[2,13] FEC did not cause FIP in coronavirus seronegative cats inoculated orally or intraperitoneally, or in coronavirus seropositive cats inoculated intraperitoneally or intratracheally. Similarly, cats previously exposed to FEC were often still susceptible to FIPV infection. In some instances, pre-existing cross-reacting humoral antibodies may actually hypersensitize cats to challenge with virulent FIPV.[1,4] Cats preimmunized with FEC-UCD or FEC-79-1683 were more sensitive to infection with the FIPV strain UDC but not FIPV strain 79-1176.[13] A mechanism proposed for hypersensitization was the enhancement of uptake of antibody-FIPV complexes by phagocytic cells, thereby accelerating virus replication and dissemination.[1,4]

IV. DIAGNOSIS

The cross-reactivity between FEC and FIPV in all the serological assays evaluated creates difficulties in the serodiagnosis of FEC and differentiation from FIPV.[2,4,5,11,12] The two disease syndromes must be distinguished primarily on the basis of clinical signs. FEC virus or viral antigen can be detected during the early stages of disease in feces by electron microscopy or by IF staining of small intestinal epithelium.[2,4] Attempts to isolate virus in cell culture are often unsuccessful.[4] Serodiagnosis of FEC using FIPV, TGEV, or CCV as the antigen source can be accomplished by demonstrating a rise in coronavirus antibody titer in convalescent serum. Serum IF antibody titers of 32 to 512 may be expected at 2 to 4 weeks after exposure compared to titers of 1600 or greater in cats infected with FIPV.[1,2]

V. PATHOGENESIS, PATHOGENICITY AND EPIDEMIOLOGY

FEC causes inapparent to moderately severe enteritis in kittens 5 to 12 weeks of age.[2] Transient vomiting may precede a mild diarrhea which persists for 2 to 5 days, accompanied by a low grade fever in some kittens.[2] A fatal hemorrhagic diarrhea has been described in one cat, but this form of the disease is uncommon.[3] In SPF kittens, clinical signs are more pronounced in kittens infected at younger ages, and usually inapparent in adult cats.[2] Kittens may become dehydrated if enteritis is severe, but mortality is usually low.

The pathogenesis of FEC resembles that of TGEV in swine and CCV infections in dogs. The target tissue is the mature villous epithelium of the small intestine and cecum, as determined by IF.[2,13] FEC-79-1683 was recovered by virus isolation in highest amounts from the ileum, jejunum, duodenum, and mesenteric lymph nodes, with lower amounts of virus present in cecum, tonsils, and thymus of infected cats.[2,3,13] Gross pathologic lesions are uncommon in FEC infections. Histopathologic lesions are usually mild and include villous atrophy, fusion of villi and crypt hyperplasia.[2] Lesions were most severe in the jejunum and ileum. This temporary loss of the absorptive villous epithelium probably accounts for the malabsorptive diarrhea associated with FEC infections.

After recovery from disease, small foci of infected cells may remain in the intestine.[4] Such foci may allow persistence of virus shedding by carrier cats, including asymptomatic antibody positive cats which may shed infectious virus for prolonged periods.[2] Epidemiologic studies of FEC infections have documented the spread of infections throughout catteries by direct contact between healthy FEC seropositive cats and susceptible kittens.[2] Kittens usually become infected at weaning (5 to 16 weeks of age), often without showing illness. Virus is also spread mechanically via dirty litter pans or animal caretakers.[2]

Seroepidemiologic surveys of FEC virus infections are complicated by the occurrence of serologically indistinguishable FIPV infections in cats. In one study it was estimated that 80 to

90% of cats within individual catteries and multiple cat households were seropositive for feline coronavirus antibodies reactive with FIPV antigen.[2]

VI. IMMUNITY, PREVENTION AND CONTROL

Most kittens do not become infected with FEC until after weaning, suggesting lactogenic immunity and possibly systemic passive immunity may provide early passive protection against infection.[4] Serum antibodies to FEC appear at 10 to 14 d postinoculation (DPI), concurrent with cessation of virus shedding in feces.[2,4] No information is available regarding local intestinal immunity to FEC, including the duration of such immunity. Serum FEC antibodies may remain persistently elevated or fluctuate over time suggestive of the loss and reacquisition of virus[4] or multiple infections with different strains of coronavirus.

Although FEC infections may not be prevented, the severity of disease may be reduced by keeping kittens segregated by ages to reduce exposure to large amounts of virus, caring for younger kittens first, and thoroughly disinfecting the premises and litter pans.[2] Symptomatic treatment of severely affected kittens may require temporary withholding of food and water, and the administration of fluid therapy to counteract dehydration.[4]

PART 3. ANTIGENIC GROUP 2 — MAMMALIAN

A. BOVINE CORONAVIRUS

I. INTRODUCTION

Bovine coronavirus (BCV) was first identified as a possible cause of calf scours in 1972 and is now recognized as one of the leading etiologic agents associated with neonatal enteritis in calves. The disease characteristically affects 3- to 21-d-old calves, producing a profuse watery diarrhea which often results in dehydration, acidosis, and death. Upper respiratory replication of the virus has recently been identified; however, clinical signs are few. The significance of respiratory BCV infection in the epidemiology and pathogenesis of the virus is unclear at this time. Distribution appears to be worldwide and prevalence rates of neonatal infection range from 11 to 81%.

II. MORPHOLOGY AND MORPHOGENESIS

Stair et al.[1] first described the morphology of a bovine coronavirus-like agent purified from calf feces and examined by negative staining electron microscopy. The viral particles ranged in size from 107 to 160 nm (average of 126 nm), with some particles displaying a "punched-out" center. The nucleocapsid was round to oblong and enveloped. A bilaminated fringe of petal-shaped surface projections averaging 11 nm extended from the surface. Doughri et al.[2] reported similar findings for the negative stained BCV, strain LY-138. Virions were enveloped, pleomorphic, and ranged in size from 70 to 120 nm, with an average of 90 nm. The envelope exhibited widely spaced petal-shaped projections of 15 to 20 nm in length averaging 25 per virion circumference.

The detailed morphology of BCV strain F15 was studied by Roseto et al. by a freeze-drying technique.[3] They showed typical pleomorphic particles with diameters ranging from 65 to 210 nm with a mean value of 120.43 +/− 15.45 nm. By the freeze-drying technique three different categories of image were observed: (1) virions which exhibited a complete corona of spikes and uneven surface, constituting the majority of particles; (2) virions with a characteristic smooth 44 nm depression over 25% of the virion surface, constituting 30% of particles; and (3) apparently broken particles clearly showing undamaged projections and the area of spike insertion at the virus membrane. A model for BCV was proposed in which a spherical particle is covered with 110 highly packed surface projections, each with a mushroom-like appearance and an average length of 25 nm.

BCV maturation occurs in the cytoplasm of mature differentiated absorptive epithelial cells and goblet cells, with no apparent nuclear involvement. As viral replication progresses, distension of smooth and rough endoplasmic reticulum with electron dense granulo-fibrillar material in the apical cytoplasm appears. The Golgi complex becomes vesiculated and viral structures appear within. In the final stages, virions are released by virus-containing vacuoles fusing with the apical plasmalemma of the infected cells.[2,3,4]

Viral morphogenesis was studied in African green monkey (vero), Madin-Darby bovine kidney (MDBK), and human rectal adenocarcinoma (HRT18) cell lines by transmission electron microscopy and summarized as follows. (1) Penetration and decapsidation — no morphological anomaly is visible in the eclipse phase which lasts about 12 h. (2) The first signs of cell infection can be noted in the cytoplasm. The endoplasmic reticulum (ER) is changed in appearance and numerous dilated zones have appeared. (3) In the hyaloplasm, dense crescent-shaped viral nucleoids adhere to the ER membrane and penetrate towards the inside by budding; the virions acquire their external membrane during this transit. (4) Surface spikes are visible on the particles located in the ER vesicles where maturation probably occurs. (5) Tubular-shaped particles are formed during viral synthesis — their diameter comparable to that of the core of the virion. These tubules are found either in the cavities of the ER or in the cytoplasmic vesicles. (6) Virus release occurs by cell lysis or ER budding. All virions remain adsorbed against the cell coat with few particles found in the medium.[6]

III. *IN VITRO* CULTIVATION

A. CELL AND ORGAN CULTURES

In 1973, Mebus et al.[7] first adapted a coronavirus-like agent isolated from a calf with diarrhea to growth in cell culture. The agent was first isolated in primary fetal bovine kidney (PFBK) cells, then adapted to secondary fetal bovine kidney cells. The virus produces CPE, and viral antigen was identified by staining with fluorescein-labeled rabbit antibodies against the virus. CPE was observed after 24 passages as a rounding, and loss of adherence of cells with syncytia formation. Takahashi et al.[8] also used primary bovine kidney cells for the isolation of the Kakegawa strain of BCV (a field strain from the feces of a cow with epizootic diarrhea). Syncytia, granularity of cells, and positive immunofluorescence by the eighth passage characterized the CPE observed.

Inaba et al.[9] and Sato et al.[10] utilized the continuous cell line, BEK-1 derived from bovine embryonic kidney, to grow the Mebus (cell culture-adapted) strain of BCV. CPE appeared at 3 d as cell rounding, which progressed to monolayer disintegration, sloughing, and cell fusion. After further passages, viral yield ranged from 10^4 to 10^5 $TCID_{50}$ (tissue culture infectious dose) per 0.1 ml. Infected cells could be detected by immunofluorescent staining and CPE could be specifically inhibited by antiserum to bovine coronavirus. Hirano et al.[11] showed that the BEK-1 cell line could also be used in a plaque assay for the Kakegawa strain of BCV.

Dea et al.[12] demonstrated that Madin-Darby bovine kidney (MDBK), Vero, and porcine kidney-15 (PK-15) continuous cell lines were satisfactory for the isolation and multiplication of the Mebus strain of BCV. Several field isolates were also grown in Vero cells and then adapted to PBFK and MDBK cells. CPE (cell rounding, detachment, and complete monolayer destruction after 96 h) and positive immunofluorescent staining were observed in all cell lines. Viral yield varied between 10^6 and 10^7 $TCID_{50}$/ml. The MDBK cell line has also been useful for plaque and plaque reduction assays.[13] Cells tested in which the virus did not produce CPE were human epidermoid carcinoma (HeLa), human epidermoid carcinoma (Hep-2), African green monkey

(BSC-1), embryonic bovine derma (EBD), embryonic bovine trachea (EBTr), Madin-Darby canine kidney (MDCK), embryonic lamb kidney (LK), and mouse L cells.

Laporte et al.[14,15] and L'Haridon et al[16] also reported the adaptation of a wild type bovine coronavirus (G110 strain) from France to primary calf kidney (PCK) cells. They also showed that the PCK adapted virus multiplied in MDBK cells, bovine skin cells, bovine lung cells, and K3 pig kidney cell lines. There was no virus growth in other pig kidney lines, baby hamster kidney (BHK21), chinese hamster ovary (CHO), or Madin-Darby canine kidney (MDCK) cell lines. High viral yields, in excess of 10^7 $TCID_{50}$/ml, were also achieved in an HRT-18 cell line following 12 passages but no CPE was observed. The HRT-18 cell line was then used for titration of the cell culture-adapted G110 and F15 strains of BCV in a plaque assay. Plaques appeared within 2 to 3 d as opalescent areas which remained colorless after neutral red or crystal violet staining.[17]

The Mebus cell culture-adapted strain of BCV has also been propagated in the MA-321 strain of human embryonic lung fibroblasts. CPE was seen as refractile, oval, or rounded cells at 4 to 5 d postinfection and complete degeneration of cell monolayers was observed at 6 to 10 d postinfection.[18]

A porcine renal cell culture-adapted strain of BCV (SC-1) was grown in a second to fifth passage of primary fetal (3 to 4 month gestation) ovine renal cells. CPE was observed at 48 to 72 h postinoculation as formation of syncytia.[19] BCV has also been shown to replicate in calf testicle cells.[20] and D2 bovine fetal spleen cells.[21]

Organ culture of BCV has been achieved in two organ systems. Stott et al.[22] showed that the Mebus (NCDV) cell culture-adapted strain would replicate in 5- to 6-month gestation fetal bovine tracheal organ culture and produce hemagglutinating activity. Hemagglutinin titers increased with viral passage and positive immunofluorescence was observed at 7 d postinoculation. Electron microscopic examination of culture fluids revealed typical coronavirus particles but no gross or histologic damage to the cilia was visible.

Bridger et al.[23] also demonstrated replication and hemagglutination by a British isolate (passaged in tracheal organ culture) of BCV in 4.5- to 6-month gestation fetal bovine intestinal organ cultures. Hemagglutinating activity peaked at 24 to 48 h postinoculation while immunofluorescence was maximal at 24 h. Typical coronavirus particles were seen in extracellular fluids harvested 18 to 96 h postinoculation. Electron microscopic examination of inoculated cultures revealed evidence of viral replication and budding. At 2 to 4 d postinoculation the columnar epithelium lost its normal morphology becoming irregular and disorganized.

B. ENHANCEMENT OF *IN VITRO* REPLICATION

Conditions for enhancing growth of cell culture-adapted strains of BCV in certain cell culture systems have been described. Dea et al.[12] showed that the factors inportant in increasing yield and appearance of CPE in Vero cells were: (1) polycation DEAE-dextran (25 μg) treatment of cells; (2) exposure of cells to hypertonic medium; (3) a slightly acidic inoculum (pH 6.5 to 7.0); (4) growth in a basic medium (pH 8.0 to 8.5); (5) incubation of cells with dactinomycin (0.01 to 0.05 μg); and (6) washing of cells with medium containing trypsin (5 μg/ml).

Storz et al.[24] noted that trypsin treatment (10 μg/ml) acccelerated CPE and size of plaques, facilitated cell fusion, improved the amount of cell-released hemagglutinin, and increased the infectivity yields in bovine fetal thyroid (BFTy) and bovine fetal brain cells (BFB). When the cell culture-adapted L9 strain of BCV was pretreated with trypsin or trypsin was present only during viral adsorption, plaque enhancement was not observed. Toth[25] showed that increasing amounts of trypsin (1, 5, or 10 μg/ml) increased the susceptibility of bovine embryonic lung cells to the Mebus strain of BCV a millionfold. Hirano et al.[11] found that when trypsin (5 μg/ml) was added to both the viral diluent and the overlay medium, plaques increased in size and numbers on BEK-1 cells infected with the Kakegawa strain of BCV. Cyr-Coats and Storz[21] demonstrated noncytopathic replication of the cell-adapted BCV L9 strain in D2 bovine fetal spleen cells only in the presence of trypsin. Bovine fetal spleen cells were nonpermissive for all wild-type BCV even in the presence of trypsin.

The mechanism(s) whereby trypsin enhances coronavirus replication are not yet fully understood. Trypsin treatment of virus might modify configurations of protein molecules in the virus envelope to render them more compatible with cellular receptor sites. Trypsin treatment of cells might promote attachment of virions by uncovering otherwise unavailable receptor sites. Alternatively, trypsin may destroy a broadly active viral inhibitor produced by cells in culture, thereby allowing multiple rounds of viral replication.[25-28]

C. ANIMAL INOCULATION

Bovine coronavirus propagation by animal inoculation may be a useful, if not necessary, method of producing large pools of field BCV strains which may not otherwise propagate well in cell culture systems. The LY-138 strain of BCV, first isolated in 1965, was maintained by oral inoculation of conventional, colostrum-deprived calves for many years by this method.[2] Akashi et al.[29] serially passed the Kakegawa strain of BCV in suckling mice, hamsters, and rats by inoculation with brain emulsions from infected laboratory animals. The Mebus strain of BCV was also adapted to suckling mouse brain by Kaye et al.[30] and later by Gerna et al.[18]

IV. PHYSICOCHEMICAL PROPERTIES

Stair et al. first described the isolation and partial characterization of a coronavirus-like agent.[1] Sharpee et al. further characterized this coronavirus-like agent and showed that it conformed to all six major properties of the coronavirus group, thereby establishing it as a true member of the Coronaviridae family.[31] Properties of BCV (Mebus strain) are as follows: (1) density of 1.18 to 1.25 g/ml; (2) complete ether and chloroform sensitivity; (3) relatively stable at pH 3.0; (4) contains ribonucleic acid; (5) thermolabile at 50°C and requires MgCl for stability; (6) inactivated by formalin, trypsin (>0.5%), deoxycholate; (7) causes hemadsorption with rat, mouse, and hamster erythrocytes at 4, 25, and 37°C but not with cat, dog, goat, sheep, cattle, man (type O), horse, goose, turkey, chicken, guinea pig, rabbit, or swine erythrocytes; (8) induces neutralizing titers as determined by hemadsorption and hemagglutination inhibition; and (9) readily filtered through 200 and 100 nm filters but not through 50 nm filters.[10,31] Physicochemical properties of an adult cow strain of BCV were similar to those above except that the virus had a density of 1.182 g/ml and could not pass through a 100-nm filter.[32]

All members of the family Coronaviridae are composed of at least three basic polypeptides: the nucleocapsid protein (N), the matrix protein (M or E1), and the peplomer protein (E2).[33] Bovine coronavirus (Mebus strain) has an internal phosphorylated nucleocapsid protein of 52-kDa mol mass. Also a small glycosylated matrix protein of 26 kDa mol mass and a glycosylated hemagglutinin of 190 kDa (normally present as subunits of 120 and 100 kDa).[34] In addition BCV has a glycosylated hemagglutinin of 140 kDa mol mass which is a disulfide-linked dimeric structure reducible to monomers of 65 kDa.[35,36]

These findings are similar to those of Storz et al. who reported glycoproteins of 180, 90, 65, and 23 kDa from BCV strain L-9 grown in fetal brain cells or fetal thyroid cells.[36] The 180-, 90-, and 65-kDa glycoproteins could be metabolically labeled with glucosamine, galactose, mannose, and fucose while the glycoprotein of 23 kDa could only be labeled with glucosamine and galactose. This new type of viral glycoprotein appears to be common to several coronaviruses.

These results compare favorably with those of Deregt et el., who reported the Quebec strain of BCV to be composed of six major structural polypeptides of mol mass 120, 100, 50, 27, and 25 kDa with the 120 kDa protein being a disulfide-bridged dimer of a 60 kDa protein. They also showed that monoclonal antibodies to the 120 kDa glycoprotein neutralized virus infectivity and that this protein is also involved in hemagglutination.[13]

However, polyacrylamide gel electrophoresis of BCV virus (F15 and wild-type strain) by other workers has shown the polypeptide sizes to be 125, 65, 50, 45, 36, 34 and 28 kDa.[38] They concluded that the BCV virus is composed of seven polypeptides and that adaptation to cell culture does not alter this.

Vautherot and Laporte[39] used a panel of 44 monoclonal antibodies to characterize three peplomer proteins, gp 105, gp 200 and gp 65/125. Glycoproteins 105 and 200 were found to derive from a common precursor p 180, whereas gp 65/125 was shown to be antigenically and structurally different. Both gp 105 and gp 65/125 had neutralizing epitopes but only monoclonal antibodies to gp 65/125 showed hemagglutination. Neutralization epitopes mapped to four domains on gp 105 and three domains on gp 65/125.

Recent work by Deregt et al.[40] has shown that the Quebec isolate of BCV contains four unique major structural proteins: (1) the matrix protein; (2) the nucleocapsid protein; (3) the peplomer protein; and (4) the hemagglutinin protein. The matrix protein(s) (E1) are a family of various *o*-linked glycosylations of p 23, which yield gp 26, gp 25 and gp 23. The nucleocapsid protein is p 53 and its trimer p 160. The peplomer (E2) protein arises from a large *N*-linked precursor glycoprotein (gp 170), which is further glycosylated (gp 190) and proteolytically cleaved to yield gp 100. The hemagglutinin protein (E3) arises from an *N*-linked glycoprotein precursor (gp 59) which rapidly dimerizes to gp 118 which is further glycosylated to gp 124. Using monoclonal antibodies, the E2 and E3 glycoproteins were shown to have neutralizing epitopes and the E1 glycoprotein did not. Neutralizing monoclonal antibodies defined two nonoverlapping antigenic domains on glycoprotein E2 and E3, plus a third site on E3 which overlapped the other two. In addition, one domain on the E3 glycoprotein could be further subdivided into two epitopes.[41]

V. GENOME

The genome of BCV consists of RNA.[31] Guy and Brian further characterized BCV genomic RNA as: (1) a homogeneous large-molecular mass species of 3.8×10^6 kDa (20-kb); (2) of constant in molecular weight after heat denaturation; (3) 80% susceptible to pancreatic RNase A digestion with the 20% resistant fraction being 4S to 7S in size; and (4) 40 to 60% polyadenylated.[42]

Lapps et al.[43] determined the RNA sequences encoding the matrix and nucleocapsid proteins using cDNA replicates of the genome. Nearest the 3′-noncoding region was the nucleocapsid gene encoding 448 amino acids giving a protein of 49,379 mol wt. Upstream to this was the matrix protein gene encoding 230 amino acids, giving a 26,376 mol wt. protein.

Little else is known about the bovine coronavirus genome or replication cycle; however, it may conform to that of the other coronaviruses. Therefore it is assumed that the viral mRNAs have a "nested set" structure with 3′ coterminal ends and sequences extending for different lengths in a 5′ direction. Each subgenomic mRNA is capped and polyadenylated and is translated independently to produce a single protein the size of which corresponds to the coding capacity of the 5′ sequences not found in the next smallest mRNA. The genome may also encode a large RNA-dependent RNA polymerase which makes full-length genomic RNA and also the subgenomic RNAs.[44] This polymerase is involved in a unique mechanism of mRNA production in which the polymerase transcribes a leader sequence and the 3′ end and then translocates to a complementary sequence along the genomic RNA, thereby giving rise to subgenomic RNAs with similar leader sequences.[45]

VI. ANTIGENIC RELATIONSHIPS

The first strain of bovine coronavirus was reported by Stair et al. and Mebus et al. who referred to it as a coronavirus-like agent, which later became known as the Mebus strain or Nebraska calf diarrhea coronavirus (NCDC).[1,7] British and Danish isolates were antigenically related to the

Mebus strain by immunofluorescent staining techniques and hemagglutination inhibition tests.[46] Four BCV isolates from Quebec were compared to the Mebus strain by counterimmunoelectrophoresis and immunodiffusion, resulting in the strains being placed in two antigenic groups.[47] An adult strain of BCV (Kakegawa strain) was identical to the Mebus strain by neutralization and immunofluorescence.[32]

Several reports have appeared describing the isolation of respiratory strains of BCV[48,49] and replication of enteric strains in the respiratory tract.[50-52] Reynolds et al.[51] and Saif et al.[52] showed that enteric strains of BCV induced respiratory shedding. In the former studies, calves which recovered from an enteric BCV infection were completely protected following challenge with either of two respiratory isolates. A panel of three hyperimmune gnotobiotic pig sera (prepared against the enteric and respiratory isolates) was shown to effectively neutralize eight heterologous cloned BCV from diverse geographic sources, indicating that the different BCV isolates belong to the same serotype despite their source.[50,51]

Monoclonal antibodies have been used to study the antigenic similarities between the Mebus strain, the Danish and British isolates, the respiratory isolate of Thomas, and two strains reported by Laporte et al. (G110 and F15 strains).[14] The respiratory isolates, G110 and F15, were similar in their reactivity with monoclonal antibodies. The G110 and F15 strains reacted with all monoclonal antibodies, while the Mebus strain and the British isolate reacted with all but one monoclonal antibody, again showing very close antigenic similarity.[53]

Several comparisons have been made between bovine coronavirus and other coronaviruses. Previous serological studies have demonstrated common antigens among BCV, human coronavirus (OC43), and mouse hepatitis virus (MHV).[54] Hogue et al. showed by immunoblotting studies that there are shared antigens among these viruses, but that BCV and human coronavirus are more closely related than are BCV and mouse hepatitis virus.[55] Several workers have further shown this antigenic relationship between BCV and human coronavirus (OC43) by neutralization, double immunodiffusion, indirect immunofluorescence tests, and immune electron microscopy.[37,56,57] Nucleic acid sequence homology and oligonucleotide fingerprint analysis have suggested a sequence homology of 96 to 98% between the BCV and OC43, indicating a recent divergence of the strains.[36,58] Lapps et al.[43] showed an overall amino-acid sequence homology of 70% with both MHV A59 and MHV JHM (72% at the nucleotide level) for the N protein and 86% for the M protein. Sequence homology of the N protein to other viruses in the Coronaviridae family were much less: 29% (37% at the nucleotide level) for transmissible gastroenteritis virus and 29% (43% at the nucleotide level) for the infectious bronchitis virus.

With a panel of monoclonal antibodies to BCV (strain G110) Vautherot and Laporte demonstrated complete cross-reactivity with human or bovine (Mebus strain, Danish, and British isolates) enteric coronavirus, and slight cross-reactivity with human respiratory coronavirus (OC43), hemagglutinating encephalitis coronavirus, and mouse hepatitis virus.[53]

Therefore, a zoonotic potential may exist for BCV and there has been one report of diarrhea in an investigator infected with BCV.[56] A human enteric coronavirus has also been reported to cause a bovine neonatal infection.[59]

VII. PATHOGENESIS AND PATHOLOGY

A. INTESTINAL REPLICATION

Stair et al. first described the experimental inoculation of calves with a coronavirus-like agent from the feces of diarrheic calves.[1] Inoculated calves developed diarrhea after a period of 18 to 36 h and become dehydrated. Diarrhea continued for 4 to 5 d and was sometimes lethal.[7] Extensive water, sodium, chloride, bicarbonate, and potassium losses occurred in the diarrheic calf.[60] Hypoglycemia, lactic acidemia, and hyperkalemia were the terminal events in BCV infection within 27 h of the onset of diarrhea.[61]

Grossly, the lesions reported were subtle. The abomasum contained curdled milk and petechia were seen on the mucosa. The intestinal tract was distended, thin, transparent, and contained a clear liquid.[4,7] Mebus described BCV infection of mainly the villous epithelium of the upper, middle, and lower small intestine, superficial and crypt colonic epithelium, and mesenteric lymph nodes.[7] A good correlation was seen between immunofluorescence and presence of virions in cells by electron microscopy. Lesions in the small intestine consisted of villous atrophy with remaining villi covered by cuboidal to squamous epithelium. The large intestinal lesions consisted of surface epithelium composed of cuboidal nonvacuolated cells, scattered dilated crypts lined by cuboidal epithelium, and a decreased number of goblet cells. Scanning electron microscopy showed small intestinal villi to be shortened, fused (Figure 7) and the villous epithelium to be composed of low cuboidal to squamous cells. The spiral colon showed atrophy of the colonic ridges and differences in the length and spacing of the microvilli on individual epithelial cells.[62]

Examination of ultrathin sections of coronavirus-infected intestinal mucosa revealed ultrastructural pathologic changes. Viral replication occurred in the cytoplasm of mature differentiated absorptive epithelial cells and goblet cells with no apparent nuclear involvement. In the early stages of infection there was an increased number of free ribosomes with distension of smooth and rough endoplasmic reticulum and electron dense granulo-fibrillar material. As viral replication progressed, well-defined membrane-bound round cytoplasmic structures filled with granulo-fibrillar; electron-dense material emerged in the apical cytoplasm. The Golgi complex became vesiculated and viral structures appeared within. In the final stages, cellular organelles become degenerate, mitochondria become swollen and dilated, and viral particles emerged after virus-containing vacuoles fused with the apical plasmalemma of the infected cell.[4,5]

Exposure of the neonate to BCV may come from the environment, infected herd mates and/or subclinically infected adult cows. Mebus et al. proposed the pathogenesis of coronaviral diarrhea after oral exposure as follows. Villous epithelial cells in the cranial part of the small intestine become infected and the infection progresses rapidly caudad. At the time most of the small intestinal villous epithelial cells are infected, the calf becomes depressed and diarrhea begins. As the infection proceeds, there is an accelerated loss of infected cells with replacement by immature squamous to cuboidal epithelial cells.[7] Increased mucosal growth may be enhanced by elevated plasma levels of neurotensin and enteroglucagon.[63]

There have been several reports of coronavirus-like agents being isolated from the feces of adult cows with "winter dysentery" or epizootic diarrhea.[8,64-67] Cows developed profuse diarrhea, became listless, lost weight, and milk production dropped. BCV particles were seen by electron microscopy in the feces, a rising serum titer to BCV was shown[67] and the virus was isolated on cell culture.[8]

B. EXTRAINTESTINAL REPLICATION

In addition to enteric replication of bovine coronavirus, several workers have recently reported respiratory infection by BCV. Thomas first reported the isolation of coronavirus from the respiratory tract of several calves showing respiratory illness.[48] Calves inoculated via respiratory routes may show no respiratory signs of illness[51,52] or they may develop a cough, nasal discharge, and increased respiratory rates.[49,68] Coronavirus antigen can be detected by immunofluorescence in lung, nasal epithelium, tracheal cells, or tracheal organ cultures of infected calves.[49,51,52,69,70] BCV particles can be seen by electron microscopy in nasal secretions from infected calves.[49,71] More recently, Saif et al.[52] have shown that some calves inoculated with fecal isolates of bovine coronavirus by an oral and/or intranasal route may develop focal interstitial emphysema in the lungs in association with coronavirus antigen detection by immunofluorescence in the lung, trachea, and nasal turbinates.

VIII. DIAGNOSIS

Methods for diagnosis of BCV infections are similar to those utilized for other viral agents.

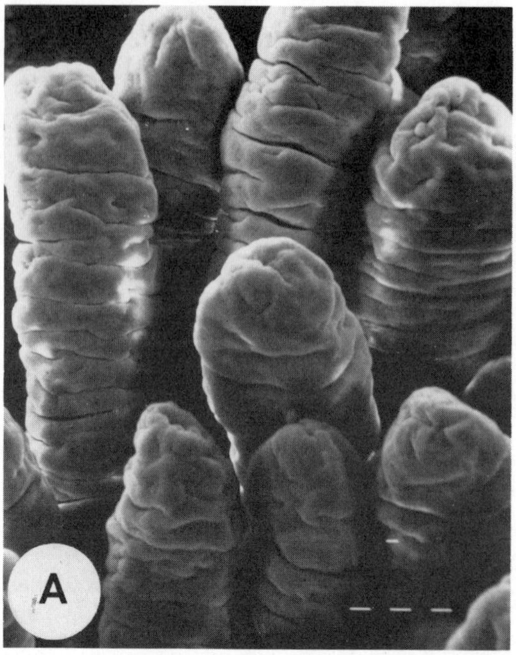

A

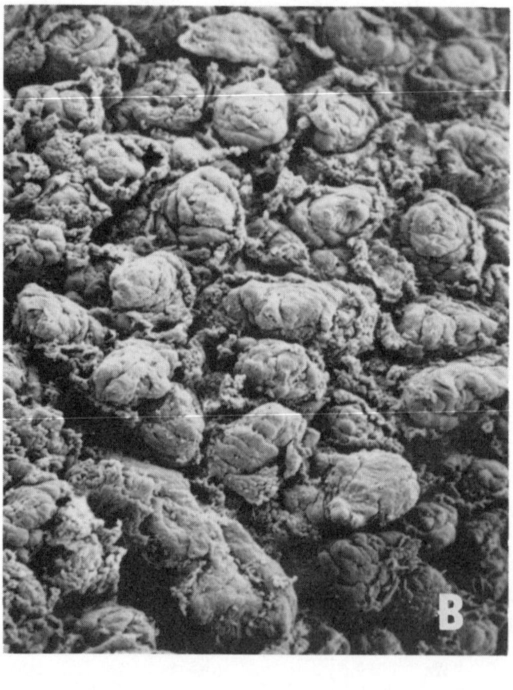

B

FIGURE 7. Scanning electron micrograph of small intestinal villi from gnotobiotic calves. (A) Intact villi in the ileum of a normal non-infected calf. (Magnification × 250.) (B) Severely atrophied villi in the ileum of a calf challenged 72 h previously with bovine coronavirus. (Magnification × 100.)

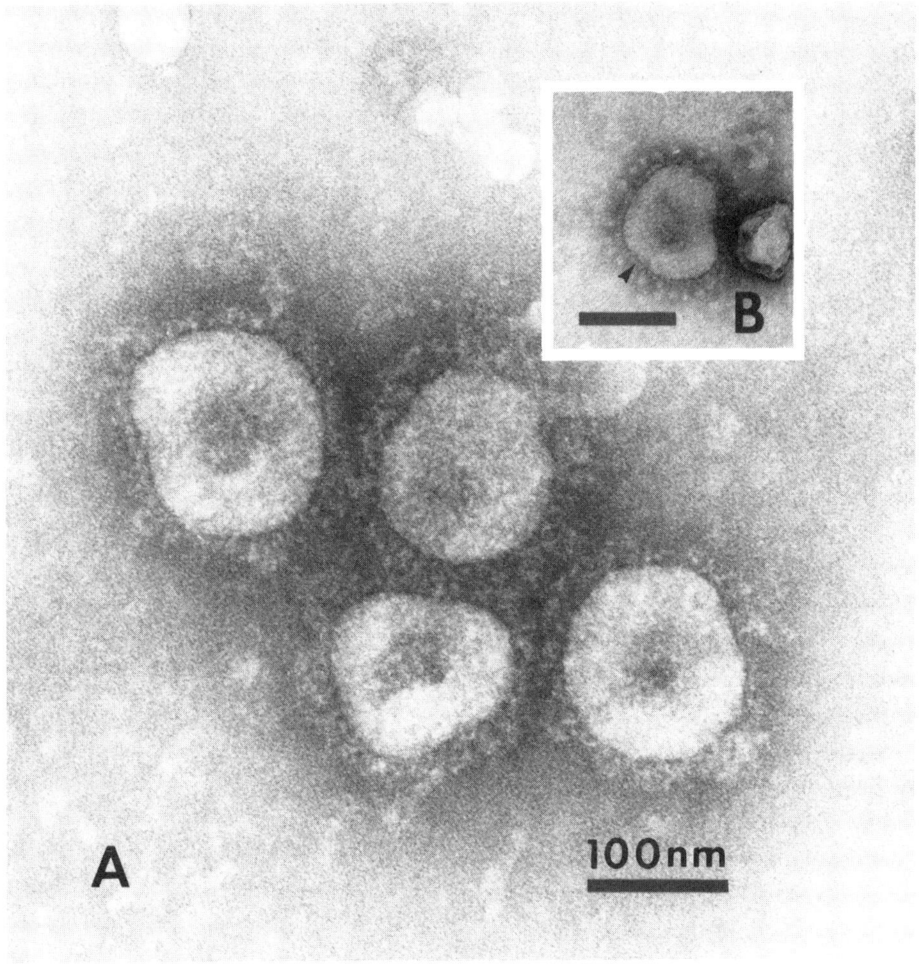

FIGURE 8. (A) Immunelectron microscopy of typical bovine coronavirus particles. (B) Direct electron microscopy of a single bovine coronavirus virion showing double fringe (short and long) of peplomers. Arrow points to short fringe. Bar = 100 nm.

These include electron microscopy, viral antigen detection, and virus isolation. Determination of a coronavirus infection was first done by electron microscopy and morphological identification of the virion.[1] This was later modified by including specific bovine coronavirus serum in the technique, producing a more sensitive and specific technique known as immune electron microscopy (Figure 8).[72] The technique has been further improved by the addition of protein A-gold to the immune complex giving a very specific and easily identifiable diagnosis (Figure 9).[71]

Many methods of BCV antigen detection have been described. Mebus et al. first showed that after necropsy, specific diagnosis of bovine coronavirus infection could be made by observation of immunofluorescence of small intestinal and colonic cells by indirect fluorescent antibody tests.[7] Saif et al. also showed direct immunofluorescence tests of the nasal epithelial cells of calves to be useful in the diagnosis of bovine coronavirus upper respiratory infections (Figure 9).[52] ELISA, utilizing polyclonal[73] or monoclonal antibodies[74] in a sandwich assay have proven useful in screening large numbers of fecal samples. The polyclonal antibody sandwich ELISA assay and electron microscopy have been found to be 82 to 100% in agreement.[75] An ELISA blocking technique based on a reduction of anti-coronavirus antibodies in a reference serum after mixing with the samples has also been described.[76] This test compared well with direct

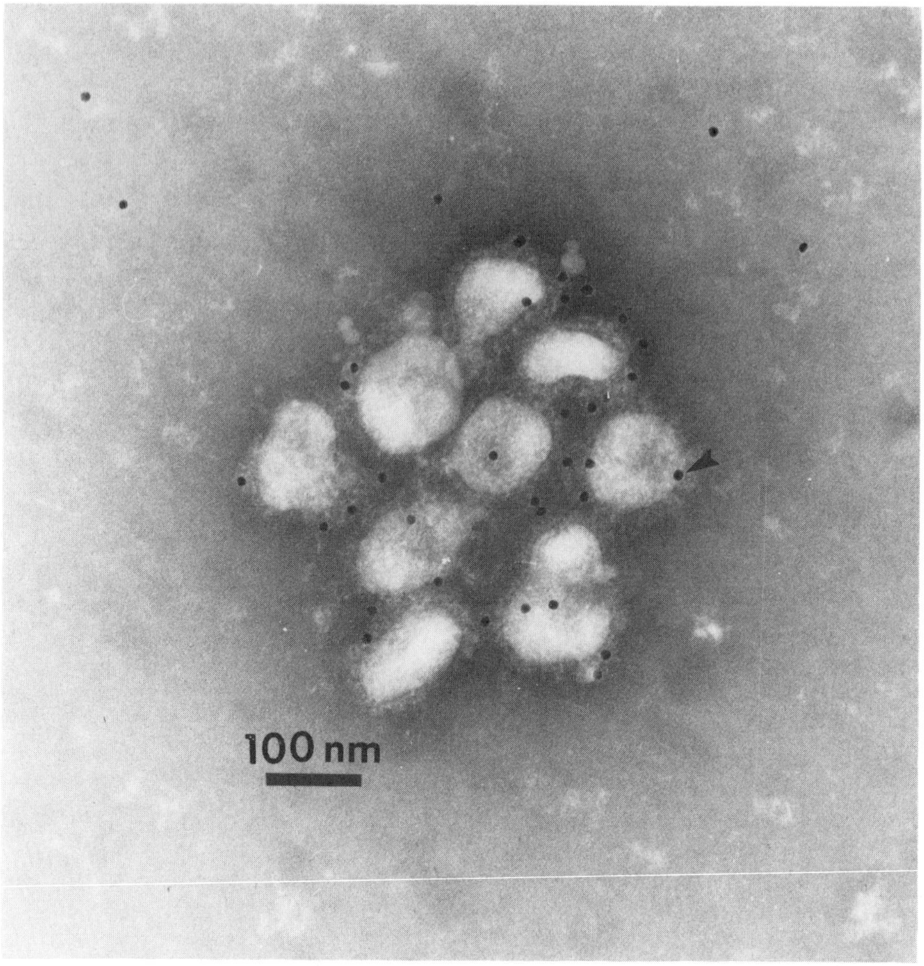

FIGURE 9. Protein A-gold immunoelectron microsopy of bovine coronavirus particles. Arrow points to a protein A-gold particle.

immunofluorescence on intestinal samples and neutralization of indirect immunofluorescence for detection of BCV. Other methods used for BCV detection include hemadsorption-elution-hemagglutination assay,[77] counterimmunoelectroosmophoresis,[78] and reversed passive hemagglutination.[79] More recently, cDNA probes labeled with ^{32}P[80] or biotin[81] have been utilized for the detection of BCV RNA in samples.

Bovine coronavirus isolation from field specimens is not routinely done, but has been achieved in fetal trachea[22] and intestinal organ culture[23] as well as using primary bovine kidney cells and Vero cells as described previously. Demonstration of a rising antibody titer to BCV in serum has been done by ELISA,[82] radioimmunoassay,[82] serum neutralization,[82] microhemagglutination-inhibition,[83] and indirect immunofluorescence.[83]

IX. EPIDEMIOLOGY

Bovine coronavirus infects dairy,[84] beef,[85] and veal calves.[86] The primary age group affected are 3- to 21-d-old calves; however BCV enteritis may occur in calves up to 3 months of age[72] and perhaps in adult cattle in association with winter dysentery or epizootic diarrhea.[8,64-67] Distribution appears to be almost worldwide.[8,46,84,86-92]

Prevalence estimates of neonatal BCV infection range widely, depending on immune status and the type of assay used. Based upon virus or virus antigen detection from diarrheic calves, prevalence ranges from 11[93] to 81%.[94] In one study BCV was not detected in the feces of healthy calves,[95] but in a second study, 8.7% of healthy calves shed BCV.[91] Based upon serology, 90[96] to 100%[82] of adult dairy cows have antibodies to BCV.

Reservoirs of infection in herds may be clinically infected calves, or chronically infected calves or cows. Reinfection of 6- to 9-month-old calves has been shown to occur based upon an antibody titer increase in a group of animals which had no contact with diseased calves, indicating chronic or recurrent infections may be possible.[83] Using ELISA, Crouch et al. showed that 5% of clinically normal cows shed BCV antigen in their feces while 70% shed BCV antigen-antibody complexes.[74,97] Whether this represents shedding of infectious virus was not determined. Collins et al. also showed that 20 to 30% of normal adult dairy cows shed BCV-like particles, as detected by electron microscopy, during the winter months and around parturition,[98] possibly resulting in exposure of the neonate at birth. However, whether these particles were truly bovine coronavirus and were infectious was not determined.

The route of transmission has historically been assumed to be fecal-oral; however with the recent evidence of BCV respiratory replication, a fecal-respiratory and/or respiratory-respiratory route may also be important in virus transmission and spread. No data are available as to the extent of respiratory shedding of BCV from calves or cows; however it has been shown to occur from calves under field conditions.[71]

X. IMMUNITY

Two approaches have been used in an attempt to provide calves with protection against coronavirus infections. The first is oral vaccination with live attenuated virus in order to stimulate active immunity in the calf. The second approach utilizes passive protection provided through lactogenic immunity, stimulated by maternal vaccination.

Oral vaccination may be theoretically effective in colostrum-deprived calves if challenge follows vaccination by 48 to 72 h.[99] However, oral vaccination was shown not to be effective in protecting colostrum-fed calves from infection, under experimental conditions[100] or in the field.[101]

Although protection through lactogenic immunity may be more promising, parenteral (intramuscular) vaccination of dams with a live attenuated coronavirus vaccine failed to significantly enhance colostrum or milk antibody titers.[102-106] In a field trial to evaluate the efficacy of such a vaccine in dairy calves, no significant difference could be shown between the vaccinated and control group in colostral antibody titers, mortality, percentage scouring, days to first scour, percentage treated for scours, days to first treatment, duration of treatment, or estimated weight gains.[104] Enhancement of colostral or milk antibody titers is, however, possible through the use of inactivated, adjuvanted coronavirus preparations[107] or by intramammary injection.[108] Alternatively, addition of incomplete freunds adjuvant to a commercial vaccine was shown to enhance colostral antibody titers in one study.[109]

Another approach to increasing neonatal immunity to coronavirus was via in utero vaccination. This has been shown to increase calf sera and gut antibody titers to BCV and to confer complete protection from BCV challenge. However a high rate of abortions and premature births precludes practical application of this procedure at this time.[110]

XI. PREVENTION AND CONTROL

To date, there has been no completely effective approach to reducing the morbidity and/or mortality from bovine coronavirus. Several factors have been shown to influence the incidence and severity of BCV infection.

Exposure of the neonate to infected fecal material should be minimized by: early removal of the calf from its dam; providing adequate air exchange to reduce airborne virus; reducing aerosol formation of fecal material; ensuring clean, dry, disinfected calving stalls or grounds; reducing contact between susceptible and scouring calves; reducing fomite transmission of virus by using foot-baths, separate feeding buckets, etc.[111] All stressors which may precipitate disease should be minimized, such as malnutrition, drafts, dampness, crowded conditions, etc.[112]

Lactogenic immunity may be important for reducing the incidence and severity of coronavirus-induced neonatal diarrhea.[105,113] It is suggested that milk containing high amounts of coronavirus specific antibodies be fed for at least the first 14 d of life to reduce the incidence of diarrhea, duration of diarrhea, and the incidence of viral shedding.[105] Ensuring that passive transfer of immunoglobulin occurs may also be an important factor for increasing neonatal protection from BCV induced scours. Hancock showed that when neonatal serum immunoglobulin levels dropped below 5 mg/ml there was also a significant increase in risk of neonatal scours.[114] It is recommended that the newborn receive at least 3 liters of colostrum via an esophageal feeder within 4 h of birth. Colostrum immunoglobulin quantity should be determined for each fresh cow and good quality colostrum frozen for feeding to calves from cows with poor colostrum.

B. MURINE CORONAVIRUS: DIARRHEA VIRUS OF INFANT MICE

I. INTRODUCTION

The diarrhea virus of infant mice was first described in 1962 by Kraft as "lethal intestinal virus of infant mice". At this time, it was thought to be mouse hepatitis virus (MHV) with altered tropism for the intestine. It was later shown to be distinct in morphology and structural polypeptide profile from mouse hepatitis virus and is now recognized as a separate virus in the Coronaviridae family, known as "Diarrhea Virus of Infant Mice" (DVIM). It classically causes epizootics of diarrhea with high mortality in 7- to 10-d-old mice, but no morbidity or mortality in adults.

II. MORPHOLOGY AND MORPHOGENESIS

Electron micrographs of DVIM show the virus to be 100 to 110 nm in diameter, excluding the surface projections, with a double fringe of radiating surface projections. The innermost corona of projections is 5 nm in length and firmly fixed in the viral envelope. The outer peplomers are club shaped, 20 nm long, and 10 nm wide at the distal end.[1]

III. CULTIVATION

DVIM replicates well in an embryonic mouse cell line (BALB/c-3T3) with maximum cytopathic effect, seen as syncytia formation, appearing at 12 to 18 h postinfection (p.i.). Syncytium formation is closely paralleled by hemagglutination titers, which peak at 11 h p.i. By immunofluorescent staining, viral antigens could only be seen in the cytoplasm and never in the nucleus.[1]

Scanning electron microscopy shows a loss of surface microvilli from BALB/c-3T3 cells with spherical buds (assumed to be budding virions) appearing from areas of syncytia formation. These pleomorphic buds absorbed mouse erythrocytes exclusively.[1]

Clinical isolates also replicate (in 2 to 4 d) to high titers (10^7 to 10^{10} 50% tissue culture infective doses per milliliter) in a mouse liver cell line (NCTC-1469).[2] However, the virus has failed to multiply in mouse L cells, mouse embryos, rhesus monkey kidney, HeLa cells, and Ehrlich ascites tumor *in vivo* and the chorioallantois of the chick embryo.[3]

IV. PHYSICOCHEMICAL PROPERTIES

Some of the physicochemical properties of cell culture passaged DVIM have been characterized. Properties of this virus are (1) hemagglutination (HA) of rat and mouse erythrocytes at 4°C but not of chicken, goose, rabbit, guinea pig or human "O" erythrocytes; (2) ether and deoxycholate lability; (3) trypsin stability; (4) heat lability (50°C for 30 min) and inactivation following storage at 4°C for 24 h; (5) unfilterable through less than 100 μm membranes; (6) elimination of HA activity and virus infectivity by bromelain treatment; (7) reduction of HA activity by pepsin treatment with infectivity unaffected; (8) reduction of HA activity and loss of infectivity by NP-40 treatment; and (9) possession of a receptor destroying activity at 20°C.[3,4,5]

Five major viral proteins constitute the DVIM virion as determined by polyacrylamide gel electrophoresis. These are VP1 (126 to 147 kDa), VP2 (65 to 72 kDa), VP3 (51 to 59 kDa), VP4 (34 to 41 kDa), and VP5 (21 to 24 kDa), with two additional smaller bands detected at 100 to 116 kDa (VP1a) and 88 to 105 kDa (VP1b).[5] VP1, VP1a, VP1b, and VP2 were removed by bromelain digestion, suggesting their location as surface proteins, whereas VP3 was unaffected. The virus is known to contain RNA,[4] but no further information about the viral genome is available.

V. ANTIGENIC RELATIONSHIPS

DVIM was initially shown to be serologically unrelated to the epizootic diarrhea of infant mice (EDIM) virus, which causes similar clinical signs, but is now known to be a rotavirus.[3] DVIM is related to infectious bronchitis virus and MHV strain 2 by complement fixation, but is clearly distinguishable from these two viruses by neutralization assays.[4] It was then shown to be partially related to mouse hepatitis virus (MHV) strains 1, 3, A59, and JHM as well as to human coronavirus OC43, and to be closely related to MHV strain S by several serological tests.[2,6] It was considered at the time to be related closely enough to MHV-S to constitute a variant of this strain and not a unique MHV strain. Recently, Sugiyama et al.[7] employed radioimmunoprecipitation to show that antisera against various individual strains of MHV immunoprecipitated all DVIM structural polypeptides, except gp69, which was not immunologically related to any of the glycoproteins of the other murine coronaviruses.

VI. PATHOGENESIS AND PATHOGENICITY

A. INTESTINAL REPLICATION

Clinical signs appear as diarrhea, dehydration, fecal staining, and in some mice neurologic signs such as scampering and jumping.[3,8] The infected infant mouse does not nurse, loses weight rapidly, becomes lethargic, and dies after a short period of cyanosis.[3] In less than 10-d-old mice, the incubation period can be as short as 36 h with death in 48 h.[3] Gross pathology consists of a shrunken stomach and a thin-walled small intestine distended by gas. Microscopically, the small intestinal villi were atrophied, reduced in number, and numerous swollen multinucleated modified epithelial cells appeared throughout the intestine. These "balloon cells" manifested a few large cytoplasmic eosinophilic inclusions but no nuclear inclusions.[2,3,8-10] By immunofluorescence, these syncytial giant cells were positive for MHV antigen.[9] Thin section electron microscopy revealed large numbers of coronavirus particles in jejunal and ileal intestinal epithelium and macrophages. Immunofluorescence on frozen sections of intestine showed viral antigen in inclusion bodies which stained equally well with MHV polyvalent antiserum, MHV-S antiserum and with antisera to five other MHV strains.[2]

Differences in age susceptibility can be explained by differences observed in gross pathology when different age groups were inoculated with DVIM. Younger animals showed a cytopathologic response with balloon cell formation, ulceration, and atrophy of villi, while in older animals, fewer balloon cells were observed and increased crypt cell proliferation compensated for cell loss.[10]

B. EXTRAINTESTINAL REPLICATION

DVIM is hepatotropic, as are all strains of MHV. Small foci of parenchymal necrosis, consisting of multinucleated giant cells and eosinophilic bodies with poor infiltration of monocytes and lymphocytes, were observed by 3 D.P.I..[9] Electron microscopy revealed virus-like particles in cytoplasmic vesicles or in the dilated endoplasmic reticulum of degenerated hepatocytes.[8] Specific immunofluorescence was demonstrated in the hepatocyte cytoplasm around large blood vessels and necrotic foci.

Histopathological lesions and virus-specific immunofluorecence were also seen in the olfactory bulb, choroid plexus of the fourth ventricle and in the cerebrum. Viral antigen was demonstrated in the lung and spleen but not in the bone marrow or other organs.[8]

VII. DIAGNOSIS

Diagnosis has traditionally been based on clinical signs of an explosive outbreak of diarrhea with high morbidity along with pathognomonic intestinal pathology.[3] Specific etiologic diagnosis can now be tentatively established by direct electron microscopy of intestinal contents and by serological testing.[2]

VIII. IMMUNITY, PREVENTION, AND CONTROL

DVIM classically appears as a rapidly spreading epizootic of diarrheal disease in the nursery of a mouse colony. The epizootic is characterized by 50 to 100% mortality in of 7- to 14-d-old mice but no morbidity or mortality in the adults.[6,8] Clinical signs are less severe in litters of multiparous dams than in primiparous ones.[9] Transmission may by fecal-oral with the adults harboring the virus and shedding the virus to the neonate.[6]

Mice surviving DVIM infections have complement fixation titers of 1:80 to 1:256.[2,9,11,12] Since litters of multiparous dams showed less severe clinical signs than litters of primiparous dams, and infected mothers developed increased antibody titers against the agent, there may be passive protection afforded the neonate through lactogenic immunity.

Epizootics have been controlled in the past by stringent use of filter tops on cages, disinfection, and physical and air barrier separation of separate mouse groups.[2] The best prevention still consists of maintaining disease-free breeder stocks and introducing only DVIM free mice into nurseries.

PART 4. ANTIGENIC GROUP 4 — AVIAN

A. CORONAVIRUS ENTERITIS OF TURKEYS: BLUECOMB DISEASE

I. INTRODUCTION

Coronavirus enteritis (CE) is an acute highly contagious disease of turkeys, characterized by wet droppings, inappetence, and weight loss. It affects turkeys of all ages, but causes highest losses in young poults. The virus does not appear to infect chickens or other avian species. The disease has also been referred to as mud fever, bluecomb, and transmissible enteritis.[1]

The disease syndrome was first described in turkeys by Peterson and Hymas in 1951,[2] but the specific etiologic agent was not identified as a coronavirus until 1973.[3,4] Although the disease caused serious economic losses in the U.S. and Canada in the 1950s and 1960s, it was subsequently eliminated from many turkey operations by depopulation and decontamination.[1] In recent years, sporadic cases of CE have been reported in the U.S., Canada, and Australia.[1,5,6] Current surveys of enteric viruses from turkeys in the U.S., based on morphologic identification of viruses by electron microscopy (EM), failed to detect coronaviruses;[7,8] however, the pleomorphic and indistinct morphology of this virus make it very difficult to identify conclusively in feces and distinguish from other pleomorphic membranous particles frequently present.[9] Although we recently detected coronavirus-like particles from flocks in two commercial operations showing severe diarrhea and high mortality, they could not be conclusively identified as coronaviruses, nor was it possible to establish their relationship with CE [20] Coronaviruses antigenically similar to CE (by ELISA and immune electron microscopy) were detected from 1982 to 1986 in Canadian turkey flocks experiencing mild to severe diarrhea.[5,6] The prevalence was 47.5% in 114 flocks from 42 commercial operations.[6]

III. MORPHOLOGY, MORPHOGENESIS, PHYSICOCHEMICAL PROPERTIES, AND GENOME

The CE virus displays typical coronavirus morphology. Virions are pleomorphic ranging in size from 50 to 150 nm in diameter (135 nm average), and enveloped, with club-shaped peplomers projecting from the envelope[3] (Figure 10). The virus has a density of 1.16 to 1.24 g/ml in sucrose and contains an RNA genome, based on sensitivity to ribonuclease. The infectivity of CE was destroyed by chloroform, but resistant to acid treatment (pH 3, 22°C for 30 min).[1,10] To date, there is no information on the protein composition or nucleic acid configuration of CE to permit comparison with other coronaviruses. The morphogenesis of CE in the intestines of infected poults and in turkey embryos resembled that seen for other coronaviruses. Small enveloped particles were observed budding into membrane-lined cisternae within the cytoplasm of epithelial cells.[11]

III. *IN VITRO* CULTIVATION

The Minnesota isolate and field isolates of CE have been successfully cultivated in embryonating turkey or chicken eggs over 16 d of age by inoculation of the amnionic cavity.[12] However, there were no gross pathological changes in the embryo, so titration of viral infectivity relied upon subsequent challenge of poults.

The CE virus had not been successfully adapted to serial propagation in any cell culture system until recently. Cell cultures tried previously included turkey, chicken, quail and monkey kidney and turkey and chicken embryo, liver, and intestine.[1] Recently, a Quebec isolate of an enteric coronavirus antigenically indistinguishable from the Minnesota isolate of CE was adapted to serial propagation in a human rectum adenocarcinoma (HRT-18) cell line by inclusion of trypsin in the inoculum and medium and stringent control of the pH.[13] Viral replication was characterized by the induction of syncytium formation and confirmed by immunofluorescence (IF), EM, and the hemagglutinating properties of the cultivated virus. The senior author had previously reported cultivation of a Quebec isolate of CE, but not the Minnesota CE isolate, in primary chicken embryo kidney cell cultures.[5] However, the concomitant infection of such cultures with reovirus made the significance of these earlier findings difficult to interpret.

IV. ANTIGENIC RELATIONSHIPS

The CE virus is antigenically unrelated to other animal enteric coronaviruses including transmissible gastroenteritis (TGE) virus, bovine coronavirus, and mouse hepatitis virus as

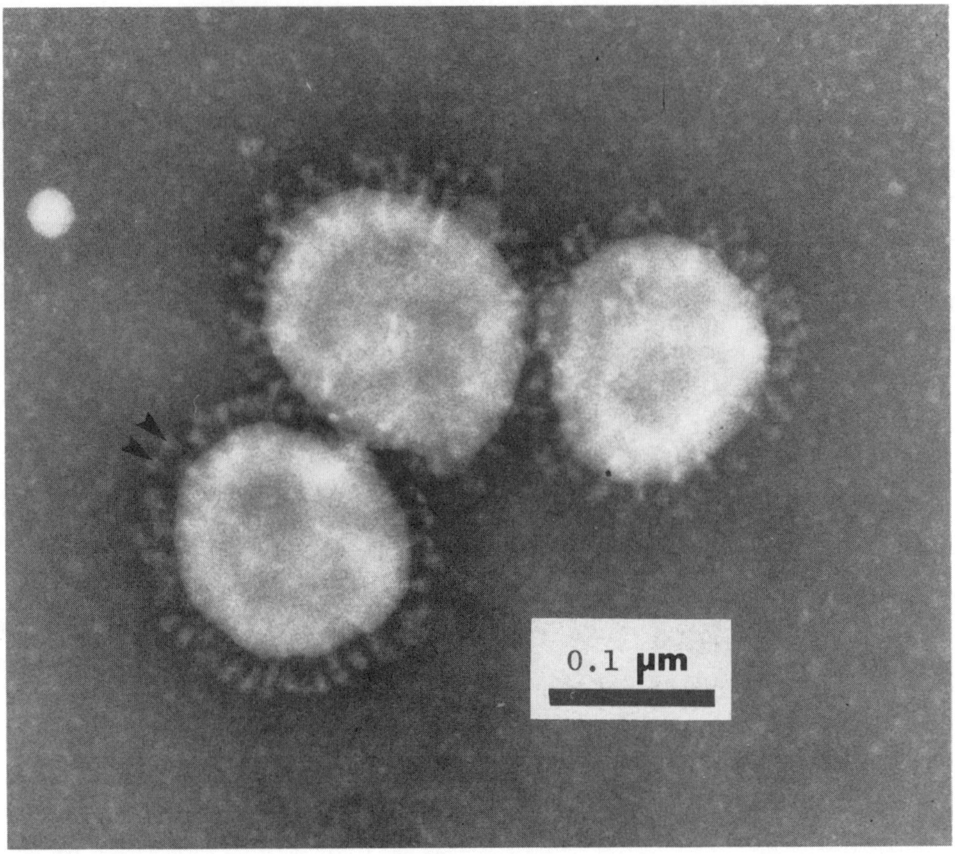

FIGURE 10. Electron micrograph of typical coronavirus particles from bluecomb disease of turkeys. Note the single layer of club-shaped surface projections (2 arrows). (Photo courtesy of Dr. Al Richie, Boone, IA.)

assessed by immune EM (IEM),[4] hemagglutination inhibition (HI), and neutralization of infectivity.[5] Use of these same techniques also failed to detect any cross-reactivity between CE and another coronavirus, avian infectious bronchitis virus.[4,5]

Different isolates of CE from diverse geographic locations were shown to be antigenically related, if not identical to one another.[1,5] These results were obtained using neutralization assays in poult challenge experiments[1] and HI assays with guinea pig hyperimmune sera produced against the Minnesota and Quebec CE isolates.[5] Such findings suggest that only a single serotype of CE may exist.

V. PATHOGENESIS, PATHOGENICITY, AND EPIDEMIOLOGY

CE infects turkeys of all ages, but not other avian species. The virus is transmitted by the fecal-oral route, with no evidence of egg transmission. The disease has a relatively short incubation period of 1 to 3 d. Affected birds become depressed, easily chilled, develop watery diarrhea often accompanied by loss of weight and dehydration, and show darkening of the head and skin (bluecomb). Morbidity is usually 100% and variable mortality (5 to 50%) is reported under field conditions.[1] Although the clinical course of CE may extend over a 2 week period, an unevenness may develop and persist in the flock for several weeks. Gross lesions are confined to the intestines with watery, gaseous gut contents. Histopathologic changes included appearance of large numbers of goblet cells at the tips of the villi and cuboidal or squamous villous epithelial cells with loss of microvilli. Villous atrophy was evident and the lamina propria was exposed and infiltrated with mononuclear cells.[1,6] Similar lesions, including a decreased villous to crypt ratio, were confirmed by scanning EM over a 3-d postinfection (D.P.I.) period[14] and correlated with a similar decline in coronavirus IF cells and reduction in shedding of viral particles by 4 D.P.I.,[15] although IF positive cells were still evident up to 28 D.P.I.

CE is probably spread mechanically among farms by personnel and equipment. Although the virus is excreted in droppings from recovered turkeys for several months, and may survive in frozen feces, it is easily destroyed by thorough cleaning and common disinfectants followed by depopulation for a 3- to 4-week period.[1]

VI. DIAGNOSIS

CE is difficult if not impossible to diagnose on the basis of clinical signs and lesions alone and must be differentiated from other enteric infections which occur in young poults. Failure to serially propagate CE virus in conventional cell culture systems has greatly hindered development of serologic tests for diagnosis. Initially, identification and titration of CE virus or neutralizing antibodies required inoculation of embryonating turkey eggs with intestinal homogenates and titration of virus or virus neutralizing antibodies in young poults.[1,12] A direct IF test was developed to detect viral antigens in intestinal epithelium of birds from field cases and confirm infection of turkey embryos and poults.[16] An indirect IF test, using frozen sections of infected intestines as a source of antigen, was used to detect antibodies from convalescing birds.[16] More recently, an ELISA test was developed to detect CE antigens. Results correlated well with direct EM or IEM detection of virus.[6] Although direct examination of intestinal contents by negative staining EM may identify the presence of coronavirus-like particles in addition to other enteric viruses, confirmatory diagnosis of CE should be made by IEM or other procedures employing CE antiserum of well-defined specificity.

In a recent report, the Minnesota and Quebec isolates of CE virus (passaged in eggs) were found to hemagglutinate rabbit and guinea pig erythrocytes.[5] Based on these findings, an HI test was developed to detect and titrate antibodies to CE in convalescent sera and confirm the antigenic similarities of the Quebec and Minnesota CE isolates.[5] Although the HI test may eventually provide a rapid, sensitive method for serologic diagnosis of CE, further work is needed to verify its sensitivity and specificity in comparison with other tests.

VII. IMMUNITY, PREVENTION, AND CONTROL

Results of field and experimental studies have indicated that turkeys which recovered from CE were immune to subsequent challenge, but some may remain carriers.[1,17] Passive immunity transferred either naturally via antibodies in the egg or artificially by subcutaneous injection of immune serum failed to provide protection against challenge.[1] SIgA antibodies to CE have been observed in bile and intestinal secretions of birds for periods up to 6 months after infection, but serum neutralizing antibody titers were low.[18] Cell-mediated immune responses, as measured by antigen-stimulated proliferation of peripheral blood lymphocytes, were also demonstrated in turkeys infected with CE.[19] However, the role these two components of the immune system play in recovery from infection and subsequent immunity was not further delineated.

The inability to serially propagate CE virus in conventional cell cultures has impeded vaccine development and no vaccines are currently available. Controlled exposure of 5- to 6-week-old poults to virulent virus has been recommended as a last resort when all other control methods fail, but only on farms with continual problems.[1] This has the disadvantage of perpetuating the infection, and abnormal losses may occur in immunocompromised flocks. At the present time, depopulation, disinfection, and security remain key factors in preventing and controlling CE.[1]

PART 5. UNCLASSIFIED

A. HUMAN ENTERIC CORONARVIRUS

I. INTRODUCTION

Coronaviruses are established causes of upper respiratory tract infections in man. In 1975 coronavirus-like particles (CVLP) were first reported as causing explosive diarrhea in young adults[1] or tropical sprue in children and adults.[2] Since then, there have been numerous reports of coronavirus-like particles being associated with outbreaks of diarrhea or necrotizing enterocolitis in neonates. The relationship between these agents and disease has been controversial, since CVLP have been found in the feces of nonclinically affected as well as clinically affected individuals. Recent molecular analysis of human enteric coronaviruses (HEC) has confirmed that they are indeed members of the Coronaviridae family.

II. MORPHOLOGY AND MORPHOGENESIS

Negative staining of CVLP from stools has revealed particles which resembled other coronaviruses. These particles were enveloped and pleomorphic with shapes ranging from round or oval to cylindrical or kidney shaped. Sizes ranged from as little as 60 nm in diameter to 1200 nm in length.[2-7] Particles have a central dense area or a distinct electron-lucent rim, depending on staining conditions.[1-3,5,6] The surface projections consist of a round or oval knob attached to a thin stalk which ranges from 20 to 46 nm in length.[2,3,5-7] Due to the pleomorphic nature of these viruses and the existence of naturally occurring pleomorphic fringed particles in stool, the identification of HEC in feces by electron microscopy is difficult and has led some to question the existence of HEC.[8]

III. CULTIVATION

In vitro propagation of HEC has been difficult and no reliable cell line has been found to grow large amounts of virus. The first attempt at cell culture propagation of HEC was by Caul et al.[3,9] using primary human embryonic kidney cells. They reported positive indirect immunofluorescence of cells using cultures inoculated with only 1 of 15 fecal specimens.[4] They were then unable to passage this isolate further in human embryonic kidney cells.

HEC also replicates in human rectal adenocarcinoma (HRT 18) cells.[6,10,11] Feces containing coronavirus-like particles from neonates with necrotizing enterocolitis produced positive immunofluorescence when inoculated onto HRT cells and stained with antiserum against bovine enteric coronavirus. No CPE was evident. The morphology of HEC grown in HRT cells, however, did not resemble those particles found in feces.

Organ culture has been the most successful method for propagating HEC. Fecal material from a patient shown to have coronavirus particles by electron microscopy was inoculated into human embryonic intestinal organ cultures. An increase in the number of extracellular virus particles, positive immunofluorescence, and the appearance of cellular changes characteristic of coronavirus replication was noted. By electron microscopy, infected epithelial cells showed a loss of the microvillous border, virus particles in the cisternae of the smooth endoplasmic reticulum, large membrane-bound virus-containing perinuclear vacuoles, and electron dense particles within cytoplasmic vesicles.[4]

Cells which were examined and found unsusceptible for the propagation of HEC were African green monkey kidney (AGMK), human embryonic fibroblasts, rabbit embryonic fibroblasts, calf embryonic fibroblasts,[12] human epidermoid oral carcinoma cells (Kb), human diploid, primary African Green monkey, Madin-Darby bovine kidney (MDBK), Madin-Darby canine kidney (MDCK)[13] a rhabdomyosarcoma line (RD), human embryonic intestine (intestine

407), a heteroploid lung line, human fetal lung (Flow 2000), primary cynomolgus monkey kidney, four strains of human embryonic tonsil fibroblasts, baby hamster kidney (BHK-21), Rhesus monkey kidney (LLC-MK2), Buffalo green monkey (SIRC),[14] and human epidermoid carcinoma (HEP-2).[15]

IV. PHYSICOCHEMICAL PROPERTIES

There are few reports of the characterization of HEC because of the difficulty in obtaining large quantities of virus. However, the following properties have been reported for HEC: (1) a buoyant density of 1.22 g/ml in CsC1, or 1.18 g/ml in glycerol-potassium tartrate; (2) stable at pH 3 to 7; (3) ether stable; (4) not thermostable at 56°C for 30 min; and (5) does not hemagglutinate goose, chicken, rat, guinea pig, rabbit, or human red blood cells.[12,15] These properties resemble those reported for other known human coronaviruses.[16]

Polyacrylamide gel electrophoresis of purified HEC has shown the virus contains four (62, 60, 34, and 32 kDa)[17] or five (190, 129, 66, 50, and 23 kDa)[15] major proteins and several minor proteins of 110, 97, 80, 77, and 56 kDa.[17] Some of these proteins react with hyperimmune guinea pig serum prepared against HEC purified from feces[17] or convalescent sera from coronavirus-infected individuals[17,15] in protein immunoblot analysis.

V. ANTIGENIC RELATIONSHIPS

Two-way cross-reactivity between human coronavirus OC43 and human enteric coronavirus (HEC), using both hyperimmune animal sera and patients' convalescent sera, was shown by immune electron microscopy.[18,19] However, no reaction between HEC antigens and antisera to OC43 could be shown in ELISA or protein immunoblots.[15,17] Antigenic relationships have also been found by direct and indirect immunofluorescence between human coronavirus strain OC43, bovine coronavirus, type 3 mouse hepatitis virus, and the 67N hemagglutinating encephalomyelitis virus of swine.[20]

It has been suggested that HEC may be related to bovine coronavirus (BCV).[10,11] Patel et al. reproduced a typical enteric BCV infection in a calf after inoculation with 5th passage HRT-18 grown CVLP, which was originally isolated from a child with necrotizing enterocolitis in France.[11] Typical coronavirus-like particles were isolated from the feces and positive immunofluorescence was observed in the colon and rectum. They concluded that HEC is either BCV or a human enteric coronavirus with specificity for at least two hosts. It has also been shown that antibodies to BCV occur in human sera from Germany and the U.S.,[21,22] and bovine sera positive for BCV have been used to identify HEC antigens in infected human cells.[10] In addition, there has been one report of a worker infected with BCV, who showed clinical signs of diarrhea and shed coronavirus particles infeces.[21]

VI. PATHOGENESIS AND PATHOGENICITY

The disease states associated with HEC generally fall into three groups: (1) acute gastroenteritis in children and adults; (2) tropical sprue in adults; and (3) necrotizing enterocolitis in infants. Gastroenteritis in children may present with one or all of the following clinical signs: fever, abdominal distention, increased stool frequency, regurgitation, abnormal gastric aspirates, and watery bloody stools,[23] with the most common signs being diarrhea (94%), fever (63%) and vomiting (51%).[14] In some cases the gastroenteritis may be fatal with severe enteritis and dehydration complicated by aspiration of vomitus or seizures.[24] Pathological findings may consist of severe dehydration, large volumes of intraluminal fluid in both the small and large bowel, with ultrastructural examination of the small intestine showing many 50- to 60-nm diameter, double-enveloped particles at the apex of epithelial cells.

Tropical sprue, a malabsorption syndrome, may present as diarrhea, steatorrhea, poor xylose absorption, and vitamin B_{12} malabsorption.[25] Intestinal biopsy of the jejunum may show patchy degeneration of epithelial cells, with ultrastructural studies showing membrane-lined vesicles containing virus-like particles of 50 to 100 nm with a distinct double outer membrane and peripheral fringe.

Necrotizing enterocolitis is the third manifestation associated with HEC infections. In one study of 11 infants (median age of 10 d, range 3 to 50) with necrotizing enterocolitis, clinical signs consisted of feeding intolerance, pneumatosis, abdominal distention with tenderness and bloody watery stools.[13,26]

It is assumed that HEC like other enteric agents is transmitted via a fecal-oral route;[13] however, no studies have been done to confirm this. It was suggested that in some infants the route of infection may be transplacental.[23] Infants born by caesarean section to mothers showing signs of gastrointestinal illness shed CVLP in their meconium, indicating infection prior to or during birth.

VII. DIAGNOSIS

Diagnosis of HEC enteritis has been almost exclusively by electron microscopy, with the exception of a few reports of cell culture propagation of the virus. Direct[2,24] and immune electron microscopy using patient's convalescent sera[13,23,27] have both been used successfully to identify HEC in the feces of children and adults.

Seroconversion to HEC (shown by agglutination titers) also occured in some newborns suffering from necrotizing enterocolitis; however, in only one of ten cases was the antibody elevation clearcut, making this an inconclusive diagnostic tool at this time.[13] Definite seroconversions and with titers of 1:100 or more in infants with necrotizing enterocolitis have been shown by ELISA, utilizing purified fecal HEC as antigen bound to the plate.[15] The ELISA should prove a useful diagnostic test in the future.

VIII. EPIDEMIOLOGY

The significance of HEC in the epidemiology of gastroenteritis in children and adults is still unclear, since most studies have found CVLP in almost as many asymptomatic individuals as symptomatic patients.[28] The prevalence of CVLP in the feces of children with gastroenteritis ranges from 0.5[5] to 39%[14] and in adults with gastroenteritis from 4[4] to 6%.[12] In asymptomatic controls the prevalence ranges from 1.4[12] to 65%[29] in children and from 5[4] to 93%[2] in adults. Many of these results are probably a reflection of the lack of reliable diagnostic tests to detect HEC and the common occurence of particles not readily morphologically distinguishable from coronaviruses in normal stools.[30] Despite this, there have been several reports of CVLP being associated with symptoms of gastroenteritis and being isolated more frequently from ill than from healthy controls.[13,18,23,21]

A seasonal incidence of HEC infections has been reported, with the peak incidence occurring in fall and early winter.[14] The primary age groups affected are those less than 3 years old and the duration of fecal virus shedding ranges from 5 to at least 25 d.

The prevalence of infection appears to be worldwide with reports of CVLP from Africa,[29] Australia,[5] Costa Rica,[32] India,[2] U.S.,[14] England,[4] France,[13] China,[27] and New Caledonia.[28]

IX. IMMUNITY, PREVENTION, AND CONTROL

Little is known about immunity to HEC infections other than they do produce seroconversions to HEC antigens[23] and related viruses. Since the significance of HEC infection is unresolved there have been no attempts at prevention or control.

B. PORCINE EPIDEMIC DIARRHEA VIRUS

I. INTRODUCTION

The disease syndrome, epidemic viral diarrhea was first reported during the 1970s in England[1,2] and Belgium.[3] Outbreaks of diarrhea were described in two different populations of swine: the first (type I) occurred only in finishing swine with no disease in suckling pigs;[1] and the second (type II) occurred in swine of all ages, including suckling pigs in which mortality was observed.[2] This latter diarrhea syndrome resembled transmissible gastroenteritis virus (TGEV) clinically, but TGEV and other known enteropathogenic agents were excluded by serology and other routine diagnostic tests. In 1978, laboratories in Belgium[4] and England[5] detected a coronavirus-like agent by negative stain electron microscopy (EM) associated with these disease outbreaks. Both groups of investigators were able to experimentally reproduce diarrhea in pigs orally inoculated with a field isolate containing the coronavirus-like particles. Both types of disease outbreaks described previously (I and II) are now known to have been caused by the same coronavirus which has been designated porcine epidemic diarrhea virus (PEDV).[3] Studies of morphology, morphogenesis, and the physicochemical characteristics of PEDV suggest tentative inclusion of this virus in the family Coronaviridae.

II. MORPHOLOGY AND MORPHOGENESIS

The morphologic characteristics of PEDV resemble those of other members of the Coronaviridae family, including the two swine coronaviruses, TGEV and hemagglutinating encephalomyelitis virus (HEV). PEDV cannot be distinguished from these latter two viruses on the basis of morphology alone. PED virions are pleomorphic, enveloped and range in size from 95 to 190 nm with a mean diameter of 130 nm.[4,5] Many particles have club-shaped peplomers, 18 to 23 nm in length radiating from the viral envelope.

The morphogenesis of PEDV in intestinal epithelial cells of naturally or experimentally infected pigs was described by a number of investigators.[5-8] The size, morphology and intracellular development of PEDV were similar to other Coronaviridae,[9] and closely resembled the morphogenesis of TGEV in intestinal cells.[7,10] Features described for PED viral morphogenesis included the following: irregular microvillus borders, presence of viral particles between the microvilli of infected cells, presence of large virus-containing intracytoplasmic vacuoles, which may represent dilated cisterna of endoplasmic reticulum, degeneration of many cellular organelles, and observed in some but not all studies, the assembly of viral particles by budding through intracytoplasmic membranes. Intracellular viral particles averaged 73 nm in diameter and were composed of a dense core often with an electron-lucent center surrounded by a double membrane, with or without surface projections. In one report, PEDV was also seen in macrophages in the lamina propria and in phagocytes in regional lymph nodes.[7]

III. *IN VITRO* CULTIVATION

To date, PEDV has not been serially propagated in porcine cell cultures, including intestinal, tracheal, lung, liver, kidney, spleen, brain, and skin explants from swine fetuses or newborn pigs.[3,11-13] Other cell types which have been tried unsuccessfully include secondary and pig kidney cell lines (PK15), secondary pig thyroid and salivary cells, and human rectal tumor cells (HRT-18).[4,11] Addition of trypsin to the growth medium was also ineffective in establishing replication of PEDV in the inoculated porcine cells.[14] Recently, PEDV was adapted to serial propagation in vero (African green monkey kidney) cells by inclusion of trypsin in the growth medium.[14] Cytopathic effects observed included syncytia formation, fusion of cells, and vacuolation. Virus replication was confirmed by immunofluorescence with FITC-conjugated anti-PEDV serum and the presence of coronavirus particles detected by electron miscroscopy from infected cell culture supernatants. PEDV failed to replicate *in vivo* in alternate species of laboratory animals (mice, rats, rabbits, and guinea pigs).

IV. PHYSICOCHEMICAL PROPERTIES AND GENOME

Failure of PEDV to replicate in conventional cell culture systems has hindered extensive biochemical characterization of the virus. The genome and replication strategy for PEDV has not been described; therefore, a comparison with other coronaviruses is not possible at the present time. Properties identified using concentrated and purified virus from gut contents of infected pigs include a density of 1.17 to 1.18 g/ml in sucrose; sensitivity to ether and chloroform, which is consistent with possession of an envelope; and failure to hemagglutinate erythrocytes from 12 different animals.[12,13,15] In addition, the structural proteins of PEDV have been described recently.[15] They consist of two clusters of proteins from 20 to 32 kDa and glycosylated proteins from 85 to 135 kDa, as well as an RNA-binding protein of 58 kDa. The properties identified for PEDV are consistent with those of other members of the Coronaviridae family.[16]

V. ANTIGENIC RELATIONSHIPS

In vitro — Immune electron microscopy (IEM) and immunofluorescence (IF) were used in two-way tests to assess the antigenic relationships between PEDV and other coronaviruses.[13,17] PEDV was antigenically distinct from two porcine coronaviruses (TGEV and HEV), canine coronavirus (CCV), bovine coronavirus (NCDV), avian infectious bronchitis virus (IBV), and feline infectious peritonitis virus (FIPV) in the two-way cross-reactivity tests. However, use of convalescent antisera to some viruses (CCV, NCDV, IBV, FIPV), but hyperimmune antisera prepared in pigs to others (PEDV, TGEV, HEV), may have resulted in a failure to detect minor antigenic differences due to lower antibody titers in convalescent sera and may require further confirmation using more sensitive tests and additional sera of well-defined specificity. The antibody titer and specificity of the serum has been found to be important in investigating antigenic relationships among coronaviruses and may account for the discrepant results reported in cross-reactivity between TGEV and CCV or FIPV.[17,18]

In vivo — Two-way cross-protection studies were conducted in seronegative (for PEDV and TGEV) conventional pigs.[19] Two pigs recovered from TGE were susceptible to PED disease and infection upon subsequent challenge exposure. Pigs recovered from PEDV developed clinical signs of TGE or seroconverted to TGE after challenge exposure with TGEV. These results thus confirm the *in vitro* findings demonstrating the lack of an antigenic relationship between PEDV and TGEV.

Only one serotype of PEDV may exist. This is based on preliminary data which has shown that strains of PEDV from Germany and France are serologically identical to the prototype strain (CV777) first described in Belgium.[3,11,13]

VI. EPIDEMIOLOGY

PEDV has been detected in swine in Belgium, England, France, West Germany, and China.[3] A disease syndrome resembling PED clinically and by the presence of coronavirus-like particles in feces, was also reported in Hungary, East Germany, Canada and Japan.[20-24] Serologic and IF examination revealed no evidence of TGEV infection, but no further characterization of the coronavirus-like agents associated with these outbreaks has been reported, except in a recent Canadian study.[23] In this latter report, the coronavirus-like particle did not react with antiserum to PEDV (CV777) by IEM or counterimmunoelectrophoresis.

A blocking ELISA test was used to conduct serologic surveys of the prevalence of PEDV in swine sera from various countries.[3,25] Antibodies to PEDV were detected in sera from Belgium, England, France, West Germany, The Netherlands, Switzerland, Bulgaria, and Taiwan. Antibodies were not detected in sera from Sweden, northern Ireland, the U.S., Australia, or Hungary. Serologic surveys were also conducted in Belgium using swine sera collected over a 15-year period, from 1969 to 1984, to try to verify whether PEDV represents occurrence of a new disease syndrome or the recognition of an existing but previously undiagnosed disease.[3] Results indicated antibodies to PEDV were not detected until 1971, which coincided with the first reported clinical outbreaks of PED. Moreover, the seroprevalence of PEDV thereafter (11 to 42% of sows positive) was similar to the seroprevalence previously reported for TGEV, for which yearly epidemic outbreaks are common.[3]

Two distinct disease patterns are often observed for PED outbreaks which occur primarily during the winter months in breeding herds.[3] In the first, the disease syndrome resembles epidemic TGE with adult animals subclinically affected or disease signs limited to depression, anorexia, and vomiting. Piglets develop watery diarrhea, possibly preceded by vomiting, and variable mortality rates (from 50 to 90%) may be observed in piglets under 1 week of age. This form is very similar to TGEV, except PEDV spreads more slowly through the herd (4 to 5 weeks), mortality rates are lower in baby pigs, and PEDV does not appear to persist or become endemic on the farm.

In the second type of outbreak, weaned pigs and adult animals may be severely affected, but suckling pigs experience only mild diarrhea and low mortality.[3] Less variation in clinical signs is seen among finishing swine, with 100% morbidity common during a PED outbreak, which may persist for only 1 week.[3] A mortality rate of 1 to 3% may occur. Finishing pigs or sows may die acutely and acute back muscle necrosis is common in necropsied animals.

PEDV is transmitted by the fecal-oral route. It is probably spread mechanically among farms by personnel and equipment, since the virus does not appear to persist after an outbreak unless susceptible animals are added.[3] Transmission of PED to non-porcine hosts has not been reported.

VII. PATHOGENESIS AND PATHOGENICITY

The pathogenesis and pathogenicity of PEDV was studied in naturally infected piglets and orally inoculated colostrum-deprived and conventional pigs ranging in age from 2 to 3 d to 6 weeks.[6-8,19,26,27] Following inoculation, vomiting and diarrhea were seen after 22 to 36 h. Of 13 inoculated pigs, 5 died from dehydration at 2 to 4 d postinoculation (DPI), 3 of which were 1-d-old at inoculation.[19] Viral replication, as assessed by IF and transmission EM on thin sections, was observed in the cytoplasm of small intestinal villous and colonic epithelial cells, and occasionally a few IF foci were seen in crypt epithelium. In one study, viral replication, as observed by EM, was seen primarily in enterocytes located on the lateral walls of villi and extending to the villus/crypt border, but not in colonic epithelium.[7] This finding agrees with results of a second study in which higher numbers of IF cells were seen at the base of the villus.[6] Viral particles also occurred in macrophages in the lamina propria and phagocytes in regional lymph nodes, which also showed IF.[6,7] However, no budding particles were observed in these cells, suggesting nonproductive infections or sites of viral degradation.

Infected enterocytes were observed by 12 to 18 h PI, reaching a maximum at 24 to 36 h PI.[6,19] This peak of fluorescent cells appeared earlier in the distal small intestine. A second peak of fluorescing cells was observed in one pig killed at 4 DPI, suggesting an infection of regenerating cells. Fluorescent cells were detected for 5 DPI, whereas virus shedding in feces persisted for 4 and 7 DPI as detected by EM and ELISA, respectively.[6,19] No extraintestinal replication of PEDV was observed in any of the aforementioned reports.

Villous atrophy was evident at 24 h PI, which coincided with the onset of diarrhea. It was most severe in the jejunum at 24 to 48 h PI in one study,[6] and in the jejunum and ileum in a second study.[7] These results were confirmed by scanning EM studies,[26] which showed desquamation of epithelial cells in the jejunum at the onset of diarrhea, and in the ileum about 6 h after onset of diarrhea. Extensive villous fusion was commonly seen after this time. No histologic changes were noted by light microscopy or scanning EM in the colon.

In histochemical studies, PEDV-infected piglets had decreased activity (lower staining intensity than the controls) of four intestinal enzymes including, alkaline phosphatase, acid phosphatase, succinic dehydrogenase, and monoamine oxydase.[29] Loss of such enzymatic activity needed to maintain digestive-absorptive capacity contributes to the malabsorptive diarrhea evident in PEDV and other coronaviral infections, such as TGEV[10,27] and canine coronavirus.[27,28]

The pathogenicity of PEDV was very similar to TGEV but with the following differences noted: TGEV infection progresses more rapidly and leads to more severe villous atrophy; TGEV infects cells throughout the entire villus, but no crypt cells, whereas PEDV infects cells mainly at the base of the villus and some crypt enterocytes; and PEDV, but not TGEV, infects colonic epithelial cells.[6,7,10,19,27]

VIII. DIAGNOSIS

A reliable diagnosis of PED cannot be made on the basis of clinical signs or histopathologic lesions alone. Occurrence of diarrhea in pigs of all ages, including baby pigs, as often seen in PED outbreaks, resembles outbreaks of TGEV. However, rotavirus and *Escherichia coli* rarely cause diarrhea in adult swine. PEDV can be diagnosed by identification of PEDV or viral antigen in feces and/or intestines, or detection of PEDV antibodies.

A. DETECTION OF VIRUS OR VIRAL ANTIGENS

Electron microscopy was the first test used for detection of PEDV in feces or gut contents.[4,5] This test relies on the morphologic identification of coronavirus particles from infected pigs; it has the further advantage of detecting the presence of other morphologically distinct enteric viruses and being applicable to specimens (feces) collected from live animals. Two problems are encountered with this procedure: first, it is difficult to identify coronavirus particles in feces, many of which may have lost their characteristic surface projections and, hence, are difficult to distinguish from other pleomorphic membranous particles commonly seen in feces.[29] Second, since PEDV and TGEV are morphologically identical, immune electron microscopy using antiserum monospecific for PEDV must be used to differentiate these two coronaviruses in feces or gut contents.[17] A more sensitive procedure recently developed to identify coronaviruses in bovine feces is the technique of protein-A gold IEM,[30] but this method has not yet been applied to PEDV detection.

A direct IF test using hyperimmune monospecific serum to PEDV prepared in hysterectomy-derived, colostrum-deprived pigs was used for detection of PEDV antigen in cryostat sections from the small intestines of infected pigs.[19] This method was more sensitive than EM for detection of PEDV, but required gut sections from pigs euthanized during the acute phase of diarrhea (less than 3 d after onset). Also, the applicability of this technique for diagnosis of PED in older finishing swine was not reported.[3]

More recently, an ELISA test was developed for detection of PEDV antigens in feces or gut contents of naturally or experimentally infected pigs.[31] This assay was more sensitive and reliable than EM for diagnosis of PEDV in both younger and older swine. Feces from experimentally inoculated pigs were consistently positive for PEDV antigen by ELISA between 2 to 5 DPI, and inconsistently positive between 6 to 8 DPI.

B. DETECTION OF VIRAL ANTIBODIES

Two methods have been reported for serologic diagnosis of PEDV. An indirect IF test was developed using PEDV-positive cryostat sections of intestines from experimentally infected pigs.[19] Whereas IF antibodies were detected in experimentally inoculated pigs after 15 DPI,[19] the practical diagnostic application of this procedure under field conditions may be limited due to the reported temperature liability of the IF antibodies.[32]

An ELISA blocking assay was also described for detection and titration of PEDV antibodies.[31] PEDV antigen for use in this assay had to be prepared in colostrum-deprived pigs since no cell culture source of virus was available. Results of testing sera collected from experimentally or naturally infected pigs indicated that PEDV antibodies were not detectable until at least 4 weeks after onset of the diarrheal outbreak, but PEDV antibodies persisted for at least 1 year.[33] Whether these findings reflect insensitivity of the ELISA for antibody detection or are an accurate indication of the immune response to PEDV, must await development of other more sensitive antibody assays.

IX. IMMUNITY, PREVENTION, AND CONTROL

The following characteristics of serum antibody responses to PEDV were defined using the ELISA blocking assay for antibody titration.[33] Whether similar responses might be observed using virus neutralizing antibody assays has not been examined, but such tests were precluded pending development of routine procedures such as recently reported for *in vitro* propagation of PEDV.[14]

PEDV antibodies were detected in serum of most animals at 3 to 4 weeks PI, reached peak titers at 4 to 6 weeks PI, then declined, but were still detectable 1 year later. Maternally derived serum antibodies disappeared in pigs between 4 to 13 weeks of age, depending upon the initial titer in serum. Five of six pigs suckling a previously infected sow showed diarrhea after challenge exposure of two piglets at 4 weeks of age. Although the sow had PED antibodies in milk at the time of challenge, the titer (80) may have been too low to be protective, and although IgA antibodies were identified they were not shown to be associated with neutralizing antibodies, which are known to be important for protection against TGEV.[34] The inability of antibodies of low titer to passively protect against PED was further demonstrated by *in vivo* neutralization studies. A pig inoculated with a PEDV-convalescent serum mixture was not protected against infection or disease, whereas another pig given PEDV and diluted hyperimmune serum remained healthy.[19] Based on preliminary findings, local protection against PED may be of short duration, since reinfection and diarrhea have been observed in seropositive animals which recovered from PED 5 months previously.[33]

Little information is available on prevention and control of PED. Since virus is introduced into farms mainly by animal or human movement, disinfection and security remain key factors for prevention and control. During a PED outbreak, measures similar to those used in TGEV outbreaks[34] may be helpful in reducing mortality, including free access to water to reduce dehydration, providing supplemental heat to young pigs, withholding feed in finishing units, and isolation of farrowing units with newly born piglets.[3] Feeding of intestinal contents from infected piglets to pregnant sows at least 2 weeks prior to farrowing may stimulate lactogenic immunity and provide some protection to their piglets. Antibiotic treatments are of no value unless secondary bacterial infections contribute to prolonged morbidity and mortality among pigs.

REFERENCES

PART 1

1. **Wege, H., Siddel, S,. and ter Meulen, V.,** The biology and pathogenesis of coronaviruses, *Curr. Top. Microbiol. Immunol.*, 99, 165, 1982.
2. **Siddell, S. G., Anderson, R., Cavanagh, D., Fujiwara, K., Klenk, H. D., Macnaughton, M. R., Pensaert, M., Stohlman, S. A., Sturman, L. and van der Zeijst, R.A.M.,** Coronaviridae, *Intervirology*, 20, 181, 1983.
3. **Sturman, L. S. and Holmes, K. V.,** The molecular biology of coronaviruses, *Adv. Virus Res.*, 28, 35, 1983.
4. **Tyrrell, D. A. J., Almeida, J. D., Berry, P. M., Cunningham, C. H., Hamre, D., Hofstad, M. S., Malluchi, L., and McIntosh, K.,** Coronaviruses, *Nature, (London)*, 220, 650, 1968.
5. **Brian, D. A., Hogue, B., Lapps, W., Potts, B., and Kapke, P.,** Comparative structure of coronaviruses, in *Proc. Fourth Int. Symp. Neonatal Diarrhea.*, Acres, S. D., Ed., VIDO, University of Saskatchewan, Saskatoon, Canada, 1984, 100.
6. **Weiss, S. R. and Leibowitz, J. L.,** Characterization of murine coronavirus RNA with virus-specific cDNA probes, *J. Gen. Virol.*, 64, 127, 1982.
7. **Lapps, W., Hogue, B. G., and Brain, D. A.,** Sequence analysis of the bovine coronavirus nucleocapsid and matrix protein genes, *Virology*, 157, 47, 1987.
8. **Jacobs, L., de Grott, R., van der Zeijst, B. A., Horzinek, M. C., and Spaan, W.,** The nucleotide sequence of the peplomer gene of porcine transmissible gastroenteritis virus: comparison with the sequence of the peplomer protein of feline infectious peritonitis virus, *Virus Res.*, 8, 363, 1987.
9. **Storz, J., Rott, R., and Kaluza, G.,** Enhancement of plaque formation and cell fusion of an enteropathogenic coronavirus by trypsin treatment, *Infect. Immun.*, 31, 1214, 1981.
10. **Bohl, E. H.,** Diagnosis of diarrhea in pigs due to transmissible gastroenteritis virus or rotavirus, *INSERM*, 90, 341, 1979.
11. **Hoffman, M. and Wyler, R.,** Propagation of the virus of porcine epidemic diarrhea in cell culture, *J. Clin. Microbiol.*, 26, 2235, 1988.
12. **Saif, L. J., Redman, D. R., Moorhead, P. D., and Theil, K. W.,** Experimental coronavirus infections in calves: Viral replication in the respiratory and intestinal tracts, *Am. J. Vet.* Res., 47, 1426, 1986.
13. **Hecket, R. A., Saif, L. J., Mohamed, T. M., and Myers, G. W.,** Epidemiology of respiratory bovine coronavirus infections detected by nasal cell immunofluorescence, Abstr. 391, Conf. of Res. Workers in Anim. Dis., Chicago, Il., November 14 to 15, 1988.
14. **Pensaert, M., Callebaut, P., and Vergote, J.,** Isolation of a porcine respiratory, non-enteric coronavirus related to transmissible gastroenteritis, *Vet. Q.*, 8, 257, 1986.
15. **Furuuchi, S., Shimizu, Y., and Kumagai, T.,** Multiplication of low and high cell culture-passaged strains of transmissible gastroenteritis virus in organs of newborn pigs, *Vet. Microbiol.*, 3, 169, 1979.
16. **Pedersen, N. C., Boyle, J. F., Floyd, K., Fudge, A., and Barker, J.,** An enteric coronavirus infection of cats and its relationship to feline infectious peritonitis, *Am. J. Vet. Res.*, 42, 368, 1981.

PART 2A

1. **Doyle, L. P. and Hutchings, L. M.,** A transmissible gastroenteritis in pigs, *J. Am. Vet. Med. Assoc.*, 108, 257, 1946.
2. **Saif, L. J. and Bohl, E. H.,** Transmissible gastroenteritis, *Diseases of Swine*, 6th ed., Leman, A. D. et al., Eds., Iowa State University Press, Ames, IA, 1986, 225.
3. **Toma, B. and Benet, J. J.,** A technique of research on microplates of the antibodies neutralizing transmissible gastroenteritis virus of swine, *Rec. Med. Vet.*, 152, 565, 1976.
4. **Egan, I. T., Harris, D. L., and Hill, H. T.,** Prevalence of swine dysentery, transmissible gastroenteritis,and pseudorabies in Iowa, Illinois and Missouri swine, Proc. 86th Annu. Meet. U. S. Anim. Health Assoc., Nashville, TN, November 7 to 12, 1982, 497.
5. **Pensaert, M., Callebaut, P. and Vergote, J.,** Isolation of a porcine respiratory, non-enteric coronavirus related to transmissible gastroenterits, *Vet. Q.*, 8, 257, 1986.
6. **Brown, I. and Cartwright, S.,** New porcine coronavirus?, *Vet. Rec.*, 119, 282, 1986.
7. **Okaniwa, A., Harada, K., and Park, D. K.,** Structure of swine transmissible gastroenteritis virus examined by negative staining, *Natl. Inst. Anim. Health Q. (Tokyo)*, 8, 175, 1968.
8. **Phillip, J. I. H., Cartwright, S. F., and Scott, A. C.,** The size and morphology of T.G.E. and vomitting and wasting disease of pigs, *Vet. Rec.*, 88, 311, 1971.
9. **Garwes, D. J. and Pocock, D. H.,** The polypeptide structure of transmissible gastroenteritis virus, *J. Gen. Virol.*, 29, 25, 1975.
10. **Siddell, S., Wege, H., and Ter Meulen, V.,** The biology of coronaviruses, *J. Gen. Virol.*, 64, 761, 1983.
11. **Thake, D. C.,** Jejunal epithelium in transmissible gastroenteritis of swine (an electron microscopic and histochemical study), *Am. J. Pathol.*, 53, 149, 1968.

12. **Pensaert, M. B., Haelterman, E. O., and Burnstein, T.,** Transmissible gastroenteritis of swine: virus-intestinal cell interactions. I. Immunofluorescence, histopathology and virus production in the small intestine through the course of infection, *Arch. Gesamte Virusforsch.*, 31, 321, 1970.
13. **Dulac, G. C., Ruckerbauer, G. M., and Boulanger, P.,** Transmissible gastroenteritis: demonstration of the virus from field specimens by means of cell culture and pig inoculation, *Can. J. Comp. Med.*, 41, 357, 1977.
14. **Harada, K., Kumagai, T., and Sasahara, J.,** Cytopathogenicity of transmissible gastroenteritis virus in pigs. *Natl. Inst. Anim. Health Q, (Tokyo)*, 3, 166, 1963.
15. **Witte, K. H.,** Isolation of the virus of transmissible gastroenteritis (TGE) from naturally infected piglets in cell cultures, *Zentralbl. Veterinaermed B*, 18, 770, 1971.
16. **McClurkin, A. W. and Norman, J. O.,** Studies on transmissible gastroenteritis of swine. II. Selected characteristics of a cytopathogenic virus common to five isolates from transmissible gastroenteritis, *Can. J. Comp. Med.*, 30, 190, 1966.
17. **Kemeny, L. J.,** Isolation of transmissible gastroenteritis virus from pharyngeal swabs obtained from sows at slaughter, *Am. J. Vet.Res.*, 39, 703, 1978.
18. **Stepanek, J., Pospisil, Z., and Mesaros, E.,** Growth activity of transmissible gastroenterits (TGE) virus in primary cultures of pig kidney cells and pig salivary gland cells, *Acta Vet. (Brno.)*, 40, 235, 1971.
19. **Rubinstein, D., Tyrrell, A. J., Derbyshire, J. B., and Collins, A. P.,** Growth of porcine transmissible gastroenteritis virus in organ cultures of pig tissue, *Nature (London)*, 227, 1348, 1970.
20. **Welter, C. J.,** TGE of swine. I. Propagation of virus in cell cultures and development of a vaccine, *Vet. Med. Small Anim. Clin.*, 60, 1054, 1965.
21. **Woods, R. D.,** Studies of enteric coronaviruses in a feline cell line, *Vet. Microbiol.*, 7, 427, 1982.
22. **Reynolds, D. J., Garwes, D. J., and Lucey, S.,** Differentiation of canine coronavirus and porcine transmissible gastroenteritis virus by neutralization with canine, porcine and feline sera, *Vet. Microbiol.*, 5, 283, 1980.
23. **Bohl, E. H.,** Diagnosis of diarrhea in pigs due to transmissible gastroenteritis virus or rotavirus, in *Viral Enteritis in Humans and Animals*, Vol. 90, Bricout, F. and Scherrer, R., Eds., INSERM, Paris, 1979, 341.
24. **Pocock, D. H. and Garwes, D. J.,** The influence of pH on the growth and stability of transmissible gastroenteritis virus in vitro, *Arch. Virol.*, 49, 239, 1975.
25. **Brian, A. B., Dennis, D.E., and Grey, J. S.,** Genome of porcine transmissible gastroenteritis virus, *J. Virol.*, 34, 410, 1980.
26. **Pike, B. V. and Garwes, D. J.,** Lipids of transmissible gastroenteritis virus and their relation to those of two different host cells, *J. Gen. Virol.*, 34, 531, 1977.
27. **Horzinek, M. C., Lutz, H., and Pedersen, N. C.,** Antigenic relationships among homologous structural polypeptides of porcine, feline, and canine coronaviruses, *Infect. Immun.*, 37, 1148, 1982.
28. **Brian, D. A., Hogue, B., Lapps, W., Potts, B., and Kapke, P.,** Comparative structure of coronaviruses, in *Proc. 4th Int. Symp. Neonatal Diarrhea*, S. Acres, Ed., VIDO, Saskatoon, Canada, 1984, 100.
29. **Hu, S., Bruszewski, J., Boone, T., and Souza, L.,** Cloning and expression of the surface glycoprotein gp 195 of porcine transmissible gastroenteritis virus, in *Modern Approaches to Vaccines*, Chanock, R. M. and Lerner, R. A., Eds., Cold Spring Harbor Laboratory, Cold Spring, Harbor, NY, 1984, 219.
30. **Garwes, D. J., Lucas, M. H., Higgins, D. A., Pike B. V., and Cartwright, S. F.,** Antigenicity of structural components from porcine transmissible gastroenteritis virus, *Vet. Microbiol.*, 3, 179, 1978/79.
31. **Laude, H., Chapsal, J. M. Gelfi, J., Labiau, S., and Grosclaude, J.,** Antigenic structure of transmissible gastroenteritis virus. I. Properties of monoclonal antibodies directed against virion proteins, *J. Gen. Virol.*, 67, 119, 1986.
32. **Jimenez, G., Correa, I., Melgosa, M. P., Brillido, M. J., and Enjuanes, L.,** Critical epitopes in transmissible gastroenteritis virus neutralization, *J. Virol.*, 60, 131, 1986.
33. **Welch, S-K. W. and Saif, L. J.,** Monoclonal antibodies to a virulent strain of transmissible gastroenteritis virus: comparison of reactivity against the attenuated and virulent virus strain, *Arch. Virol.*, 101, 221, 1988.
34. **Rasschaert, D., Delmas, B., Charley, B., Grosclaude, J., Gelfi, J., and Laude, H.,** Surface glycoproteins of transmissible gastroenteritis virus: Function and gene sequence, *Adv. Exp. Med. Biol.*, 218, 109, 1987.
35. **Wesley, R. D. and Woods, R. D.,** Identification of a 17,000 molecular weight antigenic polypeptide in transmissible gastroenteritis virus-infected cells, *J. Gen. Virol.*, 67, 1419, 1986.
36. **Delmas, B., Gelfi, J., and Laude, H.,** Antigenic structure of transmissible gastroenteritis virus. II. Domains in the peplomer glycoprotein, *J. Gen. Virol.*, 67, 1405, 1986.
37. **Woods, R. D., Wesley, R. D., and Kapke, P. A.,** Neutralization of transmissible gastroenteritis virus by complement dependent monoclonal antibodies, *Am. J. Vet. Res.*, 49, 300, 1988.
38. **Clarke, M. C.,** The effect of 5-bromodeoxyuridine and actinomycin D on the multiplication of transmissible gastroenteritis virus, *J. Gen. Virol.*, 3, 267, 1968.
39. **Dennis, D. E. and Brian, D. A.,** RNA-dependent RNA polymerase activity in coronavirus infected cells, *J. Virol.*, 42, 153, 1982.

40. **Jacobs, L., Van der Zeijst, B. A. M., and Horzinek, M. C.,** Characterization and translation of transmissible gastroenteritis virus mRNAs, *J. Virol.*, 57, 1010, 1986.
41. **Rasschaert, D., Gelfi, J., and Laude, H.,** Enteric coronavirus TGEV: partial sequence of the genomic RNA, its organization and expression, *Biochimie*, 69, 591, 1987.
42. **Kapke, P. A. and Brian, D. A.,** Sequence analysis of the porcine transmissible gastroenteritis coronavirus nucleocapsid protein gene, *Virology*, 151, 41, 1986.
43. **Kapke, P. A., Tung, F. Y., Brian, D. A., Woods, R. D., and Wesley, R.,** Nucleotide sequence of the porcine transmissible gastroenteritis coronavirus matrix protein gene, *Adv. Exp. Med. Biol.*, 218, 117, 1987.
44. **Jacobs, L., deGroot, R., van der Zeijst, B. A., Horzinek, M. C., and Spaan, W.,** The nucleotide sequence of the peplomer gene of porcine transmissible gastroenteritis (TGEV): comparison with the sequence of the peplomer protein of feline infectious peritonitis virus (FIPV), *Virus Res.*, 8, 363, 1987.
45. **Britton, P., Carmenes, R. S., Page, K. W., Garwes, D. J., and Parra, F.,** Sequence of the nucleoprotein gene from a virulent British field isolate of TGEV and its expression in *Saccharomyces cerevisiae, Mol. Microbiol.*, 2, 89, 1988.
46. **Hu, S., Bruszewski, J., Smalling, R., and Browne, J. K.,** Studies of TGEV spike protein gp 195 expressed in *E. coli* and by a TGE-vaccinia virus recombinant, in *Immunobiology of Proteins and Peptides*, Vol. III, Zauhair-Atassi, M. and Cachrach, H. L., Eds., Plenum Press, New York, 1985, 63.
47. **Kemeny, L. J.,** Antibody response in pigs inoculated with transmissible gastroenteritis virus and cross reactions among ten isolates, *Can. J. Comp. Med.*, 40, 209, 1976.
48. **Pedersen, N. C., Ward, J., and Mengeling, W. L.,** Antigenic relationship of the feline infectious peritonitis virus to coronaviruses of other species, *Arch. Virol.*, 58, 45, 1978.
49. **Pensaert, M. B. and DeBouck, P.,** A new coronavirus-like particle associated with diarrhea in swine, *Arch. Virol.*, 58, 243, 1978.
50. **Chasey, D. and Cartwright, S. F.,** Virus-like particle associated with porcine epidemic diarrhoea, *Res. Vet. Sci.*, 25, 255, 1978.
51. **Callebaut, P., Correa, I., Pensaert, M., Jimenez, G., and Enjuanes, L.,** Antigenic differentiation between transmissible gastroenteritis virus of swine and a related porcine respiratory coronavirus, *J. Gen. Virol.*, 69, 1725, 1988.
52. **Garwes, D. J., Stewart, F., Cartwright, S. F., and Brown, I.,** Differentiation of porcine coronavirus from transmissible gastroenterits virus, *Vet. Rec.*, 122, 86, 1988.
53. **Binn, L. N., Lazar, E. C., Keenan, K. P., Huxsoll, D. L., Marchwicki, R. H., and Strano, A. J.,** Recovery and characterization of a coronavirus from military dogs with diarrhea, *Proc. U. S. Anim. Health Assoc.*, 78, 359, 1974.
54. **Woods, R. D., Cheville, N. F., and Gallagher, J. E.,** Lesions in the small intestine of newborn pigs inoculated with porcine, feline and canine coronaviruses, *Am. J. Vet. Res.*, 42, 1163, 1981.
55. **Haelterman, E. O.,** Epidemiological studies of transmissible gastroenteritis of swine, *Proc. U. S. Livest. Sanit. Assoc.*, 1962, 305.
56. **McClurkin, A. W., Stark, S. L., and Norman, J. O.,** Transmissible gastroenteritis (TGE) of swine: the possible role of dogs in in the epizootiology of TGE. *Can. J. Comp. Med.*, 34, 347, 1970.
57. **Witte, K. H., Tuch, K., Dubenkropp, H., and Walther, C.,** Untersuchungen uber die Antigenverwandtschaft der Viren der Felinen Infektiosen Peritonitis (FIP) und der Transmissible Gastroenteritis (TGE) des Schweines, *Berl. Muench Tieraerztl Wochenschr.*, 90, 396, 1977.
58. **Reynolds, D. J. and Garwes, D. J.,** Virus isolation and serum antibody responses after infection of cats with transmissible gastroenteritis virus, *Arch. Virol.*, 60, 161, 1979.
59. **Underdahl, N. R., Mebus, C. A., and Torres-Medina, A.,** Recovery of transmissible gastroenteritis virus from chronically infected experimental pigs, *Am. J. Vet. Res.*, 36, 1473, 1975.
60. **Theil, K. W., Saif, L. J., Bohl, E. H., Agnes, A. G., and Kohler, E. M.,** Concurrent porcine rotaviral and transmissible gastroenteritis viral infections in a three-day-old conventional pig, *Am.J. Vet. Res.*, 40, 719, 1979.
61. **Pritchard, G. C.,** Transmissible gastroenteritis in endemically infected breeding herds of pigs in East Anglia, 1981—1985, *Vet. Rec.*, 120, 226, 1987.
62. **Cepica, A. and Derbyshire, J. B.,** Antibody-dependent and spontaneous cell-mediated cytotoxicity against transmissible gastroenteritis virus-infected cells by lymphocytes from sows, fetuses and neonatal piglets, *Can. J. Comp. Med.*, 48, 258, 1984.
63. **Hooper, B. E. and Haelterman, E. O.,** Concepts of pathogenesis and passive immunity in transmissible gastroenteritis of swine, *J. Am. Vet. Med. Assoc.*, 149, 1580, 1966.
64. **Moon, H. W.,** Mechanisms in the pathogenesis of diarrhea: A review, *J. Am. Vet. Med. Assoc.*, 172, 443, 1978.
65. **Chu, R. M., Glock, R. D., and Ross, R. F.,** Changes in gut-associated lymphoid tissues of the small intestine of eight-week-old pigs infected with transmissible gastroenteritis virus, *Am. J. Vet. Res.*, 43, 67, 1982.
66. **Butler, D. G., Gall, D. G., Kelly, M. H., and Hamilton, J. R.,** Transmissible gastroenteritis: mechanisms responsible for diarrhea in an acute enteritis in piglets, *J. Clin. Invest.*, 53, 1335, 1974.

67. **Cornelius, L. M., Hooper, B. E., and Haelterman, E. O.,** Changes in fluid and electrolyte balance in baby pigs with transmissible gastroenteritis, *Am. J. Clin. Pathol.*, 2, 105, 1968.
68. **Moon, H. W., Norman, J. O., and Lambert, G.,** Age dependent resistance to TGE of swine. I. Clinical signs and some mucosal dimensions in the small intestine, *Can. J. Comp. Med.*, 37, 157, 1973.
69. **Norman, J. O., Lambert, G., Moon, H. W., and Stark, S. L.,** Age dependent resistance to transmissible gastroenteritis (TGE). II. Coronavirus titer in tissues of pigs after exposure, *Can. J. Comp. Med.*, 37, 167, 1973.
70. **Pensaert, M. B., Haelterman, E. O., and Hinsman, E. J.,** Transmissible gastroenteritis of swine: virus-intestinal cell interactions. II. Electron microscopy of the epithelium in isolated jejunal loops, *Arch. Gesamte Virusforsch.*, 31, 335, 1970.
71. **Shepherd, R. W., Gall, D. G., Butler, D. G., and Hamilton, J. R.,** Determinates of diarrhea in viral enteritis: the role of ion transport and epithelial changes in the ileum in transmissible gastroenteritis in piglets, *Gastroenterology*, 76, 20, 1979.
72. **LaBonnardiere, C. and Laude, H.,** High interferon titer in newborn pig intestine during experimentally induced viral enteritis, *Infect. Immun.*, 32, 28, 1981.
73. **Wagner, J. E., Beamer, P. D., and Ristic, M.,** Electron microscopy of intestinal epithelial cells of piglets infected with a transmissible gastroenteritis virus, *Can. J. Comp. Med.*, 37, 177, 1973.
74. **Witte, K. H. and Walther, C.,** Age-dependent susceptibility of pigs to infection with the virus of transmissible gastroenteritis, *Proc. 4th Int. Congr. Pig Vet. Soc.*, Iowa State University, Ames, June 22 to 24, 1976, K3.
75. **Frederick, G. T., Bohl, E. H., and Cross, R. F.,** Pathogenicity of an attenuated strain of transmissible gastroenteritis virus for newborn pigs, *Am. J. Vet. Res.*, 37, 165, 1976.
76. **Hess, R. G., Bachmann, P. A., and Hanichen, T.,** Attempts to establish an immunoprophylaxis for transmissible gastroenteritis virus infection of pigs. I. Pathogenicity of the B1 strain after serial passages, *Zentralbl Veterinaermed. B*, 24, 753, 1977.
77. **Laude, H., Charley, B., and Gelfi, J.,** Replication of transmissible gastroenteritis coronavirus (TGE) in swine alveolar macrophages, *J. Gen. Virol.*, 65, 327, 1984.
78. **Furuuchi, S., Shimizu, Y., and Kumagai, T.,** Multiplication of low and high cell culture passaged strains of transmissible gastroenteritis virus in organs of newborn piglets, *Vet. Microbiol.*, 3, 169, 1979.
79. **Saif, L. J. and Bohl, E. H.,** Passive immunity to transmissible gastroenteritis virus: Intramammary viral inoculation of sows, *Ann. N. Y. Acad. Sci.*, 409, 708, 1983.
80. **Kemeny, L. J. and Woods, R. D.,** Quantitative transmissible gastroenteritis virus shedding patterns in lactating sows, *Am. J. Vet. Res.*, 38, 307, 1977.
81. **Black, J. W.,** Diagnosis of TGE by FA: evaluation of accuracy on field specimens, *Proc. U. S. Anim. Health Assoc.*, 75, 492, 1971.
82. **Saif, L. J., Bohl, E. H., Kohler, E. M., and Hughes, J. H.,** Immune electron microscopy of transmissible gastroenteritis virus and rotavirus (reovirus-like agent) of swine, *Am. J. Vet. Res.*, 38, 13, 1977.
83. **Bohac, J, Derbyshire, J. B., and Thorsen, J.,** The detection of transmissible gastroenteritis viral antigen by immunodiffusion, *Can. J. Comp. Med.*, 39, 67, 1975.
84. **Bohac, J. and Derbyshire, J. B.,** The demonstration of transmissible gastroenteritis viral antigens by immunoelectrophoresis and counterimmunoelectrophoresis, *Can. J. Microbiol.*, 21, 750, 1975.
85. **Skalinskii, E. I., Mel'nikova, L. A., Kasyuk, I. I., and Boiko, A. A.,** Diagnosis of porcine transmissible gastroenteritis virus, *Veterinaria*, 9, 104, 1977.
86. **Becker, W., Teufel, P., and Mields, W.,** Immune perioxidase method for detection of viral and chlamydial antigens. III. Demonstration of TGE antigen in pig thyroid cell cultures, *Zentralbl. Veterinaermed.*, 21, 59, 1974.
87. **Chu, R. M., Li, N. J., Glock, R. D., and Ross, R. F.,** Application of peroxidase-antiperoxidase staining technique for detection of transmissible gastroenteritis virus in pigs, *Am. J. Vet. Res.*, 43, 77, 1982.
88. **Bernard, S., Lantier, I., Laude, H., and Aynaud, J. M.,** Detection of transmissible gastroenteritis coronavirus antigens by a sandwich ELISA technique, *Am. J. Vet. Res.*, 47, 2441, 1986.
89. **Cartwright, S. F.,** Transmissible gastroenteritis of swine (TGE), *Br. Vet. J.*, 124, 410, 1968.
90. **Vannier, P., Toma, B., Madec, F., and Aynaud, J. M.,** Valuation of duration of TGE virus spread among sows of 2 infected herds by means of a serological survey of antibodies persistence, Proc. 7th Int. Congr. Pig Vet. Soc., Mexico City, Mexico, July 26 to 31, 1982, 3.
91. **Bohl, E. H. and Kumagai, T.,** The use of cell cultures for the study of TGE virus, Proc. U. S. Livest. Sanit. Assoc., Lansing, MI, October 27 to 29, 1965, 343.
92. **Benfield, D. A., Haelterman, E. O., and Burnstein, T.,** An indirect fluorescent antibody test for antibodies to transmissible gastroenteritis of swine, *Can. J. Comp. Med.*, 42, 478, 1978.
93. **Hortig, H., Roder, B., and Bruske, R.,** Vergleichende Untersuchungen zur Verwendbarkeit des Microneutralisationstests, des indirekten Immunofluoreszenztests und des Virus-Inhibitionstests zum Nachweis von Antikorpen gegen das Virus der Transmissible Gastroenteritis (TGE) des Schweines, *DTW*, 87, 192, 1980.
94. **Black, J. W.,** The detection of transmissible gastroenteritis viral antibodies by immunodiffusion, *Can. J. Comp. Med.*, 40, 161, 1976.

95. **Stone, S. S., Kemeny, L. J., and Jensen, M. T.**, Partial characterization of the principal soluble antigens associated with the coronavirus of transmissible gastroenteritis by complement fixation and immunodiffusion, *Infect. Immun.*, 13, 521, 1976.
96. **Labadie, J. P., Aynaud, J. M. Vaissaire, J., and Renault, L.**, Porcine transmissible gastroenteritis. Antibody detection by passive haemagglutination test: applications to diagnosis and epidemiology, *Rec. Med. Vet.*, 153, 931, 1977.
97. **Shimizu, M. and Shimizu, Y.**, Micro-indirect hemagglutination test for detection of antibodies against transmissible gastroenteritis vrius of pigs, *J. Clin. Microbiol.*, 6, 91, 1977.
98. **Nelson, L. D. and Kelling, C. L.**, Enzyme-linked immunosorbent assay for detection of transmissible gastroenteritis virus antibody in swine sera, *Am. J. Vet. Res.*, 45, 1654, 1984.
99. **Paul, P. S., Mengeling, W. L., Saif, L. J., and Van Deresen, R. A.**, Detection of classes of antibodies to TGE and rotavirus of swine using monoclonal antibodies to porcine immunoglobulins, Proc. 9th Int. Pig Vet. Congr., Barcelona, Spain, July 15 to 18, 1986.
100. **Hohdatsu, T., Eiguchi, Y., Ide, S., Baba, K., Yamagishi, H., Kume, T., and Matumoto, M.**, Evaluation of an enzyme-linked immunosorbent assay for the detection of transmissible gastroenteritis virus antibodies, *Vet. Microbiol.*, 13, 93, 1987.
101. **Kodama, Y., Ogata, M., and Shimizu, Y.**, Serum immunoglobulin A antibody response in swine infected with transmissible gastroenteritis virus, as determined by indirect immunoperoxidase antibody test, *Am. J. Vet. Res.*, 42, 437, 1981.
102. **Kodama, Y., Ogata, M., and Shimizu, Y.**, Characteristics of immunoglobulin A antibody in serum of swine inoculated with transmissible gastroenteritis virus, *Am. J. Vet. Res.*, 40, 740, 1980.
103. **Stepanek, J., Hampl, J., Franz, J., Mensik, P., and Skrobak, F.**, Prukax protilatek proti viru virove gastroenteritidy prasat modifikvanym autoradiografickym testem, *Vet. Med. Praha*, 27, 473, 1982.
104. **Stepanek, J., Mensik, J. Franz, J., and Hornich, M.**, Epizootiology, diagnosis and prevention of viral diarrhoea in piglets under intensive husbandry conditions, Proc. 21st World Vet. Congr., Moscow, July 1 to 7, 1979, 43.
105. **Morin, M., Turgeon, D. Jolette, J., Robinson, Y., Phaneuf, J. B., Sauvageau, R., Beauregard, M., Teuscher, E., Higgins, R., and Larivere, S.**, Neonatal diarrhea of pigs in Quebec: infectious causes of significant outbreaks, *Can. J. Comp. Med.*, 47, 11, 1983.
106. **Shimizu, M., Shimizu, Y., and Kodama, Y.**, Effects of ambient temperatures on induction of transmissible gastroenteritis in feeder pigs, *Infect. Immun.*, 21, 747, 1978.
107. **Shimizu, M. and Shimizu, Y.**, Effects of ambient temperatures on clinical and immune responses of pigs infected with transmissible gastroenteritis virus, *Vet. Microbiol.*, 4, 109, 1979.
108. **Morin, M., Solorzano, R. F., Morehouse, L. G., and Olson, L. D.**, The postulated role of feeder swine in the perpetuation of the transmissible gastroenteritis virus, *Can. J. Comp. Med.*, 42, 379, 1978.
109. **Maes, R. K. and Haelterman, E. O**, A seroepizootiological study of five viruses in a swine-evaluation station, *Am. J. Vet. Res.*, 40, 1642, 1979.
110. **Lee, K. M., Moro, M., and Baker, J. A.**, Transmissible gastroenteritis in pigs, *Am. J. Vet. Res.*, 15, 364, 1954.
111. **Kemeny, L. J., Wiltsey, V. L., and Riley, J.L.**, Upper respiratory infection of lactating sows with transmissible gastroenteritis virus following contact exposure to infected piglets, *Cornell Vet.*, 65, 352, 1975.
112. **Derbyshire, J. B., Jessett, D. M., and Newman, G.**, An experimental epidemiological study of porcine transmissible gastroenteritis, *J. Comp. Pathol.*, 79, 445, 1969.
113. **Norman, J. O., McClurkin, A. W., and Stark, S. L.**, Transmissible gastroenteritis (TGE) of swine: canine serum antibodies against an associated virus, *Can. J. Comp. Med.*, 34, 115, 1970.
114. **Klemm, R. C. and Ristic, M.**, The effect of propagation of transmissible gastroenteritis (TGE) virus in pups and the lungs of baby pigs on the immunologic properties of the virus, *Proc. 4th Int. Congr. Pig Vet. Soc.*, Iowa State University, Ames, Iowa, June 22 to 24, 1976, K11.
115. **Ferris, D. H.**, Epizootiology of porcine transmissible gastroenteritis (TGE), in *Advances in Veterinary Science and Comparative Medicine*, Vol. 17, Academic Press, New York, 1973, 57.
116. **Pilchard, E. I.**, Experimental transmission of transmissible gastroenteritis virus by starlings, *Am. J. Vet. Res.*, 26, 1177, 1965.
117. **Gough, P. M. and Jorgenson, R. D.**, Identification of porcine transmissible gastroenteritis virus in house flies (*Musca domestica* Linneaus), *Am. J. Vet. Res.*, 44, 2078, 1983.
118. **Haelterman, E. O.**, Lactogenic immunity to transmissible gastroenteritis of swine, *J. Am. Vet. Med. Assoc.*, 147, 1661, 1965.
119. **Harada, K., Furuuchi, S., Kumagai, T., and Sasahara, J.**, Pathogenicity, immunogenicity and distribution of transmissible gastroenteritis virus in pigs, *Natl. Inst. Anim. Health Q. (Tokyo)*, 9, 185, 1969.
120. **Porter, P. and Allen, W. D.**, Classes of immunoglobulins related to immunity in the pig: A review, *J. Am. Vet. Med. Assoc.*, 160, 511, 1972.
121. **Husband, A. J. and Watson, D. L.**, Immunity in the intestine, *Vet. Bull.*, 48, 911, 1978.

122. **Sprino, P. J. and Ristic, M.**, Intestinal, pulmonary, and serum antibody responses of feeder pigs exposed to transmissible gastroenteritis virus by oral and oral-intranasal routes of inoculation, *Am. J. Vet. Res.*, 43, 255, 1982.
123. **Saif, L. J.**, The Immune Response of Swine to Transmissible Gastroenteritis Virus, Ph.D. thesis, Ohio State University, Columbus, Ohio, 1974.
124. **Stone, S. S., Kemeny, L. J., and Jensen, M. T.**, Serum antibody responses of neonatal and young adult pigs to transmissible gastroenteritis coronavirus, *Vet. Immunol. Immunopathol.*, 3, 529, 1982.
125. **Frederick, G. T. and Bohl, E. H.**, Local and systemic cell-mediated immunity against transmissible gastroenteritis, an intestinal viral infection of swine, *J. Immunol.*, 116, 1000, 1976.
126. **Woods, R. D.**, Leukocyte migration-inhibition procedure for transmissible gastroenteritis viral antigens, *Am. J. Vet. Res.*, 38, 1267, 1977.
127. **Liou, P. P.**, Cellular immunity in transmissible gastroenteritis virus-infected pigs: influence of viral antigens in leukocyte migration, *J. Chin. Soc. Vet. Sci.*, 8, 135, 1982.
128. **Shimizu, M. and Shimizu, Y.**, Demonstration of cytotoxic lymphocytes to virus-infected target cells in pigs inoculated with transmissible gastroenteritis virus, *Am. J. Vet. Res.*, 40, 208, 1979.
129. **Shimizu, M. and Shimizu, Y.**, Lymphocyte proliferative response to viral antigen in feeder pigs infected with transmissible gastroenteritis, *Infect. Immun.*, 23, 239, 1979c.
130. **Welch, S-K, W., Saif, L. J., and Ram, S.**, Cell-mediated immune responses of nursing pigs inoculated with attenuated or virulent transmissible gastroenteritis virus, *Am. J. Vet. Res.*, 49, 1228, 1988.
131. **Cepica, A. and Derbyshire, J. B.**, Antibody-dependent cell-mediated cytotoxicity and spontaneous cell-mediated cytotoxicity against cells infected with porcine transmissible gastroenteritis virus, *Can. J. Comp. Med.*, 47, 298, 1983.
132. **Bohl, E. H., Frederick, G. T., and Saif, L. J.**, Passive immunity in transmissible gastroenteritis of swine: intramuscular injection of pregnant swine with a modified live-virus vaccine, *Am. J. Vet. Res.*, 36, 267, 1975.
133. **Moxley, R. A. and Olson, L. D.**, Clinical evaluation of transmissible gastroenteritis virus vaccines and vaccination procedures for inducing lactogenic immunity in sows, *Am. J. Vet. Res.*, 50, 111, 1989.
134. **Haelterman, E. O.**, Transmissible gastroenteritis of swine, Proc. 17th World Vet. Congr., Hannover, August 14 to 21, 1963, 615.
135. **Pensaert, M. B.**, Immunity in TGE of swine after infection and vaccination, in *Viral Enteritis in Humans and Animals*, Bricout, F. and Scherrer, R., Eds., INSERM, Paris, 1979, 90, 281.
136. **Saif, L. J. and Bohl, E. H.**, Role of SIgA in passive immunity of swine to enteric viral infections, in *Immunology of Breast Milk*, Ogra, P.L. and Dayton, D. H., Eds., Raven Press, New York, 1979a.
137. **Saif, L. J. and Bohl, E. H.**, Keynote address: passive immunity against enteric viral infections, *Proc. 3rd Int. Symp. Neonatal Diarrhea*, VIDO, Saskatoon, Canada, 1981.
138. **Saif, L. J.**, Passive immunity to coronavirus and rotavirus infections in swine and cattle: enhancement by maternal vaccination, in *Infectious Diarrhoea in the Young*, Tzipori, S., Ed., Elsevier Science Publ., Amsterdam, The Netherlands, 1985, 456.
139. **Bohl, E. H., Gupta, R. K. P., Olquin, M. V. F., and Saif, L. J.**, Antibody responses in serum, colostrum and milk of swine after infection or vaccination with transmissible gastroenteritis virus, *Infect. Immun.*, 6, 289, 1972.
140. **Saif, L. J., Bohl, E. H., and Gupta, R. K. P.**, Isolation of porcine immunoglobulins and determination of the immunoglobulin classes of transmissible gastroenteritis viral antibodies, *Infect. Immun.*, 6, 600, 1972.
141. **Bohl, E. H. and Saif, L. J.**, Passive immunity in transmissible gastroenteritis of swine: immunoglobulin characteristics of antibodies in milk after inoculating virus by different routes, *Infect. Immun.*, 11, 23, 1975.
142. **Stone, S. S., Kemeny, L. J., Woods, R. D., and Jensen, M. T.**, Efficacy of isolated colostral IgA and IgM(A) to protect neonatal pigs against the coronavirus of transmissible gastroenteritis, *Am. J. Vet. Res.*, 38, 1285, 1977.
143. **Saif, L. J. and Bohl, E. H.**, Passive immunity in transmissible gastroenteritis of swine: immunoglobulin classes of milk antibodies after oral-intranasal inoculations of sows with a live low cell culture-passaged virus, *Am. J. Vet. Res.*, 40, 115, 1979b.
144. **Kaji, T. and Shimizu, Y.**, Passive immunization against transmissible gastroenteritis virus in piglets by ingestion of milk of sows inoculated with attenuated virus, *Natl. Inst. Anim. Health Q. (Tokyo)*, 18, 43, 1978.
145. **Voets, M. Th., Pensaert, M., and Rondhuis, P. R.**, Vaccination of pregnant sows against transmissible gastroenteritis with two attenuated virus strains and different inoculation routes, *Vet. Q.*, 2, 211, 1980.
146. **Henning, E. R. and Thomas, P. C.**, Comparison of intramuscular and oral modified live virus TGE vaccines, *Vet. Med. Small Anim. Clin.*, 76, 1789, 1981.
147. **Matisheck, P., Emerson, W., and Searl, R. C.**, Results of laboratory and field tests of TGE vaccine, *Vet. Med. Small Anim. Clin.*, 77, 262, 1982.
148. **Shibley, G. P., Salsbury, D. L., Djurickovic, S. M., and Johnson, G.**, Application of an intramammary route of vaccination against transmissible gastroenteritis in swine, *Vet. Med. Small Anim. Clin.*, 68, 59, 1973.
149. **Bay, W. W., Doyle, L. P., and Hutchings, L. M.**, Transmissible gastroenteritis in swine, A study of immunity, *J. Am. Vet. Med. Assoc.*, 122, 200, 1953.

150. **Furuuchi, S., Shimizu, Y., and Kumagai, T.,** Comparison of properties between virulent and attenuated strains of TGE, *Natl. Inst. Anim. Health Q. (Tokyo)*, 15, 159, 1975.
151. **Hess, R. G. and Bachmann, P. A.,** In vitro differentiation and pH sensitivity of field and cell culture attenuated strains of transmissible gastroenteritis, *Infect. Immun.*, 13, 1642, 1976.
152. **Chen, K. S.,** Enzymatic and acidic sensitivity profiles of selected virulent and attenuated transmissible gastroenteritis viruses of swine, *Am. J. Vet. Res.*, 46, 632, 1985.
153. **Laude, H, Gelfi, J., and Aynaud, J. M.,** In vitro properties of low- and high-passaged strains of transmissible gastroenteritis coronavirus of swine, *Am. J. Vet. Res.*, 42, 447, 1981.
154. **Hess, R. G., Bachmann, P. A., and Mayr, A.,** Attempts to develop an immunoprophylaxis against transmissible gastroenteritis (TGE) in pigs. III. Passive immune transfer after oral vaccination with attenuated TGE virus strain B1, *Zentralbl. Veterinaermed., B*, 25, 308, 1978.
155. **Fichtner, D., Leopoldt, D., and Meyer, U.,** Untersuchungen zur Ermittlung der minimalen Antigenmenge bei der oralen Muttertierimmunisierung gegen die Transmissible Gastroenteritis der Schweine mit Riemser TGE-Vakzine, *Arch. Exp. Veterinearmed.*, 36, 577, 1982.
156. **Welter, C. J.,** Experimental and field evaluation of a new oral vaccine for TGE, *Vet. Med. Small Anim. Clin.*, 75, 1757, 1980.
157. **Saif, L. J. and Bohl, E. H.,** Experimental studies using TGE vaccines, Ohio Swine Res. Ind. Rep., Anim. Sci. Series 81-2, Ohio State Univ., Columbus, 1981, 58.
158. **Bohl, E. H., Saif, L. J., and Jones, J. E.,** Observations on the occurrence of transmissible gastroenteritis (TGE) in a vaccinated herd, Ohio Swine Res. Ind. Rep., Anim. Sci. Ser. 82-1, Ohio State University, Columbus 1982, 66.
159. **Woods, R. D.,** Efficacy of vaccination of sows with serologically related coronaviruses for control of transmissible gastroenteritis in nursing pigs, *Am.J. Vet. Res.*, 45, 1726, 1984.
160. **Woods, R. D.,** Small plaque variant transmissible gastroenteritis virus, *J. Am. Vet. Med. Assoc.*, 173, 643, 1978.
161. **Gough, P. M., Frank, C. J., Moore, D. G., Sagona, M. A., and Johnson, C. J.,** Lactogenic immunity to transmissible gastroenteritis virus induced by a subunit immunogen, *Vaccine*, 1, 37, 1983.
162. **Thorsen, J. and Djurickovic, S.,** Experimental immunization of sows with inactivated TGE virus, *Can. J. Comp. Med.*, 35, 99, 1971.
163. **Leopoldt, D. and Meyer, U.,** Transmissible gastroenteritis of swine-A model of infectious diarrhoea, *Arch. Exp. Veterinaermed.*, 32, 417, 1978.
164. **Lutter, K., Klahn, J., and Kokles, R.,** Erfahrungen bei der Sanierung der Transmissible Gastroenteritis nach dem Selektionsverfahren in einer Schweinezuchtanlage mit 1200 produktiven Sauen, *Monatsh Veterinaermed.*, 37, 121, 1982.
165. **Hooyberghs, J., Pensaert, M. B. and Callebaut, P.,** Transmissible gastroenteritis: outbreaks in swine herds previously infected with a TGE-like porcine respiratory coronavirus, Proc. 10th Int. Congr. Pig Vet. Soc., Rio de Janeiro, Brazil, August 14 to 17, 1988, 200.
166. **Pritchard, G. C.,** Observations on clinical aspects of transmissible gastroenteritis of pigs in Norfolk and Suffolk, *Vet. Rec.*, 110, 465, 1982.
167. **Saif, L. J.,** unpublished observations, 1986.

PART 2B

1. **Binn, L. N., Lazar, E. C., Keenan, K. P., Huxsoll, D. L., Marchwicki, B. S., and Strano, A. J.,** Recovery and characterization of a coronavirus from military dogs with diarrhea, *Proc. Ann. Mtg. USAHA*, 78, 359, 1974.
2. **Haelterman, E. O.,** Epidemiological studies of transmissible gastroenteritis of swine, *Ann. U. S. Livestock Sanit. Assoc.* 66, 305, 1962.
3. **Larson, D. J., Morehouse, L. G. Solorzano, R. F., and Kinden, D. A.,** Transmissible gastroenteritis in neonatal dogs: experimental intestinal infection with transmissible gastroenteritis virus, *Am. J. Vet. Res.*, 40, 447, 1979.
4. **Albrecht, G. and Lupcke, W.,** Spontaneous clinical occurrence of transmissible gastroenteritis in dogs, *Monatschr. Vet.*, 31, 865, 1976.
5. **Reynolds, D. J., Garwes, D. J., and Lucey, S.,** Differentiation of canine coronavirus and porcine transmissible gastroenteritis virus by neutralization with canine, porcine and feline sera, *Vet. Microbiol.*, 5, 283, 1980.
6. **Norman, J. O., McClurkin, A. W., and Stark, S. L.,** Transmissible gastroenteritis (TGE) of swine: canine serum antibodies against an associated virus, *Can. J. Comp. Med.*, 34, 115, 1970.
7. **Cartwright, S. F. and Lucas, M. H.,** Vomiting and diarrhoea in dogs, *Vet. Rec.*, 91, 571, 1972.
8. **Appel, M. J. G., Cooper, B. J., Greisen, H., Scott, F., and Carmichael, L. E.,** Canine viral enteritis. I. Status report on corona- and parvo-like enteritides, *Cornell Vet.*, 69, 123, 1979.
9. **Carmichael, L. E. and Binn, L. N.,** New enteric viruses in the dog, *Adv. Vet. Sci. Comp. Med.*, 25, 1, 1981.
10. **Appel, M. J.,** Canine coronavirus, in *Virus Infections of Carnivores*, Appel, M. J., Ed., Elsevier Science Publ., New York, 1987, 115.

11. **McNulty, M. S., Curran, W. L., McFerran, J. B., and Collins, D. S.,** Viruses and diarrhea in dogs, *Vet. Rec.*, 106, 350, 1980.
12. **Williams, F. P., Jr.,** Astrovirus-like, coronavirus-like, and parvovirus-like particles detected in the diarrheal stools of beagle pups, *Arch. Virol.*, 66, 215, 1980.
13. **Schnagl, R. D. and Holmes, I. H.,** Coronavirus-like particles in stools from dogs from some country areas of Australia, *Vet. Rec.*, 102, 528, 1978.
14. **Takeuchi, A., Binn, L. N., Jervis, H. R., Keenan, K. P., Hildebrandt, P. K., Valas, R. B., and Bland, F. F.,** Electron microscope study of experimental enteric infection in neonatal dogs with a canine coronavirus, *Lab. Invest.*, 34, 539, 1976.
15. **Binn, L. N., Marchwicki, R. H., and Stephenson, E. H.,** Establishment of a canine cell line: derivation, characterization and viral spectrum, *Am. J. Vet. Res.*, 41, 855, 1980.
16. **Kojima, A., Takada, H., and Okaniwa, A.,** Multiplication of canine coronavirus in CRFK cells, *Jpn. J. Vet. Sci.*, 48, 1063, 1986.
17. **Woods, R. D.,** Studies of enteric coronaviruses in a feline cell line, *Vet. Microbiol.*, 7, 427, 1982.
18. **Tuchiya, K., Kasaoka, T., Azetaka, M., Takahashi, E., and Konishi, S. I.,** Plaque assay for canine coronavirus in CRFK cells, *Jpn. J. Vet. Sci.*, 49, 571, 1987.
19. **Sturman, L. S., Holmes, K. V., and Behnke, J.,** Isolation of coronavirus envelope glycoproteins and interaction with the viral nucleoprotein, *J. Virol.*, 33, 449, 1980.
20. **Garwes, D. J. and Reynolds, D. J.,** The polypeptide structure of canine coronavirus and its relationship to porcine transmissible gastroenteritis virus, *J. Gen. Virol.*, 52, 153, 1981.
21. **Horzinek, M. C., Lutz, H., and Pedersen, N. C.,** Antigenic relationship among homologous structural polypeptides of porcine, feline, and canine coronaviruses, *Infect. Immun.*, 37, 1148, 1982.
22. **Pedersen, N. C., Ward, J., and Mengeling, W. L.,** Antigenic relationship of the feline infectious peritonitis virus to coronaviruses of other species, *Arch. Virol.*, 58, 45, 1978.
23. **Pensaert, M. B., Debouck, P., and Reynolds, D. J.,** An immunoelectron microscopic and immunofluorescent study on the antigenic relationship between the coronavirus-like agent, CCV 777, and several coronaviruses, *Arch. Virol.*, 68, 45, 1981.
24. **Woods, R. D., Cheville, N. F., and Gallagher, J. E.,** Lesions in the small intestine of newborn pigs inoculated with porcine, feline and canine coronaviruses, *Am. J. Vet. Res.*, 42, 1163, 1981.
25. **Barlough, J. E., Stoddart, G. A., Sorresso, G. P., Jacobson, R. H., and Scott, F. W.,** Experimental inoculation of cats with canine coronavirus and subsequent challenge with feline infectious peritonitis virus, *Lab. Anim. Sci.* 34, 616, 1984.
26. **Keenan, K. P., Jervis, H. R., Marchwicki, R. H., and Binn, L. N.** Intestinal infection of neonatal dogs with canine coronavirus 1-71: studies by virologic, histologic, histochemical, and immunofluorescent techniques, *Am. J. Vet. Res.*, 37, 247, 1976.
27. **Vandenbergh, J., Ducatelle, R., Debouck, P., and Hoorens, J.,** Coronavirus infection in a litter of pups, *Vet. Q.*, 2, 136, 1980.
28. **Evermann, J. F., Stann, S., DiGiacomo, R. F., Bergstrom, P. K., McKeirnan, A. J., and Giddens, W. E.,** Epizootiologic and diagnostic features of canine diarrheal disease in high and low risk dog populations, in *Proc. Am. Vet. Lab. Diag.*, 25, 229, 1982.
29. **Saif, L. J., Bohl, E. H., Kohler, E. M., and Hughes, J. H.,** Immune electron microscopy of transmissible gastroenteritis virus and rotavirus (reovirus-like agent) of swine, *Am. J. Vet. Res.*, 38, 13, 1977.
30. **Dourmashkin, R. R., Davies, H. A., Smith, H., and Bird, R. G.,** Are coronavirus-like particles seen in diarrhoea stools really viruses? *Lancet*, 2, 971, 1980.
31. **Heckert, R. A., Saif, L. J., and Myers, G.,** Development of protein A-gold immunoelectron microscopy for detection of bovine coronavirus in calves: comparison with ELISA and direct imunofluorescence of nasal epithelial cells, *Vet. Microbiol.*, 19, 217, 1989.
32. **Toma, B. and Moraillon, A.,** Infection du chien par un virus antigeniquement apparante au virus de la gastro-enterite transmissible du porc, *Rec. Med. Vet.*, 156, 464, 1980.
33. **Evermann, J. F., Foreyt, W., Maag-Miller, L., Leathers, C. W., McKiernan, A. J., and Leamaster, B.,** Acute hemorrhagic enteritis associated with canine coronavirus and parvovirus infections in a captive coyote population, *J. Am. Vet. Med. Assoc.*, 177, 784, 1980.
34. **Pensaert, M. and Callebaut, P.,** The coronaviruses: clinical and structural aspects with some practical implications, *Ann. Med. Vet.*, 122, 301, 1978.
35. **Osterhaus, A. D., Drost, G. A., Wirahadiredja, R. M., and van den Ingh, T.S.G.A.M.,** Canine viral enteritis: prevalence of parvo-, corona-, and rotavirus infections in dogs in the Netherlands, *Vet. Q.*, 2, 181, 1980.
36. **Benary, F., Kraft, W., Arens, M., Krauss, H.,** Canine coronavirus enteritis: clinical findings, diagnosis, differential diagnosis and therapy, *Kleintierpraxis*, 26, 7, 1981.

37. **Helfer-Baker, C., Evermann, J. F., McKiernan, A. J., and Morrison, W. B.,** Serological studies on the incidence of canine viruses, *Canine Pract.*, 7, 37, 1980.
38. **Martin, M. L.,** Canine coronavirus enteritis and a recent outbreak following modified live virus vaccination, *Compend. Cont. Educ. Pract. Vet.*, 7, 1013, 1985.
39. **Edwards, B. G., Fulker, R. H., Acree, W. M., and Koonse, H. J.,** Evaluation a canine coronavirus vaccine through antigen extinction and challenge studies, *Vet. Med.*, 80, 28, 1985.

PART 2C

1. **Pedersen, N. C.,** Feline infectious peritonitis virus, in *Virus Infections of Carnivores*, Appel, M. J., Ed., Elsevier Science Publ., New York, 1987, 267.
2. **Pedersen, N. C., Boyle, J. F., Floyd, K., Fudge, A. and Barker, J.,** An enteric coronavirus infection of cats and its relationship to feline infectious peritonitis, *Am. J. Vet. Res.*, 42, 368, 1981.
3. **McKeirnan, A. J., Evermann, J. F., Hargis, A., Miller, L. M., and Ott, R. L.,** Isolation of feline coronaviruses from two cats with diverse disease manifestations, *Feline Pract.*, 11, 16, 1981.
4. **Pedersen, N. C.,** Feline enteric coronavirus, in *Virus Infections of Carnivores*, Appel, M. J., Ed., Elsevier Science Publ., New York, 1987, 261.
5. **Boyle, J. F., Pedersen, N. C., Evermann, J. F., McKeirnan, A. J., Ott, R.L. and Black, J. W.,** Plaque assay, polypeptide composition and immunochemistry of feline infectious peritonitis virus and feline enteric coronavirus, *Adv. Exp. Med. Biol.*, 173, 133, 1983.
6. **Ingersoll, J. D. and Wylie, D. E.,** Identification of viral antigens that induce antibody responses on exposure to coronaviruses, *Am. J. Vet. Res.*, 49, 1467, 1988.
7. **Hayashi, T., Watanabe, Y., Nakayama, H., and Fujiwara, K.,** Enteritis due to feline infectious peritonitis virus, *Jpn. J. Vet. Sci.*, 44, 97, 1982.
8. **Hoshino, Y. and Scott, F. W.,** Coronavirus-like particles in the feces of normal cats, *Arch. Virol.*, 63, 147, 1980.
9. **Dea, S., Roy, R. S., and Elazhary, M. A. S. Y.,** Coronavirus-like particles in the feces of a cat with diarrhea, *Can. Vet. J.*, 23, 153, 1982.
10. **Pedersen, N. C., Ward, J., and Mengeling, W. L.,** Antigenic relationship of the feline infectious peritonitis virus to coronaviruses of other species, *Arch. Virol.*, 58, 45, 1978.
11. **Horzinek, M. C., Lutz, H., Petersen, N. C.,** Antigenic relationships among homologous structural polypeptides of porcine, feline and canine coronaviruses, *Infect. Immun.*, 37, 1148, 1982.
12. **Ingersoll, J. D. and Wylie, D. E.,** Comparison of serologic assays for measurement of antibody response to coronavirus in cats, *Am. J. Vet. Res.*, 49, 1472, 1988.
13. **Pedersen, N. C., Evermann, J. F., McKeirnan, A. J., and Ott, R. L.,** Pathogenicity studies of feline coronavirus isolates 79-1146 and 79-1683, *Am. J. Vet. Res.*, 45, 2580, 1984.

PART 3A

1. **Stair, E., L., Rhodes, M. B., White, R.G., and Mebus, C.A.,** Neonatal calf diarrhea: purification and electron microscopy of a coronavirus-like agent, *Am. J. Vet. Res.*, 33, 1147, 1972.
2. **Doughri, A. M., Storz, J., Hajer, I., and Fernando, H. S.,** Morphology and morphogenesis of a coronavirus infecting intestinal epithelial cells of newborn calves, *Exp. Mol. Path.*, 25, 355, 1976.
3. **Roseto, A., Bobulesco, P., Laporte, J., Gaches, D., and Peries, J.,** Bovine enteric coronavirus structure as studied by a freeze-drying technique, *J. Gen. Virol.*, 63, 241, 1982.
4. **Doughri, A. M. and Storz, J.,** Light and ultrastructural pathologic changes in intestinal coronavirus infection of newborn calves, *Zentralbl. Veterinaermed.*, B. 24, 367, 1977.
5. **Storz, J. Doughri, A. M., and Hajer, I.,** Coronaviral morphogenesis and ultrastructural changes in intestinal infection of calves, *J. Am. Med. Assoc.*, 173, 633, 1978.
6. **Tektoff, J., Dauvergne, M., Durafour, M., and Soulebot, L. P.,** Propagation of bovine enteric coronavirus in three cell systems: electron microscopic studies, in *Proc. Fourth Int. Symp. Neonatal Diarrhea*, University of Saskatchewan, Saskatoon, Canada, 1983.
7. **Mebus, C. A., Stair, E. L., Rhodes, M. B., and Twiehaus, M. J.,** Pathology of neonatal calf diarrhea included by a coronavirus-like agent, *Vet. Pathol.*, 10, 45, 1973.
8. **Takahashi, E., Inaba, Y., Sato, K., Ito, Y., Kurogi, H., Akashi, H., Satoda, K., and Omori, T.,** Epizootic diarrhoea of adult cattle associated with a coronavirus-like agent, *Vet. Microbiol.*, 5, 151, 1980.
9. **Inaba, Y., Sato, K., Kurogi, H., Takahashi, E., Ito, Y., Omori, T., Boto, Y., and Matumoto, M.,** Replication of bovine coronavirus in cell line BEK-1 culture., *Arch. Virol.*, 50, 339, 1976.
10. **Sato, K., Inaba, Y., Kurogi, H., Takahashi, E., Ito, Y., Goto, Y., Omori. T., and Matumoto M.,** Physicochemical properties of calf diarrhea coronavirus, *Vet. Microbiol.*, 2, 73, 1977.
11. **Hirano, N., Sada, Y., Tuchiya, K., Ono, K., and Murakami, T.,** Plaque assay of bovine coronavirus in BEK-1 cells, *Jap. J. Vet. Sci.*, 47, 679, 1985.

12. **Dea, S., Roy, R. S., and Begin, M. E.,** Bovine coronavirus isolation and cultivation in continuous cell lines, *Am. J. Vet. Res.*, 41, 30, 1980.
13. **Dereget, D. Crouch, C. F., Sabara, M., Gilchrist, J. E., Babiuk, L. A., and Hudson, G. R.,** Preliminary studies of a bovine coronavirus (BCV) antigen responsible for neutralization, in *Proc. Fourth Int. Symp. Neonatal Diarrhea*, University of Saskatchewan, Saskatoon, Canada, 1983.
14. **Laporte, J., L'Haridon, R., and Bobulesco, P.,** *In Vitro* culture of bovine enteritic coronavirus (BEC), in *Viral Enteritis in Humans and Animals*, Bricout, R., Pensaert, M., Flewett, T.H., and Scherrer, R., Eds., September 4 to 7, Thiverval-Grignon France, *l'Inserm*, 90, 99, 1979.
15. **Laporte, J., Bobulesco, P., and Rossi, F.,** Une lignee cellulaire particulierement sensible a la replication du coronavirus enteritique bovine: les cellules HRT 18, *C.R. Acad. Sci.*, 290, 623, 1980.
16. **L'Haridon, R., Scherrer, R., Vautherot, J. R., LaBonnardiere, C., Laporte, J., and Cohen, J.,** Adaption d'un esolement de coronavirus enterique bovine a la culture cellulaire et caracterisation de la souche obtenue, *Ann. Rech. Vet.*, 12, 243, 1981.
17. **Vautherot, J.,** Plaque assays for titration of bovine enteric coronavirus, *J. Gen. Virol.*, 56, 451, 1981.
18. **Gerna, G., Cereda, P. M., Revello, M. G., Cattaneo, E., Battaglia, M., and Gerna, M. T.,** Antigenic and biological relationships between human coronavirus OC43 and neonatal calf diarrhoea coronavirus, *J. Gen. Virol.*, 54, 91,1981.
19. **Barboi, G. H. and Pirbulescu, M.,** Le test d'inhibition de l'hemagglutination (IHA) pour le diagnostic de l'infection a coronavirus chez les bovins, *Rev. Roum. Med. - Virol.*, 37, 253, 1986.
20. **Moussa, A.,** Incurved disc form particles associated with a bovine coronavirus growing in vero and calf testicles cell cultures, in Proc. VII Int. Cong. Virol., Edmonton, Canada, August 9 to 14, 1987.
21. **Cyr-Coats, K. St. and Storz, J.,** Bovine coronavirus induced cytopathic expression and plaque formation: Host cell and virus strain determine typsin dependence, *J. Vet. Med. B*, 35, 48, 1988.
22. **Stott, E. J., Thomas, L. H., Bridger, J. C., and Jebbett N. J.,** Replication of a bovine coronavirus in organ cultures of foetal trachea, *Vet. Microbiol.*, 1, 65, 1976.
23. **Bridger, J. C., Caul, E. O., and Egglestone, S. I.,** Replication of an enteric bovine coronavirus in intestinal organ cultures, *Arch. Virol.*, 57, 43, 1978.
24. **Storz, J., Rott, R., and Kaluza, G.,** Enhancement of plaque formation and cell fusion of an enteropathogenic coronavirus by trypsin treatment, *Infect. Immun.*, 31, 1214, 1981.
25. **Toth, T. E.,**Trypsin-enhanced replication of neonatal calf diarrhea coronavirus in bovine enbryonic lung cells, *Am. J. Vet. Res.*, 43, 967, 1982.
26. **Baron, S. and McKerlie, M. L.,** Broadly active inhibitor of viruses spontaneously produced by many cell types in culture, *Infect. Immun.*, 32, 449, 1981.
27. **Hughes, T. K., Blalock, J. E., McKerlie, M. L., and Baron, S.,** Cell-produced viral inhibitor: Possible mechanism of action and chemical composition, *Infect. Immun.*, 32, 454, 1981.
28. **Debiaggi, M., Perduca, M., Romero, E., and Cereda P. M.,** Phosphatidyl-serine inhibition of OC43 and NCDV coronavirus infectivity, *Microbiologica*, 8, 313, 1985.
29. **Akashi, H., Inaba, Y., Miura, Y., Sato, K., Tokuhisa, S., Asagi, M., and Hayashi, Y.,** Propagation of the Kakegawa strain of bovine coronavirus in suckling mice, rats and hamsters, *Arch. Virol.*, 67, 367, 1981.
30. **Kaye, H. S., Yarbrough, W. B., and Reed, C. J.,** Calf diarrhoea coronavirus, *Lancet*, 2, 509, 1975.
31. **Sharpee, R. L., Mebus, C. A., and Bass, E. P.,** Characterization of a calf diarrheal coronavirus, *Am. J. Vet. Res.*, 37, 1031, 1976.
32. **Akashi, H., Inaba, Y., Miura, Y., Tokuhisa, S., Sato, K., and Satoda, K.,** Properties of a coronavirus isolated from a cow with epizootic diarrhea, *Vet. Microbiol.*, 5, 265, 1980.
33. **Siddell, S. Wege, H., and ter Meulen, V.,** The structure and replication of coronavirus, *Curr. Top. Microbiol. Immunol.*, 99, 131, 1982.
34. **King, B and Brian, D. A.,** Bovine coronavirus structural proteins, *J. Virol.*, 42, 700, 1982.
35. **King, B, Potts, J. B., and Brian, D. A.,** Bovine coronavirus hemagglutinin protein, *Virus Res.*, 2, 53, 1985.
36. **Brian, D. A., Hogue, B., Lapps, W., Potts, B., and Kapke, P.,** Comparative structure of coronaviruses, in *Proc. Fourth Int. Symp. Neonatal Diarrhea*, University of Saskatchewan, Saskatoon, Canada, Oct. 3 to 5, 1983.
37. **Storz, J., Kaluza, G., Niemann, H., and Rott, R.,** On enteropathogenic bovine coronavirus, in *Biochemistry and Biology of Coronaviruses*, ter Meulen, V., Siddell, S., and Wege, H., Eds., Plenum Press, New York, 1981, 171.
38. **Laporte, J., and Bobulesco, P.,** Polypeptide structure of bovine enteritic coronavirus: comparison between a wild strain purified from feces and a HRT 18 cell adapted strain, in *Biochemistry and Biology of Coronaviruses*, ter Meulen, V., Siddell, S., and Wege, H., Eds., Plenum Press, New York, 1981, 181.
39. **Vautherot, J. F. and Laporte, J.,** Antigenic and polypeptide structure of bovine enteric coronavirus: Characterization of the peplomers proteins, in Proc. VII Int. Cong. Vir., Edmonton Canada, August 9 to 14, 1987.
40. **Deregt, D., Sabara, M., and Babiuk, L. A.,** Structural proteins of bovine coronavirus and their intracellular processing, *J. Gen. Virol.*, 68, 2863, 1987.

41. **Deregt, D. and Babiuk, L. A.,** Monoclonal antibodies to bovine coronavirus: characteristics and topographical mapping of neutralizing epitopes on the E2 and E3 glycoproteins, *Virology*, 161, 410, 1987.
42. **Guy, J. S., and Brian, D. A.,** Bovine coronavirus genome, *J. Virol.*, 29, 293, 1979.
43. **Lapps, W., Hogue, B. G., and Brian, D. A.,** Sequence analysis of the bovine coronavirus nucleocapsid and matrix protein genes, *Virology*, 157, 47, 1987.
44. **Dennis, D. E. and Brian, D. A.,** Coronavirus cell-associated RNA-dependent RNA polymerase, in *Biochemistry and Biology of Coronaviruses*, ter Meulen, V., Siddell, S., and Wege, H., Eds., Plenum Press, New York, 1981, 155.
45. **Spaan, W., Deliis, H., Skinner, M. A., Armstrong, J., Rottier, P., Smeekens, S., Siddell, S. G., and van der Zeijst, B.,** Transcription strategy of coronaviruses: fusion of non-contiguous sequences during mRNA synthesis, in *Molecular Biology and Pathogenesis of Coronavirus*, Rottier, P. J. M., et al., Eds., Plenum Press, New York, 1984, 173.
46. **Bridger, J. C., Woode, G. N., and Meyling A.,** Isolation of coronaviruses from neonatal calf diarrhea in Great Britain and Denmark, *Vet. Microbiol.*, 3, 101, 1978.
47. **Dea, S., Roy, R. S., and Elazhary, M. A. S. Y.,** Antigenic variations among calf diarrhoea coronavirus by immunodiffusion and counterimmunoelectrophoresis, *Ann. Rech. Vet.*, 13, 351, 1982.
48. **Thomas, L. H., Gourlay, R. N., Stott, E. J., Howard, C. J., and Bridger, J. C.,** A search for new microorganisms in calf pneumonia by the inoculation of gnotobiotic calves, *Res. Vet. Sci* , 33, 170, 1982.
49. **McNulty, M. S., Bryson D. G., Allan, G. M., and Logan, E. F.,** Coronavirus infection of the bovine respiratory tract, *Vet. Microbiol.*, 9, 425, 1984.
50. **Reynolds, D. J.,** Coronavirus replication in the intestinal and respiratory tracts during infection of calves, *Ann. Rech. Vet.*, 14, 445, 1983.
51. **Reynolds, D. J., Debney, T. G., Hall, G. A., Thomas, L. H., and Parsons, K. R.,** Studies on the relationship between coronaviruses from the intestinal and respiratory tracts of calves, *Arch. Virol.*, 85, 71, 1985.
52. **Saif, L. J., Redman, D. R., Moorhead, P. D., and Theil, K. W.,** Experimental coronavirus infections in calves: viral replication in the respiratory and intestinal tracts, *Am. J. Vet. Res.*, 47, 1426, 1986.
53. **Vautherot, J. F. and Laporte, J.,** Utilization of monoclonal antibodies for antigenic characterization of coronaviruses, *Ann. Rech. Vet.*, 14, 437, 1983.
54. **Pedersen, N. C., Ward, I., and Mengeling, W. L.,** Antigenic relationship of the feline infectious peritonitis virus to coronaviruses of other species, *Arch. Virol.*, 58, 45, 1978.
55. **Hogue, B. G., King, B., and Brian, D. A.,** Antigenic relationships among proteins of bovine coronavirus, human respiratory coronavirus OC43, and mouse hepatitis coronavirus A59, *J. Virol.*, 51, 384, 1984.
56. **Storz, J. and Rott, R.,** Reactivity of antibodies in human serum with antigens of an enteropathogenic bovine coronavirus, *Med. Microbiol. Immunol.*, 169, 169, 1981.
57. **Gerna, G., Battaglia, M., Cereda, P. M., and Passarani, N.,** Reactivity of human coronavirus OC43 and neonatal calf diarrhea coronavirus membrane-associated antigens, *J. Gen. Virol.*, 60, 385, 1982.
58. **Lapps, W. and Brain, D. A.,** Oligonucleotide fingerprints of antigenically related bovine coronavirus and human coronavirus OC43, *Arch. Virol.*, 86, 101, 1985.
59. **Patel, J. R., Davies, H. A., Edington, N., Laporte, J., and Macnaughton, M. R.,** Infection of a calf with the enteric coronavirus strain Paris, *Arch. Virol.*, 73, 319, 1982.
60. **Lewis, L. D. and Phillips, R. W.,** Pathophysiologic changes due to coronavirus-induced diarrhea in the calf, *J. Am. Vet. Med. Assoc.*, 173, 636, 1978.
61. **Phillips, R. W. and Case, G. L.,** Altered metabolism, acute shock, and therapeutic response in a calf with severe coronavirus-induced diarrhea, *Am. J. Vet. Res.*, 41, 1039, 1980.
62. **Mebus, C. A., Newman, L. E., and Stair, E. L.,** Scanning electron, light, and immunofluorescent microscopy of the intestine of a gnotobiotic calf infected with calf diarrheal coronavirus, *Am. J. Vet. Res.*, 36, 1719, 1975.
63. **Hall, G. A., Parsons, K. R., and Bridger, J. C.,** Plasma enteroglucagon and neurotensin levels in gnotobiotic calves infected with enteropathogenic and non-enteropathogenic viruses, *Res. Vet. Sci.*, 38, 99, 1985.
64. **Horner, G. W., Hunter, R., and Kirkbride, C. A.,** A coronavirus-like agent present in feces of cows with diarrhea, *N. Z. Vet. J.*, 23, 98, 1975.
65. **Espinasse, J., Viso, M., Laval, A., Savey, M., and Lelayec, C.,** Winter dysentery: a coronavirus-like agent in the faeces of beef and dairy cattle with diarrhoea, *Vet. Rec.* 110, 385, 1982.
66. **Broes, A., Van Opdenbosch, E., and Wellemans, G.,**Isolement d'un coronavirus chez des bovins atteits d'enterite hemorragique hivernal (Winter dysentery) en Belgique, *Ann. Med. Vet.*, 128, 299, 1984.
67. **Saif, L. J., Redman, D. R., Brock, K. V., Kohler, E. M., and Heckert, R. A.,** Winter dysentery in adult dairy cattle: Detection of coronavirus in the faeces, *Vet. Rec.*, 123, 300, 1988.
68. **Singh, D. K., Singh, N. P., and Singh, G. K.,** Pneumoenteric syndrome in bovine neonates by bovine coronavirus, *Indian J. Vet. Med.*, 5, 55, 1985.
69. **Wellemans, G., Van Opdenbosch, E., Oudewater, J., and Pattijn, R.,** Intervention du virus corona dans un cas de pneumonie chez des bovins, *Ann. Med. Vet.*, 129, 585, 1985.
70. **Bosgiraud, C. and Nicolas, J. A.,** Pneumonie a coronavirus chez les bovins: a propos de deux observations, *Rec. Med. Vet.*, 162, 1085, 1986.

71. **Heckert, R. A., Saif, L. J., and Meyers, G. W.**, Development of protein A-gold immunoelectron microscopy for detection of bovine coronavirus in calves: comparison with ELISA and direct immunofluorescence of nasal epithelial cells, *Vet. Microbiol.*, 19, 217, 1989.
72. **Langpap, T. J., Bergeland, J. E., and Reed, D. E.**, Coronaviral enteritis of young calves: virologic and pathologic findings in naturally occurring infections, *Am. J. Vet. Res.*, 40, 1476, 1979.
73. **Ellens, D. J., Van Balken, J. A. M., and De Leeuw, P. W.**, Diagnosis of bovine coronavirus infections with hemadsorption-elution-hemagglutination assay (HEMA) and with enzyme-linked immunosorbent assay (ELISA), in *Proc. 2nd Int. Symp. on Neonatal Diarrhea*, University of Saskatchewan, Saskatoon, Canada, 1978.
74. **Crouch, C. F., Raybould, T. J. G., and Acres, S. D.**, Monoclonal antibody capture enzyme-linked immunosorbent assay for detection of bovine enteric coronavirus, *J. Clin. Microbiol.*, 19, 388, 1984.
75. **Reynolds, D. J., Chasey, D., Scott, A. C., and Bridger, J. C.**, Evaluation of ELISA and electron microscopy for the detection of coronavirus and rotavirus in bovine feces, *Vet. Rec.*, 114, 397, 1984.
76. **Van Opdenbosch, E., Wellemans, G., Wery, B., Oudewater, J., and Blondeel, H.**, A simple ELISA test for the detection of bovine coronavirus antigen in fecal and intestinal homogenates: A comparative study, *Vlaams Diergeneeskd. Tijdschr.*, 54, 385, 1985.
77. **Van Balken, J. A. M., DeLeeuw, P. W., Ellens, D. J., and Straver, P. J.**, Detection of coronavirus in calf faeces with a haemadsorption-elution-hemagglutination assay (HEHA), *Vet. Microbiol.*, 3, 205, 1978/1979.
78. **Dea, S., Roy, R. S., and Begin, M. E.**, Counterimmunoelectroosmophoresis for detection of neonatal calf diarrhea coronavirus: methodology and comparison with electron microscopy, *J. Clin. Microbiol.*, 10, 240, 1979.
79. **Sato, K., Inaba, Y., Tokuhisa, S., Miura, Y., Kaneko, N., Asagi, M., and Matumoto, M.**,Detection of bovine coronavirus in feces by reversed passive hemagglutination, *Arch. Virol.*, 80, 23, 1984.
80. **Shockley, L. J., Kapke, P. A., Lapps, W., Brian, D. A., Potgieter, L. N. D., and Woods, R.**, Diagnosis of porcine and bovine enteric coronavirus infections using cDNA probes, *J. Clin. Microbiol.*, 25, 1591, 1987.
81. **Verbeek, J. A. and Tijssen, P.**, Production of biotinylated probes to bovine coronavirus, in *Proc. VII Int. Cong. Virol.*, Edmonton, Canada, August 9 to 14, 1987.
82. **Rodak, L., Babiuk, L. A., and Acres, S. D.**, Detection by radioimmunoassay and ELISA of coronavirus antibodies in bovine serum and lacteal secretions, *J. Clin. Microbiol.*, 16, 34, 1982.
83. **Ellens, D. J., de Leeuw, P., and Straver, P. J.**, Recurrent rotavirus and coronavirus infections in cattle, *INSERM*, 90, 355, 1979.
84. **Morin, M., Lamothe, P., Gagnon, A., and Malo, R.**, A case of viral neonatal calf diarrhea in a Quebec dairy herd, *Can. J. Comp. Med.*, 38, 236, 1974.
85. **Acres, S. D., Laing, C. J., Saunders, J. R., and Radostits, O. M.**, Acute undifferentiated neonatal diarrhea in beef calves. I. Occurrence and distribution of infectious agents, *Can. J. Comp. Med.*, 39, 116, 1975.
86. **de Visser, N. A. P. C., Breukink, H. J., van Zijderveld, F. G., and de Leeuw, P. W.**, Enteric infections in veal calves: a longitudinal study of four veal calf units, *Vet. Q.*, 9, 289, 1987.
87. **Woode, G. N. and Bridger, J. C.**, Significance of bovine coronavirus infection, *Vet. Rec.*, 102, 15, 1978.
88. **Zygraich, N., Georges, A. M., and Vascoboinic, E.**, Etiologies des diarrhees neonatales du veau: resultats d'une enquete serologique relative aus virus reo-like et corona dans la population bovine Belge, *Ann. Med. Vet.*, 199, 105, 1975.
89. **Dirksen, V. G. and Bachmann, P. A.**, Zum vorkommen von rota-und coronavirus als ursache von kalberdiarrhoe in der Bundesrepublik Deutschland, *Berl. Muench. Tieraerztl. Wochenschr.*, 90, 475, 1977.
90. **Durham, P. J. K., Farquharson, F. C., and Stevenson, B. J.**, Rotavirus and coronavirus associated diarrhoea in calves, *N. Z. Vet. J.*, 27, 266, 1979.
91. **Gunenkowv, V. V., Sharabrin, O. I., and Koromyslov I. G.**, Immunofluorescence technique for the diagnosis of coronavirus infection in cattle, *Vestn. Skh. Nauki (Moscow)*, 11, 122, 1985.
92. **Valicek, L., Smid, B., Stepanek, J., and Mensik, J.**,Elektronove mikroskopicky prukaz coronaviru u telat s prujmem, *Vet. Med.*, 26, 513, 1981.
93. **Baljer, G., Eichhorn, W., Gobel, E., Wolf, M., and Bachmann, P.A.**, Prevalence and distribution of important diarrhoea pathogens in newborn calves in southern Germany in 1984-1986, *Tieraerztl. Umsch.*, 42, 56, 1987.
94. **Svoboda, I., Rodak, L., Franz, J., Stepaek, J., and Valicek, L.**, Differential diagnosis of viral diarrhoea in calves, *Vet. Med.*, 31, 497, 1986.
95. **Meyers, L. L., Firehammer, B. D., Border, M. M., and Shoop, D. S.**, Prevalence of enteric pathogens in the feces of healthy beef calves, *Am. J. Vet. Res.*, 45, 1544, 1984.
96. **Barraglia, M., Lutz, H., and Wyler, R.**, Serologische ubersichtsuntersuchung uber die verbreitung des bovinen coronavirus in der Schweiz, *Schweiz. Arch. Tierheilkd.*, 128, 213, 1986.
97. **Crouch, C. F., Bielefeldt Ohmann, H., Watts, T. C., and Babiuk, L. A.**, Chronic shedding of bovine enteric coronavirus antigen-antibody complexes by clinically normal cows, *J. Gen. Virol.*, 66, 1489, 1985.
98. **Collins, J. K., Riegel, C. A., Olson, J. D., and Fountain, A.**, Shedding of enteric coronavirus in adult cattle, *Am. J. Vet. Res.*, 48, 361, 1987.

99. **Mebus, C. A., Torres-Medina, A., and Twiehaus, M. J.,** Immune response to orally administered calf reovirus-like agent and coronavirus vaccine, in *14th Congr. Int. Assoc. Biol. Standardization, Douglas Isle of Man, 1975, Develop. Biol. Standard.*, Vol. 33, S. Karger, Basel, 1976, 396.
100. **deLeeuw, P. W., and Tiessink, W. A.,** Laboratory experiments on oral vaccination of calves against rotavirus or coronavirus induced diarrhea, *Zentralbl. Vet. Med. B.*, 32, 55, 1985.
101. **Thurber, E. T., Bass, E. P., and Bechenhauer, W. H.,** Field trial evaluation of a reo-coronavirus calf diarrhea vaccine, *Can. J. Comp. Med.*, 41, 131, 1977.
102. **Myers, L. L. and Snodgrass, D. R.,** Colostral and milk antibody titers in cows vaccinated with a modified-live rotavirus-coronavirus vaccine, *J. Am. Vet. Med. Assoc.*, 181, 486, 1982.
103. **Saif, L. J. and Smith, K. L.,** Enteric viral infections of calves and passive immunity, *J. Dairy Sci.*, 68, 206, 1985.
104. **Waltner-Toews, D., Martin, S. W., Meek, A. H., McMillan, I., and Crouch, C. F.,** A field trial to evaluate the efficacy of a combined rotavirus-coronavirus/*Escherichia coli* vaccine in dairy cattle, *Can. J. Comp. Med.*, 49, 1, 1985.
105. **Burki, F., Mostl, K., Spiegl, E., Horvath, E., and Szekely, H.,** Reduction of rotavirus-, coronavirus- and *E.coli*-associated calf-diarrheas in a large-size dairy herd by means of dam vaccination with a triple-vaccine, *J. Vet. Med. B*, 33, 241, 1986.
106. **Murakami, T. Hirano, N., Inque, A., Chitose, K., Tsuchiya, K., Ono, K., and Naito, Y.,** Transfer of antibodies against viruses of calf diarrhea from cows to their offspring via colostrum, *Jpn. J. Vet. Sci.*, 47, 507, 1985.
107. **Stepanek, J., Salajka, E., Zuffa, A., Mensik, J., and Franz, J.,** A new polyvalent vaccine against enteral infections in new-born calves, *Vet. Med.*, 32, 65, 1987.
108. **Haralambiev, H., Cisov, I., Mitov, B., Slimeonov, S., Patschev, S., and Zerkov, P.,** Antikorper in der Milch von kuhen nach uterschiedlicher Coronaviursapplikation, *Arch, Exp. Vet. Med. Leipzig*, 37, 327, 1983.
109. **Saif, L. J. and Smith, K. L.,** A review of rotavirus immunization of cows and passive protection in calves, in *Proc. Fourth Int. Symp. Neonatal Diarrhea*, Veterinary Infectious Disease Organization, University of Saskatchewan, Saskatoon, Canada, 1983.
110. **Mullaney, T. P., Newman, L. E., and Whitehair, C. K.,** Humoral immune response of the bovine fetus to in utero vaccination with attenuated bovine coronavirus, *Am. J. Vet. Res.*, 49, 156, 1988.
111. **Van Opdenbosch, E. and Wellemans, G.,** Neonatal calf diarrhea: a complex viral etiology, *INSERM*, 90, 361, 1979.
112. **Acres, S. D.,** The epidemiology of the calf scours complex in western Canada, in *Proc. Second Int. Symp. on Neonatal Diarrhea*, Veterinary Infectious Disease Organization, University of Saskatchewan, Saskatoon, Canada, 1976.
113. **Murakami, T., Hirano, N., Inoue, A., Tsuchiya, K., Chitose, K., Ono, K., and Yanagihara, T.,**Prevention of calf diarrhea with an immunoglobulin diet in beef herds, *Jpn. J. Vet. Sci.*, 48, 879, 1986.
114. **Hancock, D. D.,** Studies on the Epidemiology of Mortality and Diarrheal Morbidity in Heifer Calves in Northeastern Ohio Dairy Herds, Ph.D. dissertation, The Ohio State University, Columbus, 1984.

PART 3B

1. **Sugiyama, K. and Amano, Y.,** Morphological and biological properties of a new coronavirus associated with diarrhea in infant mice, *Arch. Virol.*, 67, 241, 1981.
2. **Hierholzer, J. D., Broderson, J. R., and Murphy F. A.,** New strain of mouse hepatitis virus as the cause of lethal enteritis in infant mice, *Infect. Immun.*, 24, 508, 1979.
3. **Kraft, L. M.,** An apparently new lethal virus disease of infant mice, *Science*, 137, 282, 1962.
4. **Sato, K., Maru, M., and Wada, T.,** Some characteristics of corona-like virus isolated from infant mice with diarrhea and inflammatory submaxillary gland of rats, *Virus*, 26, 97, 1976.
5. **Sugiyama, K. and Amano, Y.,** Hemagglutination and structural polypeptides of a new coronavirus associated with diarrhea in infant mice, *Arch. Virol.*, 66, 95, 1980.
6. **Rowe, W. P., Hartley, J. W., and Capps, W. I.,** Mouse hepatitis virus infection as a highly contagious, prevalent, enteric infection in mice, in *Proc. Soc. Exp. Biol. Med.*, 112, 161, 1963.
7. **Sugiyama, K., Ishikawa, R., and Fukuhara, N.,** Structural polypeptides of the murine coronavirus DVIM, *Arch. Virol.*, 89, 245, 1986.
8. **Ishia, T. and Fujiwara, K.,** Pathology of diarrhea due to mouse hepatitis virus in the infant mouse, *Jpn. J. Exp. Med.*, 49, 33, 1979.
9. **Ishida, T., Taguchi, F., Lee, Y., Yamada, A., Tamura, T., and Fujiwara, K.,** Isolation of mouse hepatitis virus from infant mice with fatal diarrhea, *Lab. Anim. Sci.*, 28, 269, 1978.
10. **Biggers, D. C., Kraft, L. M., and Sprinz, H.,** Lethal intestinal virus infection of mice (LVIM), *Am. J. Pathol.*, 45, 413, 1964.
11. **Broderson, J. R., Murphy, F. A., and Hierholzer, J. C.,** Lethal enteritis in infant mice caused by mouse hepatitis virus, *Lab. Anim.. Sci.*, 26, 824, 1976.

12. **Carthew, P.,** Lethal intestinal virus of infant mice is mouse hepatitis virus, *Vet. Rec.*, 101, 465, 1977.

PART 4A

1. **Pomeroy, B. S.,** Coronaviral enteritis of turkeys (Blue comb disease), In *Diseases of Poultry*, 8th ed., Hofstad, M. S., Barnes, H. J., Calnek, B. W., Reid, W. M., and Yoder, H. W., Eds., Iowa State University Press, Ames, IA, 1984.
2. **Peterson, E. H. and Hymas, T. A.,** Antibiotics in the treatment of an unfamiliar turkey disease, *Poultry Sci.*, 30, 466, 1951.
3. **Panigrahy, B., Naqi, S. A., and Hall, C. F.,** Isolation and characterization of viruses associated with transmissible enteritis (bluecomb) of turkeys, *Avian Dis.*, 17, 430, 1973.
4. **Ritchie, A. E., Deshmukh, D. R., Larsen, C. T., and Pomeroy, B. S.,** Electron microscopy of coronavirus-like particles characteristic of turkey bluecomb disease, *Avian Dis.*, 17, 546, 1973.
5. **Dea, S., Marsolais, G., Beaubien, J. and Ruppanner, R.,** Coronaviruses associated with outbreaks of transmissible enteritis of turkeys in Quebec: hemagglutination properties and cell cultivation, *Avian Dis.*, 30, 319, 1986.
6. **Dea, S. and Tijssen, P.,** Viral agents associated with outbreaks of diarrhea in turkey flocks in Quebec, *Can. J. Vet. Res.*, 52, 53, 1988.
7. **Saif, L. J., Saif, Y. M., and Theil, K. W.,** Enteric viruses in diarrheic turkey poults, *Avian Dis.*, 29, 798, 1985.
8. **Reynolds, D. L., Saif, Y. M., and Theil, K. W.,** A survey of enteric viruses of turkey poults, *Avian Dis.*, 31, 89, 1987.
9. **Almeida, J. D.,** Uses and abuses of diagnostic electron microscopy, *Curr. Top. Microbiol. Immunol.*, 104, 147, 1983.
10. **Deshmukh, D. R. and Pomeroy, B. S.,** Physicochemical characterization of a bluecomb coronavirus of turkeys, *Am. J. Vet. Res.*, 35, 1549, 1974.
11. **Adams, N. R., Ball, R. A., Annis, C. L., and Hofstad, M. S.,** Ultrastructural changes in the intestines of turkey poults and embryos affected with transmissible enteritis, *J. Comp. Pathol.*, 82, 187, 1972.
12. **Adams, N. R. and Hofstad, M. S.,** Isolation of the agent of transmissible enteritis of turkeys in avian embryos, *Avian Dis.*, 15, 426, 1971.
13. **Dea, S. Garzon, S., and Tijssen, P.,** Isolation and trypsin-enhanced propagation of turkey enteric (bluecomb) coronavirus in a continuous human rectal adenocarcinoma cell line, *Am. J. Vet. Res.*, 50, 1310, 1989.
14. **Gonder, E., Patel, B. L., and Pomeroy, B. S.,** Scanning electron, light and immunofluorescent microscopy of coronaviral enteritis of turkeys (bluecomb), *Am. J. Vet. Res.*, 37, 1435, 1976.
15. **Pomeroy, K. A., Patel, B. L., Larsen, C. T., and Pomeroy, B. S.,** Combined immunofluorescence and transmission electron microscopic studies of sequential intestinal samples form turkey embryos and poults infected with turkey enteritis virus, *Am. J. Vet. Res.*, 39, 1348, 1978.
16. **Patel, B. L., Gonder, E., and Pomeroy, B. S.,** Detection of turkey coronaviral enteritis (bluecomb) in field epiornithics, using the direct and indirect fluorescent antibody test, *Am. J. Vet. Res.*, 38, 1407, 1977.
17. **Pomeroy, B. S., Larsen, C. T., Deshmukh, and Patel, B. L.,** Immunity to transmissible (coronaviral) enteritis of turkeys (bluecomb), *Am. J. Vet. Res.*, 36, 553, 1975.
18. **Nagaraja, K. V. and Pomeroy, B. S.,** Secretory antibodies against turkey coronaviral enteritis, *Am. J. Vet. Res.*, 39, 1463, 1978.
19. **Nagaraja, K. V. and Pomeroy, B. S.,** Cell-mediated immunity against turkey coronaviral enteritis (bluecomb), *Am. J. Vet. Res.*, 41, 915, 1980.
20. **Saif, L. J. and Saif, Y. M.,** unpublished observations, 1988.

PART 5A

1. **Caul, E. O., Paver, W. K., and Clarke, S. K. R.,** Coronavirus particles in faeces from patients with gastroenteritis, *Lancet*, 1, 1192, 1975.
2. **Mathan, M, Mathan, V. I., Swaminathan, S. P., Yesudoss, S., and Baker, S. J.,** Pleomorphic virus-like particles in human faeces, *Lancet*, 1, 1068, 1975.
3. **Caul, E. O. and Egglestone, S. I.,** Further studies on human enteric coronavirus, *Arch. Virol.* 54, 107, 1977.
4. **Clarke, S. K. R., Caul, E. O., and Egglestone, S. I.,** The human enteric coronaviruses, *Postgrad. Med. J.*, 55, 135, 1979.
5. **Schnagl, R. D., Morey, R., and Homes, I. H.,** Rotavirus and coronavirus-like particles in aboriginal and non-aboriginal neonates in Kalgoorlie and Alice Springs, *Med. J. Aust.*, 2, 178, 1979.
6. **Sureau, C., Amiel-Tison, C., Moscovici, O., Lebon, P., Laporte, J., and Chany, C.,** Une epidemic d'enterocolitis ulcernoecrosantes en maternite. Arguments en faveur de sou orinine virale, *Bull. Acad. Natl. Med.*, 164, 286, 1980.
7. **Caul, E. O., Ashley, C. R., and Egglestone, S. I.,** Recognition of human enteric coronaviruses by electron microscopsy, *Med. Lab. Sci.*, 34, 259, 1977.

8. **Dourmashkin, R. R., Davies, H. A., Smith, H., and Bird, R. G.,** Are coronavirus-like particles seen in diarrhoea stools really viruses?, *Lancet*, 2, 971, 1980.
9. **Caul, E. O. and Clark, S. K. R.,** Coronavirus propagated from patient with nonbacterial gastroenteritis, *Lancet*, 2, 953, 1975.
10. **Macnaughton, M. R. and Davies, H. A.,** Human enteric coronaviruses, *Arch Virol.*, 70, 301, 1981.
11. **Patel, J. R., Davies, H. A., Edington, N., Laporte, J., and Macnaughton, M. R.,** Infection of a calf with the enteric coronavirus strain paris, *Arch Virol.*, 73, 319, 1982.
12. **Maass, G. and Baumeister, H. G.,** Coronavirus-like particles as aetiological agents of acute non-bacterial gastroenteritis in humans, in *Int Symp. Enteric Infection in Man and Animals: Stand. of Immunol. Proc.*, Dublin, Ireland, 1982, *Develop. Biol. Standard.*, Vol. 5, (S. Karger, Basel, 1983), 319.
13. **Chany, C., Moscovici, O., Lebon, P., and Rousset, S.,** Association of coronavirus infection with neonatal necrotizing enterocolitis, *Pediatrics*, 69, 209, 1982.
14. **Mortensen, M. L., Ray, C. G., Payne, C. M., Friedman, A. D., Minnich, L. L., and Rousseau, C.,** Coronaviruslike particles in human gastrointestinal disease, *Am. J. Dis. Child.*, 139, 928, 1985.
15. **Resta, S., Luby, J. P., Rosenfeld, C. R., and Siegel, J. D.,** Isolation and propagation of a human enteric coronavirus, *Science*, 229, 978, 1985.
16. **Tyrrell, D. A. J., Alexander, D. J., Almeida, J. D., Sunningham, C. H., Easterday, B. C., Garwes, D. J., Hoerholzer, J. C., Kapikian, A. Z., Macnaughton, M. R., and MacIntosh, K.,** Coronaviridae, second report, *Intervirology*, 10, 321, 1978.
17. **Battaglia, M., Passarani, N., Di Matteo, A., and Gerna, G.,** Human enteric coronaviruses: further characterizations and immunoblotting of viral proteins, *J. Infect. Dis.*, 155, 140, 1987.
18. **Gerna, G., Passarani, N., Battaglia, M., Revello, M. G., Torre, D., and Cereda, P. M.,** Coronaviruses and gastroenteritis: evidence of antigenic relatedness between human enteric coronavirus strains and human coronavirus OC43, *Microbiologica*, 7, 315, 1984.
19. **Gerna, G., Passerani, N., Cereda, P. M., and Battaglia, M.,** Antigenic relatedness of human enteric coronavirus strains to human coronavirus OC43: a preliminary report, *J. Infect. Dis.*, 150, 618, 1984.
20. **Pederson, N. C., Ward, J., and Mengeling, W. L.,** Antigenic relationship of the feline infectious peritonitis virus to coronavirus of other species, *Arch. Virol.*, 58, 45, 1978.
21. **Storz, J. and Rott, R.,** Reactivity of antibodies in human serum with antigens of an enteropathogenic bovine coronavirus, *Med. Microbiol. Immunol.*, 169, 169, 1981.
22. **Sharpee, R. and Mebus, C. A.,** Letter to the editor, *Lancet*, 1, 639, 1975.
23. **Vaucher, Y. E., Ray, C. G., Minnich, L. L., Payne, C. M., Beck, D., and Lowe, P.,** Pleomorphic, enveloped, virus-like particles associated with gastrointestinal illness in neonates, *J. Infect. Dis.*, 145, 27, 1982.
24. **Rettig, P. J. and Altshuler, G. P.,** Fetal gastroenteritis associated with coronavirus like particles, *Am. J. Dis. Child.*, 139, 245, 1985.
25. **Baker, S. J., Mathan, M., Mathan, V. I., Jesudoss, S., and Swaminathan, S. P.,** Chronic enterocyte infection with coronavirus, *Dig. Dis. Sci.*, 27, 1039, 1982.
26. **Siegel, J. D., Luby, J. P., Laptook, A. R., and Butler, S.,** Identification of coronavirus (CRNV) in a premature nursery during an outbreak of necrotizing enterocolitis (NEC) and diarrhea (D), *Pediatr. Res.*, 17(Suppl.), 181A, 1983.
27. **Yongnian, H., Wang, N. L., Lo, H. N., Nie, A. G., and Li, A.,** A finding of coronavirus particles in feces of patients with diarrhea, *Chin. J. Epidemiol.*, 8, 25, 1987.
28. **Marshall, J. A., Blirch, C. J., Williamson, H. G., Bowden, D. K., Boveington, C. M., Kuberski, T., Bennett, P. H., and Gust, I. D.,** Coronavirus-like particles and other agents in the faeces of children in Efate, Vanuatu, *J. Trop. Med. Hyg.*, 85, 213, 1982.
29. **Sitbon, M.,** Human-enteric-coronaviruslike particles (CVLP) with different epidemiological characteristics, *J. Med. Virol.*, 16, 67, 1985.
30. **Almeida, J. D.,** Uses and abuses of diagnostic electron microscopy, *Curr. Top. Microbiol. Immunol.*, 104, 147. 1983.
31. **Weindling, A. M., Walker-Smith, J. A., and Bird, R.,** Micro-organisms in outpatient infantile gastroenteritis, *Arch. Dis. Childhood*, 55, 185, 1980.
32. **Simhon, A. and Mata, L.,** Fecal rotaviruses, adenoviruses, coronavirus-like particles, and small round viruses in a cohort of rural Costa Rican children, *Am. J. Trop. Med. Hyg.*, 34, 931, 1985.

PART 5B

1. **Anon.,** Epidemic diarrhoea, *Vet. Rec.*, 90, (Suppl.(95), 49, 1972.
2. **Wood, E. N.,** An apparently new syndrome of porcine epidemic diarrhoea, *Vet. Rec.*, 100, 243, 1977.
3. **Pensaert, M.,** Porcine epidemic diarrhea, in *Diseases of Swine*, 6th ed., Leman, A. D., et al., Eds., Iowa State University Press, Ames, IA, 402, 1986.
4. **Pensaert, M. B. and DeBouck, P.,** A new coronavirus-like particle associated with diarrhea in swine, *Arch. Virol.*, 58, 243, 1978.

5. **Chasey, D. and Cartwright, S. F.,** Virus-like particles associated with porcine epidemic diarrhoea, *Res. Vet. Sci.*, 25, 255, 1978.
6. **DeBouck, P., Pensaert, M., and Coussement, W.,** The pathogenesis of an enteric infection in pigs experimentally induced by the coronavirus-like agent CV777, *Vet. Microbiol.*, 6, 157, 1981.
7. **Pospischil, A., Hess, R. G., and Bachmann, P. A.,** Light microscopy and ultrahistology of intestinal changes in pigs infected with enzootic diarrhoea virus (EVD): comparison with transmissible gastroenteritis (TGE) virus and porcine rotavirus infections, *Zentralbl. Veterinaermed. B*, 28, 564, 1981.
8. **Ducatelle, R., Coussement, W., Charlier, G., DeBouck, P., and Hoorens, J.,** Pathology of experimental CV777 coronavirus enteritis in piglets. II. Electron microscopic study, *Vet. Pathol.*, 19, 57, 1982.
9. **Takeuchi, A., Binn, L. N., Jervis, H. R., Keenan, K. P., Hildebrand, P. K., Valas, T. R. B., and Bland, F. F.,** Electron microscope study of experimental enteric infection in neonatal dogs with a canine coronavirus, *Lab. Invest.*, 34, 539, 1976.
10. **Thake, D. C.,** Jejunal epithelium in transmissible gastroenteritis of swine, an electron microscopic and histological study, *Am. J. Pathol.*, 53, 149, 1968.
11. **Hess, R. G., Bollwahn, W., Pospischil, A., Heinritzi, K., and Bachmann, P. A.,** Neue aspekte der Virusaetiologie bei Durchfallerkrankungen des Schweines: Vorkommen von Infektionen mit dem epizootischen virusdiarrhoe-(EVD-) virus, *Berl. Muench. Tieraertzl. Wochenschr*, 93, 445, 1980.
12. **Callebaut, P. and DeBouck, P.,** Some characteristics of a new porcine coronavirus and detection of antigen and antibody by ELISA, Proc. 5th Int Congr Virol., Strasbourg, France, August 2 to 7, 1981, 420.
13. **Witte, K. H., Pragner, D., Ernst, H., and Neuhoff, H.,** Die epizootische virusdiarrhoe (EVD), *Tieraerztl Umsch.*, 36, 235, 1981.
14. **Hoffman, M. and Wyler, R.,** Propagation of the virus of porcine epidemic diarrhea in cell culture, *J. Clin. Microbiol.*, 26, 2235, 1988.
15. **Egberink, H. F., Ederveen, J., Callebaut, P., Horzinek, M.C.,** Characterization of the structural proteins of porcine epizootic diarrhea virus, strain CV777, *Am. J. Vet. Res.*, 49, 1320, 1988.
16. **Siddel, S., Wege, H., and ter Meulen, V.,** The structure and replication of coronaviruses, *Curr. Top. Microbiol. Immunol.*, 99, 131, 1983.
17. **Pensaert, M. B., DeBouck, P., and Reynolds, D. J.,** An immunoelectron microscopic and immunofluorescent study on the antigenic relationship between the coronavirus-like agent CV777 and several coronaviruses, *Arch. Virol.*, 68, 45, 1981.
18. **Pedersen, N. C., Ward, J., and Mengeling, W. L.,** Antigenic relationship of the feline infectious peritonitis virus to coronaviruses of other species, *Arch. Virol.*, 58, 45, 1978.
19. **DeBouck, P. and Pensaert, M.,** Experimental infection of pigs with a new porcine enteric coronavirus CV777, *Am. J. Vet. Res.*, 41, 219, 1980.
20. **Horvath, I. and Moscari, E.,** Ultrastructural changes in the small intestinal epithelium of suckling pigs affected with a transmissible gastroenteritis (TGE)-like disease, *Arch. Virol.*, 68, 103, 1981.
21. **Leopoldt, D., Lehnert, C., Koitzsch, R., Tesmer, S., Granzow, H., and Heinrich, H. W.,** Neue virusbedingte Durchfallerkrankungen des Schweiner (TGE-aehnliche Erkrankungen), *Monatsh. Veterinaermed.*, 36, 411, 1981.
22. **Turgeon, D. C., Morin, M., Golette, J., Higgins, R., Marsolais, G., and DeFranco, E.,** Coronavirus-like particles associated with diarrhea in baby pigs in Quebec, *Can. Vet. J.*, 21, 100, 1980.
23. **Dea, S., Vaillancourt, J., Elazhary, Y., and Martin, G. P.,** An outbreak of diarrhea in piglets caused by a coronavirus antigenically distinct from TGE virus, *Can. Vet. J.*, 26, 108, 1985.
24. **Takahashi, K., Okada, K., and Oshshima, K.,** An outbreak of swine diarrhea of a new type associated with coronavirus-like particles in Japan, *Jpn. J. Vet. Sci.*, 45, 829, 1983.
25. **DeBouck, P., Callebaut, P., and Pensaert, M.,** Prevalence of the porcine epidemic diarrhea (PED) virus in the pig population of different countries, Proc. 7th Int. Congr. Pig Vet. Soc., Mexico City, July 26 to 31, 1982, 53.
26. **Ducatelle, R., Coussement, W., Charlier, G., DeBouck, P., and Hoorens, J.,** Three-dimensional sequential study of the intestinal surface in experimental porcine CV777 coronavirus enteritis, *Zentralbl. Veterinaermed. B*, 28, 483, 1981.
27. **Ducatelle, R., Coussement, W., DeBouck, P., and Hoorens, J.,** Pathology of experimental CV777 coronavirus enteritis in piglets, I. Histological and histochemical study, *Vet. Pathol.*, 19, 46, 1982.
28. **Keenan, K. P., Jervis, H. R., Marchwicki, R. H., and Binn, L. N.,** Intestinal infection of neonatal dogs with canine coronavirus 1-71: studies by virologic, histologic, histochemical and immunofluorescent techniques, *Am. J. Vet. Res.*, 37, 247, 1976.
29. **Almeida, J. D.,** Uses and abuses of diagnostic electron microscopy, *Current Top. Microbiol. Immunol.*, 104, 147, 1983.
30. **Heckert, R. A., Saif, L. J., and Myers, G. W.,** Development of protein A-gold immunoelectron microscopy for detection of bovine coronavirus in calves: Comparison with ELISA and direct immunofluorescence of nasal epithelial cells, *Vet. Microbiol.* 19, 217, 1989.

31. **Callebaut, P., DeBouck, P., and Pensaert, M.,** Enzyme-linked immunosorbent assay for the detection of the coronavirus-like agent and its antibodies in pigs with porcine epidemic diarrhea, *Vet. Microbiol.*, 7, 295, 1982.
32. **Prager, D. and Witte, K.,** Die serologische diagnose der epizootischen virusdiarrhoea (EVD) des schweines mit hilfe der indireckten immunofluoreszenztechnik (IIFT), II. Antikorper-antwort nach experimenteller infektion, *Tieraerztl. Umsch.*, 36, 477, 1981.
33. **DeBouck, P. and Pensaert, M.,** Porcine epidemic diarrhea: kinetics of actively and passively acquired serum antibodies and the effect of reinfection, Proc. 8th Int. Congr. Pig Vet. Soc., Ghent, Belgium, August 27 to 31, 1984, 53.
34. **Saif, L. J. and Bohl, E. H.,** Transmissible gastroenteritis virus, in *Diseases of Swine*, 6th ed., Leman, A. D. et al., Eds., Iowa State University Press, Ames, IA, 1986, 225.

Chapter 10

TOROVIRUSES

Marian C. Horzinek and Marianne Weiss

TABLE OF CONTENTS

I. INTRODUCTION

In 1982, Woode et al.[1] reported studies with a hitherto unclassified virus isolated from calves with diarrhea in Breda, Iowa. Breda virus (BRV) could be propagated in gnotobiotic calves and caused agglutination of rat erythrocytes. Indications for the existence of two serotypes were obtained including identification of a second serotype from diarrheic calves in Ohio and Iowa.[2] In 1972, a virus was isolated during routine laboratory diagnostic work from a horse under observation at the surgery clinic in Berne, Switzerland. The animal died about 1 week after the sample had been collected showing pseudomembranous enteritis and miliary granulomas and necrosis in the liver upon post-mortem examination. *Salmonella lille* was considered as the causative agent. Berne virus (BEV; laboratory designation of the strain: P138/72) was not neutralized by diagnostic antisera against notorious equine viruses and was shown to possess a unique morphology and substructure.[3] More recently, particles resembling the Berne/Breda viruses in morphology were described in the stools of children and adults with gastroenteritis in Birmingham, England and Bordeaux, France (BIV);[4] the human viruses were found physico-chemically and antigenically related with BRV.[5]

When discussing electron micrographs of negatively stained, purified BEV with colleagues experienced in the identification of enteric viruses, it appeared that similar particles are not uncommon in fecal material from humans and animals. Because of their ill–defined morphology, however, they were often regarded as nonviral in nature; others were inspired by the surface projections to make comparisons with coronaviruses. The resemblance, albeit superficial, has brought forth designations such as "corona-like" viruses, "minicorona" viruses, etc. We were able to show, however, that BEV has a unique pattern of structural polypeptides. For coronaviruses, the molecular weight of about 50 kDa of the nucleocapsid protein[6] is a taxonomic criterion; BEV, however, possesses a major capsid polypeptide of 20 kDa.[7] Also, replication of coronaviruses is independent from host cellular functions as they can be grown in enucleated cells and in the presence of actinomycin D.[8] This is in contrast to BEV, where indications for a nuclear involvement during virus replication have been obtained.[9]

On the basis of the available evidence, the establishment of a new virus family has been proposed at the International Congress of Virology at Sendai/Japan. Pending the acknowledgment by the International Committee on Taxonomy of Viruses, the name "Toroviridae" will be used for the Berne/Breda group of viruses.[10] The latin word "torus" refers to the circular convex moldings found at the bases of columns and alludes to the shape of the nucleocapsid in extracellular virus. A Toroviridae Study Group has been established whose first report has summarized data relevant for classification;[11] other reviews have appeared recently.[12–14]

II. MORPHOLOGY AND MORPHOGENESIS

Since most reports of toroviruses are based on observations of particles in feces, the negatively stained virion has been extensively described.[1–5, 15] Torovirions are spherical, oval, elongated, or kidney-shaped particles measuring 120 to 140 nm in diameter. In preparations of purified virions we have observed a sausage-like internal structure with transverse striations (estimated periodicity 4.5 nm[3] to 4.8 nm[5]) which appeared tightly attached to the membrane and did not leave the particle when the membrane was damaged, e.g., by detergent treatment. Depending upon the preparation, virions are either bald or studded with projections (peplomers).[3,7] Figure 1 shows a preparation of purified BEV particles negatively stained with uranyl acetate which, in our hands, is superior to phosphotungstate contrasting. Figure 2 illustrates the morphology of BRV particles from calves and an unclassified Breda–like virus from pigs.

There is some controversy about the morphology and dimensions of the peplomers. We have described them for BEV as "drumsticks", consisting of a thin stalk carrying a distal spherule

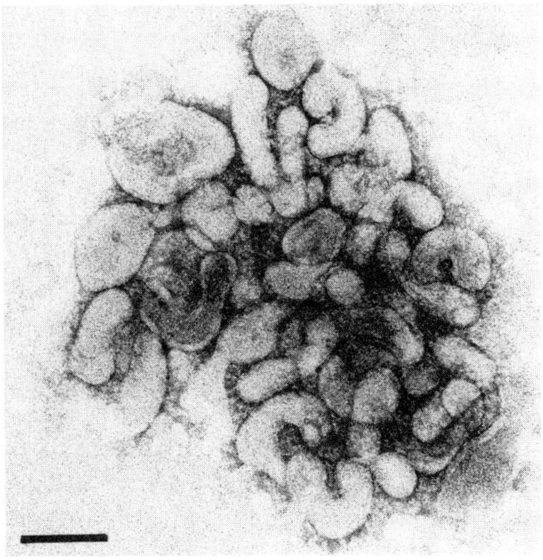

FIGURE 1. Particles of Berne virus grown in cells of the E. derm. (equine dermis) line and purified by ultracentrifugation into a 15%/50% sucrose interphase; negative staining with 1% unbuffered uranyl acetate. Bar represents 100 nm.

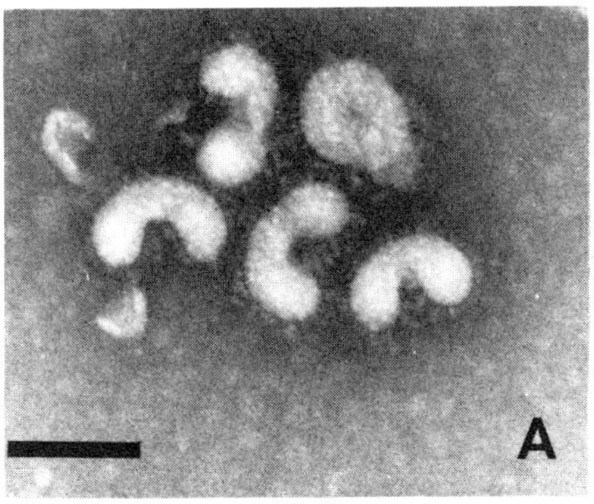

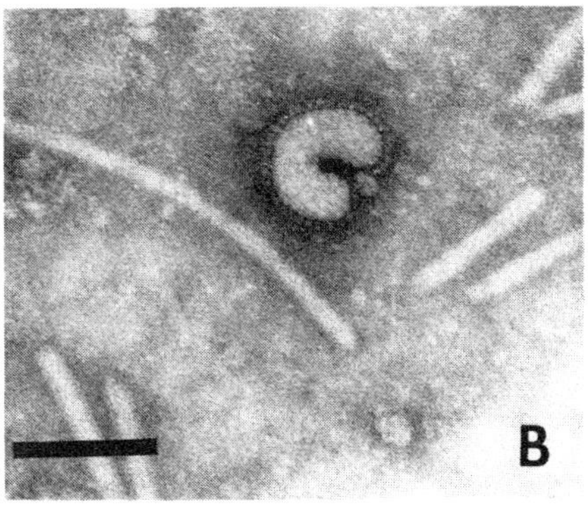

A B

FIGURE 2. Electron micrograph of negatively stained (3% PTA, PH7) Ohio Breda virus from gnotobiotic calf intestinal contents (A) and Breda–like virus particles (B) from conventional pig diarrheic feces. The particles in (A) are aggregated following incubation with antiBreda virus 2 serum. No coating or clumping of the particles in (B) was observed using the same or Breda virus 1 antiserum. Bar represents 200 nm. (Photos and results courtesy of Dr. Linda J. Saif.)

(total length about 20 nm).[3] Woode et al. have recorded surface projections of 7.6 to 9.5 nm in BRV.[1] The particles of BIV were described as carrying peplomers 7 to 9 nm in length. Occasionally what appeared to be a second ring of smaller peplomers was seen, partly superimposed upon the first.[4] The longer peplomers (17 to 24 nm) are believed by Woode et al. to be of doubtful specificity[1] and have been reported only occasionally on particles in samples of human feces.[4] The differences in peplomer length between BEV (20 nm) and BRV/BIV (7 to 9 nm) have been recently confirmed in a comparative study.[5] The distance between the viral and cytoplasmic membrane measured in thin sections through BEV particles adsorbed to the cell surface corroborates the longer estimates,[13, 16]

The thin section morphology of toroviruses was studied so far only with BEV (in infected equine cells)[3,17] and BRV (in gut epithelium from infected calves).[18,19] BEV-infected cells revealed densely staining spherical, elliptical, and elongated particles accumulating at the cytoplasmic membrane and in vacuoles. At higher magnification, a clear distinction can be made between an electron-lucent envelope and a dark core. Rod- and crescent shaped cores are prevalent in virions in the extracellular space. Twin circular structures with a conspicuous light center, which we interpret as cross-sections through a hollow, tubular nucleocapsid (diameter 23 nm) bent into an open torus are regularly seen; they are surrounded by a tightly fitting membrane. The length of the tubular structure can only be approximated; the calculated average value (104 ± 16 nm) is meaningless in this respect, since it only reflects the frequency distribution of orientations in space of particles. A better estimate is our maximum value of 171 nm, but the true length of the capsid may be slightly greater (180 nm, when assuming a nonoverlapping 23-nm tube filling the space within a discoidal envelope with an inner diameter of 80 nm). From the asymmetric distribution of the diameters of all opaque internal structures of circular or elliptical shape in thin-section electron micrographs, we concluded that the virus particle is not spherical in shape.[3] Figure 3 gives an impression of the thin-section morphology of BEV.

In cells of the intestinal mucosa of BRV-infected calves, elongated enveloped virions with rounded ends were detected, with average dimensions of 35 × 80 nm. The particles were described as being pleomorphic and varying in length.[18, 19] The crescent shape described above for BEV was not reported.

The morphological data are best explained by assuming a nucleocapsid of helical symmetry, which is tightly coiled into a hollow tube — not unlike in rhabdoviruses; this structure is either straight or bent into an open torus. A tightly fitting envelope surrounds the core. The virion may assume the morphology of a biconcave disk — like an erythrocyte — as visualized in our cross-sections. Alternatively, the envelope could follow the smaller curvature of the torus, thereby creating a sausage or kidney shape.[12]

A particle with this degree of pleomorphism — bacilliform, kidney-shaped, or discoidal — is quite uncommon in virology; different orientations of the particles with respect to the electron beam tend to increase the heterogeneity of the picture still further. This is certainly one reason why electron microscopists were reluctant to accept these structures as viral in nature, when encountering them, e.g., in fecal specimens.

In thin sections through equine dermis (E. derm.) cells infected with BEV, particles are visible 10 h after infection. Virions are encountered in all parts of the Golgi system and, infrequently, in the rough endoplasmic reticulum. Budding of virus particles single or in groups occurs predominantly in the Golgi system. From images obtained at different stages of the process, the following sequence of events can be reconstructed: the preformed intracytoplasmic nucleocapsid approaches the membrane with one of its rounded poles upon which the membrane progressively attaches to the capsid unilaterally. The part of the nucleocapsid which is already enveloped has a higher electron density and a uniform diameter, whereas the part still embedded in the cytoplasmic matrix is of lesser electron density and varying width. The result of this morphopoietic process is an entirely enveloped tubulus, which is encountered free in the lumen of Golgi cisternae. The particles are, without exception, rod-like with two rounded ends and a regularly structured core of uniform electron density.[17] Their bacilliform shape is emphasized by occasional packaging into bundles, an observation also made with BRV.[18,19] In cross-sections, the virion appears as three concentric circles; the innermost circle of highest electron density probably constitutes a transversal section through the nucleocapsid. Its electron lucent center emphasizes the tubular nature of the capsid. The diameters of the circles are 24, 37, and 47 nm, respectively.[17]

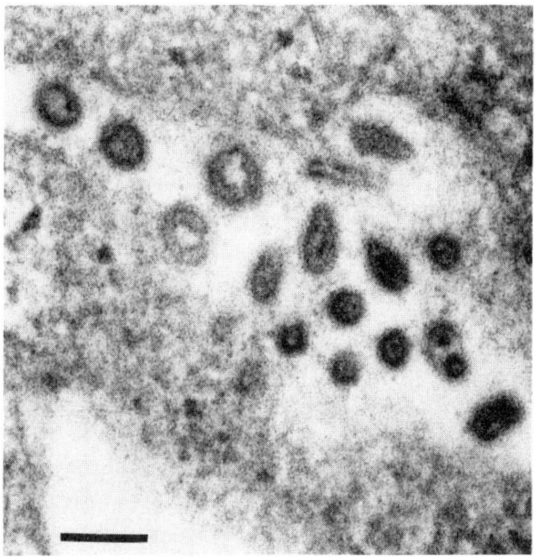

Figure 3. Thin section through cells of the EMS (embryonic mule skin) line infected with Berne virus; note variation in morphology described in the text. Bar represents 100 nm.

Virion morphogenesis in the Golgi system is observed in the lateral vesicles as well as the flat cisternae. Sometimes, virions are seen in several cisternae of the same stack. Late in infection, budding occurs also into smooth surfaced vesicles in locations distant from the nucleus, where the correlation to the Golgi system can no longer be established. In addition, budding into the rough endoplasmic reticulum and the perinuclear space has occasionally been encountered.[17]

Virus-containing vesicles merge with the peripheral cytoplasmic membrane and release their contents. Virus is found flattened against the cell surface where it may show the typical twin circular structures indicative for the torus morphology of the nucleocapsid. Aberrant virion forms of extreme length (250 nm) are occasionally observed in vacuoles. The proper morphology is seen in extracellular particles and in vacuoles near the cell surface, but not in the cytoplasm.[17]

III. PHYSICOCHEMICAL PROPERTIES

Toroviruses possess a lipoprotein membrane, as evidenced by electron microscopy, by their low buoyant density in sucrose (1.14 to 1.18 g/ml), and their sensitivity to organic solvents.[3,5,20,21] Sedimentation coefficients of about 400 S have been determined for BEV and BRV.[20] Four classes of structural polypeptides have been found in BEV[9,7,22] and are probably also present in BRV.[20]

Accounting for about 84% of the protein mass, a 20 kDa polypeptide (N) is most prevalent in the BEV virion. It is detected in gradient fractions with maximum infectivity (1.16 g/ml), but also in an intermediate peak (1.11 g/ml) which contains about 10% of the infectivity of the virion peak.[3] The 20 kDa protein is part of a substructure which can be liberated from the virion by treatment with Triton X-100 and which, subsequently, can be pelleted through sucrose of virion density. In infected cells, it was present in a component of nucleoprotein density (1.36 g/ml in CsCl) which contained hot TCA-soluble phosphate label indicative of nucleic acid.[23] In blotting experiments, the 20 kDa protein was the only RNA-binding polypeptide species detectable in the infected cell lysate; its isoelectric point is in the neutral to basic range, as indicated by two-dimensional PAGE.[23] These properties qualify the 20 kDa species as the main protein constituent of the nucleocapsid.[7]

Second in abundance (about 13% of the virion protein mass) is a 22 kDa protein (E) which occurs in a noninfectious structure of low density (1.07 g/ml) present in media from BEV-infected cultures; treatment with Triton X-100 results in its conversion into slowly sedimenting material, an observation indicative of its membrane nature. When virions were submitted to the same treatment, material with a similar sedimentation behavior was generated. These observations support the conclusion that the 22 kDa protein is part of the envelope of BEV.[7,9]

A phosphorylated polypeptide of 37 kDa (M) is probably also associated with the viral envelope, since it can be demonstrated in gradients both in the low density and the virion peak; it is devoid of RNA binding properties.[7] Further studies are needed for its localization; however, it can be radioiodinated together with the major envelope protein E in intact virions where the N protein remains unlabeled and must, therefore, be exposed.[13] Under the assumption that it spans the membrane, it may have a role in nucleocapsid recognition during budding. The membrane proteins effectively shield the lipid bilayer, thereby contributing to the extreme deoxycholate stability of BEV.[21]

Using ^{3}H-glucosamine and ^{3}H-manose label, two virus-specific glycosylated polypeptide species with M_r values of about 200 kDa and in the 70- to 100- kDa range, respectively, were recognized in BEV-infected embryonic mule skin cells. In purified virions only the latter glycoprotein (P) occurred; it was heterogenous in its electrophoretic behavior and relative abundance, with major bands of 96 and 75 kDa and a minor 86 kDa species. Concanavalin A was bound to the virion surface, as evidenced by reduction in infectivity. The viral glycoprotein signals coincided with the maximum of infectivity and hemagglutinating activity in an isokinetic sucrose gradient. Polyclonal rabbit immune serum and a neutralizing and hemagglutination-inhibiting monoclonal antibody raised against BEV recognized both the 96/86/75 kDa and the "200 kDa" glycoproteins. Using tunicamycin, a concentration-dependent inhibition of synthesis of infectious virus was noted. However, particles were released in small quantities which lacked the glycoproteins; these were also absent from cytoplasmic extracts but novel polypeptides of about 150 and 115 kDa were identified instead. Translation of polyA-selected intracellular RNA from infected cells in rabbit reticulocyte cell-free system also resulted in the appearance of a high M_r polypeptide (173 kDa). Using pulse-chase labeling and radioimmunoprecipitation, a precursor-product relationship between the intracellular "200 kDa" and the virion glycoproteins has been demonstrated. These experiments identify the N-glycosylated proteins in the 75 to 96 kDa range as constituents of the peplomeric envelope projection of BEV; they probably arise by post-translational processing of a precursor molecule involving glycosylation and subsequent cleavage.[22] Using surface radioiodination of BRV particles purified from feces of calves with diarrhea, a 105-kDa and a 85-kDa polypeptide were identified in SDS-PAGE. Also, neutralizing sera obtained after immunization of mice with one serotype recognized these proteins in preparations of another serotype of BRV.[20]

Thermal inactivation of BEV proceeds at a linear rate in the 31 to 43° C range; similar observations have been made with other RNA viruses (picornaviruses, retroviruses, togaviruses, etc.). Compared to transmissible gastroenteritis virus of swine (TGEV), a coronavirus, it appears that BEV is more readily heat inactivated. Storage at temperatures lower than –20°C preserves the infectivity, while at 4°C appreciable loss occurs between 92 and 185 h. Freeze-drying or desiccation at 22°C causes only insignificant loss of infectivity.[21]

A high pH stability was noted for BEV, with inactivation occurring only at values <2.5 and >10.3; this is quite unusual for an enveloped virus and allows for a gastric passage during pathogenesis. Another remarkable stability feature is the resistance of BEV to sodium deoxycholate and phospholipase C, trypsin, and chymotrypsin.[21]

IV. GENOME

The presence of an RNA genome was inferred from the observation that growth of BEV is unaffected by IUDR under conditions where a DNA virus is inhibited.[3] Recent unpublished experiments using uridine labeling have shown that the genome of BEV consists of a single RNA species with an apparent molecular weight of about 9×10^6 Da. This value was estimated from gels in which genomic RNAs of mouse hepatitis virus and TGEV were coelectrophoresed. The molecule is polyadenylated to an extent comparable with that of TGEV; it is single-stranded, of positive polarity, and infectious when assayed under transfection conditions. A population of particles with a slower sedimentation than the virion has been regularly observed in preparations of BEV; these particles are noninfectious and may contain RNAs of smaller size.[23]

V. CULTIVATION

Original isolation of BRV was made by oral inoculation into gnotobiotic calves. Attempts to grow the virus in cell or organ cultures or in embryonated eggs were unsuccessful.[1]

The P138/72 strain of BEV was isolated in secondary horse kidney cells, inoculated with material from a rectal swab; the same material yielded the virus on two subsequent reisolations. Significantly, this has remained the only equine torovirus strain until now, although some hundred isolation attempts were made. Rapid adaptation to cells of equine dermis and embryonic mule skin was observed where the virus caused a lytic cytopathic effect. In these lines, however, virus yields tended to decrease with higher cell passage levels; attempts to establish permanent lines of equine cells using transfection with adeno-, papilloma-, herpes-, and retroviral genes have been without success so far.[23] In view of the wide-spread occurrence of antibodies against toroviruses in cattle,[24-26] fetal calf sera to be used in torovirus growth and isolation must be pretested.

We have studied some 20 established cell lines of different origin for their ability to support multiplication of BEV; all attempts were negative. Also BRV could not be adapted to growth in culture until now and must be propagated in calf-to-calf passages using gnotobiotic or colostrum deprived animals.[1,2,18]

The study of toroviruses is obviously hampered by the lack of suitable culture systems. The requirement by BRV for highly differentiated cells which are difficult to maintain or grow *in vitro* is illustrated by its tropism *in vivo* for the epithelial cells of domes in the ileum.[27] In animals, toroviruses may reach extremely high concentrations. BRV has been reported to attain hemagglutinin titers of 10^7,[1,28] which would correspond to about 10^{12} physical particles when using the ratios established for orthomyxo- and togaviruses.

VI. DIAGNOSIS

The diagnostic methods for BRV have been reviewed;[28] they include hemagglutination (HA) and hemagglutination-inhibition (HAHI), immunofluorescence (IF), enzyme-linked immunosorbent assay (ELISA), and electron microscopy. All isolates possess common antigens, which can be detected by IF of intestinal sections and by ELISA. An antigen-independent method for detecting toroviruses in fecal material would be electron microscopy of thin sections through ultracentrifuge pellets. The twin circular structures resulting from cross-section through the toroidal capsid are unmistakable and would confirm a preliminary diagnosis from negatively stained preparations. The high concentrations of particles in feces make this approach feasible.

VII. ANTIGENIC RELATIONSHIPS

The Breda agent (Iowa 1 isolate) is antigenically unrelated to bovine corona-, parainfluenza, rota-, parvo, and pestiviruses (bovine viral diarrhea/mucosal disease).[1] Two further isolations of BRV were reported: the Ohio 1 strain from 5-month-old diarrheic calves[29] and the Iowa 2 isolate from a 2-day-old experimental calf.[2] An HA activity of BRV for rat and mouse erythrocytes[1] and of BEV for human group 0, rabbit and guinea pig red cells,[16] was discovered; also BIV (isolate 1690) agglutinated rat erythrocytes.[5] On the basis of HAHI tests, ELISA, and immune electron microscopy, the three BRV isolates were assigned to serotype 1 (BRV1) being represented by the first Iowa isolate and serotype 2 (BRV2) including the Ohio 1 and Iowa 2 isolates.[2]

Using radioimmunoprecipitation followed by SDS polyacrylamide gel electrophoresis, we have localized the epitope(s) responsible for the cross-reactions between BRV serotypes on the P protein.[20] The P polypeptide of BEV is also recognized by monoclonal antibodies which had been selected on the basis of their neutralizing and HAHI activity.[30] Guinea pig immune sera prepared against both BRV serotypes inhibited HA of the homologous viruses to high titers and also neutralized BEV; similarly prepared highly neutralizing anti-BEV serum did not inhibit HA of either BRV serotype.[5]

VIII. EPIDEMIOLOGY

The BEV is widespread in the Swiss horse population (81%) and has been so during the last decade; rises in antibody titers were noted in 9% of paired sera sampled at random. Positive reactions were also obtained in serum neutralization tests and ELISA using small numbers of horse sera from Germany, France, and the U.S.[24] Using BRV as ELISA antigen, 57% of horse sera selected at random in the U.S. had antibodies.[25] The results of neutralization tests and ELISA were correlated in 83% of the random samples tested; 13% were neutralization-positive and ELISA-negative and in 4% the inverse was observed. These latter results may indicate that more distantly related toroviruses occur in the horse.[24]

Neutralizing activity for BEV was found in the sera of cattle (86%), goats (69%), sheep (34%), pigs (81%), laboratory rabbits (17%), and two species of wild mice (*Clethrionomys glareolus* and *Apodemus sylvaticus*); a third murine species *(Apodemus flavicollis)* was seronegative. Inconclusive results were obtained with feline and human sera; those from dogs and foxes *(Vulpes vulpes)* were consistently negative.[24]

Some experience using ELISA for the seroepidemiology of BRV has been obtained. Quesada et al.[25] have found 88.5% of randomly collected cattle sera in the U.S. to be antibody-positive; we have determined a proportion of 90.8% in the German Federal Republic. Most of the seronegative animals were calves; the fraction of seronegative adult cattle (4.8%) comprised bulls in artificial insemination units, zoo animals, and cows from isolated farms.[26]

IX. PATHOGENESIS

Pathogenetic mechanisms have been studied only with BRV. After experimental inoculation of eight gnotobiotic calves via the oral route, three developed severe diarrhea with lesions in the small and large intestine. Using IF, viral antigen was demonstrated in the jejunum, ileum, and spiral colon; infection of both the crypt and villus epithelial cells was observed. Virus excretion correlated with the onset of diarrhea or a change in the appearance of the feces.[1]

Intranasal inoculation of gnotobiotic calves within 2 h after caesarean section resulted in diarrhea 48 to 60 h later. Consistent microscopic findings in the intestinal mucosa were changes in enterocytes of the lower small intestine and the large intestine; the lesions were characterized by cell volume expansion, duplication of the Golgi complex with many small vesicles associated with both faces, distension and fragmentation of the cytocavitary network, and occurrence of phagolysosomes containing cellular debris and viral structures. Most affected cells contained tubular aggregates, also in the nucleoplasm. The dome epithelial cells showed alterations.[18]

X. IMMUNITY

In a horizontal study of BEV seropositive mares and their offspring, a decline of maternal neutralizing antibodies and a sudden synchronous seroconversion in all foals were observed, again without clinical symptoms.[24] Colostral antibodies to BRV in calves (as measured by ELISA) waned during the first 7 weeks of age; antibodies following infection developed from the 6th month of age onward.[26]

XI. PREVENTION AND CONTROL

Since the importance of toroviruses as agents causing enteric infections and sometimes disease in humans and animals only begins to emerge, control measures have not yet been developed.

REFERENCES

1. **Woode, G. N., Reed, D. E., Runnels, P.L., Herrig, M. A., and Hill, H. T.,** Studies with an unclassified virus isolated from diarrheic calves, *Vet. Microbiol.,* 7, 221, 1982.
2. **Woode, G. N., Saif, L. J., Quesada, M., Winand, N. J., Pohlenz, J. F., and Gourley, N. K.,** Comparative studies on three isolates of Breda virus of calves, *Am. J. Vet. Res.,* 45, 1003, 1985.
3. **Weiss, M., Steck, F., and Horzinek, M. C.,** Purification and partial characterization of a new enveloped RNA virus (Berne virus), *J. Gen. Virol.,* 64, 1849, 1983.
4. **Beards, G. M., Hall, C., Green, J., Flewett, T. H., Lamouliate F., and Du Pasquier, P.,** An enveloped virus in stools of children and adults with gastroenteritis that resembles the Breda virus of calves, *Lancet,* 2, 1050, 1984.
5. **Beards, G. M., Brown, D. W. G., Green, J., and Flewett, T. H.,** Preliminary characterization of torovirus–like particles of humans: comparison with Berne virus of horses and Breda virus of calves, *J. Med. Virol.,* 20, 67, 1986.
6. **Siddell, S., Wege, H., and Ter Meulen, V.,** The structure and replication of coronaviruses, *Curr. Top. Microbiol. Immunobiol.,* 99, 131, 1982.
7. **Horzinek, M. C., Ederveen, J., and Weiss, M.,** The nucleocapsid of Berne virus, *J. Gen. Virol.,* 66, 1287, 1985.
8. **Wilhelmsen, K. C., Leibowitz, J. L., Bond, C. W., and Robb, J. A.,** The replication of murine coronaviruses in enucleated cells, *Virology,* 110, 225, 1981.
9. **Horzinek, M. C., Weiss, M., and Ederveen, J.,** Berne virus is not 'coronavirus–like', *J. Gen. Virol.,* 65, 645, 1984.
10. **Horzinek, M. C.,** Nonarbo animal togaviruses and control perspectives, in *Control of Virus Diseases,* Kurstak, E., Ed., Academic Press, New York, 1984, 163.
11. **Horzinek, M. C., Flewett, T. H., Saif, L. J., Spaan, W. J. M., Weiss, M., and Woode, G. N.,** A new family of vertebrate viruses, Toroviridae, *Intervirology,* 27, 17, 1987.
12. **Horzinek, M. C. and Weiss, M.,** Toroviridae: a taxonomic proposal, *Zentralbl. Vet. Med. B,* 31, 649, 1984.
13. **Weiss, M. and Horzinek, M. C.,** The proposed family Toroviridae: agents of enteric infections, *Arch. Virol.,* 92, 1, 1987.
14. **Horzinek, M. C., Weiss, M., and Ederveen, J.,** Toroviridae: a proposed new family of enveloped RNA viruses, *Ciba Symp. Series,* 1987.
15. **Moussa, A., Dannacher, G., and Fedida, M.,** Nouveaux virus intervenant dans l'étiologie des entérites néonatales des bovins, *Rec. Med. Vet,* 159, 185, 1983.
16. **Zanoni, R., Weiss, M., and Peterhans, E.,** The hemagglutinating activity of Berne virus, *J. Gen. Virol.,* 67, 2485, 1986.

17. **Weiss, M. and Horzinek, M. C.,** Morphogenesis of Berne virus (proposed family toroviridae), *J. Gen. Virol.*, 67, 1305, 1986.
18. **Pohlenz, J. F. L., Cheville, N. F., Woode, G. N., and Mokresh, A. H.,** Cellular lesions in intestinal mucosa of gnotobiotic calves experimentally infected with a new unclassified bovine virus (Breda virus), *Vet. Pathol.*, 21, 407, 1984.
19. **Fagerland, J. A., Pohlenz, J. F. L., and Woode, G. N.,** A morphologic study of the replication of Breda virus (proposed family Toroviridae) in bovine intestinal cells, *J. Gen. Virol.*, 67, 1986.
20. **Koopmans, M., Ederveen, J., Woode, G. N., and Horzinek, M. C.,** The surface proteins of Breda virus, *Am. J. Vet. Res.*, 47, 1896 1986.
21. **Weiss, M., and Horzinek, M. C.,** Resistance of Berne virus to physical and chemical treatment, *Vet. Microbiol.*, 11, 41, 1986.
22. **Horzinek, M. C., Ederveen, J., Kaeffer, B., de Boer, D., and Weiss, M.,** The peplomers of Berne virus, *J Gen. Virol.*, 67, 2475, 1986.
23. **Horzinek, M. C. and Ederveen, J.,** unpublished observations, 1986.
24. **Weiss, M., Steck, F., Kaderli, R., and Horzinek, M. C.,** Antibodies to Berne virus in horses and other animals, *Vet. Microbiol.*, 9, 523, 1984.
25. **Quesada, M., Kelso Gourley, N. and Woode, G. N.,** unpublished observations, 1986.
26. **Van den Boom, U.,** Entwicklung eines ELISA zum Nachweis von Toroviren und Torovirus–Antikoerpern beim Rind, epidemiolog Untersuchungen ueber Torovirusinfektionen beim Rind, Inaug. Diss Vet. School, Hannover, Fed. Rep. Germany, 1986.
27. **Woode, G. N., Pohlenz, J. F., Kelso Gourley, N., and Fagerland, J. A.,** Astrovirus and Breda virus infections of dome cell epithelium of bovine ileum, *J. Clin. Microbiol.*, 19, 623, 1984.
28. **Woode, G. N., Mohammed, K. A., Saif, L. J., Winand, N. J., Quesada, M., Kelso, N. E., and Pohlenz, J. F.,** Diagnostic methods for the discovered "Breda" group of calf enteritis inducing viruses, *Proc. 3rd Int. Symp. World Assoc. Vet. Lab. Diagn.*, 2, 533, 1983.
29. **Saif, L. J., Redman, D. R., Theil, K. W., Moorhead, P. D., and Smith, C. K.,** Studies of an enteric "Breda" virus in calves, Abstr. 236, *62nd Annu. Meet. Conf. Res. Workers in Anim. Dis.*, 1981.
30. **Kaeffer, B., Ederveen, J., and Horzinek, M. C.,** unpublished observations, 1986.

Chapter 11

BOVINE VIRAL DIARRHEA VIRUS

Kenny V. Brock

TABLE OF CONTENTS

I. INTRODUCTION

Bovine viral diarrhea virus (BVDV) is widespread and a major contributor of economic losses to the cattle industry.[1] BVDV was historically recognized as an enteric pathogen associated with mucosal disease and outbreaks of diarrhea in susceptible herds.[2] However, the most common outcome of BVDV infection was subclinical infection. The virus also is fetopathic and causes abortions and congenital malformations.[3,4,5] In addition, it has recently been recognized that BVDV is a major factor in the pathogenesis of respiratory tract disease in cattle and may be associated synergistically with a number of disease pathogens.[6] This synergism may result from the transient immunosuppression produced by BVDV infection.[6-8] Strains of BVDV vary in antigenicity, pathogenicity, and cytopathogenicity and little is known about the molecular mechanisms responsible for this variation. The significance of BVDV as a pathogen has been recognized for over 30 years; however, the complex pathogenesis and biology of this virus remains unclear.

II. MORPHOLOGY AND MORPHOGENESIS

Cytopathic BVDV was isolated in 1957 by Underdahl et al. from two instances of mucosal disease.[9] Following the isolation of BVDV by Underdahl and others, the virus was further characterized by Hermodsson. BVDV was ether sensitive, contained RNA, and virions were approximately 40 nm in diameter.[10] Darbyshire[11] first reported a cross-reaction of BVDV with hog cholera virus by agar gel precipitation, which was later confirmed by Dinter using cross-neutralization studies[12] and Mengeling by immunofluorescence.[13] The buoyant density of BVDV was first reported by Fernelius[14] to be 1.15 g/cm^3 and in 1968, Hafez et al. described the morphology of BVD virions as 40 to 60 nm in size, depending on preparation procedures.[15] These findings led to the proposal that BVDV be included in the Togaviridae family.[16,17]

As BVDV was studied and characterized further, there remained some disagreement over the buoyant density of BVDV, which was important for purification of the virus and also necessary in its classification. Contradictory to Fernelius' findings,[14] Parks et al. determined that the buoyant density was approximately 1.115 g/cm^3.[18] The buoyant density was later confirmed by others to be approximately 1.12 g/cm.[3,19-21] Different estimations of these values were attributed to heterogeneity in size and buoyant density of virions.[20] Various reports have been published on the morphology of BVDV. Virion size estimates have ranged from 40 to 120 nm.[15,17,19] Chu reported that BVDV particles were pleomorphic and measured 120 (+/– 30) nm.[19] Based on morphologic characteristics, sedimentation coefficient, buoyant density, and antigenic relationship, BVDV, hog cholera virus, and Border disease virus were taxonomically placed in the Togaviridae family as members of a separate genus, *Pestivirus*.[22,23] The Oregon C24V strain of BVDV was selected as the prototype virus for the Pestivirus genus. Virions from the latter BVDV strain were pleomorphic, without any discernable symmetry, and measured 50 to 60 nm in diameter.[23]

III. *IN VITRO* CULTIVATION

BVDV was first isolated as a cytopathogenic agent in bovine kidney cell cultures.[9] BVDV has been isolated in numerous types of bovine cell cultures, such as bovine fetal kidney, bovine turbinate cells, bovine testicular cells, and bovine endothelial cells.[24,25] Several cytopathic and noncytopathic biotypes of BVDV have been isolated. Most cytopathic strains produce visible cytopathic effects in cell cultures within 36 to 72 h of inoculation. Noncytopathic virus may require 5 to 7 d of growth in cell cultures before antigen can be detected by fluorescent antibody.

BVDV also seems to replicate in most cell types of other species including human cell culture lines.[26] This should be of concern to all investigators using fetal bovine serum as a cell culture medium supplement, since practically all FBS is contaminated with from 10^1 to 10^6 $CCID_{50}$/ml of serum. Toth and Hesse reported that VERO cells were refractory to BVDV infection, thus allowing the elimination of BVDV from bovine respiratory syncytial virus stock.[27] However, some VERO cell lines are known by the author to be contaminated with noncytopathic BVDV. Heat inactivation was ineffective for eliminating BVDV from bovine fetal serum.[28] A method using beta-propiolactone treatment and heat inactivation to remove adventitious BVDV from fetal bovine serum has been reported by Brock and Potgieter.[29]

BVDV contamination of biologicals is recognized as a major concern. The major source of adventitious contamination of biologics by BVDV is fetal bovine serum, which is used universally in cell culture media.[28,30] At present, it is difficult to guarantee that products are free of BVDV, since it is difficult to detect BVDV in fetal bovine serum without exhaustive virus isolation procedures. Cloned BVDV sequences used as DNA hybridization probes could be an important method of detecting BVDV contamination in fetal bovine serum, because of its sensitivity and specificity. Hybridization probes also would enable detection of BVDV contamination in other viral stocks that have been contaminated during subculture or isolation. The extent of the presence of BVDV in these materials also may be underestimated, since it is difficult to detect BVDV contamination of viruses more aggressively cytopathogenic than BVDV.

IV. PHYSICOCHEMICAL PROPERTIES

Early studies identified three to four separate species of virus-specific polypeptides.[31,32] Frost identified three structural proteins in BVDV-infected cells by polyacrylamide gel electrophoresis.[31] Four BVDV proteins, with molecular weights of 110,000, 93,000, 70,000, and 23,000 Da (110, 93, 70, and 23 kDa), were described by Pritchett.[33] Matthaeus also reported finding three BVDV structural proteins by immunoprecipitation of radiolabeled polypeptides with molecular weights of 57, 44, and 34 kDa. He indicated that the two larger proteins were glycosylated (gp57 and gp44).[32] Coria in 1983 detected four major proteins of the cytopathic Singer strain of BVDV with molecular weights of approximately 75, 66, 54, and 26 kDa and determined that the gp75 and gp54 were glycosylated proteins.[34] Cell-free translation of BVDV RNA resulted in the *in vitro* synthesis of two species of proteins ranging from 50 to 150 kDa.[35] Protein synthesis with isolated polysomes from infected cell cultures resulted in polypeptides of 80 and 115 kDa.[35]

More recently, BVDV proteins have been further characterized. Three major proteins with molecular weights 115, 80, and 55 kDa and two minor proteins, 45 and 38 kDa, have been described by Purchio, 1984.[36] Wit5out using immunoprecipitation, Donis identified 12 viral polypeptides with molecular weights of 165, 135, 118, 80, 75, 62, 56 to 58, 48, 37, 35, 32, and 19 kDa in infected cell cultures.[37] Since the sum of the molecular weights of the polypeptides identified by Donis et al. was greater than the coding capacity of the BVDV genome, some of the polypeptides must have been precursor polypeptides.[37] The polypeptides were identified from infected cell cultures by radiolabeling in the presence of a hypertonic initiation block and SDS-polyacrylamide gel electrophoresis analysis.[37] Six specific glycoproteins were identified by Donis including two major species of 48 kDa and 56 to 58 kDa and less abundant species of 118, 75, 65, and 25 kDa.[38]

V. GENOME

BVDV RNA from infected cell culture supernate was characterized by Pritchett in 1975.[20] The major RNA component isolated had a sedimentation rate of 38S which represents a molecular weight of 3.22×10^6 Da. Two minor species of genomic RNA sedimented at 31S and 24S which corresponded to molecular weights of 2.09×10^6 and 1.22×10^6 Da, respectively.[20] Purchio later identified only one species of BVDV intracellular RNA in infected cell cultures.[21,35] The BVDV genome consisted of single-stranded RNA because it lacked resistance to high concentrations of RNase and because base composition analysis indicated it contained uridine.[21] Viral RNA also lacked a poly-A tail.[21]

The molecular characteristics, replication strategy, biotype differences, and the pathogenesis of BVDV are largely unknown.[39] This deficit in our knowledge concerning BVDV is due, in part, to difficulty in manipulation of BVDV in the laboratory and the lack of appropriate biologicals for the identification of BVDV. The organization of BVDV genomic RNA differed from other members of the Togaviridae family.[40] BVDV genomic RNA was not poly-adenylated as were most members of the Togaviridae family.[20] From preliminary unpublished sequence data, it was speculated that the genome contained not one but two open reading frames.[41] Subgenomic species of viral RNA have not been detected in cells infected with BVDV which differs from the life cycle of members of the Togaviridae family.[21] Also speculative arguments have been presented that the structural proteins probably were encoded from the 5′ end of the genome RNA whereas the 3′ end encoded the nonstructural proteins which was opposite to that of members of the Togaviridae family.[41]

It has been proposed that either the *Pestivirus* genus belongs in a separate family or that its members belong in the Flaviviridae family.[39,41,42] Flaviviruses lacked a 3' poly-A tail, had similar genome organization as BVDV (5′ structural and 3′ nonstructural), and produced viral proteins by post-translational cleavage of one polyprotein.[42] Evidence suggested that BVDV replication was similar to that of the Flaviviridae and Picornaviridae in that the genome was translated into a polyprotein with subsequent post-translational processing.[38,41] Because BVDV genomic RNA possibly contained two open reading frames, it may deserve separate family status.[39,41]

Renard was first to report the successful cloning of BVDV sequences in 1985.[43] BVDV genomic RNA was 12.5 kilobases (kb) long and was a single-stranded molecule.[43] BVDV genomic RNA was purified from bovine embryonic kidney cells infected with a high passage Osloss strain of cytopathic BVDV. Complementary DNA (cDNA) was synthesized after denaturation of BVDV RNA with methylmercury hydroxide and addition of a poly-A tail with poly-A polymerase. First- and second-strand synthesis was done according to the techniques of Gubler and Hoffman. A library of clones was obtained ranging in size from 0.5 to 1.5 kb representing five overlapping groups of clones.[43] Brock et al. also have reported on the molecular cloning of cDNA from a pneumotropic strain of BVDV, strain 72.[29]

VI. ANTIGENIC RELATIONSHIPS

BVDV strains were antigenically related but some cytopathic and noncytopathic strains were differentiated antigenically by serum virus neutralization[44,45] and indirect immunofluorescence.[46] Gillespie et al. however, reported that in a study of three strains, neutralizing antibody to heterologous as well as the homologous strain of BVDV virus was produced which was later confirmed by Castrucci et al.[47,48] Grouping according to serotype has not existed, but strains were distinguished by their ability to produce cytopathic effects in cell culture. Grouping of strains within noncytopathic or cytopathic biotypes has been the major means of biological differentiation.[32] Noncytopathic strains did not cause microscopically visible cytopathic changes in cell culture and were identified by indirect methods such as immunofluorescence or viral interference.[49,50] Analysis of glycoproteins induced by various isolates of BVDV suggested marked heterogeneity between virus strains.[51,52,53] Cytopathic BVDV strains synthesized an 80 kDa polypeptide that was absent in noncytopathic BVDV-infected cells.[51,54] It has been postulated that the 118 kDa polypeptide present in cells infected with both biotypes of BVDV may be a precursor protein of the 80 kDa polypeptide and the proteolytic cleavage site may not be present in the 118 kDa polypeptide of noncytopathic strains.[51] This molecular difference may be a factor determining the biological differences that exist between different biotypes of BVDV.[51] Dubovi et al. used monoclonal antibodies to distinguish between noncytopathic and cytopathic BVDV by their recognition of the presence or absence of epitopes between biotypes.[52] The observations of these authors confirmed an earlier report by Peters et al. who indicated that differences between cytopathic and noncytopathic strains were recognized by certain monoclonal antibodies.[46] Peters was able to identify epitopes present on polypeptides of cytopathic virus that were not present in cell cultures infected with noncytopathic virus.[46] This monoclonal antibody reacted with an epitope present on a 79 kDa polypeptide present in virions of cytopathic strains of BVDV.[46]

VII. PATHOGENESIS AND PATHOGENICITY

Bovine viral diarrhea virus (BVDV) has been recognized as a common pathogen and a major cause of disease in cattle.[8,55,56] However, the most common outcome of primary postnatal BVDV infection is subclinical disease.[8,57] BVDV is responsible for three separate disease syndromes in cattle: bovine viral diarrhea, mucosal disease, and fetal disease.[8,57] In 1946, Olafson et al. described a disease affecting cattle that was characterized by high morbidity and low mortality.[2] The disease was called bovine viral diarrhea and its most important features were fever, excess salivation, diarrhea, and erosions of the oral and gastrointestinal mucosa.[2]

Mucosal disease of cattle was described by Ramsey and Chivers in 1953.[58] It was a sporadic disease in herds with a high mortality rate.[58] Mucosal disease was difficult to reproduce experimentally, but it was obvious that BVDV was involved in the pathogenesis of the disease.[59] It was distinguished from bovine viral diarrhea by its more sporadic occurrence, lower morbidity, longer course (up to 90 d), higher mortality, the absence of detectable antibodies, and persistent viremia.[1,8] The pathogenesis of mucosal disease and BVD differed.[60–62] Mucosal disease was a late manifestation of *in utero* infection, since it occurred only in persistently infected, viremic, antibody-negative cattle (immunotolerant as a result of *in utero* infection) after superinfection with a different strain of virus.[60,63,64] Bovine viral diarrhea, on the other hand, resulted from a primary postnatal infection.[62,65]

Kahrs et al. first reported that BVDV readily crossed the placenta and resulted in fetal disease.[3] Fetal death usually occurred when the fetus was infected in the first trimester of gestation.[5] Fetal death as a result of BVDV infection in the second trimester was not uncommon either.[5] Fetal infections also may result in various congenital malformations such as ocular, musculo-skeletal, nervous system, and skin lesions.[3-5] *In utero* infection also results in the development of persistent, immunotolerant infections in the postnatal animal.[24] A major clinical consequence of BVDV infection is the enhancement of disease by various pathogens and facilitating infections by opportunistic organisms.[56,59] Enhancement of respiratory tract disease by BVDV infection was first recognized by Dinter et al. in 1961.[66,67] Clinical and experimental observations of synergism of BVDV with bacterial and viral pathogens has been reported on several occasions.[68-70] *In vitro* and *in vivo* experimental evidence have indicated that the synergism of BVDV infection with other pathogens is a consequence of the immunosuppression induced by BVDV infection in cattle.[71,72]

BVDV was first isolated and identified by Underdahl, et al. in 1957, but its relationship with mucosal disease was not established.[9] The etiological connection of BVDV was established by Gillepsie et al. when a cytopathic strain was isolated from a case of bovine viral diarrhea.[73] The virus was classified in the Togaviridae family as a member of the *Pestivirus* genus.[17] BVDV was closely related to Border disease virus of sheep but more distantly related to hog cholera virus.[11]

A. PRENATAL INFECTIONS

BVDV crossed the placenta efficiently (approaching 100% experimentally) and therefore resulted in fetal infections when susceptible dams were infected during gestation.[5,24,74,75] Disease in pregnant dams usually was subclinical, as were most postnatal infections, but fetal infection resulted in abortion, congenital malformations, weak, stunted, neonates, and persistently infected, apparently immunotolerant, calves.[4,24,73,75] The response of the fetus to infection primarily depended on the stage of gestation, but other determinants of disease may have been the biological variation of virus strain, immune status of the host, and breed differences.[24,63,72,74]

An increase in numbers of repeat breeders and infertility resulted from infection of cows in early gestation.[24] Infection of the conceptus occurred when BVDV-containing semen was used in artificial insemination.[76] Infection before 100 d of gestation often resulted in death of the fetus.[24] Hypomyelinogenesis, evident as tremors at birth, occurred also as a result of infection before 100 d of gestation.[8] Hypomyelinogenesis was a major characteristic of *in utero* infections in sheep with Border disease virus.[77–79]

Infection of the fetus between 100 and 150 d of gestation during the final stages of organogenesis led to destruction of differentiating cells.[8] Sequelae such as retardation of tissue growth and lowered birth weights, cerebellar hypoplasia and dysplasia, cerebral cavitation, and retinal hypoplasia occurred.[3–5]

Persistently infected immunotolerant calves were a result of infection with certain biotypes of the virus before complete development of immunological competence.[60,63] Most fetuses obtained immunocompetence to BVDV at approximately 126 d of gestation.[8] Therefore, infection must have occurred before or during the development of self-recognition.[8] Transplacental infection with a noncytopathic strain of BVDV between 60 to 120 d of gestation led to a persistent infection in fetuses and postnatal animals.[8,80,81] Infection of pregnant cows during this time with a cytopathic virus resulted in abortion, not persistent infections, even from modified-live cytopathic strains after vaccination.[80,82] "Immunotolerance", the phenomenon of failure of the immune response to recognize an antigen as non-self, apparently developed to the antigens of the noncytopathic strains.[8,24,83] Animals persistently infected with BVDV excreted large amounts of virus.[72] Approximately 1.7% of animals going to slaughter in Europe were persistently infected with BVDV.[8] Bolin et al. identified persistently infected animals in 9% of herds examined and 1.7% of the total number tested.[84] These animals were the primary source of virus within herds and were responsible for spreading the virus to other susceptible herd members.[84] Some persistently infected animals appeared normal but up to 50% died prematurely due to wasting and debilitation during the first year of life.[8] Lymphocytes and neutrophils were functionally impaired, since neutrophil phagocytosis was reduced and the blastogenic response of lymphocytes to mitogens was attenuated.[72,85] Serum from persistently infected animals and BVDV-infected cell culture fluids decreased the blastogenic response of lymphocytes to phytohemagglutinin and concanavalin A.[86,87] Immunosuppression in persistently infected animals differed from the transient immunosuppression associated with primary postnatal BVDV infections in cattle, since normal numbers of circulating lymphocytes were present and neutrophil dysfunction was not as severe in the former.[72,88,89] Lesions in persistently infected animals consisted of degeneration and satellitosis of neurons, glomerular basement membrane thickening, and antigen deposition in kidneys, neurons, lymph nodes, spleen, intestinal epithelial cells, and endothelial cells.[90,91] The presence of immune-complex deposition indicated that there may have been some degree of specific BVDV antibody production in persistently infected animals.[90,91] Persistently infected animals had a high probability of developing mucosal disease.[60]

Infection of the fetus after organogenesis and after the development of immunocompetence was usually of no consequence, because most fetuses developed an immune response *in utero* and subsequently eliminated the virus.[24]

B. POSTNATAL INFECTIONS

Several consequences of primary postnatal infections in susceptible cattle include: subclinical infections, bovine viral diarrhea, and transient immunosuppression.[8,57] Subclinical infections were the most common result of primary postnatal BVDV infections. BVDV infection in cattle was characterized by a transient but profound and multifactorial immunosuppression.[6–8] Decreased responsiveness of lymphocytes to mitogens, depressed levels of cellular interferon production, and sensitivity of exogenous interferon, decreased antibody production, depressed monocyte phagocytosis, impaired neutrophil iodination reaction, and reduced monocyte chemotaxis as a result of BVDV infection have been described.[86,88,92–95] In addition, BVDV-infected animals had reduced numbers of lymphocytes, macrophages, and neutrophils. Vaccination with some strains of modified-live BVDV vaccines were immunosuppressive.[96,97] Modified-live BVDV vaccination resulted in decreased lymphocyte bastogenesis, decreased neutrophil chemotaxis and phagocytosis, and decreased numbers of circulating lymphocytes and neutrophils.[97] This immunosuppression may have resulted in severe secondary infections after vaccination of stressed, feeder calves from virus-free herds.[98]

Recently, it has been recognized that BVDV infections were a pivotal event in the pathogenesis of respiratory tract disease in cattle.[68,99] Several observations have led to this conclusion.[6,66,100] BVDV was the most common virus isolated from cattle with shipping fever.[6] Infection with BVDV enhanced respiratory tract disease in cattle by causing an increased susceptibility to *Pasteurella haemolytica*,[68,101] infectious bovine rhinotracheitis virus,[7,100,102] and malignant catarrhal fever virus.[69] Edwards et al. reported enhanced shedding of bovine herpesvirus 1 (BHV-1) after virulent virus challenge inoculation of animals that had been vaccinated with BHV-1 while infected with BVDV.[7]

Enhanced disease developed in animals after concurrent infections of BVDV and infectious bovine rhinotracheitis virus,[7,103] papular stomatitis virus,[104] bovine leukemia virus,[105] or malignant catarrhal fever virus.[69] Wray and Roeder reported the exacerbation of *Salmonella* infections and persistence of fecal bacterial excretion in calves as a result of BVDV infection.[70] Bacteremia due to impaired clearance of endogenous bacteria was reported in calves experimentally inoculated with BVDV.[106]

BVDV infections have been common since 60 to 95% of animals in herds were seropositive.[8,107] Serologic data indicated that fewer than 10% of herds consisted entirely of animals with no previous exposure to BVDV.[108] BVDV infection was characterized by transient fever, leucopenia, and viremia (which persisted for up to 15 d).[8] Bovine virus diarrhea occurred in some susceptible herds as an explosive outbreak of diarrhea with high morbidity and no mortality.[1,8] Affected animals had mild to severe diarrhea, erosions of the oral mucosa, and a drop in milk production. Animals developed lifelong immunity to BVDV after infection.[8]

C. MUCOSAL DISEASE

Mucosal disease was characterized by fever, anorexia, dehydration, profuse watery diarrhea, serous catarrhal exudate, and oral and nasal mucosal ulceration.[58] Erosions of the coronary band and the interdigital cleft often occurred and resulted in lameness.[8,57,62,65] Death sometimes occurred within 2 to 3 d but more commonly the course was 2 to 3 weeks.[8,62] A chronic form occasionally developed that lasted several months before death.[76] Pathological lesions in cattle with mucosal disease were extensive mucosal erosions and submucosal hemorrhages throughout the digestive tract, inflammation and necrosis of the Peyer's patches, dehydration, and lymphoid depletion from lymphoid organs.[1,2,55,62]

Mucosal disease occurred in persistently infected, antibody-negative cattle.[59,62] Evidence indicated that super-infection with certain cytopathic strains of BVDV in cattle persistently infected with certain noncytopathic strains will cause mucosal disease.[62,63,09] It may have been a function of a complex relationship of antigenic determinants between different BVDV strains.[71] Persistently infected animals do respond with neutralizing antibody production to heterologous strains, but not to the homologous strain.[71] A certain combination of two strains of BVDV were required to precipitate mucosal disease, because not all combinations of noncytopathic and cytopathic BVDV caused mucosal disease.[63,71] In one study, the development of neutralizing antibody in calves from vaccination resulted in a delayed onset of mucosal disease after challenge inoculation.[71] Tolerance to BVDV in persistently infected cattle may have been terminated by an immunological response following exposure to certain strains of the virus.[71] Certain combinations of virus infections result in a more chronic disease than others.[17] The tolerance present in persistently infected cattle apparently was not complete since there were microscopic lesions suggestive of immune-mediated pathology (type III hypersensitivity reaction) involving antigen-antibody complexes in the kidney glomeruli.[71,90,91]

Vaccination of persistently infected cattle with modified-live and killed BVDV also has been associated with the subsequent development of mucosal disease.[71,96,110] Most modified-live BVDV vaccines have been cytopathic and have been incriminated in the development of mucosal disease in calves persistently infected with a noncytopathic strain.[71,98] Vaccination of pregnant cows with modified-live BVDV vaccines led to birth of persistently infected, immunotolerant calves that were chronic virus shedders.[98] However, a temperature-sensitive mutant of BVDV, used as a modified-live vaccine, did not result in detectable postvaccinal sequelae in pregnant cows.[111] Chronic mucosal disease in a persistently infected bull was associated, in a retrospective study, with inoculation of killed-BVDV vaccine.[72]

The exact mechanism of the pathogenesis of mucosal disease is not understood at present.[112] Some investigators feel that mucosal disease may result from conversion of noncytopathic BVDV in the persistently infected animal to a cytopathic biotype, without requiring superinfection with a homologous cytopathic strain.[54,104] *In vitro* observations have suggested that interferons and tumor necrosis factor may have a cytopathic-inducing effect on noncytopathic BVDV in cell culture.[113]

VIII. DIAGNOSIS

BVDV infection usually has been suspected when disease such as mucosal disease, abortions, or congenital malformations were present in animals of a herd.[1] Birth of unthrifty animals, excessive neonatal deaths, and episodes of enteric or respiratory tract disease (often after the introduction of new animals into the herd) have also indicated the introduction of BVDV into a herd.[62,82] Laboratory diagnosis was essential when classical signs of disease were not present.[1] Seroconversion measured in paired serum samples by serum neutralization,[47,48,114] immunofluorescence,[49,115] immunoperoxidase test,[116] and ELISA indicated active BVDV infections.[117–120] Serum samples also have been suitable for isolation of virus from persistently infected animals.[120,121] The identification of viral antigen or antibody in fetal tissues or fluids from an aborted fetus was evidence of fetal infection, but did not indicate BVDV fetopathy since some transplacental BVDV infections, especially in late gestation, were harmless. Abortion occurred weeks after infection of the dam at which time the latter likely had developed an antibody response.[8] Therefore an increase of antibody titer in paired sera taken after abortion was not likely to occur. Virus isolation, serum virus neutralization for specific antibodies, and the detection of viral antigen (ELISA and immunofluorescence) have been the primary methods of determining BVDV infections.[110,120] The separation of noncytopathic and cytopathic isolates in mixed infections required special methods, such as reverse plaque titration as reported by Itoh et al.[45,122] Terminal dilutions of noncytopathic BVDV grew to higher titer than cytopathic BVDV by plaque purification.[123] Itoh et al. reported differentiation of a noncytopathic virus present in cytopathic stocks of BVDV using the exaltation of Newcastle disease virus assay.[50]

IX. EPIZOOTIOLOGY

Because classical clinical signs of BVDV have been the exception rather than the rule in BVDV-infected cattle, most infections have been surreptitious, which made it difficult to identify and attribute a herd problem to BVDV.[1] Prevention of BVDV fetal infections, especially in early gestation, and the identification of persistently infected animals have been identified as the major focus in prevention and control of BVDV infection.[124]

Introduction of a persistently infected animal in a susceptible herd has resulted in the occurrence of several types of clinical disease, such as enteritis, respiratory disease, decreased weight gain, reduced milk production, early embryo loss, abortions, congenital malformations, and neonatal deaths.[1,82] A variable number of progeny from dams which were infected during pregnancy would be persistently infected and would remain a source of the virus in the herd.[1,82] Persistently infected heifers and cows always produced persistently infected offspring as a result of transplacental transmission.[125] Virus spread among cattle by inhalation or ingestion of materials contaminated by oculonasal discharge, saliva, urine, and feces.[57,110] Semen, uterine secretions, amniotic fluid, and placental tissues containing virus also have been responsible for transmission.[8,126] Exposure of preimplantation embryos *in vitro* did not cause adverse effects on *in vitro* development of embryos, but the *in vivo* effects of BVDV were not examined.[127] Infection also has occurred by transmission from persistently infected reservoir animals, such as sheep or wild ruminants.[110,128] However, the frequency of this occurrence is unknown.[125] BVDV has infected sheep, goats, swine, and wild ruminants.[107,110,129] Both BVDV and Border disease virus strains have infected cattle and sheep experimentally.[110, 128]

X. IMMUNITY

Persistently infected animals are capable of developing a neutralizing antibody response to heterologous BVDV strains of cytopathic and noncytopathic biotypes. In most cattle (80%) with mucosal disease, there was no detectable antibody response to BVDV.[54] Most animals infected postnatally respond to BVDV infection by producting neutralizing antibody that give lifelong protection against the homologous virus.[44-46,50] Neutralizing antibody provides protection against infection with homologous virus and some, but not all, heterologous strains of BVDV.[50,53,54] Virus can be isolated from lymphocytes following the development of neutralizing antibody.[59] Therefore, it is unknown if virus is completely cleared from an animal infected postnatally.

XI. PREVENTION AND CONTROL

Most modified-live vaccines contain cytopathic biotypes of BVDV, strain NADL, Oregon C24V, or Singer strain.[98] Little substantive information has been reported on the antigenic differences among of strains of BVDV vaccines.[98] It has been recommended that only inactivated vaccines be used in closed herds where BVDV has not been diagnosed.[98,130] BVDV vaccination prior to conception has been effective in protecting fetuses from *in utero* BVDV infection.[110]

There are many facets of the pathogenesis of BVDV infection that still are not clear. The mechanism by which persistence is established as a result of *in utero* infection and that by which postnatal reinfection with certain BVDV strains precipitates mucosal disease should be elucidated for the rational control of virus spread and to prevent BVDV-induced disease. It is likely that specific genomic differences amoung BVDV strains exist that correlate with the different consequences of BVDV infection and other differences in the natural history of these viruses. Strain-specific hybridization probes could be useful in tracking the molecular epidemiology and natural history of BVDV. Once identified, these genomic differences could be studied further by gene sequencing and identifying and characterizing gene products from these sequences.

BVDV is considered to be a major contributor to economic loss in the cattle industry as a result of disease and loss of production, primarily due to *in utero* infection and its capacity to induce immune dysfunction. Persistently infected animals are the major source of virus in the environment, but many of these animals appear healthy and are difficult to identify.[131,132] Therefore, the detection of the immunotolerant, persistently infected animals is important. The most common method of herd screening for BVDV infection is by seroconversion, but seronegative, persistently infected animals cannot be identified in this manner. Serum virus neutralization is commonly used for identification and assaying BVDV antibodies in bovine sera. It is expensive, relatively insensitive, and laborious. Neutralizing antibody titers to homologous BVDV strains are often much higher than to heterologous strains, which complicates diagnostic serology based on serum virus neutralization.[49] Many of these disadvantages also exist when virus isolation is attempted, which is commonly used to identify infected animals.[57] The presence of virus in persistently infected calves may be masked by the presence of passive antibody.[57]

It is important, therefore, to use rapid, sensitive, and specific methods of identifying the presence of BVDV infection in herds. Dot–blot hybridization for BVDV in serum samples spotted on nitrocellulose membranes could allow simultaneous screening of many animals and may facilitate the detection of persistently infected animals in a herd.

REFERENCES

1. **Perdrizet, J. A., Rebhun, W. C., Dubovi, E. J., and Donis, R. O.,** Bovine virus diarrhea-clinical syndromes in dairy herds, *Cornell Vet.*, 77, 46, 1987.
2. **Olafson, P., MacCallum, A. C., and Fox, F. H.,** An apparently new transmissible disease of cattle, *Cornell Vet.*, 36, 205, 1946.
3. **Kahrs, R. F.,** The relationship of bovine viral diarrhea-mucosal disease to abortion in cattle, *J. Am. Vet. Med. Assoc.*, 153, 1652, 1968.
4. **Kahrs, R. F., Scott, F. W., and de Lahunta, A.,** Bovine viral diarrhea-mucosal disease, abortion, and congenital cerebellar hypoplasia in a dairy herd, *J. Am. Vet. Med. Assoc.*, 156, 851, 1970.
5. **Kahrs, R. F.,** Effects of bovine viral diarrhea on the developing fetus, *J. Am. Vet. Med. Assoc.*, 163, 877, 1973.
6. **Reggiardo, C.,** Role of BVD virus in shipping fever of feedlot cattle: case studies and diagnostic considerations, *Proc. Am. Assoc. Vet. Lab. Diagn.*, 22, 315, 1979.
7. **Duffell, S. J. and Harkness, J. W.,** Bovine virus diarrhoea-mucosal disease infection in cattle, *Vet. Rec.*, 117, 240–245, 1985.
8. **Edwards, S. J., Wood, L., Hewitt-Taylor, C., and Drew, T. W.,** Evidence for an immunocompromising effect of bovine pestivirus on bovid herpesvirus 1 vaccination, *Vet. Res. Comm.*, 10, 2, 1986.
9. **Underdahl, N. R., Grace, O. D., and Hoerlein, A. B.,** Cultivation in tissue-culture of cytopathic agent from bovine mucosal disease, *Proc. Soc. Exp. Biol. Med.*, 94, 795, 1957.
10. **Hermodsson, S. and Dinter, Z.,** Properties of bovine virus diarrhoea virus, *Nature (London)*, 194, 893, 1962.
11. **Darbyshire, J. H.,** Agar gel diffusion studies with a mucosal disease of cattle. II. A serological relationship between a mucosal disease and swine fever, *Res. Vet. Sci.*, 3, 125, 1962.
12. **Dinter, Z.,** Relationship between bovine virus diarrhoea virus and hog cholera virus. *Zentralbl. Bakteriol. Parasitkde. I*,188, 475, 1962.
13. **Mengeling, W. L., Gutekunst, D. E., Fernelius, A. L., and Pirtle, E. C.,** Demonstration of an antigenic relationship between hog cholera and bovine viral diarrhea viruses by immunofluorescence, *Can. J. Comp. Med.*, 27, 162, 1963.

14. **Fernelius, A. L.,** Characterization of bovine viral diarrhea viruses, *Arch. Virol.*, 25, 211, 1968.
15. **Hafez, S. M., Petzoldt, K., and Reczko, E.,** Morphology of bovine viral diarrhoea virus, *Acta. Virol.*, 12, 471, 1968.
16. **Femner, F., Bauchmann, P. A., Gibbs, E. P. J., Murphy, F. A., Studdert, M. J., and White, D. O.,** *Veterinary Virology*, Academic Press, Orlando, FL, 1987.
17. **Horzinek, M. C., Maess, J., and Laues, R.,** Studies on the substructure of togaviruses, *Arch. Virol.*, 33, 306, 1971.
18. **Parks, J. B., Pritchett, R. F., and Zee, Y. C.,** Buoyant density of bovine viral diarrhea virus (36511), *Proc. Soc. Exp. Biol. Med.*, 140, 595, 12.
19. **Chu, H.-J. and Zee, Y. C.,** Morphology of bovine viral diarrhea virus, *Am. J. Vet. Res.*, 45, 845, 1984.
20. **Pritchett, R., Manning, J. S., and Zee, Y. C.,** Characterization of bovine viral diarrhea virus RNA, *J. Virol.*, 15, 1342, 1975.
21. **Purchio, A. F., Larson, R., and Collett, M. S.,** Characterization of virus-specific RNA synthesized in bovine cells infected with bovine viral diarrhea virus, *J. Virol.*, 48, 320, 1983.
22. **Laude, H.,** Nonarbo-Togaviridae comparative hydrodynamic properties of the pestivirus genus, *Arch. Virol.*, 62, 347, 1979.
23. **Westaway, E. G., Brinton, M. A., Gaidamovich, S. Y., Horzinek, M. C., Igarashi, A., Kaariainen, L., Lvov, D. K., Porterfield, J. S., Russel, P. K., and Trent, D. W.,** Togaviridae, *Intervirology*, 24, 125, 1985.
24. **Done, J. T., Sterlecki, S., Richardson, C., Harkness, J. S., Sands, J. J., Patterson, D. S. P., Sweasey, D., Shaw, I. G., Winkler, C. E., and Duffell, S. J.,** Bovine virus diarrhoea-mucosal disease virus: pathogenicity for the fetal calf following maternal infection, *Vet. Rec.*, 106, 473, 1980.
25. **Johnson, J.,** *In vitro* Characteristics of Bovine Viral Diarrhea Virus Infection in Two Bovine Cell Types: Effect of Host Cell, M. S. thesis, Iowa State University, Ames, 1988.
26. **Tessler, J., Stewart, W. C., Kresse, J. I., and Snyder, M. L.,** Tween 80: a marker for differentiation of hog cholera and bovine viral diarrhea viruses, *Can. J. Comp. Med.*, 41, 127, 1977.
27. **Toth, T. E. and Hesse, R. A.,** Elimination of contaminating bovine viral diarrhea virus from bovine respiratory respiratory syncytial virus stock, *J. Virol. Methods*, 6, 241, 1983.
28. **Rossi, C. R., Brigdman, C. R., and Kiesel, G. K.,** Viral contamination of bovine fetal lung cultures and bovine fetal serum, *Am. J. Vet. Res.*, 41, 1680, 1980.
29. **Brock, K. V., Brian, D. A., Rouse, B. T., and Potgieter, L. N. D.,** Molecular cloning of a pneumopathic strain of bovine viral diarrhea virus and its diagnostic application, *Can. J. Vet. Res.*, 52, 451, 1988.
30. **Hassan, A. K. M. and Scott, G. R.,** A technique to obviate the risk of inadvertent infection of cell cultures with bovine viral diarrhoea virus, *J. Comp. Pathol.*, 96, 241, 1986.
31. **Frost, J. W. and Liess, B.,** Separation of structural proteins of bovine viral diarrhea virus after degradation by different splitting techniques, *Arch. Virol.*, 42, 2, 1973.
32. **Matthaeus, W.,** Detection of three polypeptides in preparation of bovine viral diarrhoea virus, *Arch. Virol.*, 59, 299, 1979.
33. **Pritchett, R. F. and Zee, Y. C.,** Structural proteins of bovine viral diarrhoea virus, *Am. J. Vet. Res.*, 36, 1731, 1975.
34. **Coria, M. F., Schmerr, M. J. F., and McClurkin, A. W.,** Characterization of the major structural proteins of purified bovine viral diarrhea virus, *Arch. Virol.*, 76, 335, 1983.
35. **Purchio, A. F., Larson, R., Torborg, L. L., and Collett, M. S.,** Cell-free translation of bovine viral diarrhea virus RNA, *J. Virol.*, 52, 3, 1984.
36. **Purchio, A. F., Larson, R., and Collett, M. S.,** Characterization of bovine viral diarrhea virus proteins, *J. Virol.*, 50, 666, 1984.
37. **Donis, R. O. and Dubovi, E. J.,** Characterization of bovine viral diarrhoea-mucosal disease virus-specific proteins in bovine cells, *J. Gen. Virol.*, 68, 1597, 1987.
38. **Donis, R. O and Dubovi, E. J.,** Glycoproteins of bovine viral diarrhoea-mucosal disease virus in infected bovine cells, *J. Gen. Virol.*, 68, 1607, 1987.
39. **Horzinek, M. C. and Van Berlo, M. F.,** The pestiviruses: where do they belong?, *Ann. Rech. Vet.*, 18, 115, 1987.
40. **Horzinek, M. C.,** *Non-Arthropod-Borne Togaviruses*, Academic Press, London, 1981.
41. **Renard, A., Schmetz, D., Guiot, C., Brown-Shimmer, S., Dagenais, L., Pastoret, P. P., Dina, D., and Martial, J. A.,** Molecular cloning of the bovine viral diarrhea virus genomic RNA, *Ann. Rech. Vet.*, 18, 121, 1987.
42. **Rice, C. M., Lenches, E. M., Eddy, S. R., Shin, S. J., Sheets, R. L., and Strauss, J. H.,** Nucleotide sequence of yellow fever virus: implications for flavivirus gene expression and evolution, *Science*, 229, 726, 1985.
43. **Renard, A., Guiot, C., Schmetz, D., Dagenais, L., Pastoret, P. P., Dina, D., and Martial, J. A.,** Molecular cloning of bovine viral diarrhea viral sequences, *DNA*, 4, 429, 1985.
44. **Fernelius, A. L., Lambert, G., and Booth, G. D.,** Bovine viral diarrhea virus-host cell interactions: serotypes and their relationship to biotypes by cross neutralization, *Am. J. Vet. Res.*, 32, 229, 1971.

45. **Coria, M. F., Schmerr, M. J. F., McClurkin, A. W., and Bolin, S. R.,** Differentiation of cytopathic and noncytopathic isolates of bovine viral diarrhea virus by virus neutralization, *Am. J. Vet. Res.,* 45, 2129, 1984.
46. **Peters, W., Greiser-Wilke, I., Moennig, V., and Liess, B.,** Preliminary serological characterization of bovine viral diarrhoea virus strains using monoclonal antibodies, *Vet. Microbiol.,* 12, 195, 1986.
47. **Castrucci, G., Avellini, G., Cilli, V., Pedini, B., McKercher, D. G., and Valente, C.,** A study of immunologic relationships among serologically heterologous strains of bovine viral diarrhea virus by cross immunity tests, *Cornell Vet.,* 65, 65, 1975.
48. **Gillespie, J. H., Coggins, L., Thomson, J., and Baker, J. A.,** Comparison by neutralization tests of strains of virus isolated from virus diarrhea and mucosal disease, *Cornell Vet.,* 51, 155, 1961.
49. **Fernelius, A. L.,** Noncytopathogenic bovine viral diarrhea viruses detected and titrated by immunofluorescence, *Can. J. Comp. Med.,* 28, 121, 1964.
50. **Itoh, O., Sasaki, H., and Hanaki, T.,** A study of serologic relationships among non–cytopathic strains of bovine viral diarrhea-mucosal disease virus by reverse plaque technique, *Jpn. J. Vet. Sci.,* 46, 669, 1984.
51. **Donis, R. O., and Dubovi, E. J.,** Differences in virus-induced polypeptides in cells infected by cytopathic and noncytopathic biotypes of bovine virus diarrhea-mucosal disease virus, *Virology,* 158, 168, 1987.
52. **Dubovi, E. J., Corapi, W., and Donis, R. O.,** Probing the antigenic diversity of cytopathic and noncytopathic BVD virus with monoclonal antibodies, *Am. Assoc. Vet. Lab. Diagnosticians 29th Annu. Proc.,* Louisville, KY, October 19 to 21, 1986, 259.
53. **Popcock, D. H., Howard, C. J., Clarke, M. C., and Browelie, J.,** Variation in the intracellular polypeptides profiles from different isolates of bovine virus diarrhoea virus, *Arch. Virol.,* 94, 43, 1987.
54. **Donis, R. O. and Dubovi, E. J.,** Molecular specificity of the antibody responses of cattle naturally and experimentally infected with cytopathic and noncytopathic bovine viral diarrhea virus biotypes, *Am. J. Vet. Res.* 48, 1549, 1987.
55. **Armstrong, D. A., McClurkin, A. W., Potgieter, L. N. D., and Reggiardo, C.,** Focus: bovine virus diarrhea, Part 1, Clinical signs and diagnosis, *Bov. Vet. Forum,* 1, 1985.
56. **Potgieter, L. N. D.,** Current concepts on the role of viruses in respiratory tract disease of cattle, *Bov. Pract.,* 12, 75, 1977.
57. **Baker, J. C.,** Bovine viral diarrhea virus: a review, *J. Am. Vet. Med. Assoc.,* 190, 1449, 1987.
58. **Ramsey, F. K. and Chivers, W. H.,** Mucosal disease of cattle, *North Am. Vet.,* 34, 629, 1953.
59. **Malmquist, W. A.,** Bovine viral diarrhea-mucosal disease: etiology, pathogenesis, and applied immunity, *J. Am. Med. Assoc.,* 152, 763, 1986.
60. **Brownlie, J., Clarke, M. C., and Howard, C. J.,** Experimental production of fatal mucosal disease in cattle, *Vet. Rec.,* 114, 535, 1984.
61. **Brownlie J., Clarke, M. C., Howard, C. J., and Pocock, D. H.,** Pathogenesis and epidemiology of bovine virus diarrhoea virus infection of cattle, *Ann. Rech. Vet.,* 18, 157, 1986.
62. **Nagele, M. J.,** Outbreak of mucosal disease among apparently immunotolerant heifers, *Vet. Rec.,* 115, 496, 1984.
63. **Bolin, S. R., McClurkin, A. W., Cutlip, R. C., and Coria, M. F.,** Severe clinical disease induced in cattle persistently infected with noncytopathic bovine viral diarrhea virus by superinfection with cytopathic bovine viral diarrhea virus, *Am. J. Vet. Res.,* 46, 573, 1985.
64. **Brownlie, J., Clarke, M. C., and Howard, C. J.,** Aetiology and pathogenesis of mucosal disease: current concepts, observations and speculation, *Aust. Vet. J.,* 62, 142, 1985.
65. **Harkness, J. W.,** Aetiology and pathogenesis of mucosal disease, *Aust. Vet. J.,* 62, 395, 1985.
66. **Dinter, Z. and Bakos, K.,** Viruses associated with acute respiratory and enteric diseases in cattle. *Bull. Off. Int. Epizool.,* 56, 29, 1961.
67. **Dinter, Z., Hansen, H. J., and Roneus, O.,** Untersuchungen uber mucosal disease. I. Isolieriung und identifizierung des virus der virusdiarrhoe. *Zentralbl. Vet. Med.,* 9, 739, 1962.
68. **Potgieter, L. N. D., McCracken, M. D., Hopkins, F. M., and Guy, J. S.,** Comparison of the pneumopathogenicity of two strains of bovine viral diarrhea virus, *Am. J. Vet. Res.,* 46, 151, 1985.
69. **Sharpe, R. T., Bicknell, S. R., and Hunter, A. R.,** Concurrent malignant catarrhal fever and bovine virus diarrhoea virus infection in a dairy herd, *Vet. Rec.,* 120, 545, 1987.
70. **Wray, C. and Roeder, P. L.,** Effect of bovine virus diarrhoea-mucosal disease virus infection on salmonella infection in calves, *Res. Vet. Sci.,* 42, 213, 1987.
71. **Bolin, S. R., McClurkin, A. W., Cutlip, R. C., and Coria, M. F.,** Response of cattle persistently infected with noncytopathic bovine viral diarrhea virus to vaccination for bovine viral diarrhea and to subsequent challenge exposure with cytopathic bovine viral diarrhea virus, *Am. J. Vet. Res.,* 46, 2467, 1985.
72. **Bolin, S. R., Roth, J. A., Uhlenhopp, E. K., and Pohlenz, J. F.,** Immunologic and virologic findings in a bull chronically infected with noncytopathic bovine viral diarrhea virus, *J. Am. Vet. Med. Assoc.,* 190, 1015, 1987.
73. **Gillespie, J. H., Baker, J. A., and McEntee, K.,** A cytopathic strain of diarrhea virus, *Cornell Vet.,* 50, 73, 1960.
74. **Orban, S., Leiss, B., Hafez, S. M., Frey, H. F., Blindow, H., and Passe-Patzer, B.,** Studies on transplacental transmissibility of a bovine virus diarrhoea (BVD) vaccine virus. I. Inoculation of pregnant cows 15 to 90 days before parturition (190th to 265th day of gestation), *Zentralbl. Vet. Med. B,* 30, 619, 1983.

75. **Sweat, R. L.,** Comments on bovine viral diarrhea-mucosal disease, *J. Am. Vet. Med. Assoc.*, 152, 867, 1968.
76. **Barlow, R. M., Nettleton, P. F., Gardiner, A. C., Greig, A., Campbell. J. R., and Bonn, J. M.,** Persistent bovine virus diarrhoea virus infection in a bull, *Vet. Rec.*, 118, 321, 1986.
77. **Done, J. T., Wooley, J., Barnard, V. V., Upcott, D. H., Herbert, C. N., and Terlecki, S.,** Border disease of sheep: spinal cord morphometry, *J. Comp. Path.*, 95, 325, 1985.
78. **Lim, C. F. and Carnegie, P. R.,** A survey of hairy shaker disease (border disease, hypomyelinogenesis congenita) in sheep, *Aust. Vet. J.*, 61, 174, 1984.
79. **Plant, J. W., Walker, K. H., Acland, H. M., and Gard, G. P.,** Pathology in the ovine foetus caused by an ovine pestivirus, *Aust. Vet. J.*, 60, 137, 1983.
80. **Liess, B. S., Orban, S., Frey, H. R., Trautwein, G., Weifel, W., and Blindow, H.,** Studies on transplacental transmissibility of a bovine virus diarrhoea (BVD) vaccine virus in cattle. II. Inoculation of pregnant cows without detectable neutralizing antibodies to BVD virus 90 to 229 days before parturition (51st to 190th day of gestation), *Zentralbl. Vet. Med. B*, 30, 669, 1984.
81. **McClurkin, A. W., Littledyke, E. T., Cutlip, R. C., Frank, G. H., Coria, M. F., and Bolin, S. R.,** Production of cattle immunotolerant to bovine virual diarrhea virus, *Can. J. Comp. Med.*, 48, 156, 1984.
82. **Cranwell, M. P. and Drew, T. W.,** Mucosal disease in cattle, *Vet. Rec.*, 229, 1984.
83. **Coria, M. F. and McClurkin, A. W.,** Specific immune tolerance in an apparently healthy bull persistently infected with bovine viral diarrhea virus, *J. Am. Vet. Med. Assoc.*, 172, 449, 1978.
84. **Bolin, S. R. and McClurkin, A. W.,** Frequency of persistent bovine viral diarrhea virus infection in selected cattle herds, *Am. J. Vet. Res.*, 46, 2385, 1985.
85. **Roth, J. A., Bolin, S. R., and Frank, D. E.,** Lymphocyte blastogenesis and neutrophil function in cattle persistently infected with bovine viral diarrhea virus, *Am. J. Vet. Res.*, 47, 1139, 1986.
86. **Markham, R. J. F. and Ramnaraine, M. L.,** Release of immunosuppressive substances from tissue culture cells infected with bovine viral diarrhea virus, *Am. J. Vet. Res.*, 46, 879, 1985.
87. **Steck, F., Lazary, S., and Fey, H.,** Immune responsiveness in cattle fatally affected by bovine virus diarrhea-mucosal disease, *Zentralbl. Veterinarmed, B*, 27, 429, 1980.
88. **Bolin, S. R., McClurkin, A. W., and Coria, M. F.,** Effects of bovine viral diarrhea virus on the percentages and absolute numbers of circulating B and T lymphocytes in cattle, *Am. J. Vet. Res.*, 46, 884, 1985.
89. **Muscoplat, C. C., Johnson, D. W., and Teuscher, E.,** Surface immunoglobulin of circulating lymphocytes in chronic bovine diarrhea: abnormalities in cell populations and cell function, *Am. J. Vet. Res.*, 33, 1101, 1973.
90. **Cutlip, R. C., McClurkin, A. W., and Coria, M. F.,** Lesions in clinically healthy cattle persistently infected with the virus of bovine viral diarrhea–glomerulonephritis, *Am. J. Vet. Res.*, 41, 1938, 1980.
91. **Hewicker, M., Trautwein, G., Stahl, C., and Liess, B.,** Kidney lesions in cattle persistently infected with bovine viral diarrhoea virus, *J. Vet. Med.*, 34, 1, 1987.
92. **Atluru, D., Notowidjojo, W., Johnson, D. W., and Muscoplat, C. C.,** Suppression of in vitro immunoglobulin biosynthesis in bovine spleen cells by bovine viral diarrhea virus, *Clin. Immunol. Immunopathol.* 13, 254, 1979.
93. **Johnson, D. W. and Muscoplat, C. C.,** Immunologic abnormalities in calves with chronic bovine viral diarrhea, *Am. J. Vet. Res.*, 34, 1139, 1973.
94. **Ketelsen, A. T., Johnson, D. W., and Muscoplat, C. C.,** Depression of bovine monocyte chemotaxis by bovine viral diarrhea virus, *Infect. Immun.*, 25, 565, 1979.
95. **Muscoplat, C. C., Johnson, D. W., and Stevens, J. B.,** Abnormalities of in vitro lymphocytic responses during bovine viral diarrhea infection, *Am. J. Vet. Res.*, 34, 753, 1973.
96. **Peter, C. P., Tyler, D. E., and Ramsey, F. K.,** Characteristics of a condition following vaccination with bovine virus diarrhea vaccine, *J. Am. Vet. Med. Assoc.*, 150, 46, 1967.
97. **Roth, J. A. and Kaeberle, M. L.,** Suppression of neutrophil function induced by a vaccinal strain of bovine viral diarrhea virus with and without the administration of ACTH, *Am. J. Vet. Res.*, 44, 2366, 1983.
98. **Neaton, M. J.,** Which BVD vaccine should I use?, *Vet. Med.*, 81, 876, 1986.
99. **Roth, J. A., Corstvet, R. E., and Fulton, R. W.,** Bovine respiratory disease, *Proc. of a Seminar, Western Vet. Conf.*, 1987, 1.
100. **Potgieter, L. N. D., McCracken, M. D., Hopkins, F. M., Walker, R. D., and Guy, J. S.,** Experimental production of bovine respiratory tract disease with bovine viral diarrhea virus, *Am. J. Vet. Res.*, 45, 1582, 1984.
101. **Turk, J. R., Corstvet, R. E., McClure, J. R., Gossett, K. A., Enright, F. M., and Pace, L. W.,** Synergism of bovine virus diarrhea virus and *Pasteurella haemolytica* serotype 1 in bovine respiratory disease complex. I. Leukocyte alterations and pulmonary lesion volumes, *Am. Assoc. Vet. Lab. Diagn.*, 28, 67, 1985.
102. **Potgieter, L. N. D., McCracken, M. D., Hopkins, F. M., and Walker, R. D.,** Effect of bovine viral diarrhea virus infection on the distribution of infectious bovine rhinotracheitis virus in calves, *Am. J. Vet. Res.*, 45, 687, 1984.
103. **Greig, A., Gibson, I. R., Nettleton, P. F., and Herring, J. A.,** Disease outbreak in calves caused by a mixed infection with infectious bovine rhinotracheitis virus and bovine virus diarrhoea virus, *Vet. Rec.*, 108, 480, 1981.
104. **Bohac, J. G. and Yates, W. D. G.,** Concurrent bovine virus diarrhea and bovine papular stomatitis infection in a calf, *Can. Vet. J.*, 21, 310, 1980.

105. **Roberts, D. H., Lucas, M. H., Wibberley, G., and Westcott, D.,** Response of cattle persistently infected with bovine virus diarrhoea virus to bovine leukosis virus, *Vet. Rec.*, 122, 293, 1988.
106. **Reggiardo, C. and Kaeberle, M. L.,** Detection of bacteremia in cattle inoculated with bovine viral diarrhea virus, *Am. J. Vet. Res.*, 42, 218, 1981.
107. **Elazhary, M. A. S. Y., Silim, A., and Dea, S.,** Prevalence of antibodies to bovine respiratory syncyticial virus, bovine viral diarrhea virus, bovine herpesvirus-1, and bovine parainfluenza-3 virus in sheep and goats in Quebec. *Am. J. Vet. Res.*, 45, 1660, 1984.
108. **Edwards, S.,** Prevalence of bovine virus diarrhoea virus viraemia, *Vet. Rec.*, 120, 71, 1987.
109. **McClurkin, A. W., Bolin, S. R., and Coria, M. F.,** Isolation of cytopathic and noncytopathic bovine viral diarrhea virus from the spleen of cattle acutely and chronically affected with bovine viral diarrhea, *J. Am. Vet. Med. Assoc.*, 186. 568, 1985.
110. **Duffel, S. J., Sharp, M. W., Winkler, C. E., Terlecki, S., Richardson, C., Done, J. T., Roeder, P. L., and Herbert, C. N.,** Bovine virus diarrhoea-mucosal disease virus-induced fetopathy in cattle: efficacy of prophylactic maternal pre-exposure, *Vet. Rec.*, 114, 558, 1984.
111. **Lobmann, M., Charlier, P., Florent, G., and Zygraich, N.,** Clinical evaluation of a temperature-sensitive bovine viral diarrhea vaccine strain, *Am. J. Vet. Res.*, 45, 2498, 1984.
112. **Ohmann, H. B. and Babiuk, L. A.,** Viral infections in domestic animals as models for studies for viral immunology and pathogenesis, *J. Gen. Virol.*, 66, 1, 1986.
113. **Ohmann, H. B. and Babiuk, L. A.,** Influence of interferons α_1 and of tumour necrosis factor on persistent infection with bovine viral diarrhoea virus in vitro, *J. Gen. Virol.*, 69, 1399, 1988.
114. **Chen, K. S. and Johnson, D. W.,** Neutralization kinetics of bovine viral diarrhea virus by hyperimmune seum: one or multi-hit mechanism, *Comp. Immunol. Microbiol. Infect. Dis.*, 9, 37, 1986.
115. **Smithies, L. K. and Robertson, S. M.,** The laboratory diagnosis of bovine virus diarrhea by fluorescent antibody, *Proc. Annu. Meet. U.S. Anim. Health Assoc.*, 1969, 539.
116. **Ward, A. C. S. and Kaeberle, M. L.,** Use of an immunoperoxidase stain for the detection of bovine viral diarrhea virus by light and electron microscopies, *Am. J. Vet. Res.*, 45, 165, 1984.
117. **Bock, R. E., Burgess, G. W., and Douglas, I. C.,** Development of an enzyme linked immunosorbent assay (ELISA) for the detection of bovine serum antibody to bovine viral diarrhea virus, *Aust. Vet. J.*, 63, 406, 1986.
118. **Chu, H.-J., Zee, Y, C., Ardans, A. A., and Dai, K.,** Enzyme-linked immunosorbent assay for the detection of antibodies to bovine viral diarrhea virus in bovine sera. *Vet. Microbiol.*, 10, 325, 1985.
119. **Heuschele, W. P.,** BVD virus strain variation and laboratory diagnostic problems, *Proc. Am. Assoc. Vet. Lab. Diagn.*, 18, 91, 1975.
120. **Howard, C. J., Clarke, M. C., and Brownlie, J.,** An enzyme-linked immunosorbent assay (ELISA) for the detection of antibodies to bovine viral diarrhoea virus (BVDV) in cattle sera, *Vet. Microbiol.*, 10, 359, 1985.
121. **Rae, A. G., Sinclair, J. A., and Nettleton, P. F.,** Survival of bovine virus diarrhoea virus in blood from persistently infected cattle, *Vet. Rec.*, 120, 504, 1987.
122. **Itoh, O., Sassaki, H., and Hanaki, T.,** Reverse plaque formation method for titration of non-cytopathogenic bovine viral diarrhea–mucosal disease virus, *Natl. Inst. Anim. Health O.*, 23, 27, 1983.
123. **Itoh, O., Sugiyama, M., Nakamura, S., and Sasaki, H.,** Characterization of a non-cytopathogenic agent isolated from cytopathogenic bovine viral diarrhea-mucosal disease virus stock, *Microbiol. Immunol.*, 28, 1163, 1984.
124. **Roeder, P. L. and Harkness, J. W.,** BVD virus infection: prospects for control, *Vet. Rec.*, 118, 143, 1986.
125. **Duffel, S. J., Sharp, M. W., Winkler, C. E., Terlecki, S., Richardson, C., Done, J. T., Roeder, P. L., and Hebert, C. N.,** Bovine virus diarrhoea-mucosal disease virus-induced fetopathy in cattle: efficacy of prophylactic maternal pre-exposure. *Vet. Rec.*, 114, 558, 1984.
126. **Meyling, A. and Jensen, A. M.,** Transmission of bovine virus diarrhoea virus (BVDV) by artificial insemination (AI) with semen from a persistently-infected bull, *Vet. Microbiol.*, 17, 97, 1988.
127. **Potter, M. L., Corstvet, R. E., Looney, C. R., Fulton, R. W, Archbald, L. F., and Godke, R. A.,** Evaluation of bovine viral diarrhea virus uptake by preimplantation embryos, *Am. J. Vet. Res.*, 45, 1778, 1984.
128. **Nettleton, P. F.,** Pathogenesis and epidemiology of border disease, *Ann. Rech. Vet.*, 18, 147, 1986.
129. **Lamontagne, L. and Roy, R.,** Presence of antibodies to bovine viral diarrhea-mucosal disease virus (border disease) in sheep and goat flocks in Quebec, *Can. J. Comp. Med.*, 48, 225, 1984.
130. **Armstrong, D. A., McClurkin, A. W., Potgieter, L. N. D., and Reggiardo, C.,** Focus: bovine virus diarrhea, Part 2, Vaccination programs and management practices, *Bov. Vet. Forum,* 1, 1, 1986.
131. **Harkness, J. W.,** The control of bovine viral diarrhoea virus infection, *Ann. Rech. Vet.*, 18, 167, 1987.
132. **McClurkin, A. W., Coria, M. F., and Cutlip, R. C.,** Reproductive performance of apparently healthy cattle persistently infected with bovine viral diarrhea virus, *J. Am. Vet. Med. Assoc.*, 174, 1116, 1979.

V. Mixed Enteric Infections

Chapter 12

MIXED INFECTIONS IN THE INTESTINAL TRACT

David R. Snodgrass

I. INTRODUCTION

Diarrhea in young animals and children can be caused by any one of a variety of infectious microorganisms acting alone. The outcome of such enteric infections in neonates is the result of complex interactions between host, infectious agent, and the environment. Superimposed on this complexity is the common observation that two or more enteropathogens may simultaneously infect the same individual with the possibility of an interactive effect. Enteropathogens may also interreact with other agents not regarded as primarily enteropathogenic in their own right.

II. OCCURRENCE OF COMBINED INFECTIONS

A. CALVES

Although many infectious agents have the potential to cause diarrhea in young calves, only relatively few are known to do so commonly. Microorganisms of wide distribution and proven enteropathogenicity are rotavirus, coronavirus, enterotoxigenic (K99 + STa +) strains of *Escherichia coli* (ETEC), *Salmonella* spp, and *Cryptosporidium.* [1] Discussion in this chapter will focus on the natural and experimental observations made on combined infections with these enteropathogens.

Few detailed epidemiological and microbiological surveys on diarrhea in young calves have been performed, but there is general agreement that combined infections with more than one enteropathogen are common.[2,3] Surveys in the U.K. indicate that 15 to 20% of diarrheic calves and 2% of healthy calves concurrently excrete more than one enteropathogen[4,5] (Table 1). An additional question as to whether primary infection with one agent leads to subsequent infection with another has not been investigated in the field.

The nature of combined infections varies with the prevalence of the different agents in a given region. In North America, combined infections of rotavirus and either enterotoxigenic *E. coli* or coronavirus are widespread.[2,3,6–8] This reflects the high prevalence of all three of these agents in the calf population. However in Europe, ETEC infection occurs as a clinically distinct and comparatively uncommon watery diarrhea in very young calves of 1 to 2 d old only, and coronavirus infection also appears less common.[4,5,9,10] It is, therefore, unusual for ETEC infection to occur in combination with any other agent. On the other hand, combined infections of rotavirus with either coronavirus or more particularly with *Cryptosporidium* occurred in 4% and 11% of diarrheic calves, respectively.[4] Coronavirus infections were observed to be 3 times more common in the presence of rotavirus infection.[5]

B. PIGLETS

The range of common and proven enteropathogens in piglets contains a variety of viruses, bacteria, and protozoa remarkably similar to that occurring in calves — coronaviruses (transmissible gastroenteritis [TGE] virus and epidemic diarrhea virus), rotavirus, ETEC (mainly K88 + LT +), and *Isospora suis.* Reports on combined infections in suckling piglets with rotavirus and TGE virus,[11] rotavirus and *I. suis,*[12] and rotavirus with other enteric viruses[13,14] exist. In a comprehensive survey in Canada of neonatal piglet diarrhea for a range of infectious agents in a number of individuals and outbreaks, one enteropathogen only was detected in 78% of piglets, and the predominant agent involved was TGE virus, which infected more than half of all piglets.[15] Combined infections occurred in 12% of the piglets examined. However, in Taiwan, multiple infections particularly involving coronavirus and ETEC, occurred in 60% of 317 diarrheic piglets.[16]

The syndrome of postweaning diarrhea in piglets has been associated with ETEC infection and predisposing factors which may be either infectious, nutritional, immunological, or environmental.[17-19] In particular, Lecce et al.[18] showed that ETEC infections were benign in

TABLE 1
Detection of Multiple Infections (% Incidence)

Survey	Calf status	No. of enteropathogens per calf 0	1	2	3
Moredun[a]	Healthy	65	33	2	0
	Diarrhea	29	56	15	< 1
Compton[b]	Healthy	77	21	2	0
	Diarrhea	31	49	17	3

[a] Snodgrass, D. R., et al., *Vet. Rec.*, 119, 31, 1986.
[b] Reynolds, D. J., et al., *Vet. Rec.*, 119, 34, 1986.

weaned pigs unless they were concurrently infected with rotavirus, in which case typical postweaning diarrhea occurred.

C. CHILDREN

Many viral, bacterial, and protozoal infections can cause diarrhea in young children.[20,21] However, reports of the occurrence of mixed infections in childhood diarrhea are surprisingly scarce. In developing countries, hygienic considerations suggest that endemic infections must frequently coinfect children, as is often the case with young animals. A comprehensive investigation of the occurrence of mixed intestinal infections in a developing country was carried out in Brazil by Guerrant et al.[20] They found that the commonest infections in children also were rotavirus and ETEC, but 14 of 35 children with enterotoxic coliforms were infected with other enteropathogens, and 12 of 24 children with rotavirus also simultaneously excreted other enteropathogens. Other reports of mixed infections from developing countries have been made from Rwanda, where *Cryptosporidium* was diagnosed in combined infections with either *Salmonella* or *Campylobacter* in 2% of diarrheic children,[22] and from Taiwan where Echeverria et al [23] observed that 7/11 ETEC infections were accompanied by rotavirus.

Concurrent intestinal infections with more than one enteropathogen are also common in industrialized western countries. In surveys conducted in the mid-1970s, infection with two or more viruses was detected in 14% of patients,[24] and 13% of children with rotavirus were also found to be infected with other intestinal pathogens.[25] However in a more recent large survey of children under 2 years of age hospitalized with acute diarrhea, two or more intestinal pathogens were detected in 28% of patients.[26] There was no difference in disease severity as measured by duration of diarrhea and vomiting, occurrence of dehydration, character of stools, clinical and biochemical features, and general condition, between those who excreted single or multiple organisms.[26]

D. POULTRY

Viruses are of considerable importance in enteritis of poultry. In a study of diarrheic turkey poults, rotaviruses, astroviruses, reoviruses, enteroviruses, and adenoviruses, were all detected, with combined infections occurring more commonly than single-agent infections.[27]

III. EXPERIMENTAL STUDIES

The classical concept of primary viral infection predisposing to secondary bacterial infection has predominated in the design of experimental studies to investigate interactions between neonatal enteric pathogens. Hence, the infectious agents examined have been almost exclusively rotavirus and ETEC, and calves in particular have been used as experimental animals.

The first demonstration of interaction between rotavirus and ETEC was made by Gouet et al.[28] They showed that in colostrum-deprived newborn calves a dose of 10^{10} ETEC produced a fatal diarrhea, but that infection with either 10^8 ETEC or rotavirus produced a nonfatal diarrhea. However, when rotavirus infection at a few hours of age was followed by 10^8 ETEC 24 h later, a severe fatal diarrhea occurred. In calves a few days old, simultaneous infection with rotavirus and ETEC caused fatal diarrhea, but no interaction was observed when ETEC infection was followed by rotavirus. Thus the basic concepts of the nature of rotavirus/ETEC interaction were determined: that interaction does occur, and that in particular rotavirus enables an ETEC infection to establish in circumstances where, for reasons of age of animal or titer of inoculum, it would not otherwise do so; and that this interaction can be most readily demonstrated when rotavirus is given prior to or simultaneous with the ETEC inoculum.

Subsequent studies have added to these observations by repeating and extending them, using calves of differing status from gnotobiotic to conventional suckling, using different ages of calves, and using different strains and doses of the infectious agents. In general, those experimenters who, by reason of dose of inoculum or age of calf, were able to produce a diarrhea with ETEC infection alone did not show significant synergistic effects by superimposing rotavirus infection. For example, Runnels et al.[29] using 5 to 8 d-old gnotobiotic calves measured fecal dry weight, bacterial counts, viral shedding, and histological lesions. Combined infections tended to be slightly more severe than either rotavirus or ETEC infections separately. However, the same group later extended the period of observation and produced a severe fatal diarrhea in some dually infected calves which, they concluded, demonstrated synergistic interaction.[30] Experiments in neonatal gnotobiotic piglets also demonstrated a more severe clinical disease in dually infected animals, but without general enhancement of bacterial titers.[31]

On the other hand, those studies in which the ETEC infection alone was either abortive or subclinical, were often able to show that concurrent rotavirus infection produced a marked enhancement of the clinical disease. In experiments with calves over 1 week of age, Tzipori et al.[32] were unable to produce diarrhea with either rotavirus or ETEC alone, but combined infections caused clinical diarrhea in either gnotobiotic or microbiologically conventional calves. Similarly, Snodgrass et al.[33] showed that ETEC were not able to colonize the intestine of 6 d old conventional calves unless rotavirus was also present. Bacterial colonization was shown to be greatly increased by rotavirus infection. The most detailed evidence for the mechanism of interaction was provided in experiments by Hess et al.[34] A small dose of 10^7 ETEC in specific pathogen-free calves under 24 h old produced no clinical signs, and rotavirus alone caused a mild diarrhea. A severe diarrhea ensued when calves were infected with the two agents simultaneously or when the rotavirus was given prior to the ETEC. However, if the calf was infected with ETEC before rotavirus, no exacerbation of disease was evident. With the combined infections, the incubation period was reduced, titers of both rotavirus and ETEC in the intestine were increased, and the severity of histological lesions was increased.

In experiments with postweaning diarrhea in pigs, ETEC were shown to be unable to establish infection in 4-week-old animals. However, if the pigs were infected first with rotavirus, the ETEC infection became established and severe diarrhea of long duration occurred.[18] This ability of rotavirus infection to enhance ETEC infection in postweaning pigs has been demonstrated by Tzipori et al. also.[17]

Experimental observations on rotavirus/ETEC interactions have also been made in other species. Increased mortality with combined infections has been observed in mice[35] and lambs,[36] and diarrhea in foals could be caused by combined infections in situations where neither infection alone caused diarrhea.[37]

There are few published records of experimental observations into combined infections other than those with rotavirus and ETEC. A synergistic action with both increased morbidity and mortality in precolostral calves infected with more than one agent, including rotavirus, coronavirus, and BVD virus has been demonstrated.[38] Tzipori et al.[39] infected two lambs with rotavirus and *Cryptosporidium* and did not show any exacerbation compared with the already severe cryptosporidial diarrhea. Because of the common occurrence in calves of coinfections with rotavirus and either *Cryptosporidium* or coronavirus, further experimental investigations in this area are warranted. Concurrent coccidiosis and the stress of weaning caused recrudescence of bovine parvovirus with resulting severe diarrhea in calves, with possible implications for postweaning diarrhea.[40]

Using a mouse model of infection, it was shown that a normal intestinal microflora did not modify rotavirus diarrhea.[40] However, rotavirus infection in gnotobiotic mice facilitated intestinal colonization by potentially pathogenic anaerobic bacteria, indicating a potential for rotavirus infection to enhance opportunistic bacterial infections.

IV. MECHANISMS OF INTERACTION

There is a considerable amount of knowledge on the pathogenic mechanisms of the common enteric pathogens, but very little on the mechanisms of their interaction. Rotavirus and coronavirus multiply in the intestinal epithelial enterocytes. Although they infect primarily the small intestine, the area infected is not consistent — some reports suggest jejunum is most severely affected,[42,43] while others found the lesions to be largely in the ileum.[44-46] Infection in the large intestine has also been noted occasionally.[44] Infected cells desquamate, producing villus atrophy and an epithelium either temporarily devoid of cells or lined with immature cuboidal cells produced from hyperplastic and hypertrophic crypts. These lesions produce defects in digestion, particularly of disaccharides, and impaired absorption.[47] There is also evidence that dysfunction in the intestine may continue for at least 2 weeks after rotaviral infection.[48]

Enteric bacteria have a variety of virulence mechanisms in their repertoire. Even within *E. coli*, strains that can cause diarrhea can be categorized as enterotoxigenic, enteroinvasive, and enteropathogenic.[49] ETEC produce a heat-labile enterotoxin, which attaches to enterocytes through specific membrane-bound gangliosides, and/or heat-stable enterotoxins. ETEC colonize principally the ileum, with lesser colonization in the jejunum and none in the duodenum.[57] The most notable invasive bacteria are *Shigella* and *Salmonella*, although these organisms may also produce enterotoxin. *Salmonella* infect principally the ileum and large bowel.

Cryptosporidia invade the apical border of epithelial cells in the ileum and, to a lesser extent, the large intestine, causing widespread villus atrophy and fusion.[52]

Because of these widely differing pathogenic mechanisms, the potential exists for a variety of effects when two enteropathogens coinfect the same intestinal tract. For example, it is conceivable that a jejunal rotavirus infection and an ileal cryptosporidial infection would have an additive pathophysiological effect with exacerbation of the clinical problem. Similarly, a malabsorptive viral infection with a secretory bacterial infection could compound the clinical problem. Where both agents are infecting the epithelium in the same region of the intestine, these interactions are probably more complex and more specific.

Although there is clear experimental evidence that rotavirus enhances the ability of ETEC to infect the intestine, it is not known whether this is due to altered specific or nonspecific immune mechanisms or to enhanced receptor attachment. Bovine astrovirus has been shown to infect and damage the dome cell epithelium on intestinal Peyer's patches,[53] which may result in abnormalities of immune function. It is not known what factors control the marked age restriction of ETEC infections in calves. It is possible that receptors present on a relatively mature epithelium at birth are important for bacterial attachment, but again it seems unlikely that the immature cuboidal epithelium produced after rotavirus infection would contain these same receptors. Altered bacterial adherence has been demonstrated after viral infection in the respiratory tract, with influenza virus increasing the adherence of several bacterial species to pharyngeal cells *in vitro*.[54] An alternative possibility is that rotavirus impairs nonspecific immune factors such as the mucus layer or intestinal motility.

With these interactive mechanisms unknown, it is also possible that two enteropathogens could have mutually antagonistic effects, either through stimulation of nonspecific immune factors such as interferon, or through destruction or alteration of target receptors or cells. No such reactions have so far been reported or investigated.

V. CONCLUSIONS

With many enteropathogenic infections endemic in the environment of the young animal or child, it is not surprising that mixed infections of the intestinal tract are frequently observed. This phenomenon has been reported much more commonly from veterinary medicine, and its significance in human medicine may well be underestimated. In most of the infections discussed, the presence of mixed infections is merely an added complication to the successful resolution of the case. However, in postweaning diarrhea of piglets the principal enteropathogen (ETEC) is not capable of initiating the disease on its own, but requires the presence of predisposing factors, one of which may be a concurrent rotavirus infection. In this case a combined infection is an essential prerequisite for disease to occur.

Experimental studies with combined infections have successfully confirmed the interactive effects of rotavirus and ETEC, but without ascertaining the underlying mechanisms. No experimental information is available on the potentially equally important virus-protozoa or virus-virus interactions exemplified by rotavirus infections combined with either *Cryptosporidium* or coronavirus, and these areas deserve a higher research priority.

It is also true that the microbial etiology of diarrhea in the young has not been fully elucidated, and that as information on newly described enteropathogens becomes available new interactions will also be described, perhaps sometimes involving agents not normally recognized as enteropathogens in their own right.

REFERENCES

1. **Tzipori, S.**, The aetiology and diagnosis of calf diarrhoea, *Vet. Rec.*, 108, 510, 1981.
2. **Morin, M., Lariviere, S., and Lallier, R.**, Pathological and microbiological observations made on spontaneous cases of acute neonatal calf diarrhea., *Can. J. Comp. Med.*, 40, 228, 1975.
3. **Moon, H. W., McClurkin, A. W., Isaacson, R. E., Pohlenz, J., Skartvedt, S. M., Gillette, K. G., and Baetz, A. L.**, Pathogenic relationships of rotavirus, *Escherichia coli*, and other agents in mixed infections in calves, *J.Am.Vet.Med.Assoc.*, 173, 577, 1978.
4. **Snodgrass, D. R., Terzolo, H. R., Sherwood, D., Campbell, I., Menzies, J. D., and Synge, B. A.**, Aetiology of diarrhoea in young calves, *Vet. Rec.*, 119, 31, 1986.
5. **Reynolds, D. J., Morgan, J. H., Chanter, N., Jones, P. W., Bridger, J. C., Debney, T. G., and Bunch, K. J.**, The microbiology of calf diarrhoea in southern Britain, *Vet. Rec.*, 119, 34, 1986.
6. **Acres, S. D., Saunders, J. R., and Radostits, O. M.**, Acute undifferentiated neonatal diarrhoea of beef calves. The prevalence of enterotoxigenic *E. coli*, reo-like (rota) virus and other enteropathogens in cow-calf herds, *Can. Vet. J.*,18, 113, 1977.
7. **Bulgin, M. S., Anderson, B. C., Ward, C. S., and Evermann, J. F.**, Infectious agents associated with neonatal calf disease in southwestern Idaho and eastern Oregon, *J.Am.Vet.Med.Assoc.*, 180, 1222, 1982.
8. **Marsolais, G., Assaf, R., Montpetit, P., and Marois, P.**, Diagnosis of viral agents associated with neonatal calf diarrhea, *Can. J. Comp. Med.*, 42, 168, 1978.
9. **Sherwood, D., Snodgrass, D. R., and Lawson, G. H. K.**, Prevalence of enterotoxigenic *Escherichia coli* (ETEC) in calves in Scotland and northern England, *Vet. Rec.*, 113, 208, 1983.

10. **Moerman, A., de Leeuw, P. W., van Zijderveld, F. G., Baanvinger, T., and Tiessink, J. W. A.**, Prevalence and significance of viral enteritis in Dutch dairy calves, in Proc. XII World Congr. on Diseases of Cattle, Amsterdam, The Netherlands, 1982, 228.
11. **Theil, K. W., Saif, L .J., Bohl, E. H., Agnes, A. G., and Kohler, E. M.**, Concurrent porcine rotaviral and transmissible gastroenteritis viral infections in a three-day-old conventional pig, *Am. J. Vet. Res.* 40, 719, 1979.
12. **Roberts, L. and Walker, E.J.**, Field study of coccidial and rotaviral diarrhea in unweaned piglets, *Vet. Rec.*, 110, 11, 1982.
13. **Bridger, J. C.**, Detection by electron microscopy of caliciviruses, astroviruses and rotavirus-like particles in the faeces of piglets with diarrhea, *Vet. Rec.*, 107, 532, 1980.
14. **Saif, L. J., Bohl, E. H., Theil, K. W., Cross, R. F., and House, J. A.**, Rotavirus-like, calicivirus-like, and 23-nm virus-like particles associated with diarrhoea in young pigs, *J. Clin. Microbiol.*, 12, 105, 1980.
15. **Morin, M., Turgeon, D., Jolette, J., Robinson, Y., Phaneuf, J. B., Sauvageau, R., Beauregard, M., Teuscher, E., Higgins, R., and Lariviere, S.** , Neonatal diarrhea of pigs in Quebec: infectious causes of significant outbreaks, *Can. J. Comp. Med.* 47, 11, 1983.
16. **Chu, R. M., Yang, P. C., Chang, W. F.**, Neonatal diarrhoea of pigs in Taiwan in aetiology, epidemiology, and treatment, in *Infectious Diarrhea in the Young*, Tzipori, S., Ed., Excerpta Medica, Amsterdam, 1985.
17. **Tzipori, S., Chandler, D., Makin, T., and Smith, M.**, *Escherichia coli* and rotavirus infections in four-week-old gnotobiotic piglets fed milk or dry food, *Aust. Vet. J.*, 56, 279, 1980.
18. **Lecce, J. G., Balsbaugh, R. K., Clare, D. A., and King, M. W.**, Rotavirus and hemolytic enteropathogenic *Escherichia coli* in weanling diarrhea of pigs, *J.Clin. Microbiol.*, 16, 715, 1982.
19. **Miller, B. G., Newby, C. R., Stokes, C. R., and Bourne, F. J.**, Influence of diet on postweaning malabsorption and diarrhoea in the pig, *Res. Vet. Sci.*, 36, 187, 1984.
20. **Guerrant, R. L., Kirchhoff, L. V., Shields, D. S., Nations, M. K., Leslie, J., de Sousa, M. A., Araujo, J. G., Correia, L. L., Sauer, K. T., McClelland, K. E., Trowbridge, F. L., and Hughes, J. M.**, Prospective study of diarrheal illnesses in northeastern Brazil: patterns of disease, nutritional impact, etiologies and risk factors, *J. Infect. Dis.*, 148, 986, 1983.
21. **Black, R. E.**, Relative importance of enteropathogens effecting humans, in *Infectious Diarrhoea in the Young*, Tzipori, S., Ed., Excerpta Medica, Amsterdam, 1985.
22. **Bogaerts, J., Lepage, P., Rouvroy, D., and Vandepitte, J.**, *Cryptosporidium* spp., a frequent cause of diarrhea in Central Africa, *J. Clin. Microbiol.*, 20, 874, 1984.
23. **Echeverria, P., Ho, M. T., Blacklow, N. R., Quinnan, G., Portnoy, B., Olson, J. G., Conklin, R., DuPont, H. L., and Cross, J. H.**, Relative importance of viruses and bacteria in the etiology of pediatric diarrhea in Taiwan, *J. Infect. Dis.*, 136, 383, 1977.
24. **Madeley, C. R., Cosgrove, B. P., Bell, E. J., and Fallon, R. J.**, Stool viruses in babies in Glasgow. *J. Hyg.*,78, 261, 1977.
25. **Rodriguez, W. J., Kim, H. W., Arrobio, J. O., Brandt, C. D., Chanock, R. M., Kapikian, A. Z., Wyatt, R. G., and Parrott, R. H.**, Clinical features of acute gastroenteritis associated with human reovirus-like agent in infants and young children, *J. Pediatr.*, 91, 188, 1977.
26. **Ellis, M. E., Watson, B., Mandal, B. K., Dunbar, E. M., Craske, J., Curry, A., Roberti, J., and Lomax, J.**, Microorganisms in gastroenteritis.,*Arch. Dis. Child.*, 59, 848, 1984.
27. **Saif, L. J., Saif, Y. M., and Theil, K. W.**, Enteric viruses in diarrheic turkey poults, *Avian Dis.*, 29, 798, 1985.
28. **Gouet, Ph., Contrepois, M., Dubourguier, H. C., Riou, Y., Scherrer, R., Laporte, J., Vautherot, J. F., Cohen, J., and L'Haridon, R.**, The experimental production of diarrhea in colostrum deprived axenic and gnotoxenic calves with enteropathogenic *Escherichia coli*, rotavirus, coronavirus and in a combined infection of rotavirus and *E. coli*, *Ann. Rech. Vet.*, 9, 433, 1978.
29. **Runnels, P. L., Moon, H. W., Whipp, S. C., Matthews, P. J., and Woode, G. M.**, Interaction of rotavirus and enterotoxigenic *Escherichia coli* (ETEC) in gnotobiotic calves, in Proc. II Int. Symp. on Neonatal Diarrhea, Acres, S. D., Forman, A. J., and Fast, H., Eds., Saskatoon, Canada.
30. **Runnels, P. L., Moon, H. W., Matthews, P. J., Whipp, S. C., Woode, G. N.**, Effects of microbial and host variables on the interaction of rotavirus and *Escherichia coli* infections in gnotobiotic calves, *Am. J. Vet.*. Res., 47, 1524, 1986.
31. **Benfield, D. A., Francis, D. H., McAdaragh, J. P., Johnson, D. D., Bergeland, M. E., Rossow, K., and Moore, R.**, Combined rotavirus and K99 *Escherichia coli* infection in gnotobiotic pigs, *Am. J. Vet. Res.*, 49, 330, 1988.
32. **Tzipori, S., Makin, T. J., Smith, M. L., and Krautil, F. L.**, Clinical manifestations of diarrhea in calves infected with rotavirus and enterotoxigenic *Escherichia coli*, *J. Clin. Microbiol.*, 13, 1011, 1981.
33. **Snodgrass, D. R., Smith, M. L., and Krautil, F. L.**, Interaction of rotavirus and enterotoxigenic *Escherichia coli* in conventionally-reared dairy calves, *Vet.Microbiol.*, 7, 51, 1982.
34. **Hess, R. G., Bachmann, P. A., Baljer, G., Mayr, A., Pospischil, A., and Schmid, G.**, Synergism in experimental mixed infections of newborn colostrum-deprived calves with bovine rotavirus and enterotoxigenic *Escherichia coli* (ETEC), *Zentralbl. Vet. Med. B*, 31, 585, 1984.

35. **Newsome, P. M. and Coney, K. A.**, Synergistic rotavirus and *Escherichia coli* diarrheal infection of mice, *Infect. Immun.*, 47, 573, 1985.
36. **Wray, C., Dawson, M., Afshar, A., and Lucas, M.**, Experimental *Escherichia coli* and rotavirus infection in lambs, *Res. Vet. Sci.*, 30, 379, 1981.
37. **Tzipori, S., Makin, T., Smith, M., and Krautil, F.**, Enteritis in foals induced by rotavirus and enterotoxigenic *Escherichia coli*, *Aust. Vet. J.*, 58, 20, 1982.
38. **Van Opdenbosch, E., Wellemans, G., and Oudewater, J.**, Interaction of BVD, corona and rotavirus in neonatal calf diarrhoea: experimental infections in newborn calves, *Vlaams Diergeneeskd. Tijdschr.*, 50, 163, 1981.
39. **Tzipori, S., Sherwood, D., Angus, K. W., Campbell, I., and Gordon, M.**, Diarrhea in lambs: experimental infections with Enterotoxigenic *Escherichia coli*, rotavirus and *Cryptosporidium* sp. *Infect. Immun.*, 33, 40l, 1981.
40. **Durham, P. K. J., Johnson, R. H., and Parker, R. J.**, Exacerbation of experimental parvoviral enteritis in calves by coccidia and weaning stress, *Res. Vet. Sci.*, 39, 16, 1985.
41. **Moreau, M., Corthier, G., Muller, M., Dubos, F., and Raibaud, P.**, Relationships between rotavirus diarrhea and intestinal microflora establishment in conventional and gnotobiotic mice, *J. Clin. Microbiol.*, 23, 863, 1986.
42. **Davidson, G. P., Gall, D. G., Petric, M., Butler, D. G., and Hamilton, J. R.**, Human rotavirus enteritis induced in conventional piglets, *J. Clin. Invest.*, 60, 1402, 1977.
43. **Pearson, G. R., McNulty, M. S., and Logan, E. F.**, Pathological changes in the small intestine of neonatal calves naturally infected with reo-like virus (rotavirus), *Vet. Rec.*, 102, 454, 1978.
44. **Snodgrass, D. R., Angus, K. W., and Gray, E. W.**, Rotavirus infection in lambs: Pathogenesis and pathology, *Arch. Virol.*, 55, 263, 1977.
45. **Crouch, C. F. and Woode, G. N.**, Serial studies of virus multiplication and intestinal damage in gnotobiotic piglets infected with rotavirus, *J. Med. Microbiol.*, 11, 325, 1978.
46. **Theil, K.W., Bohl, E.H., Cross, R.F., Kohler, E.M., and Agnes, A.G.**, Pathogenesis of porcine rotaviral infection in experimentally inoculated gnotobiotic pigs, *Am. J. Vet. Res.*, 39, 213, 1978.
47. **Hamilton, J. R. and Gall, D. G.**, Pathophysiological and clinical features of viral enteritis, in *Virus Infections of the Gastrointestinal Tract*, Tyrrell, D. A. J. and Kapikian, A. Z., Eds., Marcel Dekker, New York, 1982, 227.
48. **Snodgrass, D. R., Ferguson, A., Allan, F., Angus, K. W., and Mitchell, B.**, Small intestinal morphology and epithelial cell kinetics in lamb rotavirus infections, *Gastroenterology*, 76, 277, 1979.
49. **Gross, R. J. and Rowe, B.**, *Escherichia coli* diarrhoea, *J. Hyg. Camb.*, 95, 531, 1985.
50. **Sack, R. B.**, Acute diarrheal diseases in humans caused by bacteria, in *Virus Infections of the Gastrointestinal Tract*, Tyrrell, D. A. J. and Kapikian, A. Z., Eds., Marcel Dekker, New York, 1982, 239.
51. **Bellamy, J. E. and Acres, S. D.**, Enterotoxigenic colibacillosis in colostrum-fed calves: pathologic changes, *Am. J. Vet. Res.*, 40, 1391, 1979.
52. **Angus, K. W.**, Cryptosporidiosis in man, domestic animals and birds: a review, *J. R. Soc. Med.*, 76, 62, 1983.
53. **Woode, G. N., Pohlenz, J. C., Gourley, N. E. K., and Fagerland, J. A.**, Astrovirus and Breda virus infections of dome cell epithelium of bovine ileum, *J. Clin. Microbiol.*, 19, 623, 1984.
54. **Fainstein, V., Musher, D. M., and Cate, T. R.**, Bacterial adherence of pharyngeal cells during viral infection, *J. Infect. Dis.*, 141, 172, 1980.

VI. Vaccine Strategies and Immunity to Enteropathogenic Viral Infections

Chapter 13

THE INTESTINAL IMMUNE SYSTEM

Alan J. Husband

TABLE OF CONTENTS

I. INTRODUCTION

The ingestion of potentially pathogenic material ensures a continual barrage of antigens against which intestinal immune mechanisms present the first line of defense. These defenses have evolved a highly sophisticated network of effector activities involving specific antibodies and antigen-reactive effector cells, which not only eliminate infections of the mucosa, but also shield the systemic immune system from unnecessary stimulation which would otherwise create chronic inflammation.

The defenses in the gut, as in systemic immunity, depend on both molecular and cellular effectors, but it is a feature characteristic of mucous surfaces that the immunoglobulin produced during the molecular or humoral response is predominantly of the immunoglobulin A (IgA) isotype. There is also evidence that even the cellular effectors form a specialized population at mucosal sites. The mechanism by which these responses are enlisted and their subsequent deployment and engagement with antigen will be reviewed in this chapter.

II. ANTIGEN UPTAKE AND PROCESSING IN THE INTESTINE

The anatomy and composition of the intestinal lymphatic system and gut-associated lymphoid tissue are discussed in detail by Yoffey and Courtice[1] and are summarized in Figure 1. The lymphatic system of the intestine originates from small lacteals in the villi and begin blindly under the epithelium at the tip of each villus. In Peyer's patches (PP) there are dense networks of large, thin-walled lymphatic capillaries surrounding lymphoid nodules. The lacteals and lymphatic capillaries anastomose at the base of the lamina propria, forming a plexus which branches through the muscularis mucosae via the mesenteric ducts to the mesenteric lymph nodes (MLN). The lymphatics efferent from the MLN drain to a common intestinal duct, which enters the cisterna chyli reaching the venous circulation via the thoracic duct. The role of the intestinal lymphatics is to return interstitial fluid and transudated serum proteins to the blood, transport fat absorbed from the intestine, and provide a mechanism for transport and recirculation of cells. In addition they also transport antigenic material absorbed from the intestinal lumen, and an understanding of the way in which antigens are absorbed and processed in the intestine is vital to the development of intestinal immunization procedures.

It is well documented that small amounts of orally administered antigens are absorbed intact from the intestinal lumen.[2-5] Indeed, Shreeve and Thomlinson[6] have reported that after oral dosage of neonatal pigs with *Escherichia coli* extracts, intact antigen was detected in the superficial cells of the small intestine after 1 min. and in the spleen after 5 min. Other studies have revealed that at least part of the absorbed antigen is transported from the intestinal lumen by the lymphatic system.[7,8]

The mechanism and site of uptake of lumenal antigens has aroused considerable debate. Volkheimer and Schulz[9] suggested that particulate antigens pass between epithelial cells as the result of a kneading action and proposed the term "persorption" to describe this process. Volkheimer et al.[10] used this theory to explain his observation that iron particles up to 52 μm in diameter are absorbed from the intestinal lumen and Raettig[11] suggested that microorganisms could also gain entry in this way.

But while some antigen may gain direct entry across the normal intestinal epithelium, the primary sites for antigen uptake in gut-associated lymphoid tissue (GALT) are the PP.[12] In rats, engagement of intestinal IgA-producing cell precursors with antigen occurs entirely within the PP, since removal of the MLN, the only lymphoid tissue other than PP which the precursors encounter before reaching the intestinal lamina propria, did not affect the IgA cell response to orally administered antigen.[13]

The ultrastructure of PP has been described in detail in rats,[14] mice,[15] humans,[16] nonhuman primates,[17] and pigs.[18] They consist of lymphoid aggregates in the lamina propria and submucosa

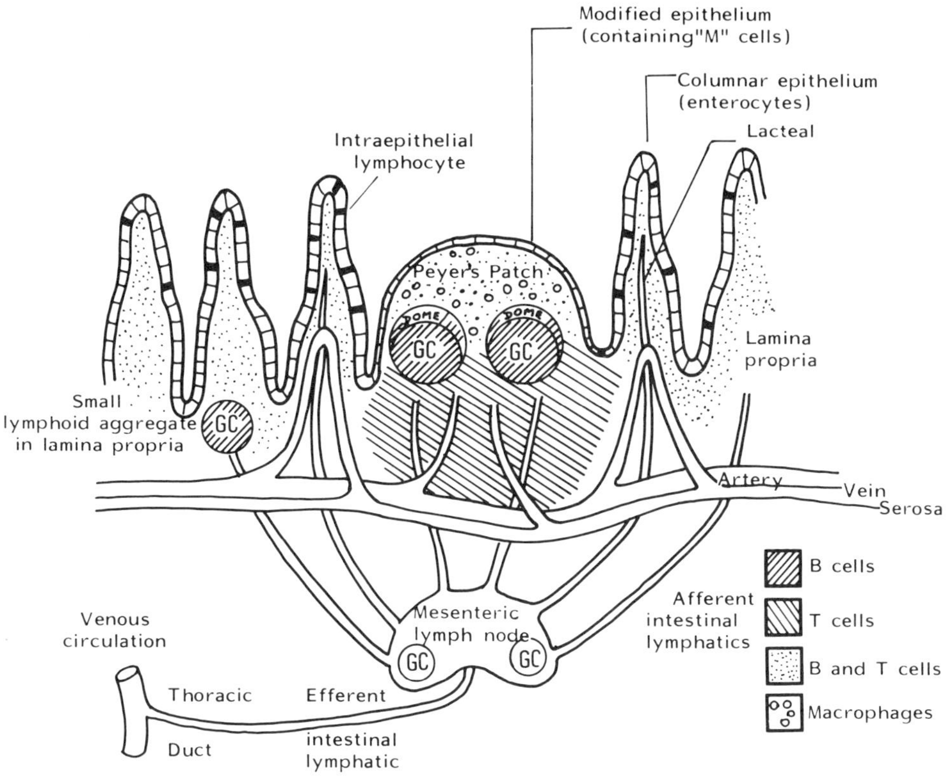

FIGURE 1. Structure of gut-associated lymphoid tissue. GC = germinal center.

and appear as nodules on the lumenal wall of the gut, occurring predominantly in the small intestine.[14] They have a layer of cuboidal epithelial cells, lacking the organized arrangement of the villous epithelium and are free of goblet cells. Interspersed among the microvillus-covered PP epithelial cells are numerous membranous (M) cells, which are specialized for antigen uptake,[12] that are distinguished by their lack of microvilli, having instead widely spaced microfolds on their lumenal surface. They often embrace one or more intraepithelial lymphocytes (IEL) and have interdigitating processes at their junction with neighboring epithelial cells.[18] Antigen is taken up by pinocytosis and carried through the M cells.[12] Immediately beneath, and sometimes interposed between, the epithelial cells, are macrophages which engulf the antigens taken up in this way, transport them from intercellular spaces into the follicle dome of the PP (a prominent area of lymphocyte traffic[19]), and present them to lymphoid cells to initiate an immune response.[20] This region contains a mixture of T lymphocytes (which are associated with immune regulation and cell-mediated immunity), B lymphocytes (which are responsible for antibody production), macrophages, and plasma cells (mature antibody-producing cells). Within the PP follicles, germinal centers occur consisting primarily of B lymphocytes, while T lymphocytes occupy the regions between follicles. In the interfollicular region (the T cell region), a population of interdigitating, dendritic cells have been observed which also trap and present antigen.[21]

Several authors have noted in human and animal species the presence within the intestinal mucosa of microscopic lymphoid nodules often lacking the specialized epithelial morphology characteristic of PP.[22-24] Whether lumenal antigen processing occurs in these unaggregated lymphoid follicles is unknown, but they may play a role similar to that of PP.

Although the natural physiological mode of PP stimulation is via the intestinal lumen, Pierce and Gowans[25] described a method of intraperitoneal immunization using antigen emulsified in Freund's complete adjuvant (FCA), which resulted in increased numbers of antigen-specific immunoblasts appearing in lymph draining the PP.[13] This procedure is thought to result in stimulation of the PP via the intestinal serosa.[26,27]

III. MOLECULAR EFFECTORS OF IMMUNITY IN THE INTESTINE

As early as 1919, Besredka[28] described immunity to *Shigella* infection after oral immunization, which was not correlated with serum antibody levels, and suggested that immunity in the gut operated independently of systemic immunity. It was not until much later that Heremans[29] detected a unique class of immunoglobulins in mucosal secretions that was poorly represented in serum.[30,31] This immunoglobulin isotype was designated IgA and its distribution could be accounted for by local production in plasma cells within the tissues underlying mucosal epithelia.[32-36] Indeed, it has been calculated that the human intestinal lamina propria contains up to 200,000 IgA plasma cells per cubic millimeter or 7.5×10^{10} cells in the entire gut.[37,38]

Hanson[39] detected a subtle difference between the IgA molecules in milk compared to serum, and it soon became apparent that IgA occurred in two distinct forms. The nonsecretory form[40] is a dimer of 150,000 mol wt subunits linked by a joining (J) chain yielding a molecule of approximately 330,000 mol wt, whereas the secretory form is approximately 420,000 mol wt and contains an additional glycoprotein moiety,secretory component (SC).[40] In the intestine, the SC molecule is produced by mucosal epithelial cells,[41] especially in the glandular region,[42] and is expressed on their basal membrane,[43] where it acts as a specific receptor for IgA molecules released into the subepithelial interstitial fluid by plasma cells in the submucosa.[44] The J chain endows IgA with structural characteristics which are crucial to binding to SC.[45,46] After the SC-IgA complex forms, it is transferred through the cell by reverse pinocytosis to the lumen and secreted into the mucus layer.[47] By virtue of its association with SC, the secretory IgA molecule remains relatively resistant to proteolysis by intestinal proteases.[48]

In most species the bulk of IgA in mucosal secretions, particularly in the intestine, is locally produced[49] (Figure 2). This has been confirmed in sheep using radiolabeled purified immunoglobulins which revealed that, whereas all of the IgG2 and most of the IgG1 in intestinal secretion is derived by transudation from plasma, the bulk of the IgA is of local origin.[50,51] Similar conclusions were obtained by Bull, Bienenstock, and Tomasi[52] using perfused loops of human intestine. These findings are not surprising in view of the high density of IgA-producing plasma cells in the lamina propria.[53] In contrast to most monogastric species, ruminants also have substantial numbers of IgG1-producing plasma cells in the gut.[54,55] In sheep, there is little evidence for transport of locally produced IgG into intestinal secretions where, in conformity with most other species, IgA is the predominant isotype.[50,56,57] This is probably either due to the lack of a specific transport mechanism for IgG1 and IgG2 or to the distribution of IgG-producing cells deeper in the submucosa and, hence, more distant from the surface epithelium than IgA-producing cells.[58,59] Therefore, IgG is removed from interstitial fluid via lymph rather than via the mucosal surface. Antibody produced in the regional MLN is predominantly of the IgG class, but almost all of this is drained via intestinal lymph.[59]

Some immunoglobulins reach intestinal secretion from plasma. IgG of plasma origin appears in intestinal secretion by passive transudation[57] (Figure 2) and despite the presence of an SC-mediated transport mechanism for IgA, evidence for selective transport of serum-derived IgA into intestinal secretion is lacking.[51] This is probably explained on the basis of SC receptor saturation by the large amounts of locally produced IgA at this site. Serum-derived IgA is selectively transported into saliva,[60] respiratory tract secretions,[61] and, in some species, into bile[62,63] and milk.[51,64-66]

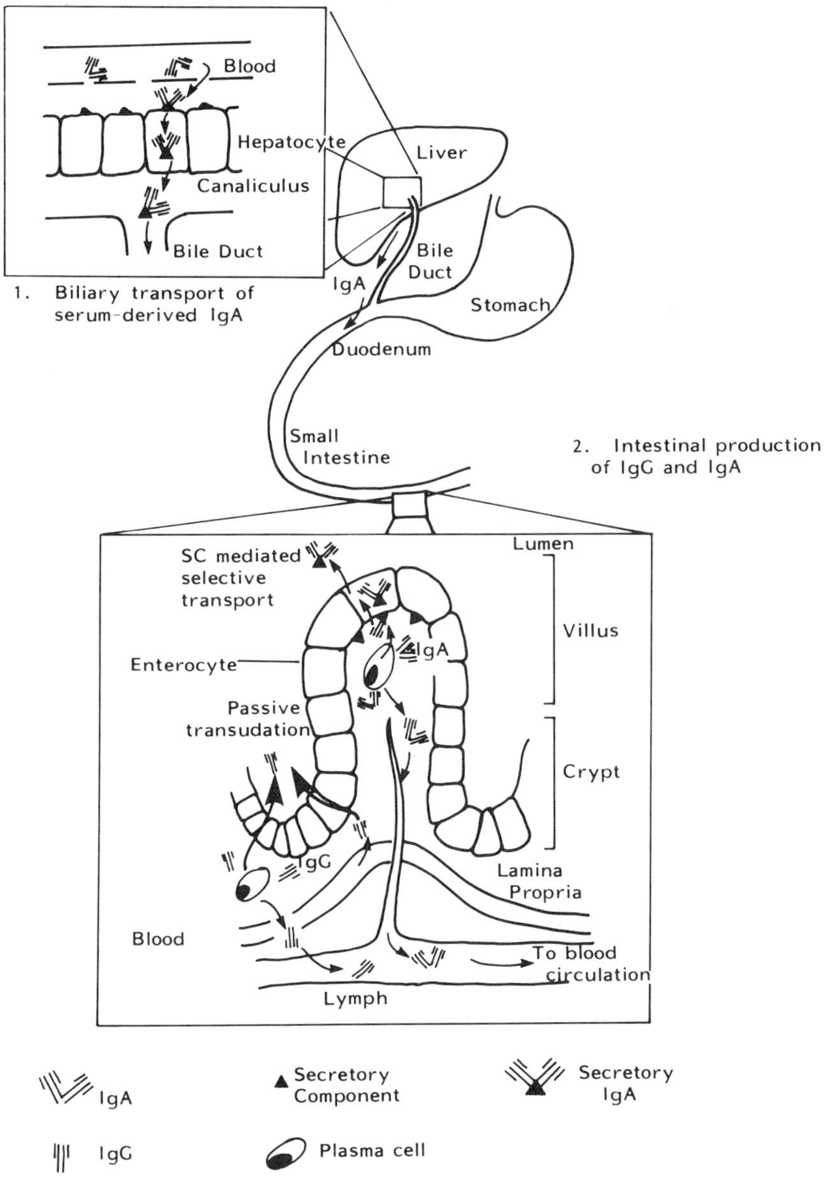

FIGURE 2. Origin of immunoglobulins in the intestinal lumen. IgA is locally produced by plasma cells underlying the lamina propria and is transported into the lumen by SC-mediated selective transport. IgA in blood also gains access to the intestine via the bile after SC-dependent selective transport through hepatocytes. IgG is present in the lamina propria interstitial fluid either by leakage from blood vessels or local production and although a small amount of IgG enters the intestinal lumen by passive transudation the bulk is returned to the circulation via the lymph.

Although the mucosa contributes large amounts of immunoglobulin to the intestinal secretion via local production and/or serum transudation, in some species bile is an additional source of intestinal IgA (Figure 2). In rats, bile contributes as much as 5 to 12 mg to intestinal secretion per day,[67,68] and this source of intestinal IgA may be quantitatively more important than IgA produced locally in the intestine in this species.[69] Concentration of IgA in bile occurs by selective removal of dimeric IgA molecules from the circulation by hepatocytes which express surface SC[62,63] in a manner analogous to mucosal epithelial cells. The IgA is transported through the hepatocytes in vesicles, exocytosed by the canalicular membrane, and then released into bile which in turn is secreted into the lumen of the gut.[67,70] While this occurs in rats and rabbits,[71] only small amounts of serum-derived IgA enter bile in this way in humans,[60] guinea pigs,[71] and sheep.[72-75] Thus, these species are dependent almost entirely on the secretion of IgA produced by plasma cells in the intestinal lamina propria to maintain the lumenal IgA antibody levels. In most species there is substantial local production of IgA within the biliary tree and in sheep about half of the IgA in bile arises by local production.[245]

While large amounts of IgA are produced in the intestine, not all is transported into the lumen, probably because of SC receptor saturation. A substantial amount of IgA of intestinal origin reaches the circulation via draining intestinal lymph. Indeed in sheep, 90% of the IgA in intestinal lymph originates from GALT[59] and this constitutes the major source of plasma IgA[75,76] The mammary gland[76] and respiratory tract[77] mucosa also contribute some IgA to plasma. This explains the predominance of IgA in the dimeric form in ruminant[78] and rodent [79] plasma and probably also accounts for the predominance of IgA dimers in chicken[80,81] and porcine[82] plasma. By contrast, in humans the majority of IgA in serum is monomeric and is principally derived from bone marrow cells.[83,84]

The protective functions of IgA are paradoxically quite limited and these have been extensively reviewed elsewhere.[85] IgA is not normally able to activate complement via the classical pathway[86,87] and inhibits complement activation by other classes of antibody.[88,89] It also inhibits complement deposition and consumption by antigen-antibody complexes.[90,91] These features explain the ability of IgA antibodies to react with antigen without initiating inflammatory reactions. IgA immune complexes are capable of complement activation *in vitro* [92] and, in conjunction with complement and lysozyme, secretory IgA may participate in bacteriolysis.[93]

Despite the fact that IgA antibodies can promote opsonization of bacteria facilitating their ingestion and destruction by monocytes,[94] they are relatively poor opsonins[95] and may even inhibit opsonization by antibodies of other isotypes.[96] IgA plays a role in neutralization of toxins[97-99] and viruses,[100,101] participates in complement independent ADCC,[94,102] enhances IgG-dependent cell mediated cytotoxicity,[103] and augments cytotoxic activities of lymphocytes of GALT origin but not peripheral lymph node origin.[104] In the spleen, both macrophages and lymphocytes participate in IgA-dependent ADCC, but among GALT cells only lymphocytes are involved.[105]

However, the major role for secretory IgA appears to be in preventing colonization of the intestinal tract by pathogenic organisms by blocking adhesion to the mucosal epithelium[106-108] and by reducing absorption of antigen by the intestine.[109-111] Walker et al.[109] observed that the intestinal absorption of a macromolecular antigen was reduced in animals that previously had been orally immunized with that antigen. Williams and Gibbons [107] demonstrated that specific IgA antibody in the intestinal lumen was responsible for this phenomenon and proposed that it functions as an antigen disposal mechanism, since preparations of IgA antibody to *Streptococcus* organisms specifically inhibited the adherence of these organisms to the intestinal epithelial cells. Andre´et al.[110] showed that a single intragastric administration of human serum albumin decreased the absorption of subsequent doses and that this interference with absorption was a result of the combination of antigen with secretory antibodies. Similar findings were reported by Stokes and colleagues.[111]

While some authors have seen this function of secretory IgA antibody as a barrier to absorption of antigen, an alternative explanation could be that the presence of IgA antibody facilitates digestion of antigen by the intestinal epithelial cell after combination with antigen at the cell surface and subsequent pinocytotic uptake of the antigen-antibody complex.[112] That is, the specific IgA antibody acts as an antigen receptor at the epithelial cell surface and the antigen-antibody complex is more readily degraded within the cell. There is as yet no strong evidence to support this contention but it is a possibility which requires further investigation.

IV. CELLULAR EFFECTORS OF IMMUNITY IN THE GUT

A. B CELLS

Most IgA antibody in intestinal secretions is produced locally by subepithelial plasma cells. Study of the origin and migration patterns of these cells has led to an understanding of the induction of IgA responses and the potential which the gut represents in providing a pool of cellular effectors for extraintestinal mucosal sites, now referred to as the common mucosal immune system. In 1964 Gowans and Knight[113] observed that large dividing cells entering blood from thoracic duct lymph, most of which originated from GALT, displayed a predilection for the intestinal wall after their intravenous injection.[113] The later observation, that IgA is the predominant immunoglobulin isotype secreted by plasma cells in the intestine,[37] encouraged speculation that these thoracic lymphoblasts were indeed the precursors of intestinal IgA plasma cells. Several other findings supported this hypothesis. Firstly, Guy-Grand et al.[114] showed in mice that radiolabeled thoracic duct large lymphocytes, accumulating in the intestine after intravenous injection, contain IgA. This was later confirmed in rats by Husband et al.[115] who showed that about 70% of thoracic duct large lymphocytes which home to the intestine contain IgA. Secondly, Williams and Gowans[116] showed that the bulk of lymphoblasts in thoracic duct lymph in rats express cytoplasmic and/or surface IgA.[116]

The PP origin of the IgA plasma cell precursors found in thoracic duct lymph was established by Craig and Cebra[117] by demonstrating IgA plasma cells of donor origin in the intestine of recipient rabbits after adoptive transfer of PP cells. This experiment clearly demonstrated that PP are an enriched source of potential IgA-producing cells. Further evidence for the PP origin of IgA precursor cells was provided by Husband and Gowans[13] using rats bearing Thiry Vella isolated intestinal loops. After intraperitoneal priming with antigen in FCA, a procedure which primes GALT for an IgA-specific antibody-containing cell response, antigen was administered locally into loops from which PP had been surgically removed. In these animals the IgA antibody-containing cell response was abrogated, whereas the response was virtually unaffected if MLN were removed prior to lumenal challenge. Similar data obtained from loops with or without PP have been reported by Cebra,[118] but, in contrast to these findings, Keren et al.[22] found that loops lacking PP were able to mount a vigorous IgA response to an invasive bacteria although this was delayed in onset compared to loops containing PP.

Thus, the population of predominantly IgA-secreting cells in the intestine is maintained by a supply of large lymphocytes from PP which, after stimulation by antigen from the gut lumen, enter the blood via the thoracic duct and migrate from blood into the intestinal lamina propria,[115] as shown in Figure 3. The observation that both thoracic duct lymphocytes and MLN cells home preferentially to the intestine, while cells of peripheral lymph node origin migrate preferentially to peripheral nodes,[114,119] substantiates the concept that there is a pool of cells displaying mucosally-targeted migration behavior. With respect to lymphoblasts, the immediate precursors of plasma cells, this represents a one-way traffic, and recirculation from the gut back into the blood does not occur,[113] although small lymphocytes are able to recirculate through the gut.[13,120] This migration pathway for IgA-specific antibody-containing cells has been confirmed in ruminants,[26] but in pigs both small recirculating lymphocytes and blast cells responding to intestinal antigen enter the blood circulation within the MLN and not via thoracic duct lymph.[121,122] More recently, Tseng[123] proposed a modification to this pathway in mice with evidence that there is an additional population of IgA precursor cells present in PP which migrate initially to the spleen and do not reach the gut lamina propria until after about 12 to 15 d.

The factors determining the site of localization of migrating IgA plasma cell precursors have been intensively studied because of the implications to localization of a response to infection and the possibilities of exploitation of the common mucosal system to provide protection of sites other than the intestine by oral immunization. Many theories have been proposed in this regard.

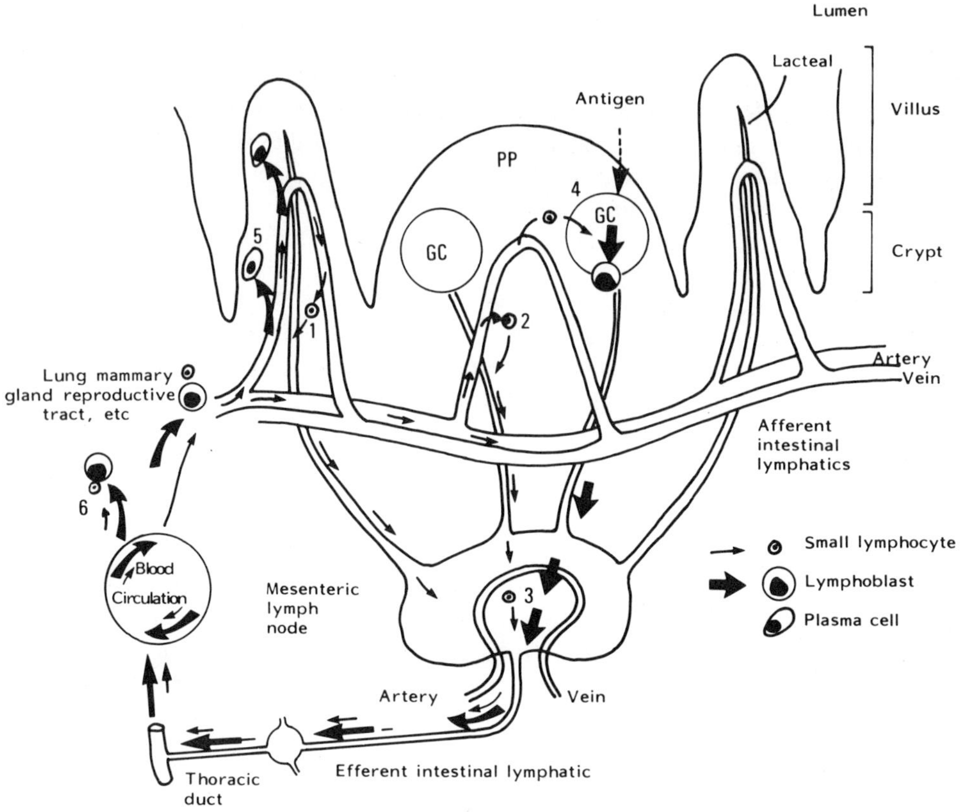

FIGURE 3. Cell migration in the intestine and the common mucosal immune system. A proportion of the small lymphocytes in blood supplying the gut leave the capillaries in the lamina propria (1) and Peyer's patches (PP) (2). These either remain in intestinal lymphoid tissue or leave the gut via afferent lymphatic drainage. Some of the small lymphocytes in blood supplying the mesenteric nodes also leave the circulation in the node (3). All cells which are not retained in the GALT are returned to the blood circulation via efferent intestinal and thoracic duct lymph and continue to recirculate through gut or other mucosal tissues. Lymphoblasts are generated in response to absorbed luminal antigen in germinal centers (GC) (4) and migrate to the blood circulation via afferent lymphatics, mesenteric lymph nodes, efferent lymphatics and thoracic duct. These cells may then leave the circulation in the laminae propriae of the intestine (5) and other mucosal organs (6) where they reside as antibody-producing cells and do not reenter the circulation.

While it seemed likely that these cells migrate to mucosal sites by virtue of their capacity to secrete IgA,[119,124-126] this has been discounted on the basis of the selective migration of GALT-derived T cells to the gut in mice[127,128] and the similar migration of mucosal IgG-secreting cells.[129] Furthermore, treatment of mice with large doses of anti-IgA antisera did not affect homing of IgA precursor cells.[119] The attraction of antigen for cells migrating to the intestine was suggested by Gowans and Knight,[113] but the migration of intestinal lymphoblasts to the gut in neonatal rats[130] and to antigen-free grafts of fetal gut in mice[131] were interpreted as evidence for antigen independence of lymphoblast migration. On the other hand, Ogra and Karzon[132] presented evidence in favor of antigen dependent migration by demonstrating IgA antibody confined to specifically immunized segments of human intestine.

This apparent conflict was later explained in experiments showing that precursor cells, generated by intestinal challenge of rats 2 weeks after intraperitoneal priming with antigen in FCA, localized equally in challenged or nonchallenged isolated intestinal loops during the first 6 h after their intravenous injection.[13,133] However, by 12 h (by which time extravazation of injected cells was complete) numbers in the nonchallenged loops had diminished, whereas in challenged loops they continued to expand reaching a peak at 36 h after injection. These experiments suggest that the predilection of IgA precursor cells for mucosal tissues is not determined by antigen, since extravazation occurs randomly at these sites. However, retention and proliferation after extravazation is dependent on the presence of specific antigen. Subsequent experiments have confirmed this hypothesis and further indicate that extravazation of IgA-containing cells in the intestine is not influenced by the level of small intestine from which the precursor cells originated.[134]

Butcher et al.[135] have identified determinants on mucosal-derived lymphocytes for which there are complementary receptors on endothelial cells in PP. However, while this may explain the tissue specificity of small recirculating cells of mucosal origin for PP tissues, it does not satisfactorily explain the tissue-specific homing of GALT blast cells to the intestinal lamina propria since the endothelial receptors identified in PP do not appear in the lamina propria.

An alternative view proposed by Elson et al.[136] is that perhaps the mucosal selectivity is only an apparent effect. The selective accumulation of IgA-specific B cells in mucosal tissues may be only a reflection of differential proliferation of IgA precursor cells in response to the high ratio of IgA helper to suppressor T cells present at these sites.

B. T CELLS

T cells orchestrate most of the molecular immune effector activities in the gut and play a primary role in cellular reactivity to enteric antigens and the pathological changes associated with hypersensitivity diseases in the gut.

Regarding regulation, IgA responses are characteristically highly thymus dependent,[137-139] and a population of helper T cells in GALT specific for IgA responses has been identified in mice,[140,141] rats[142] and humans.[143] In mice, these cells bear Fc receptors for IgA molecules.[144] Whether this isotype-specific regulation occurs as a result of directed isotype switching[145] or by induced selective clonal expansion of preswitched B cells[146,147] is unclear. Isotype-specific suppression of IgA responses by murine T cells with surface receptors for IgA has also been described.[148] Suppressor T cells specific for IgG responses are also produced in response to enteric antigens and these function to limit the systemic IgG response,[149-152] reducing the potential for inflammatory responses to mucosally presented antigens.

During acute enteric viral infections, subsets of T lymphocytes cytotoxic for virus-infected cells appear.[153,154] In small animals cytotoxic T cell activity in the gut was demonstrated by Kagnoff[155] in response to feeding tumor cells and by Husband et al.[156] after feeding allogeneic lymphocytes. In pigs, Frederick and Bohl[157] and Huntley et al.[158] demonstrated T cell lymphokine production in the gut in response to oral administration of antigens. Indeed, the pathology associated with local hypersensitivity reactions in the gut, especially in food allergies[159] and intestinal parasitic infection,[160] is no doubt the result of local cell-mediated immunity and lymphokine production.

Kinetic studies suggest that T cells generated in GALT in response to antigen in the intestine also display a tissue-specific homing pattern and migrate via thoracic duct lymph and blood circulation to preferentially populate mucosal sites.[127,156,161] Their distribution in the gut mucosa, however, is unusual in that in addition to their appearance in lamina propria there are numerous cells of apparent T lymphocyte lineage interspersed between epithelial cells.[18,128] These intraepithelial T lymphocytes are probably derived from PP; 85% or more in the mouse and man express the phenotype corresponding to the cytotoxic or suppressor T cell population.[162-165] The role of intraepithelial T cells is unclear, but it is possible that they are able to enter the lumen sampling its antigenic contents and then reenter the lamina propria to generate T effector cells.[166,167]

C. NK CELLS

Among the IEL population, are other cells which express ADCC and natural killing (NK) activity.[168] NK cells show spontaneous cytotoxicity and, unlike cytotoxic T cells, do not require the target cell to share their own histocompatibility antigens and do not require prior sensitization.[169] The properties of memory and target specificity of classical immune reactions are thus not exhibited by NK cells. The available evidence would suggest that gut NK cells are similar, but not identical, to splenic NK cells.[164,170] The main effector cells of NK activity in human, rat, and mouse IEL populations have large granular lymphocyte morphology,[164] but represent only about 15% of the total granulated IEL population.[246] NK cell activity is also associated with mononuclear cells in the intestinal lamina propria, but only as a minority population.[171] Circulating precursors of NK cells in mice appear to display selective localization patterns with some populations having a predilection for the gut epithelium.[172] In pigs, NK activity against transmissible gastroenteritis virus-infected target cells has been described among intestinal IEL.[173] A similar predilection for the gut epithelium has been described following adoptive transfer to neonatal recipients of mononuclear cells which conferred spontaneous cell-mediated cytotoxic activity and resistance to transmissible gastroenteritis.[174]

Natural anti-bacterial activity observed in the early phase of intestinal defense against enteric bacterial infection may be attributable to gut NK cell activity.[175] This may explain the observation that higher doses of bacteria are required to cause lethal infections if given orally than if given systemically.[176]

V. THE DEVELOPMENT OF INTESTINAL IMMUNITY

The intestinal immune system in most animals is anatomically well developed at birth, but parameters of local immunity are slow to reach adult levels and full-scale secretion of IgA antibody is delayed. This creates a paradox whereby there is a critical period between decline of maternally derived passive protection and onset of endogenous immunity. Thus defenses are low at a time when the need to establish protection against enteric pathogens is paramount.

By the end of gestation, most components of GALT are well differentiated. The PP in calves[177] and pigs[178] are present in numbers equivalent to adults by the time of birth and are morphologically well formed. MLN are also well formed with small follicles appearing, and lymphocyte infiltration is observed in the intestinal lamina propria at this time.[179]

Animals of most species are born virtually devoid of plasma cells in the gut, but there is a rapid appearance of these cells after birth following exposure to environmental antigens, and GALT tissues undergo rapid proliferative changes in response to the massive onslaught of environmental antigens. Studies in mice have indicated that by the time of birth, B cells are committed with respect to isotype and bear surface IgM and IgD in addition to the isotype to which they are committed.[180,181] Upon exposure to antigen, cytoplasmic immunoglobulin of this isotype appears and surface IgD and IgM are lost. In PP, precursors committed to IgA synthesis are apparently selectively accumulated from the recirculating pool of B lymphocytes, since there is unusual uniformity of surface IgA expression among PP cells in mice[182] and IgA is the predominant immunoglobulin produced by cells arising from PP precursors.[117] In young sheep, however, IgM is the predominant surface immunoglobulin isotype on ileal PP cells,[183] and only about half the cells generated from PP precursors in response to gut antigens are IgA-specific, the remainder producing IgM or IgG antibody.[27]

In neonatal calves the infiltration of the gut lamina propria with IgA- and IgM-producing plasma cells is observed as early as 4 d after birth, with IgM cells predominating for the first few weeks of life but IgA-cells predominating after 5 weeks.[184] Antibody-producing cells of all classes appear in the duodenum of pigs as early as 1 week of age, with a predominance of IgM cells up to 3 weeks of age, but IgA cells predominating thereafter.[185] By 12 weeks of age approximately 90% of the immunoglobulin producing cells in the intestine of pigs are of the IgA class.[186]

The rapid appearance of plasma cells in the gut of neonatal animals is dependent on stimulation by environmental antigens and is absent from germ-free animals.[34] Whether this proliferation is entirely due to antigen-specific responses or whether nonspecific components of the environmental challenge are involved is controversial. Bacterial lipopolysaccharide (LPS) is a well documented B lymphocyte mitogen and oral immunization with LPS before birth has been shown to accelerate the appearance of immunoglobulin-producing cells in GALT,[187] suggesting that bacterial endotoxins may play a role in amplification of the gut response during early ontogeny. It has also been suggested that components of colostrum may preferentially stimulate IgA synthesis [188]

The development of cell mediated immunity in GALT is an important aspect of the immune repertoire that has received relatively little attention. There are some indications that cell-mediated immune responsiveness is less well developed than antibody responses in neonatal animals. Evidence for this has come mainly from studies of parasitized animals, where cell-mediated immunity plays a key role in parasite expulsion, which show that neonatal sheep and rats are more susceptible to infection with helminth parasites than adults.[189-191] In rats, this neonatal susceptibility occurs despite levels of reaginic and protective antibody equivalent to those recorded in adults.[192] These findings are consistent with the observation in piglets that lymphocyte and NK-cell cytotoxicity against transmissible gastroenteritis virus are absent during the first week of life,[193] and the cell-mediated response to porcine enterovirus in the intestine of 5-week-old pigs is weak and localized,[194] whereas adult pigs mount a rapid cell-mediated response to orally administered antigen both in the intestine and at remote sites.[195]

Newborn rats have reduced numbers of total and granulated IEL and those expressing T cell markers are predominantly of suppressor phenotype, with cells of helper phenotype absent until 2 weeks of age.[196] In these studies, adult numbers and subset distribution of IEL were still not attained by the time of weaning. In PP, the lymphocyte subset distribution reached mature levels by 4 weeks of age.

In calves[197] and pigs,[198] IgA levels in intestinal secretion are slow to reach adult levels. In humans, measurments of plasma IgA, which presumably reflects mucosal activity with respect to IgA production, has revealed that whereas the concentration of IgA does not reach adult values until about 10 years of age, IgA antibodies to specific antigens to which children are regularly exposed reach adult levels by about 5 years of age.[247] This indicates that the slow development of IgA is probably related to limited antigenic experience rather than an inherent deficiency in IgA responsiveness. One of the other reasons for the delay in endogenous IgA synthesis in neonatal animals appears to be a feedback inhibition effect mediated by passively acquired maternal antibody, since in colostrum-deprived animals the onset of IgA synthesis occurs much earlier.[197,199,200] Hormonal factors have also been implicated in delayed development of local immunity. The use of corticosteroids to induce parturition caused a delay in IgA production in calves,[210] and since endogenous corticosteroids are involved in the birth process, it is possible that the relative incompetence of the immune system at birth is related to the perinatal hormonal environment. In addition, the increased susceptibility to enteric disease during times of neonatal stress, such as immediately after weaning, may also be accounted for by increased adrenal cortex hormonal activity.

VI. THE ROLE OF THE GUT IN EXTRAINTESTINAL DEFENSE

Initial observations regarding an entero-mammary axis with respect to IgA antibodies were made in studies with swine.[202,203] Subsequently, there have been numerous demonstrations of common mucosal immunity between the gut and other mucosal sites. This phenomenon has opened new possibilities in the control of diseases at extraintestinal mucosal surfaces because of the ability of the gut to contribute cellular and molecular effectors of immunity to distant sites. This is particularly relevant because the gut is the largest mucosal organ and the best endowed with inductive lymphoid tissue necessary for the generation, of an IgA response to local antigen.

In contrast, bronchus-associated lymphoid tissue (BALT) is relatively less abundant,[85] especially in ruminants,[77] and the mammary gland and genital tracts are virtually devoid of such tissues to the extent that local responses at these sites are relatively difficult to elicit.

GALT has been shown to play a major role in IgA responses at distant mucosal sites by contribution of GALT-derived IgA plasma cell precursors. These cells, reflecting the antigenic experience of the gut, relocate to other mucosal sites, subject to those controlling factors discussed previously. Their relocation to the mammary gland has been confirmed in a number of monogastric species.[204-206] In ruminants, despite the evidence for enhanced milk antibody responses after intestinal immunization,[207] there is no evidence for migration of IgA plasma cell precursors from the gut to the mammary gland.[208,209] IgA-containing cells of GALT origin can also migrate to the respiratory tract, genital tract, salivary glands, and intestine[129,210] (see Figure 3). Conversely, cells of BALT migrate to both the intestine and respiratory tract.[129] Migration of cells to the cervix and mammary gland appears to be under hormonal control.[211,212]

Experiments in sheep have confirmed the ability of GALT immunization to contribute to local IgA immunity in the respiratory tract by translocation of cells from GALT. Intraperitoneal immunization of sheep with antigen emulsified in FCA produces an IgA antibody-containing cell response in the intestine.[27] When an intratracheal dose of antigen was given in conjunction with intraperitoneal intestinal priming, an enhanced IgA-specific antibody-containing cell response in the upper respiratory tract was obtained which exceeded the response to either intratracheal or intraperitoneal immunization alone. This enhanced response did not occur if the supply of GALT-derived precursors was diverted from the circulation by lymphatic cannulation.[213]

Evidence for common mucosal immunity is also available in humans. Volunteers given an enteric-coated oral vaccine containing killed bacteria produced specific antibodies in saliva of the IgA class;[214,215] in other experiments, oral administration of enteric coated capsules containing adenovirus resulted in total protection against epidemic respiratory disease.[216] Similarly, oral vaccination with herpes simplex virus enhanced the protection against intravaginal challenge,[217] and ingestion of enteric capsules containing *Streptococcus mutans* led to the appearance of specific IgA antibodies in salivary and lacrimal secretions in the absence of detectable serum antibody.[218]

It is of interest to note that the well described gut-mammary axis may not only be a one-way process, but mammary gland immunization may in turn result in an IgA response in the gut. Sheep given a single intramammary dose of soluble antigen mounted an IgA-specific response in the intestinal lamina propria.[208] Evidence was obtained in these experiments that antigen is translocated from the mammary gland to GALT via the blood circulation, stimulating production of IgA precursor cells from GALT which migrate to the gut lamina propria via the previously described pathway. Similar conclusions were drawn by Saif and Bohl.[219] After inoculation of lactating mammary glands of pigs with transmissible gastroenteritis virus, local IgA antibody production was observed in noninfected glands. This was explained on the basis of relocation from the gut of specific precursors following a gut response to the intramammary inoculation. The role of circulating macrophages in the translocation of antigen between mucosal sites and the possibility of a mucosally targeted macrophage population requires further investigation, particularly in view of the finding that there is substantial macrophage traffic in GALT[220] and that some macrophages possess surface receptors for IgA.[221]

The preceding data has addressed the phenomenon by which GALT has been shown to play a major role in providing a pool of IgA-producing cells which can disseminate to a variety of mucosal sites. However, the great majority of IgA cells generated in GALT return to the lamina propria of the intestine,[113,124] where they produce antibodies specific to those antigens encountered in the lumen of the intestine which stimulated their production from GALT. As discussed previously, some of the IgA molecules produced in the intestine enter the circulation via lymph. Since IgA in this form is readily transported at mucosal surfaces via the SC receptor mechanism, this represents an alternative avenue by which the gut may contribute to distant mucosal immune responses. In rodents, the bulk of this circulating IgA appears to be removed and transported into bile through hepatocytes,[67] but in species other than rats and rabbits there is considerable variation in the extent of this avenue of transport.[71]

In sheep only about 2 to 5% of the IgA in plasma is removed via bile,[72,222,223] but considerably larger amounts of IgA are selectively transported from serum into the respiratory tract[61,77] and mammary gland[51,72] secretions. Studies using injections of radiolabeled immunoglobulins indicated that during early and midlactation, IgA in mammary secretions is almost wholly serum-derived, but during involution the bulk of the IgA in secretion is locally produced. At this latter time, there is no evidence of selective transport of IgA into secretion. Similarly in intestinal secretions, where local production is the predominant source of IgA, there is no evidence for the selective transport of serum IgA.[51] These studies suggest the existence of an inverse relationship between selective transport and local production of IgA, probably reflecting relative SC availability. The threefold elevation of IgA transport into bile in lactating sheep[72] suggests that hormonal changes associated with lactation may enhance IgA transport across other mucosal epithelia, a possibility which requires further investigation. These studies indicate that at sites of limited local IgA production there is a considerable contribution to the IgA content in secretions from GALT-derived IgA molecules entering the circulation via intestinal lymph.

Similarly, in humans there is selective transport of serum-derived IgA into saliva,[60] but in mice there is some conflict as to whether IgA is selectively transported into milk[65-66] and there is no evidence for its transport into milk in rats.[224]

VII. DEFENSE MECHANISMS AGAINST VIRAL ENTERIC INFECTIONS

Gastroenteric viral infections commonly occur after oral inoculation, the target cells for viruses causing diarrhea being the epithelium of the small intestine.[225] The intestinal mucosal defenses are therefore vital to the course of and recovery from gastroenteric disease and the repertoire of defenses encompasses both antigen-specific and nonantigen-specific mechanisms.

A. ANTIGEN-SPECIFIC IMMUNE MECHANISMS

These have been discussed in detail in preceding sections and are summarized in Figure 4. Only a brief description is presented here.

Antibody response — Large amounts of virus-specific antibodies, particularly of the IgA class, are generated in response to enteric viral infection. These antibodies play a role in protection against enteric viral disease probably both prior to infection of epithelial cells and after infection when virus is released into the intestinal lumen. Free virus in the lumen can be neutralized by specific antibody as a result of virus agglutination, opsonisation, and inhibition of absorption. Specific antibody is also required in ADCC reactions.

Cell-mediated immune mechanisms — The cell-mediated antigen-specific immune mechanisms, which have been shown to protect against viral infections, include T lymphocyte cytotoxicity and ADCC. Antigen-specific T lymphocytes are stimulated by viral antigens expressed on the surface of infected cells and are able to directly lyse the target cells by membrane interactions.[154] ADCC is effective against antibody-coated virus-infected cells and in the intestine lymphocytes are the predominant effector cell in this activity.[105]

B. NONANTIGEN-SPECIFIC DEFENSE MECHANISMS

There are a number of additional nonspecific defense mechanisms in the intestine which either prevent viral entry into cells or inhibit intracellular replication. Perhaps the most important nonimmune mechanisms are the low pH of the gastric contents (although all viruses causing gastroenteric disease in man and animals are resistant to low pH conditions[225]), the secretions

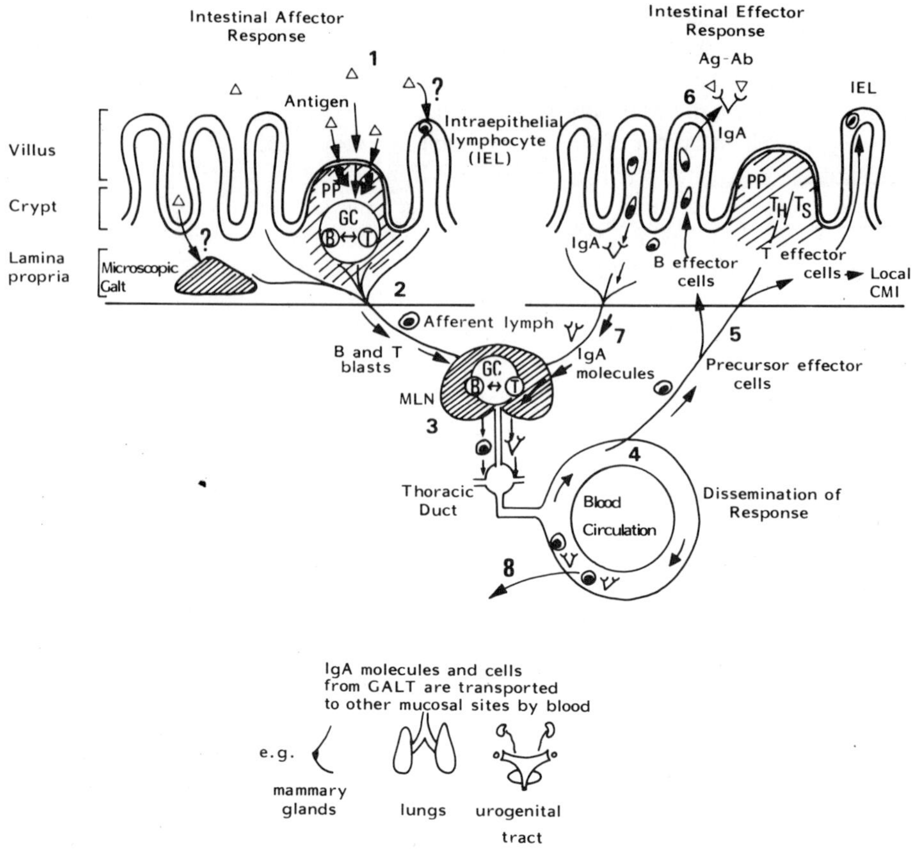

FIGURE 4. The uptake processing and response to antigen in the intestine. The predominant site for antigen uptake by GALT is the modified Peyer's patch (PP) epithelium (1), although the extent of antigen interaction via the normal epithelium with either microscopic lymphoid nodules or intraepithelial lymphocytes (IEL) is unclear. Antigen presenting cells in PP stimulate antigen-reactive B and T cells in germinal centers which then leave via afferent lymphatics (2) and enter mesenteric nodes (MLN). There is some amplification of the response in MLN (3), adding to the population of migrating blast cells which leave the node and enter the blood circulation via efferent and thoracic duct lymph (4). These effector cell precursors may then return to the intestine via the circulation (5), where they mature into effectors providing populations of IEL, B cells involved in antibody production, and T cells involved in cell-mediated immunity (CMI) and regulation. IgA antibodies either enter the intestinal lumen to bind with antigen (6) or enter blood circulation via the lymph (7). These circulating IgA molecules together with blood-borne T and B blasts of GALT origin may also provide specific immunity at other mucosal sites (8).

in the intestine (proteolytic enzymes, bile, pancreatic secretions), the effects of peristalsis in removing lumenal contents, and the integrity of the mucosal epithelium. In addition to presenting a physical barrier, the epithelium is an important structure determining the outcome of viral infection by the nature of receptors on the epithelial cell surface, since viruses have specific receptor requirements as a prerequisite to infection.[226] The mucus layer overlying the epithelium has been shown to play a role in defense by inhibiting penetration and adhesion of pathogens[227,228] and providing a depot for secreted antibody.[20] In immune animals, mucus secretion from goblet cells is increased in the presence of antigen,[229] immune complexes,[230] and IgE-mediated intestinal anaphylaxis reactions.[231]

There are also a range of nonantigen-specific immune mechanisms mediated by various granulocytic cells in the intestine. NK cells show spontaneous cytotoxicity for virus-infected cells and their activities have been discussed in a previous section. Macrophages are found in both PP and lamina propria of the gut, and their distribution and function in the intestine have been reviewed in detail by LeFevre, Hammer, and Joel.[232] Accumulations of mononuclear cells in the intestinal lamina propria are particularly pronounced during infections with transmissible gastroenteritis virus in piglets[233] and enteric viral infection of mice.[234] Once viruses have gained entry into the epithelial cells, macrophages provide nonspecific defense by spontaneous cell-mediated cytotoxicity against infected cells[235] and by ingestion and removal of free virus. Antigen ingested by macrophages in the PP is presented to lymphoid cells *in situ* to initiate a mucosal response[20] and macrophages from both the lamina propria and PP may migrate to mesenteric lymph nodes via afferent mesenteric lymph where an IgG response can occur.[220,236,237] Macrophages may also dispose of ingested material by migration through the epithelium to be excreted in faeces.[237] The role of lumenal phagocytes is controversial but may assist in the elimination of lumenal virus.[167,238,239]

The dramatic increase in the number of mucosal mast cells in response to infection is another contributor to nonspecific defense.[240] The rapid elimination of organisms from the bowel lumen is facilitated by the release of vasoactive substances from mast cells in response to degranulating signals received following antigen encounter.[241,242]

Once epithelial invasion occurs, nonspecific soluble mediators are released from a number of cell types in the gut. One of the most potent of these is a lymphokine, interferon, produced by lymphocytes, macrophages, epithelial cells, and fibroblasts. It has a local effect preventing viral replication in surrounding cells. Interferon production in the intestine has been described in association with viral gastroenteritis in piglets.[243] The observation that in infected piglets epithelial cells associated with PP are more resistant to infection than those from normal villi[233] could be due to local interferon production in PP. Human fetal intestine has also been shown to markedly inhibit enteric viral replication after treatment with interferon, but endogenous interferon production could not be demonstrated in these studies, possibly because of its rapid proteolytic inactivation.[244] Other lymphokines are important in modulating the antiviral functions of macrophages, such as macrophage migration inhibition factor and macrophage arming factor both of which are products of antigen-stimulated T lymphocytes.

REFERENCES

1. **Yoffey, J. M. and Courtice, F. C.,** *Lymphatics , Lymph and Lymphomyeloid Complex,* Academic Press, London, 1970.
2. **Cooper, G. N. and Thonard, J. C.,** Serum antibody responses to intestinal implantation of antigens in rats, *J. Pathol. Bacteriol.,* 93, 213, 1967.
3. **Rothberg, R. M., Kraft, S. C., and Michalek, S. M.,** Systemic immunity after local antigenic stimulation of the lymphoid tissue of the gastrointestinal tract, *J. Immunol..,* 111, 1906, 1973.
4. **Warshaw, A. L., Walker, W. A., and Isselbacher, K. J.,** Protein uptake by the intestine: evidence for absorption of intact macromolecules, *Gastroenterology,* 66, 987, 1974.
5. **Halsey, J. F. and Benjamin, D. C.,** Induction of immunologic tolerance in nursing neonates by absorption of tolerogen from colostrum, *J. Immunol.,* 116, 1204, 1976.
6. **Shreeve, J. and Thomlinson, J. R.,** Absorption of *Escherichia coli* endotoxin by the neonatal pig, *J. Med . Microbiol.,* 5, 55, 1972.
7. **Carter, P. B. and Collins, F. M.,** The route of enteric infection in normal mice, *J.Exp. Med.,* 139, 1189, 1974.
8. **Warshaw, A. L., Walker, W. A., Cornell, R., and Isselbacher, K. J.,** Small intestinal permeability to macromolecules. Transmission of horseradish peroxidase into mesenteric lymph and portal blood, *Lab Invest.,* 25, 675, 1971.
9. **Volkheimer, G. and Schulz, F. H.,** The phenomenon of persorption, *Digestion,* 1, 213, 1968.

10. **Volkheimer, G., Schulz, F. H., Lindeneau, A., and Beitz,U.,** Persorption of metallic iron particles, *Gut,* 10, 32, 1969.
11. **Raettig, H.,** Mechanisms of oral immunization with inactivated microorganisms, *Progr. Immunobiol. Stand.,* 4, 337, 1970.
12. **Owen, R. L.,** Ultrastructure of antigen trapping epithelia of mucosal lymphoid organs, in *Regulation of the Immune Response,* Ogra, P. L. and Jacobs, D. M., Eds., S. Karger, Basel, 1983, 88.
13. **Husband, A. J. and Gowans, J. L.,** The origin and antigen-dependent distribution of IgA-containing cells in the intestine, *J. Exp. Med.,* 148, 1146, 1978.
14. **Owen, R. L. and Jones, A. L.,** Scanning electron microscope evaluation of Peyer's patches in rats and humans, *Anat. Rec.,* 175, 404, 1973.
15. **Owen, R. L.,** Sequential uptake of horseradish peroxidase by lymphoid follicle epithelium of Peyer's patches in the normal unobstructed mouse intestine: an ultrastructural study, *Gastroenterology,* 72, 440, 1977.
16. **Owen, R. L. and Jones, A. L.,** Epithelial cell specialization within human Peyer's patches. An ultrastructural of intestinal lymphoid follicles, *Gastroenterology,* 66, 189, 1974.
17. **Owen, R. L. and Jones, A. L.,** Specialized lymphoid follicle epithelial cells in the human and non-human primate. A possible antigen uptake site, in *Scanning Electron Microscopy/1974, Part III. Proc. of the Workshop on Advances in Bio-Medical Applications of the Scanning Electron Microscope,* I.I.T. Research Institute, Chicago, IL, 1974, 697.
18. **Chu, R. M., Glock, R. D., and Ross, R. F.,** Gut-associated lymphoid tissues of young swine with emphasis in dome epithelium of aggregated lymph nodules(Peyer's patches) of the small intestine, *Am. J. Vet. Res.*, 40, 1720, 1979.
19. **Sousa, M.de,** *Lymphocyte Circulation: Experimental and Clinical Aspects,* J. Wiley & Son, New York, 1981.
20. **Richman, L. K., Graeff, A. S., and Strober, W.,** Antigen presentation by macrophage-enriched cells from the mouse Peyer's patch, *Cell. Immunol.,* 62, 110, 1981.
21. **Sminia, T., Janse, E. M., and Wilders, M. M.,** Antigen-trapping cells in Peyer's patches of the rat, *Scand. J. Immunol.,* 16, 481, 1982.
22. **Keren, D. F., Holt, P. S., Collins, H. H., Gemski, P., and Formal, S. B.,** The role of Peyer's patches in the local immune response of rabbit ileum to live bacteria, *J. Immunol.,* 120, 1892, 1978.
23. **Arey, L. B.,** *Human Histology,* W.B. Saunders, Philadelphia, 1974, 222.
24. **Burbige, E. J. and Sobky, R. Z. F.,** Endoscopic appearance of colonic lymphoid nodules: a normal variant, *Gastroenterology,* 72, 524, 1977.
25. **Pierce, N. F. and Gowans, J. L.,** Cellular kinetics of the intestinal response to cholera toxin in rats, *J. Exp. Med.,* 142, 1550, 1975.
26. **Husband, A. J., Beh, K. J., and Lascelles, A. K.,** IgA-containing cells in the ruminant intestine following intraperitoneal and local immunization, *Immunology,* 37, 597, 1979.
27. **Beh, K. J., Husband, A. J., and Lascelles, A. K.,** Intestinal response of sheep to intraperitoneal immunisation, *Immunology,* 37, 385, 1979.
28. **Besredka, A.,** De la vaccination contre des etats typhoides par le voie buccale, *Ann. Immunol. Paris,* 33, 882, 1919.
29. **Heremans, J. F., Heremans, M. T., and Schultz, H. E.,** Isolation and description of a few properties of the B_{2a}-globulin of human serum, *Clin. Chim. Acta,* 4, 96, 1959.
30. **Chodirker, W. B. and Tomasi, T. B.,** Gamma globulins: quantitative relationships in human serum and non-vascular fluids, *Science,* 142, 1080, 1936.
31. **Tomasi, T. B. and Zigelbaum, S.,** The selective occurrence of $gamma_{1a}$ globulins in certain body fluids, *J. Clin. Invest.,* 42, 1552, 1963.
32. **Crabbe´, P. A. and Heremans, J. F.,** The distribution of immunoglobulin-containing cells along the human gastrointestinal tract, *Gastroenterology,* 51, 305, 1966.
33. **Tourville, D., Adler, R., Bienenstock, J., and Tomasi, T. B.,** The human secretory immunoglobulin system: Immunohistological localization of gamma A, secretory "piece", and lactoferrin in normal human tissues, *J. Exp. Med.,* 129, 411, 1969.
34. **Crabbe´, P. A., Nash, D. R., Bazin, H., Eyssen, H., and Heremans, J. F.,** Antibodies of the IgA type in intestinal plasma cells of germ-free mice after oral stimulation or parenteral immunization with ferritin, *J. Exp. Med.,* 130, 723, 1969.
35. **Tomasi, T. B.,** Structure and function of mucosal antibodies, *Annu. Rev. Med.,* 21, 281, 1970.
36. **Rubin, W., Fauci, A. S., Sleisenger, M. H., and Jeffries, G. H.,** Immunoflourescent studies in adult celiac disease, *J. Clin. Invest.,* 44, 475, 1965.
37. **Crabbe´, P. A., Carbonara, A. D., and Heremans, J. F.,** The normal human intestinal mucosa as a major source of plasma cells containing gamma-immunoglobulin, *Lab. Invest.,* 14, 235, 1965.
38. **Heremans, J. F.,** The secretory immune system. A critical appraisal, in *The Immune System and Infectious Diseases,* Neter, E. and Milgrom, F., Eds., S. Karger, Basel, 1975, 376.

39. **Hanson, L. A.,** Comparative immunological studies of the immune globulins of human milk and blood serum, *Int. Arch. Allergy,* 18, 241, 1961.
40. **Tomasi, T. B. and Bienenstock, J.,** Secretory immunoglobulins, *Adv. Immunol.,* 9, 1, 1968.
41. **Poger, M. E. and Lamm, M. E.,** Localization of free and bound secretory component in human intestinal epithelial cells. A model for the assembly of secretory IgA, *J. Exp. Med.,* 139, 629, 1974.
42. **Brandtzaeg, P.,** Human secretory immunoglobulin M. An immunochemical and immunohistochemical study, *Immunology,* 29, 559, 1975.
43. **Brandtzaeg, P.,** Mucosal and glandular distribution of immunoglobulin components. Immunohistochemistry with a cold ethanol fixation technique, *Immunology,* 26, 1101, 1974.
44. **Kuhn, L. and Kraehenbuhl, J.-P.,** Role of secretory component, a secreted glycoprotein, in the specific uptake of IgA dimer by epithelial cells, *J. Biol. Chem.,* 254, 11072, 1979.
45. **Eskeland, T. and Brandtzaeg, P.,** Does J chain mediate the combination of 19S IgM and dimeric IgA with their polymerization?, *Immunochemistry,* 11, 161, 1974.
46. **Brandtzaeg, P.,** Complex formation between secretory component and human immunoglobulins related to their content of J chain, *Scand. J. Immunol.,* 5, 411, 1976.
47. **Brandtzaeg, P. and Baklien, K.,** Intestinal secretion of IgA and IgM: a hypothetical model, *Ciba Found. Symp.,* 46, 77, 1977.
48. **Ghetie, V. and Mota, G.,** The decrease of human colostral immunoglobulin A resistance to papain action after gradual release of the secretory component, *Immunochemistry,* 10, 839, 1973.
49. **Plaut, A. G. and Keonil, P.,** Immunoglobulins in human small intestinal fluid, *Gastroenterology,* 56, 522, 1969.
50. **Cripps, A. W., Husband, A. J., and Lascelles, A. K.,** The origin of immunoglobulins in intestinal secretion of sheep, *Aust. J. Exp. Biol. Med. Sci.,* 52, 711, 1974.
51. **Sheldrake, R. F., Husband, A. J., Watson, D. L., and Cripps, A. W.,** *J. Immunol.,* 132, 363, 1984.
52. **Bull, D. M., Bienenstock, J., and Tomasi, T. B.,** Studies on human intestinal immunoglobulin A, *Gastroenterology,* 60, 370, 1971.
53. **Tomasi, T. B., Tan, E. M., Solomon, A., and Prendergast, R. A.,** Characteristics of an immune system common to certain external secretions, *J. Exp. Med.,* 121, 101, 1965.
54. **Curtain, C. C., Clark, B. L., and Dufty, J. H.,** The origins of the immunoglobulins in the mucous secretions of cattle, *Clin. Exp. Immunol.,* 8, 151, 1971.
55. **Lee, C. S. and Lascelles, A. K.,** Antibody-producing cells in antigenically stimulated mammary glands and in the gastro-intestinal tract of sheep, *Aust. J. Exp. Biol. Med. Sci.,* 48, 525, 1970.
56. **Lascelles, A. K. and McDowell, G. H.,** Localized humoral immunity with particular reference to ruminants, *Transplant. Rev.,* 19, 170, 1974.
57. **Sheldrake, R. F. and Husband, A. J.,** Immune defences at mucosal surfaces in ruminants, *J. Dairy Res.,* 52, 599, 1985.
58. **Brandtzaeg, P., Baklien, K., Fausa, O., and Hoel, P.,** Immunohistochemical characterization of local immunoglobulin formation in ulcerative colitis, *Gastroenterology,* 66, 1123, 1974.
59. **Quin, J. W., Husband, A. J., and Lascelles, A. K.,** The origin of the immunoglobulins in intestinal lymph of sheep, *Aust. J. Exp. Biol. Med. Sci.,* 53, 205, 1975.
60. **Delacroix, D. L., Hodgson, H. J. F., McPherson, A., and Dive, C.,** Selective transport of immunoglobulin A in bile. Quantitative relationships of monomeric and polymeric immunoglobulin A, immunoglobulin M, and other proteins in serum, bile and saliva, *J. Clin. Invest.,* 70, 230, 1982.
61. **Scicchitano, R., Sheldrake, R. F., and Husband, A. J.,** Origin of immunoglobulins in respiratory tract secretion and saliva of sheep, *Immunology,* 58, 315, 1985.
62. **Orlans, E., Peppard, J., Fry, J. F., Hinton, R. H., and Mullock, B. M.,** Secretory component as the receptor for polymeric IgA on rat hepatocytes, *J. Exp. Med.,* 150, 1577, 1979.
63. **Socken, D. J., Jeejeebhoy, K. N., Bazin, H., and Underdown, B. J.,** Identification of secretory component as an IgA receptor on rat hepatocytes, *J. Exp. Med.,* 150, 1538, 1979.
64. **Russell, M. W., Brown, T. A., and Mestecky, J.,** Preferential transport of IgA and IgA-immune complexes to bile compared with other external secretions, *Mol. Immunol.,* 19, 677, 1982.
65. **Halsey, J. F. Johnson, B. H., and Cebra, J. J.,** Transport of immunoglobulin from serum into colostrum, *J. Exp. Med.,* 151, 767, 1980.
66. **Halsey, J. F., Mitchell, C., Meyer, R., and Cebra, J. J.,** Metabolism of immunoglobulin A in lactating mice: origins of immunoglobulin A in milk, *Eur. J. Immunol.,* 12, 107, 1982.
67. **Jackson, G. D. F., Lemaitre-Coelho, I., Vaerman, J.-P., Bazin, H., and Beckers, A.,** Rapid disappearance from serum of intravenously injected rat myeloma IgA and its secretion into bile, *Eur. J. Immunol.,* 8, 123, 1978.
68. **Lemaitre-Coelho, I., Jackson, G. D. F., and Vaerman, J. -P.,** Rat bile as a convenient source of secretory IgA and free secretory component, *Eur. J. Immunol.,* 8, 588, 1977.

69. **Lemaître-Coelho, I., Jackson, G. D. F., and Vaerman, J.-P.,** Relevance of biliary IgA antibodies in rat intestinal immunity, *Scand. J. Immunol.,* 8, 459, 1978.
70. **Renston, R. H., Jouvs, A. L., Christiansen, W. D., Hradek, G. T., and Underdown, B. J.,** Evidence for a vesicular transport mechanism in hepatocytes for biliary secretion of immunoglobulin A, *Science,* 208, 1276, 1980.
71. **Vaerman, J.-P., Lemaître-Coelho, I., Limet, J. N., and Dalacroix, D. L.,** Hepatic transfer of polymeric IgA from plasma to bile in rats and other mammals: a survey, in *Recent Advances in Mucosal Immunity,* Strober, W., Hanson, L. A. and Sell, K. W., Eds., Raven Press, New York, 1982, 233.
72. **Sheldrake, R. F., Scicchitano, R., and Husband, A. J.,** The effect of lactation on the transport of serum-derived IgA into bile of sheep, *Immunology,* 54, 471, 1985.
73. **Orlans, E., Peppard, J. V., Payne, A. W. R., Fitzharris, B. M., Mullock, B. M., Hinton, R. H., and Hall, J. G.,** Comparative aspects of the hepatobiliary transport of IgA, *Ann. N.Y. Acad. Sci.,* 409, 411, 1983.
74. **Hall, J. G., Gyure, L. A., and Payne, A. W. R.,** Comparative aspects of the transport of immunoglobulin A from blood to bile, *Immunology,* 41, 899, 1980.
75. **Heremans, J. F. and Vaerman, J. P.,** Biologic significance of IgA antibodies in serum and secretions, in *Progress in Immunology,* Vol. 1, Amos, B., Ed., Academic Press, New York, 1971, 875.
76. **Beh, K. J., Watson, D. L., and Lascelles, A. K.,** Concentrations of immunoglobulins and albumin in lymph collected from various regions of the body of sheep, *Aust. J. Exp. Biol. Med. Sci.,* 52, 81, 1974.
77. **Scicchitano, R., Husband, A. J., and Cripps, A. W. C.,** Immunoglobulin-containing cells and the origin of immunoglobulins in the respiratory tract of sheep, *Immunology,* 52, 529, 1984.
78. **Vaerman, J.-P.,** Studies on IgA Immunoglobulins in Man and Animals, Ph.D. thesis, Universite´ Catholique de Louvain, Belgium, 1970.
79. **Vaerman, J.-P., Andre´, C., Bazin, H., and Heremans, J. F.,** Mesenteric lymph as a major source of serum IgA in guinea pig and rats, *Eur. J. Immunol.,* 3, 580, 1973.
80. **Leslie, G.A. and Martin, L.N.,** Studies on the secretory immunologic system of the fowl. III. Serum and secretory IgA of the chicken, *J. Immunol.,* 110, 1, 1973.
81. **Lebacq-Verheyden, A. M., Vaerman, J.-P., and Heremans, J. F.,** Quantification and distribtuion of chicken immunoglobulins IgA, IgM and IgG in serum and secretions, *Immunology,* 27, 683, 1974.
82. **Porter, P. and Allen, W. D.,** Classes of immunoglobulins related to immunity in the pig, *J. Am. Vet. Med. Ass.* 160, 511, 1972.
83. **Heremans, J. F.,** Immunoglobulin A, in *The Antigens,* Vol. 2, Sela, M., Ed., Academic Press, New York, 1974 365.
84. **Kutteh, W. H., Prince, S. J., and Mestecky, J.,** Tissue origins of human polymeric and monomeric IgA, *J Immunol.,* 128, 990, 1982.
85. **McDermott, M. R., Befus, A. D., and Bienenstock, J.,** The structural basis for immunity in the respiratory tract, *Int. Rev. Exp. Pathol.,* 23, 47, 1982.
86. **Ishizaka, T., Ishizaka, K., Borsos, T., and Rapp, H.,** C´1 fixation by human isoagglutinins: fixation of C´1 by gamma G and gamma M but not gamma A antibody, *J. Immunol.,* 97, 716, 1966.
87. **Colten, H. R. and Bienenstock, J.,** Lack of C_3 activation through classical or alternate pathways by human secretory IgA anti-blood group A antibody, *Adv. Exp. Med. Biol.,* 45, 305, 1974.
88. **Hall, W. H., Manion, R. E., and Zinneman, H. H.,** Blocking serum lysis of *Brucella abortus* by hyperimmune rabbit immunoglobulin A, *J. Immunol.,* 107, 41, 1971.
89. **Griffiss, J. McL. and Bertram, M. A.,** Immunoepidemiology of meningoccoccal disease in military recruits II. Blocking of serum bactericidal activity by circulating IgA early in the course of invasive disease, *J. Infect Dis.,* 136, 733, 1977.
90. **Russell-Jones, G. J., Ey, P. L., and Reynolds, B. L.,** The ability of IgA to inhibit the complement-mediated lysis of target red blood cells sensitized with IgG antibody, *Mol. Immunol.,* 17, 1173, 1980.
91. **Russell-Jones, G. J., Ey, P. L., and Reynolds, B. L.,** The ability of IgA to inhibit complement consumption by complement-fixing antigens and antigen-antibody complexes, *Aust. J. Exp. Biol. Med. Sci.,* 62, 1, 1984
92. **Johnson, K. J., Wilson, B. S., Till, G. O., and Ward, P. A.,** Acute lung injury in rat caused by immunoglobulin A immune complexes, *J. Clin. Invest.,* 74, 358, 1984.
93. **Adinolfi, M., Glynn, A. A., Lindsay, M., and Milne, C. M.,** Serological properties of gamma A antibodies to *Escherichia coli* present in human colostrum, *Immunology,* 10, 517, 1966.
94. **Lowell, G. H., Smith, L. F., Griffiss, J. M., and Brandt, B. L.,** IgA-dependent, monocyte-mediated antibacterial activity, *J. Exp. Med.,* 152, 452, 1980.
95. **Reynolds, H. Y. and Thompson, R. E.,** Pulmonary host defences, II. Interaction of respiratory antibodies with *Pseudomonas aeruginosa* and alveolar macrophages, *J. Immunol.,* 111, 369, 1973.
96. **Musher, D. M., Goree, A., Baughn, R. E., and Birdsall, H. H.,** Immunoglobulin A from bronchopulmonary secretion blocks bactericidal and opsonizing effects of antibody to nontypable *Haemophilus influenzae, Infect Immun.,* 45, 36, 1984.

97. **Kaur, J., McGhee, J. R., and Burrows, W.,** Immunity to cholera: The occurrence and nature of antibody-active immunoglobulins in the lower ileum of the rabbit, *J. Immunol.*, 108, 387, 1972.
98. **Yardley, J. H., Keren, D. F., Hamilton, S. R., and Brown, G. D.,** Local (immunoglobulin A) immune response by the intestine to cholera toxin and its partial suppression with combined systemic and intra-intestinal immunization, *Infect. Immun.*, 19, 589, 1978.
99. **Pierce, N. F., Cray, W. C., and Engel, P. F.,** Antitoxic immunity to cholera in dogs immunized orally with cholera toxin, *Infect. Immun.*, 27, 632, 1980.
100. **Ogra, P. L., Karzon, D. T., Righthand, F., and MacGillivray, M.,** Immunoglobulin response in serum and secretion after immunization with live and inactivated poliovaccine and natural infection, *N. Engl. J. Med.*, 279, 893, 1968.
101. **Ogra, P. L. and Karzon, D. T.,** Poliovirus antibody response in serum and nasal secretions following intranasal inoculation with inactivated poliovaccine, *J. Immunol.*, 102, 15, 1969.
102. **Lowell, G. H., Smith, L. F., Griffiss, J. McL., Brandt, B. L., and MacDermott, R. P.,** Antibody-dependent mononuclear cell-mediated antimeningococcal activity. Comparison of the effects of convalescent and postimmunization immunoglobulins G, M, and A, *J. Clin. Invest.*, 66, 260, 1980.
103. **Shen, L. and Fanger, M. W.,** Secretory IgA antibodies synergize with IgG in promoting ADCC by human polymorphonuclear cells, monocytes, and lymphocytes, *Cell Immunol.*, 59, 75, 1981.
104. **Tagliabue, A., Nencioni, L., Villa, L., Keren, D. F., Lowell, G. H., and Boraschi, D.,** Antibody-dependent cell-mediated antibacterial activity of intestinal lymphocytes with secretory IgA, *Nature* (London), 306, 184, 1983.
105. **Tagliabue, A., Boraschi, D., Villa, L., Keren, D. F., Lowell, G. H., Rappuoli, R., and Nencioni, L.,** IgA-dependent cell-mediated activity against enteropathogenic bacteria: distribution, specificity, and characterization of the effector cells, *J. Immunol.*, 133, 988, 1984.
106. **McClelland, D. B. L., Samson, R. R., Parkin, D. M., and Shearman, D. J. C.,** Bacterial agglutination studies with secretory IgA prepared from human gastrointestinal secretions and colostrum, *Gut*, 13, 450, 1972.
107. **Williams, R. C. and Gibbons, R. J.,** Inhibition of bacterial adherence by secretory immunoglobulin A: a mechanism of antigen disposal, *Science*, 177, 697, 1972.
108. **Parry, S. H. and Porter, P.,** Immunological aspects of cell membrane adhesion demonstrated by porcine enteropathogenic *Escherichia coli*, *Immunology*, 34, 41, 1978.
109. **Walker, W. A., Isselbacher, K. J., and Bloch, K. J.,** Intestinal uptake of macromolecules: Effect of oral immunization, *Science*, 177, 608, 1972.
110. **Andre´, C., Lambert, R., Bazin, H., and Heremans, J. F.,** Interference of oral immunization with the intestinal absorption of heterologous albumin, *Eur. J. Immunol.*, 4, 701, 1974.
111. **Stokes, C. R., Soothill, J. F., and Turner, M. W.,** Immune exclusion is a function of IgA, *Nature (London)*, 255, 745, 1975.
112. **Green, F. H. Y. and Freed, D. L. J.,** Antibody-facilitated digestion and the consequences of its failure, in *Antigen Absorption by the Gut*, Hemmings, W. A., Ed., MTP Press, Lancaster, 1978, 189.
113. **Gowans, J. L. and Knight, E. J.,** The route of re-circulation of lymphocytes in the rat, *Proc. R. Soc. B.*, 159, 257, 1964.
114. **Guy-Grand, D., Griscelli, C., and Vassalli, P.,** The gut-associated lymphoid system: nature and properties of the large dividing cells, *Eur. J. Immunol.*, 4, 435, 1974.
115. **Husband, A. J., Monié, H. J., and Gowans, J. L.,** The natural history of the cells producing IgA in the gut, *Ciba Found. Symp.*, 46, 29, 1977.
116. **Williams, A. F. and Gowans, J. L.,** The presence of IgA on the surface of rat thoracic duct lymphocytes which contain internal IgA, *J. Exp. Med.*, 141, 335, 1975.
117. **Craig, S. W. and Cebra, J. J.,** Peyer's patches: an enriched source of precursors for Iga-producing immunocytes in the rabbit, *J. Exp. Med.*, 134, 188, 1971.
118. **Cebra, J. J., Kamat, R., Gearhart, P., Robertson, S. M., and Tseng, J.,** The secretory IgA system of the gut, *Ciba Found. Symp.*, 46, 5, 1977.
119. **McWilliams, M., Phillips-Quagliata, J. M., Lamm, and M. E.,** Characteristics of mesenteric lymph node cells homing to gut-associated lymphoid tissue in syngeneic mice, *J. Immunol.*, 115, 54, 1975.
120. **Gowans, J. L. and Steer, H. W.,** The function and pathways of lymphocyte recirculation, *Ciba Found. Symp.*, 71, 113, 1980.
121. **Bennell, M. A. and Husband, A. J.,** Route of lymphocyte migration in pigs. I. Lymphocyte circulation in gut-associated lymphoid tissues, *Immunology*, 42, 469, 1981.
122. **Bennell, M. A. and Husband, A. J.,** Route of lymphocyte migration in pigs. II. Migration to the intestinal lamina propria of antigen-specific cells generated in response to intestinal immunization in the pig, *Immunology*, 42, 475, 1981.
123. **Tseng, J.,** A population of resting IgM-IgD double-bearing lymphocytes in Peyer's patches: the major precursor cells for IgA plasma cells in the gut lamina propria, *J. Immunol.*, 132, 2730, 1984.

124. **Griscelli, C., Vassalli, P., and McClusky, R. T.**, The distribution of large dividing lymph node cells in syngeneic recipient rats after intravenous injection, *J. Exp. Med.*, 130, 1427, 1969.
125. **Hall, J. G., Parry, D. M., and Smith, M. E.**, The distribution and differentiation of lymph-borne immunoblasts after intravenous injection into syngeneic recipients, *Cell. Tissue Kinet.*, 5, 269, 1972.
126. **Scollay, R., Hopkins, J., and Hall, J. G.**, Possible role of surface immunoglobulin in non-random recirculation of small lymphocytes, *Nature(London)*, 260, 528, 1976.
127. **Rose, M. L., Parrott, D. M. V., and Bruce, R. G.**, Migration of lymphoblasts to the small intestine. I. Effect of *Trichinella spiralis* infection on the migration of mesenteric lymphoblasts and mesenteric T lymphoblasts in syngeneic mice, *Immunology*, 31, 723, 1976.
128. **Sprent, J.**, Fate of H2-activated T lymphocytes in syngeneic hosts. I. Fate in lymphoid tissues and intestines traced with ^{3}H-thymidine, ^{125}I-deoxyuridine and ^{51}chromium, *Cell. Immunol.*, 21, 278, 1976.
129. **McDermott, M. R. and Bienenstock, J.**, Evidence for a common mucosal immunologic system. I. Migration of B immunoblasts into intestinal respiratory and genital tissues, *J. Immunol.*, 122, 1892, 1979.
130. **Halstead, T. E. and Hall, J. G.**, The homing of lymph-borne immunoblasts to the small gut of neonatal rats, *Transplantation*, 14, 339, 1972.
131. **Ferguson, A. A. and Parrott, D. M. V.**, Growth and deveopment of 'antigen-free' grafts of foetal mouse intestine, *J. Pathol.*, 106, 95, 1972.
132. **Ogra, P. L. and Karzon, D. T.**, Distribution of poliovirus antibody in serum, nasopharynx and alimentary tract with poliovaccine, *J. Immunol.*, 102, 1423, 1969.
133. **Husband, A. J.**, Kinetics of extravasation and redistribution of IgA-specific antibody-containing cells in the intestine, *J. Immunol.*, 128, 1355, 1982.
134. **Husband, A. J. and Dunkley, M. L.**, Lack of site of origin effects on distribution of IgA antibody-containing cells, *Immunology*, 54, 215, 1985.
135. **Butcher, E. C., Scollay, R. G., and Weissman, I. L.**, Organ specificity of lymphocyte migration: mediation by highly selective lymphocyte interaction with organ-specific determinants on high endothelial venules, *Eur. J. Immunol.*, 10, 556, 1980.
136. **Elson, C. O., Heck, J. A., and Strober, W.**, T cell regulation of IgA synthesis, in *Immunology of Breast Milk*, Ogra, P. L. and Dayton, D. H., Eds., Raven Press, New York, 1979, 37.
137. **Clough, J. D., Mims, L. H., and Strober, W.**, Deficient IgA antibody responses to arsanilic acid bovine serum albumin (BSA) in neonatally thymectomized rabbits, *J. Immunol.*, 106, 1624, 1971.
138. **Crewther, P. and Warner, N. L.**, Serum immunoglobulins and antibodies in congenitally athymic (nude) mice, *Aust. J. Exp. Biol. Med. Sci.*, 50, 625, 1972.
139. **Ebersole, J. L., Taubman, M. A., and Smith, D. J.**, Thymic control of secretory antibody responses in the rat, *J. Immunol.*, 123, 19, 1979.
140. **Kiyono, H., McGhee, J. R., Mosteller, L. M., Eldridge, J. H., Koopman, W. J., Kearney, J. F., and Michalek, S. M.**, Murine Peyer's patch T cell clones. Characterization of antigen-specific helper T cells for immunoglobulin A responses, *J. Immunol.*, 156, 1115, 1982.
141. **Elson, C. O., Heck, J. A., and Strober, W.**, T-cell regulation of murine IgA synthesis, *J. Exp. Med.*, 149, 632, 1979.
142. **Husband, A. J. and Dunkley, M. L.**, T-cell regulation of the mucosal IgA immune response, *J. Leukocyte Biol.*, 36, 417, 1984.
143. **Mayer, L., Fu, S. M., and Kunkel, H. G.**, Human T cell hybridomas secreting factors for IgA-specific help, polyclonal B cell activation, and cell proliferation, *J. Exp. Med.*, 156, 1860, 1982.
144. **Kiyono, H., Phillips, J. O., Colwell, D. E., Michalek, S. M., Koopman, W. J. and McGhee, J. R.**, Isotype specificity of T cell clones: Fc(alpha) receptors regulate T and B cell collaboration for IgA responses, *J. Immunol.*, 133, 1087, 1984.
145. **Kawanishi, H., Saltzman, L. E., and Strober, W.**, Characteristics and regulatory function of murine con A-induced, cloned T cells obtained from Peyer's patches and spleen: mechanisms regulating isotype-specific immunoglobulin production by Peyer's patch B cells, *J. Immunol.*, 129, 475, 1982.
146. **Cebra, J. J., Fuhrman, J. A., Gearhart, P. J., Hurwitz, J. L., and Shahin, R. D.**, B lymphocyte differentiation leading to a commitment to IgA expression may depend on cell division and may occur during antigen-stimulated clonal expansion, in *Recent Advances in Mucosal Immunity*, Strober, W., Hanson, L. A. and Sell, K. W., Eds., Raven Press, New York, 155.
147. **Gearhart, P. J., Hurwitz, J. L., and Cebra, J. J.**, Successive switching of antibody isotypes expressed within the lines of a B-cell clone, *Proc. Natl. Acad. Sci. U.S.A.*, 77, 5424, 1980.
148. **Hoover, R. G. and Lynch, R. G.**, Isotype-specific suppression of IgA: Suppression of IgA responses in Balb/c mice by T(alpha) cells, *J. Immunol.*, 130, 521, 1983.
149. **Challacombe, S. J. and Tomasi, T. B.**, Systemic tolerance and secretory immunity after oral immunization, *J. Exp. Med.*, 152, 1459, 1980.

150. **Mattingly, J. A. and Waksman, B. H.**, Immunologic suppression after oral administration of antigen. I. Specific suppressor cells found in rat Peyer's patches after oral administration of sheep erythrocytes and their systemic migration, *J. Immunol.,* 121, 1878, 1978.
151. **Ngan, J. and Kind, L. S.**, Suppressor T cells for IgE and IgG in Peyer's patches of mice made tolerant by the oral administration of ovalbumin, *J. Immunol.,* 120, 861, 1978.
152. **Richman, L. K., Graeff, A. S., Yarchoan, R., and Strober, W.**, Simultaneous induction of antigen-specific IgA helper T cells and IgG suppressor T cells in the murine Peyer's patch after protein feeding, *J. Immunol.,* 126, 2079, 1981.
153. **Liew, F. Y.**, Regulation of delayed-type hypersensitivity to pathogens and alloantigens, *Immunol. Today,* 3, 18, 1982.
154. **Klein, E.**, Natural and activated cytotoxic T lymphocytes, *Immunol. Today,* 1, iv, 1980.
155. **Kagnoff, M. F.**, Effects of antigen-feeding on intestinal and systemic immune responses. I. Priming of precursor cytotoxic T cells by antigen feeding, *J. Immunol.,* 120, 395, 1978.
156. **Husband, A. J., Dunkley, M. L., Cripps, A. W., and Clancy, R. L.**, Antigen-specific response among T lymphocytes following intestinal administration of alloantigens, *Aust. J. Exp. Biol. Med. Sci.,* 62, 687, 1984.
157. **Frederick, G. T. and Bohl, E. H.**, Local and systemic cell-mediated immunity against transmissible gastroenteritis, an intestinal infection of swine, *J. Immunol.,* 116, 1000, 1976.
158. **Huntley, J., Newby, T. J., and Bourne, F. J.**, The cell-mediated immune response of the pig to orally administered antigen, *Immunology,* 37, 225, 1979.
159. **Mowat, A. M. and Ferguson, A.**, Induction and expression of mucosal cell mediated immunity, in *The Mucosal Immune System,* Bourne, F. J., Ed., Martinus Nijhoff, The Hague, 1980, 107.
160. **Ferguson, A. and McDonald, T. T.**, Effects of local delayed hypersensitivity on the small intestine, *Ciba Found. Symp.,* 46, 305, 1977.
161. **Howard, J. C., Hunt, S. V., and Gowans, J. L.**, Identification of marrow-derived and thymus-derived small lymphocytes in the lymphoid tissue and thoracic duct lymph of normal rats, *J. Exp. Med.,* 135, 200, 1972.
162. **Guy-Grand, D., Griscelli, C., and Vassalli, P.**, The mouse gut T lymphocyte, a novel type of T cell. Nature, origin and traffic in mice in normal and graft-versus-host conditions, *J. Exp. Med.,* 148, 1661, 1978.
163. **Lyscom, N. and Brueton, M. J.**, Intraepithelial, lamina propria and Peyer's patch lymphocytes of the rat small intestine: isolation and characterization in terms of immunoglobulin markers and receptors for monoclonal antibodies, *Immunology,* 45, 775, 1982.
164. **Tagliabue, A., Befus, A. D., Clark, D. A., and Bienenstock, J.**, Characteristics of natural killer cells in the murine intestinal epithelium and lamina propria, *J. Exp. Med.,* 155, 1785, 1982.
165. **Flexman, J. P., Shellam, G. R., and Mayrhofer, G.**, Natural cytotoxicity, responsiveness to interferon and morphology of intra-epithelial lymphocytes from the small intestine of the rat, *Immunology,* 48, 733, 1983.
166. **Heyworth, M. F. and Owen, R. L.**, Harvesting and characterization of lymphocytes from the mouse small intestinal lumen after experimental infection, *Immunobiology,* 163, 243, 1982.
167. **Heatley, R. V. and Bienenstock, J.**, Luminal lymphoid cells in the rabbit intestine, *Gastroenterology,* 82, 268, 1982.
168. **Arnaud-Battandier, F., Bundy, B. M., O'Neill, M., Bienenstock, J., and Nelson, D. L.**, Cytotoxic activities of gut mucosal lymphoid cells in guinea pigs, *J. Immunol.,* 121, 1059, 1978.
169. **Koren, H. and Herberman, R.**, The cryptic orphan killer cells, *Immunol. Today,* 4, 97, 1983.
170. **Tagliabue, A., Luini, W., Soldateschi, D., and Boraschi, D.**, Natural killer activity of gut mucosal lymphoid cells in mice, *Eur. J. Immun.,* 11, 919, 1981.
171. **Gibson, P. R., Dow, E. L., Selby, W. S., Strickland, R. G., and Jewell, D. P.**, Natural killer cells and spontaneous cell-mediated cytotoxicity in human intestine, *Clin. Exp. Immunol.,* 56, 438, 1984.
172. **Bienenstock, J., Befus, A. D., McDermott, M., Mirski, S., Rosenthal, K., and Tagliabue, A.**, The mucosal immunological network: compartmentalization of lymphocytes, natural killer cells, and mast cells, *Ann. N.Y. Acad. Sci.,* 409, 164, 1983.
173. **Cepica, A. and Derbyshire, J. B.**, Antibody-dependent cell-mediated cytotoxicity against cells infected with transmissible gastroenteritis virus, *Can. J. Comp. Med.,* 47, 298, 1984.
174. **Cepica, A. and Derbyshire, J. B.**, The effect of adoptive transfer of mononuclear lymphocytes from an adult donor on spontaneous cell-mediated cytotoxicity and resistance to transmissible gastroenteritis in neonatal piglets, *Can. J. Comp. Med.,* 48, 360, 1984.
175. **Nencioni, L., Villa, L., Boraschi, D., Berti, B., and Tagliabue, A.**, Natural and antibody-dependent cell-mediated activity against *Salmonella typhimurium* by peripheral and intestinal lymphoid cells in mice, *J. Immunol.,* 130, 903, 1983.
176. **Tagliabue, A., Nencioni, L., Villa, L., and Boraschi, D.**, Genetic control of *in vitro* natural cell-mediated activity against *Salmonella typhimurium* by intestinal and splenic lymphoid cells in mice, *Clin. Exp. Immunol.,* 56, 531, 1984.

177. **Doughri, A. M., Altera, K. P., and Kainer, R. A.,** Some developmental aspects of the bovine fetal gut, *Zentralbl. Vet. Med. A.,* 19, 417, 1972.
178. **Chapman, H. A., Johnson, J. S., and Cooper, M. D.,** Ontogeny of Peyer's patches and immunoglobulin-containing cells in pigs, *J. Immunol.,* 112, 555, 1974.
179. **Cole, G. J. and Morris, B.,** The lymphoid apparatus of the sheep: Its growth, development and significance in immunologic reactions, *Adv. Vet. Sci. Comp. Med.,* 17, 225, 1973.
180. **Cooper, M. D., Kearney, J. F., Lawton, A. R., Abney, E. R., Parkhouse, R. M. E., Preud'homme, J.L., and Seligman, M.,** Generation of immunoglobulin class diversity in B cells: a discussion with emphasis on IgD development, *Ann. Immunol. Inst. Pasteur,* 127C, 573, 1976.
181. **Parkhouse, R. M. E. and Cooper, M. D.,** A model for the differentiation of B lymphocytes with implications for the biological role of IgD, *Immunol. Rev.,* 37, 105, 1977.
182. **Butcher, E. C., Rouse, R. V., Coffman, R. L., Nottenberg, C. N., Hardy, R. R., and Weissman, I. L.,** Surface phenotype of Peyer's patch germinal center cells: implications for the role of germinal centers in B cell differentiation, *J. Immunol.,* 129, 2698, 1982.
183. **Miyasaka, M., Dudler, L., Bordmann, G., Leiserson, W. M., Gerber, H. A., Reynolds, J., and Trnka, Z.,** Differentiation of B lymphocytes in sheep. I. Phenotypic analysis of ileal Peyer's patch cells and the demonstration of a precursor population for sIg^+cells in the ileal Peyer's patches, *Immunology,* 53, 515, 1984.
184. **Allen, W. D. and Porter, P.,** Localization of immunoglobulins in intestinal mucosa and the production of secretory antibodies in response to intraluminal administration of bacterial antigens in the preruminant calf, *Clin. Exp. Immunol.,* 21, 407, 1975.
185. **Brown, P. J. and Bourne, F. J.,** Development of immunoglobulin-containing cell populations in intestine, spleen, and mesenteric lymph node of the young pig, as demonstrated by peroxidase-conjugated antiserums, *Am. J. Vet. Res.,* 37, 1309, 1976.
186. **Allen, W. D. and Porter, P.,** The relative frequencies and distribution of immunoglobulin-bearing cells in the intestinal mucosa of neonatal and weaned pigs and their significance in the development of secretory immunity, *Immunology,* 32, 819, 1977.
187. **Husband, A. J. and McDowell, G. H.,** Immunity to experimental enteritis in lambs vaccinated prenatally, *Res. Vet. Sci.,* 25, 343, 1978.
188. **Pittard, W. B. and Bill, K.,** Immunoregulation by breast milk cells, *Cell. Immunol.,* 42, 437, 1979.
189. **Manton, V. J. A., Peacock, R., Poynter, D., Silverman, P. H., and Terry, R. J.,** The influence of age on naturally acquired resistance to *Haemonchus contortus* in lambs, *Res. Vet. Sci.,* 3, 308, 1962.
190. **Urquhart, G. M., Jarrett, W. F. H., Jennings, F. W., McIntyre, W. I. M., and Mulligan, W.,** Immunity to *Haemonchus contortus* infection: relationship between age and successful vaccination with irradiated larvae, *Am. J. Vet. Res.,* 27, 1645, 1966.
191. **Ogilvie, B. M. and Hockley, D. J.,** Effects of immunity on *Nippostrongylus brasiliensis* adult worms. Reversible and irreversible changes in infectivity, reproduction and morphology, *J. Parasitol.,* 54, 1073, 1968.
192. **Jarrett, E. E. E., Urquhart, G. M., and Douthwaite, R. M.,** Immunological unresponsiveness to helminth parasites. II. Antibody response and hypersensitivity reaction associated with *Nippostrongylus brasiliensis* infection in young rats, *Exp. Parasitol.,* 24, 270, 1969.
193. **Cepica, A. and Derbyshire, J. B.,** Antibody-dependent and spontaneous cell-mediated cytotoxicity against transmissible gastroenteritis virus infected cells by lymphocytes from sows, fetuses and neonatal pigs, *Can. J. Comp. Med.,* 48, 258, 1984.
194. **Brundage, L. J., Derbyshire, J. B., and Wilkie, B. N.,** Cell mediated immune responses in a porcine enterovirus infection in piglets, *Can. J. Comp. Med.,* 44, 61, 1980.
195. **Huntley, J., Newby, T. J., and Bourne, F. J.,** The cell-mediated immune response of the pig to orally administered antigen, *Immunology,* 37, 225, 1979.
196. **Lyscom, N. and Brueton, M. J.,** The development of intraepithelial and Peyer's patch lymphocyte sub-types in the small intestine of newborn rats, *Clin. Exp. Immunol.,* 54, 158, 1983.
197. **Husband, A. J. and Lascelles, A. K.,** Antibody responses to neonatal immunization in calves, *Res. Vet. Sci.,* 18, 201, 1975.
198. **Bourne, F. J., Pickup, J., and Honour, J. W.,** Intestinal immunoglobulins in the pig, *Biochim. Biophys. Acta.,* 229, 18, 1971.
199. **Watson, D. L., Bennell, M. A., and Chaniago, T. D.,** The effect of circulating, maternally-derived antibody on the development of a local immune response in the intestine of the neonatal pig, *Am. J. Vet. Res.,* 40, 61, 1977.
200. **Logan, E. F., Stenhouse, A., Ormród, D. J., and Penhale, W. J.,** The role of colostral immunoglobulins in intestinal immunity to enteric colibacillosis in the calf, *Res. Vet. Sci.,* 17, 290, 1974.
201. **Husband, A. J., Brandon, M. R., and Lascelles, A. K.,** The effect of corticosteroid on absorption and endogenous production of immunoglobulins in calves, *Aust. J. Exp. Biol. Med. Sci.,* 51, 707, 1973.
202. **Saif, L. J., Bohl, E. H., and Gupta, R. K. P.,** Isolation of porcine immunoglobulin and determination of the immunoglobulin classes of transmissible gastroenteritis virus, *Infect. Immun.,* 6, 600, 1972.

203. **Bohl, E. H., Gupta, R. K. P., Olquin, M. V. F., and Saif, L. J.,** Antibody responses in serum, colostrum, and milk of swine after infection or vaccination with transmissible gastroenteritis virus, *Infect. Immun.,* 6, 289, 1972.
204. **Roux, M. E., McWilliams, M., Phillips-Quagliata, J. M., Weisz-Carrington, P., and Lamm, M. E.,** Origin of IgA-secreting plasma cells in the mammary gland, *J. Exp. Med.,* 146, 1311, 1977.
205. **Rose, M. L., Parrott, D. M. V., and Bruce, R. G.,** The accumulation of immunoblasts in extravascular tissues including mammary gland, peritoneal cavity, gut and skin, *Immunology,* 35, 415, 1978.
206. **Weisz-Carrington, P., Roux, M. E., McWilliams, M., Phillips-Quagliata, J. M., and Lamm, M. E.,** Organ and isotype distribution of plasma cells producing specific antibody after oral immunization: evidence for a generalized secretory immune system, *J. Immun.,* 123, 1705, 1979.
207. **Lascelles, A. K., Beh, K. J., and Husband, A. J.,** Origin of antibody in mammary secretion with particular reference to the IgA system, *Adv. Exp. Biol. Med.,* 137, 493, 1980.
208. **Sheldrake, R. F., Husband, A. J., and Watson, D. L.,** Specific antibody-containing cells in the mammary gland of non-lactating sheep following intraperitoneal and intramammary immunization, *Res. Vet. Sci.,* 38, 312, 1985.
209. **Sheldrake, R. F., Husband, A. J., Watson, D. L., and Cripps, A. W.,** The effect of intraperitoneal and intramammary immunization of sheep on numbers of antibody-containing cells in the mammary gland and antibody titres in blood serum and mammary secretions, *Immunology,* 56, 605, 1985.
210. **Montgomery, P. C., Rosner, B. R., and Cohn, J.,** The secretory antibody response. Anti-DNP antibodies induced by dinitrophenylated type III pneumococcus, *Immunol. Commun.,* 3, 143, 1974.
211. **Weisz-Carrington, P., Roux, M. E., and Lamm, M. E.,** Plasma cells and epithelial immunoglobulins in the mouse mammary gland during pregnancy and lactation, *J. Immunol.,* 119, 1306, 1977.
212. **McDermott, M. R., Clark, D. A., and Bienenstock, J.,** Evidence for a common mucosal immunologic system. II. Influence of the estrus cycle on B immunoblast migration into genital and intestinal tissues, *J. Immunol.,* 124, 2536, 1980.
213. **Scicchitano, R., Husband, A. J., and Clancy, R. L.,** Contribution of intraperitoneal immunization to the local immune response in the respiratory tract of sheep, *Immunology,* 53, 375, 1984.
214. **Clancy, R. L., Cripps, A. W., Husband, A. J., and Buckley, D.,** Specific immune response in the respiratory tract after administration of an oral polyvalent baterial vaccine, *Infect. Immun.,* 39, 491, 1983.
215. **Puigdollers, J. M., Serna, G. R., Rey, I. H., Barruffet, M. T. T., and Torroella, J. J.,** Immunoglobulin production in man stimulated by an orally administered bacterial lysate, *Respiration,* 40, 142, 1980.
216. **Edmondson, W. P., Purcell, R. H. and Gundelfinger, B. F.,** Immunisation by selective infection with type 4 adenovirus grown in human diploid tissue culture. II. Specific protective effect against epidemic disease, *J. Am. Med. Assoc.,* 195, 453, 1966.
217. **Sturn, B. and Schneweis, K. E.,** Protective effect of an oral infection with herpes simplex virus type 1 against subsequent genital infection with herpes simplex virus type 2, *Med. Microbiol. Immunol.,* 165, 119, 1978.
218. **Mestecky, J., McGhee, J. R., Arnold, R. R., Michalek, S. M., Prince, S. J., and Babb, J. L.,** Selective induction of an immune response in human external secretions by ingestion of bacterial antigen, *J. Clin. Invest.,* 61, 731, 1978.
219. **Saif, L. J. and Bohl, E. H.,** Passive immunity to transmissible gastroenteritis virus: intramammary viral inoculation of sows, *Ann. N.Y. Acad. Sci.,* 409, 708, 1983.
220. **MacPherson, G. G. and Steer, H. W.,** Properties of mononuclear phagocytes derived from the small intestinal wall of rats, in *Function and Structure of the Immune System,* Muller-Ruchholtz, W. and Muller-Hermelink, H. K., Eds., Plenum Press, New York, 1979, 433.
221. **Lowell, G. H., Smith, L. F., Griffiss, J. M., and Brandt, B. L.,** IgA-dependent, monocyte-mediated antibacterial activity, *J. Exp. Med.,* 152, 452, 1980.
222. **Orlans, J. V., Peppard, A. W. R., Fitzharris, B. M., Mullock, B. M., Hinton, R. H., and Hall, J. G.,** Comparative aspects of the hepatobiliary transport of IgA, *Ann. N.Y. Acad. Sci.,* 409, 411, 1983.
223. **Scicchitano, R., Husband, A. J., and Cripps, A. W.,** Biliary transport of serum IgA in sheep, *Immunology,* 53, 121, 1984.
224. **Dahlgren, U., Ahlstedt, S., Hedman, L., Wadsworth, C., and Hanson, L. A.,** Dimeric IgA in the rat is transferred from serum into bile but not into milk, *Scand. J. Immunol.,* 14, 95, 1981.
225. **Flewitt, T. H.,** The virology of acute infectious diarrhoea, in *Microbes and Infections of the Gut,* Goodwin, C. S., Ed., Blackwell, Melbourne, 1984, 159.
226. **Holland, J. J.,** Enterovirus entrance into specific host cells, and subsequent alterations of cell protein and nucleic acid synthesis, *Bacteriol. Rev.,* 28, 3, 1964.
227. **Lee, G. B. and Ogilvie, B. M.,** The mucus layer in intestinal nematode infections, in *The Mucosal Immune System in Health and Disease,* Report of the 81st Ross Conf. on Pediatric Research, Ross Laboratories, Columbus, 1981, 69.
228. **Schrank, G. D. and Verwey, M. F.,** Distribution of cholera organisms in experimental *Vibrio cholerae* infections: Proposed mechanisms of pathogenesis and anti-bacterial immunity, *Infect. Immun.,* 13, 195, 1976.

229. **Lake, A. M., Bloch, K. J., Neutra, M. R., and Walker, W. A.**, Intestinal goblet cell mucus release. II. *In vivo* stimulation by antigen in the immunized rat, *J. Immunol.*, 122, 834, 1979.
230. **Walker, W. A., Wu, M., and Bloch, K. J.**, Stimulation by immune complexes of mucus release from goblet cells of the rat small intestine, *Science*, 197, 370, 1977.
231. **Lake, A. M., Bloch, K. J., Sinclair, K. J., and Walker, W. A.**, Anaphylactic release of intestinal goblet cell mucus, *Immunology*, 39, 173, 1980.
232. **LeFevre, M. E., Hammer, R., and Joel, D. D.**, Macrophages of the mammalian small intestine: a review, *J. Reticuloendothel. Soc.*, 26, 553, 1979.
233. **Hooper, B. E. and Haelterman, E. O.**, Lesions of the gastrointestinal tract of pigs infected with transmissible gastroenteritis, *Can. J. Comp. Med.*, 33, 29, 1969.
234. **Biggers, D. C., Kraft, L. M., and Sprinz, H.**, Lethal intestinal virus infection of mice, *Am. J. Pathol.*, 45, 413, 1964.
235. **Gibson, P. R., Dow, E. L., Selby, W. S., Strickland, R. G., and Jewell, D. P.**, Natural killer cells and spontaneous cell-mediated cytotoxicity in human intestine, *Clin. Exp. Immunol.*, 56, 438, 1984.
236. **Beh, K. J.**, The antibody-containing cell response of the lamina propria of the sheep and the uptake of soluble antigen into afferent ileal lymph following intra-intestinal infusion of soluble or particulate antigen, *Immunology*, 54, 479, 1985.
237. **LeFevre, M. E., Olivio, R., Vanderhoff, J. W., and Joel, D. D.**, Accumulation of latex in Peyer's patches and its subsequent appearance in villi and mesenteric lymph nodes, *Proc. Soc. Exp. Biol. Med.*, 159, 298, 1978.
238. **Lipscomb, M. F., Toews, G. B., Lyons, C. R., and Uhr, J. W.**, Antigen presentation by guinea pig alveolar macrophages, *J. Immunol.*, 126, 286, 1981.
239. **Bellamy, J. E. C. and Nielsen, N. O.**, Immune mediated emigration of neutrophils into the lumen of the small intestine, *Infect. Immun.*, 9, 615, 1974.
240. **Befus, A. D. and Bienenstock, J.**, Immunologically mediated intestinal mastocytosis in *Nippostrongylus brasiliensis* -infected rats, *Immunology*, 38, 95, 1979.
241. **Barth, E. E. E., Jarrett, W. F. H., and Urquhart, G. M.**, Studies on the mechanisms of the self-cure reaction in rats infected with *Nippostrongylus brasiliensis*, *Immunology*, 10, 459, 1966.
242. **Nawa, Y. and Miller, H. R. P.**, Adoptive transfer of the intestinal mast cell response in rats infected with *Nippostrongylus brasiliensis*, *Cell. Immunol.*, 42, 225, 1979.
243. **Bonnardiere, C. La and Laude, H.**, Interferon induction in rotavirus and coronavirus infections: a review of recent results, *Ann. Rech. Vet.*, 14, 507, 1983.
244. **Albright, D. J., Whalen, R. A., and Blacklow, N. R.**, Sensitivity of human foetal intestine to interferon, *Nature (London)*, 247, 218, 1974.
245. **Scicchitano, R. and Husband, A. J.**, unpublished
246. **Petit, A., Ernst, P. B., Befus, A. D., Clark, D. A., Rosenthal, K. L., Ishizaka, T., and Bienenstock, J.**, unpublished observations, 1985.
247. **Gleeson, M., Clancy, R. L., and Cripps, A. W.**, personal communication.

Chapter 14

ENTERIC VIRUS VACCINES: THEORETICAL CONSIDERATIONS, CURRENT STATUS, AND FUTURE APPROACHES

Linda J. Saif and Daral J. Jackwood

TABLE OF CONTENTS

I. INTRODUCTION

To date, few if any commercial vaccines have been developed which demonstrate a high degree of efficacy for prevention of neonatal viral diarrheas in man or animals. Although live oral poliovirus vaccines are commonly used in infants and are highly efficacious, the mechanism of protection against poliovirus differs from that needed to prevent viral diarrheas. Polioviruses replicate initially in intestinal lymphoid cells and do not induce lesions in intestinal epithelial cells.[1] Either live attenuated oral poliovirus vaccines or parenterally administered inactivated vaccines are effective in stimulating circulating antibodies which prevent the systemic spread of poliovirus to the central nervous system.[2] Moreover, following natural infection, the individual's intestinal tract is highly resistant to reinfection.[3] By comparison, enteropathogenic viruses replicate and induce lesions only in the gastrointestinal tract and the susceptible target cell is the villous enterocyte.[4,5] Thus, active immunity against enteropathogenic viral infections depends on locally produced SIgA and IgM antibodies, with local cell-mediated immunity possibly playing a role in final elimination of the virus.[6] However, the latter point is unresolved. Researchers studying rotavirus infections have proposed that the presence of specific neuralizing antibodies in the gut may be necessary to completely prevent or overcome virus replication, basing their conclusions on the ability of passively administered antibodies (polyclonal or monoclonal) to prevent infection and diarrhea.[7-10] In attempts to further define the role of cell-mediated immunity against rotavirus infections, suckling athymic nude mice (T-cell deficient) recovered from rotavirus in an identical manner to infected normal mice,[9,11] whereas SCID mice (no functional T or B cells or antibody production) and an immunodeficient patient with X-linked agammaglobulinemia (no antibodies) developed chronic symptomatic rotavirus infections.[9,12] Although T-cells cytotoxic for rotavirus have been identified in mice,[13] the time course required for their activation after antigen exposure mitigates against their playing a role in prevention of infection: their role in recovery from enteropathogenic viral infections is unclear at present. Currently, only oral vaccines containing live replicating organisms, as opposed to inactivated vaccines, have been highly efficient in evoking mucosal immune responses, particularly SIgA antibodies.[2] Furthermore, oral use of nonreplicating soluble or killed antigens has often induced systemic tolerance[14,15] or immunity of only short duration. Whether the problems encountered in oral administration of soluble protein antigens might be overcome by use of improved mucosal adjuvants (muramyl dipeptide, cholera toxin, etc.) or new delivery systems (liposomes, microencapsulation, and sustained release systems) is uncertain.

The presence of specific serum antibodies to enteropathogenic viruses usually does not correlate with resistance to intestinal infection,[7,16,17,28] although some moderation of the severity of rotavirus diarrhea was seen in ruminants when very high titers of colostrally-derived serum antibodies were present.[5,7,18] Transudation of maternally derived (via colostrum, egg yolk, or transplacentally) high titered serum IgG antibodies to the intestine may thus provide an explanation for the infrequent occurrence of viral diarrheas in weaned dairy calves or poultry under a week of age,[5,18-20] or the asymptomatic rotavirus infections seen in bottle-fed infants under a month of age.[21]

Neonatal viral diarrheas occur most frequently in mammals during the nursing period (and again in pigs at weaning) and in poultry from 1 to 4 weeks of age.[4,5,18,19,22] In man, their peak occurrence worldwide is in children under 2 years of age.[22,23] Little is known about the duration of immunity following natural infection with enteropathogenic viruses. In experimental studies, animals that recover from enteric viral infections resist challenge with the virulent strain of virus on a short-term basis.[5,24,25] However, natural infection of children with one serotype of rotavirus may not prevent reinfection with the same or different serotypes, but frequently prevents clinically significant diarrhea upon reinfection.[21,26]

II. ROTAVIRUS AND CORONAVIRUS INFECTIONS

These two enteric viruses account for a high percentage of viral diarrheas in neonatal animals. Both viruses infect the small intestinal villous epithelial cells causing villous atrophy and consequently malabsorptive diarrhea and dehydration.

Pigs of all ages are susceptible to the swine enteric coronavirus, transmissible gastroenteritis virus (TGEV), but fatal disease primarily occurs in the neonate.[27] TGE causes both epidemics in seronegative herds, often resulting in 100% mortality in newborn piglets, and endemics in seropositive herds, resulting in high morbidity but lower mortality among young pigs. Rotavirus and coronavirus infections occur as endemics in young animals in immune herds and are often associated with high morbidity, but lower mortality. In young children (6- to 18-months old), rotaviruses are one of the leading causes of dehydrating diarrhea worldwide.[21,22]

Because rotaviruses and coronaviruses are well-characterized enteropathogens and their epidemiologic distribution is known, they are important models for studies of immunity to enteric viruses. Moreover, effective vaccines are generally not currently available for their prevention or control, or for the prevention and control of any other viral-induced diarrheas.

III. ACTIVE IMMUNITY TO ROTAVIRUS

A. VACCINES TO STIMULATE ACTIVE IMMUNITY IN ANIMALS

Two approaches have been devised to try to prevent rotavirus infections in pigs and calves. The first approach involves oral inoculation of neonatal calves or pigs with live attenuated rotavirus vaccines to stimulate active immunity. While this approach provided effective protection based on challenge studies in colostrum-deprived calves,[28,29] its efficacy in double-blind field trials has been questioned.[30-34] No significant differences in the incidence of diarrhea morbidity or mortality or rotavirus-associated diarrhea were observed among calves from five vaccinated and five nonvaccinated dairy herds.[33] Similarly, in studies in swine, oral vaccination of nursing pigs with a live attenuated rotavirus vaccine failed to prevent rotavirus-associated diarrhea in the pigs at weaning.[35] No significant differences in weight gains, rotavirus shedding, and rotavirus-associated diarrhea were observed among vaccinated and nonvaccinated pigs. Several reasons are possible for these observed failures of the commercial vaccines including: (1) specific colostral antibodies may neutralize the vaccine virus in the intestine; (2) calves and pigs may become infected with virulent rotaviruses at birth, prior to vaccination; and (3) there may be a lack of cross-protection due to infection by different rotavirus serotypes. The latter possibility has been confirmed in experimental studies in pigs and calves.[24,25] However, most investigators have suggested that under field conditions, vaccine failures may be largely mediated by interference by maternal antibodies.[31,34,35] Widespread occurrence of these antibodies and the degree of viral attenuation needed to insure innocuousness in the neonate currently limit the feasibility of this approach. In addition, the usefulness of the active immunization approach to prevent epidemic TGE viral infections in pigs is questionable since pigs are highly susceptible to TGEV at birth, but development of active immunity following oral vaccination requires at least 4 to 5 days.[36]

B. VACCINES TO STIMULATE ACTIVE IMMUNITY IN INFANTS AND CHILDREN

The current approach to immunization against rotavirus in children is focused on stimulation of active immunity. Various vaccine strategies are under consideration, including use of live attenuated animal rotavirus strains, animal-human rotavirus reassortants, and rDNA engineered live replicating or subunit vaccines.[37,38]

Previous investigators evaluating the use of a live-attenuated bovine rotavirus (RIT 4237) as a vaccine in infants reported efficacy against rotavirus diarrhea when tested in developed countries,[39] but a lack of efficacy when tested in developing countries.[40,41] Reasons proposed for vaccine failure include the following: (1) higher rotavirus challenge doses may be common in developing countries surpassing vaccine-induced immunity; (2) other enteric microorganisms present in the gut may interfere with vaccine replication; (3) higher levels of breast milk or placentally derived antibodies in infants may neutralize vaccine virus and thereby prevent replication; and (4) vaccine-induced immunity may be poor in malnourished infants. A less-attenuated strain of bovine rotavirus (WC3) is currently under test in vaccine trials in infants in a developed country to access its efficacy against natural infection with serotypically distinct human rotaviruses.[42]

A live rhesus rotavirus (RRV) vaccine tested in infants showed efficacy based on preliminary results in two field trials, one in a developed country[43] and the second in a developing country.[44] More extensive field trials are needed to confirm and extend these findings. Additional research is also needed to investigate reasons for current or prior vaccine failures to improve the design of new vaccines for future efficacy testing.

IV. PASSIVE IMMUNITY

A. IMMUNOGLOBULINS (IgS) IN MAMMARY SECRETIONS

In man, primates, rabbits, guinea pigs, and poultry, passive serum Igs are acquired transplacentally or via the yolk sac.[45] In infants, only small amounts of IgA antibodies are absorbed via the colostrum and their significance in passive immunity is uncertain.[46] Rodents represent an intermediate position in that they acquire maternal antibodies both transplacentally and also selectively absorb IgG from the milk for periods up to 17 to 20 d after birth.[45] In contrast, ungulates (calves, lambs, pigs, and foals) are born agammaglobulinemic and acquire passive serum Igs entirely from the colostrum, nonselectively via the gut, for approximately 24 h after birth.[45,47-49] In these latter species, IgG which is predominant in colostrum is almost entirely serum derived.

Predominance of serum-derived IgG in porcine and bovine colostrum (Figure 1) assures transfer of serum antibody from the dam into the circulation of the neonate via transient intestinal absorption after nursing.[47-49] Passively acquired circulating antibodies provide protection against system infections, but generally not against enteric infections,[16,17] except for possible resecretion of serum antibodies onto mucosal surfaces in ruminants.[5,18,47,50,51]

In swine, the concentration of colostral IgG decreases ~30-fold during the first week of lactation while SIgA declines only ~3-fold, becoming the predominant Ig in milk.[48,49] This is analogous to the predominance of SIgA in milk of other monogastrics including man,[52,53] as illustrated in Figure 1. Milk SIgA antibodies are intestinally absorbed, only minimally, and thus play an important role locally in protection of the gut.

An important question for developing efficacious maternal vaccination procedures in monogastrics concerns the origin of SIgA antibodies in milk. In porcine milk, more than 90% of SIgA is locally produced by mammary gland plasma cells.[49] Extensive direct and indirect evidence now exists that the precursors of mammary gland IgA plasmacytes originate from the intestinal tract.[53-57] This gut-mammary SIgA axis, initially proposed in studies of passive immunity to TGEV in swine,[54,55,58,59] has now been elaborated upon and confirmed, using a variety of antigens in other monogastrics including man.[53,56-57] Furthermore, this gut-mammary immunologic relationship may also extend to T-lymphocytes, since T-lymphocytes from mammary secretions respond in antigen-specific lymphocyte proliferation assays against

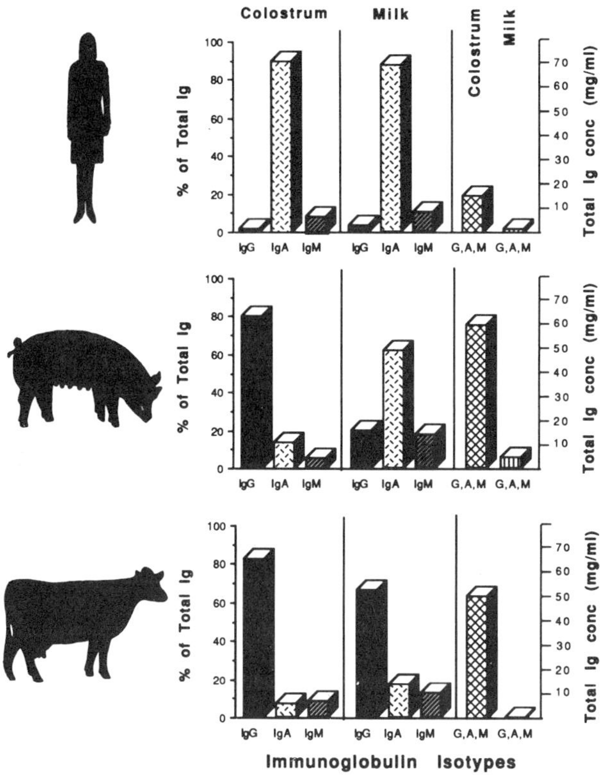

FIGURE 1. Relative percentage and total concentration of immunoglobulins in mammary secretions of human beings, pigs, and cows.

enteric antigens, whereas circulating lymphocytes do not. Maternal vaccination to elicit lactogenic immunity, thus, is contingent upon stimulation of local intestinal immunity with subsequent transfer of sensitized lymphocytes, IgA antibodies, or both to the mammary gland. This has been done most effectively by colonizing the gut with live organisms such as enteric viruses[54,55,58,59] or *Escherichia coli*.[49,53,61,62] The gut-mammary immunologic relationship assures that the neonate receives milk antibodies against enteric pathogens endemic in the population. Studies in mice have shown IgA lymphoblasts home from the mesenteric lymph node to mammary glands under the influence of pregnancy-associated hormones.[56] Thus, sexual maturity of the dam, stage of pregnancy, and lactation are additional variables to consider in maternal vaccination programs aimed at enhancing passive immunity.

An additional mechanism proposed for occurrence of milk SIgA antibodies with specificity for enteric antigens is the selective transport of serum polymeric IgA (of intestinal origin) into mammary secretions,[63] similar to its transport into bile.[64] Although this mechanism was demonstrated in ruminants,[65] there are conflicting reports on its existence in rodents,[65] and no reports on its occurrence or significance in other species.

In ruminants, IgG_1 predominates in colostrum and milk, its concentration decreasing ~100-fold in the transition to milk (Figure 1).[47,51,65,66] Whereas IgG in porcine colostrum represents mainly a serum transudate,[48,49] in the cow IgG_1 is selectively transported from serum to colostrum and milk.[66] Serum dimeric IgA is also selectively transported into mammary secretions during early and midlactation but not during involution.[65] Knowledge of these aspects of lactogenic immunity is important for designing effective maternal vaccination regimes in cows.

Evidence for the role of milk antibodies in passive protection against enteric viral infections in monogastrics (swine) and ruminants (cattle) will be extensively reviewed in the following sections, since immune milk has been evaluated for control of a number of neonatal infectious diseases in veterinary medicine. The importance of breast milk has also been recognized in a number of human diseases[67,68] and many studies have documented an association between breast feeding and a lower incidence of gastroenteritis in infants.[69] Moreover, human milk containing rotavirus antibodies has been used therapeutically in immunodeficient patients with rotavirus infections[12] and was shown to reduce the severity of clinical symptoms in infants who acquired nosocomial rotavirus infections.[70] However, to date there are few reports describing attempts to enhance passive immunity in breast milk by vaccination of pregnant or lactating women. The demonstrated feasibility of using attenuated or inactivated enteric viruses to increase SIgA anitbodies in milk of naturally primed animals suggests studies of this approach in lactating women may be warranted, particularly in developing countries where long-term breast feeding is practiced.

The mechanism of mucosal immune protection mediated by SIgA or IgG_1 milk antibodies has not been well-characterized. Mechanisms have evolved in systemic immunity to maintain sterility within this system by complete destruction and elimination of microorganisms via complement-mediated lysis or opsonization, both of which evoke strong inflammatory responses. On the other hand, the normal microflora and food antigens constantly present in the intestine represent a massive antigenic challenge which could evoke extreme damage if inflammatory-type immune responses were repeatedly stimulated. Thus, the intestinal immune system has evolved with its major role to prevent the uptake and penetration of pathogens and food antigens through the mucosal barrier. The IgA system fulfills this functional role in many ways. IgA is inefficient at activating complement and although it does so by the alternative pathway, the biological significance of this may be limited, since mucosal secretions lack certain complement components.[71] SIgA antibodies are effective in virus neutralization and may thereby prevent attachment of viruses to the intestinal mucosa similar to their prevention of adhesion of *E. coli* to enterocytes.[72] However, in one study, SIgA antibodies to TGEV did not prevent virus attachment to cells *in vitro,* but did prevent penetration of TGE virus into the cells; whether a similar mechanism occurs *in vivo* is unclear.[73]

Another activity attributed to SIgA antibodies involves regulation of mucous secretion from goblet cells.[74] Thus, prevention of viral attachment or penetration combined with increased flow of mucous and normal intestinal motility may contribute to the rapid removal of enteropathogens from the epithelial cell surfaces. In addition, recent evidence suggests IgA may participate directly in antibody dependent cell-mediated cytotoxicity, either alone or by increasing the activity of IgG,[75,76] thereby providing another possible method for elimination of mucosal pathogens.

B. EPIDEMIC VIRAL INFECTIONS: VACCINES TO STIMULATE PRIMARY ANTIBODY RESPONSES IN MAMMARY SECRETIONS OF SERONEGATIVE SWINE

Results of maternal TGEV vaccination and mechanisms of passive immunity to TGEV have been reviewed[27,58,59,77,78] and are summarized in Table 1. Parenteral vaccination of pregnant sows with attenuated TGEV led to elevated IgG TGEV antibodies in serum and colostrum, but generally only low levels of IgG antibodies in milk. Protection was poor (56% litter mortality) except in litters of intramammarily (IMm) vaccinated sows (14% litter mortality) which had high IgG TGEV antibody titers in colostrum and early milk. These results were confirmed by others using attenuated strains of TGE virus.[77] Oral administration of virulent virus or natural infection led to elevated SIgA TGEV milk antibodies which persisted throughout lactation and provided the greatest protection (6% litter mortality). This confirms early observations that sows which recover from TGE transmit protection to their piglets.[16] The failure of orally or

TABLE 1
Vaccination of Sows with TGE Virus: Summary of Milk Antibody Titers, Isotypes and Passive Protection in Suckling Piglets

		TGE antibody[a]			
		3-5 DPP milk		Litter mortality[b]	
Virus form	Route	GMT	Isotype	No./Total	Percent
Live					
Virulent	Oral	330	IgA	2/35	6
Attenuated	IM	125	IgG	39/70	56
	IMm	975	IgG	2/14	14
	Oral/IN	124	IgG(IgA)	5/20	25
None	—	<1	—	60/68	88

[a] GMT = geometric mean virus neutralization antibody titer; DPP = days post-partum; antibody isotypes were determined by virus neutralization tests on IgA or IgG gel filtration fractions.

[b] Piglets were challenged at 2 to 4 d of age with virulent TGE virus.

intranasally-administered attenuated TGE viruses (25 to 100% mortality)[27,58,59,77,78] may relate to inability of the attenuated virus to adequately infect or stimulate IgA antibodies in the adult intestine. Limited intestinal IgA responses would provide seeding of few IgA TGE-sensitized lymphocytes to mammary tissue. Administration of attenuated virus in enteric-coated capsules or direct inoculation into the gut lumen also failed to induce SIgA milk antibodies or passive protection.[77] However, oral or IMm vaccination with a small plaque variant strain of TGEV, led to high titers of IgA and IgG milk antibodies and increased litter protection (14 to 24% litter mortality).[79] The positive correlation between optimal protection and predominance of SIgA antibodies in milk may be due to several factors including: (1) the persistence of SIgA milk antibodies throughout lactation[48,49] (2) the stability of SIgA to proteolytic intestinal enzymes;[80] and (3) a possible selective binding of milk SIgA to villous enterocytes[81] thereby preventing viral attachment. However, IgG TGEV milk antibodies were also protective if milk antibody titers were extremely high and persisted during the challenge period, as in litters of IMm-vaccinated sows.[58,82]

C. ENDEMIC VIRAL INFECTIONS

1. Vaccines to Boost Antibodies in Mammary Secretions of Seropositive Swine

Most adult swine are seropositive for rotavirus antibodies.[25,78] TGEV and other enteric viruses also occur frequently as endemics in immune herds. Thus optimization of lactogenic immunity may often involve boosting rather than priming for milk antibody responses. There is little information on methods to boost lactogenic immunity in previously exposed animals, particularly with rotaviruses in regard to homotypic and heterotypic immunity.

Enhancement of rotavirus milk antibody titers can occur in at least two ways: first, natural infection of the litter with rotavirus may lead to a subclinical reinfection of the dam, evoking an anamnestic response in sow serum and milk. Second, parenteral vaccination of pregnant or lactating sows with live attenuated rotaviruses also enhances rotavirus antibody titers in serum and milk.[59,78]

In naturally infected sows, rotavirus milk antibodies were mainly associated with SIgA based on isotype-specific ELISA done on whole milk or virus neutralization done on Biogel-fractionated milk. After natural infection of the litter with rotavirus, increases in rotavirus antibody titers

TABLE 2
Detection of Rotavirus Shedding and Diarrhea in Naturally Infected Nursing Piglets of Vaccinated or Nonvaccinated Sows

Rotavirus vaccination groups		Total no. of pigs	Diarrhea	Rotavius shedding		
No.	Vaccine[a]		% positive	% positive	Onset (DPP)[b] x, (range)	Duration (d) x, (range)
1	OSU Type I	50	53	62	37 (28—45)	4 (1—7)
2	OSU Type I + Gottfried Type II	16	63	100	37 (33—44)	4 (3—5)
3	None	38	70	61	23 (17—28)	5.4 (1—9)

[a] Live attenuated porcine rotaviruses were injected either intramuscularly or intramammarily shortly before or after farrowing.
[b] DPP = days post-partum.

in milk were mainly associated with SIgA, with smaller increases in IgG.[78] Milk antibody titers increased to a similar degree to two serotypes of porcine rotavirus (OSU and Gottfried). Because the serotype of rotavirus infecting the pigs was not identified in this study, it was not possible to determine if the increased antibody responses in milk were homotypic, heterotypic, or both. The increased milk antibody titers could be due to : (1) subclinical infection of the sow's gut leading to translocation of additional sensitized lymphocytes to the mammary gland; or (2) a lacteal entrance of virus leading to clonal expansion of mammary gland IgA precursor cells (intestinally derived) as proposed in experimental studies of TGEV in swine.[82]

Parental vaccination with live attenuated rotavirus or TGEV of previously orally-primed sows led to boosting of both SIgA and IgG antibodies in milk, but titers declined by the end of lactation.[78,82] In sows vaccinated with OSU rotavirus, serum and milk neutralizing antibody titers to a heterotypic porcine rotavirus (Gottfried) also increased, but the increase was less than against the homotypic virus.

Results of our studies corroborate reports which showed challenge of seropositive adults with one rotavirus serotype led to increases in both homotypic and heterotypic rotavirus antibody titers.[83,84] However, titers increased only to heterotypic viruses to which vaccinates had preexposure antibodies,[83,84] as was also noted in our studies.[78] Our observations in sows also concur with others who noted similar increases in breast milk SIgA antibodies in women orally primed to cholera (endemic exposure) and then parenterally boosted with cholera vaccine.[85]

A summary of preliminary studies of protection against diarrhea and natural infection with rotavirus in litters from vaccinated or control sows is shown in Table 2. Maternal vaccination did not prevent natural infection with rotavirus in nursing piglets. However, it did successfully delay the onset and shorten the duration of both rotavirus shedding and diarrhea. There was also less diarrhea among litters of vaccinated dams compared with control litters. Reasons for failure of these increased milk antibodies to effectively mediate longer or more complete passive immunity are uncertain, but could relate to massive challenge doses or additional antigenic differences between the vaccine serotypes and the infecting rotavirus. However, maternal vaccination may be useful in decreasing rotavirus morbidity, especially in herds where rotavirus infection may have an early onset (<3 weeks of age) resulting in high morbidity and increased mortality. Such findings need to be verified in additional studies.

2. Vaccines to Boost Antibodies in Mammary Secretions of Seropositive Cows

Attempts to significantly enhance antibody titers in mammary secretions by parenteral vaccination of cattle with rotavirus vaccines have met with various degrees of success. The results of such studies, including our own findings, have been reviewed recently.[78,86] Two factors have probably greatly influenced any positive results achieved. First, bovine IgG_1 is selectively transported from serum to colostrum[87] and also to milk, although the magnitude is reduced after colostrum formation.[88,89] Second, rotavirus infections in cattle are enzootic; hence most cattle are serpositive and have primarily IgG_1 rotavirus antibodies in normal colostrum.[86,90,91] Although these colostral rotavirus antibodies may be protective if fed undiluted on a frequent basis,[5,8] such antibody titers normally decline to low or nondetectable levels in milk after a few days of lactation.[5,28,32,90] Therefore, since IgG_1 bovine rotavirus antibodies are prevalent in colostrum and are mainly serum-derived, parenteral vaccination of cows represents a rational approach to boosting these antibodies in mammary secretions.

Significantly enhanced antibody titers in mammary secretions were first reported by Snodgras et al.[90] after parenteral vaccination of cows using inactivated bovine rotavirus. However, in their initial studies, suckling calves born to vaccinated dams were only partially protected against challenge with bovine rotavirus.[90] Possible explanations for this lack of protection included: high challenge dose of virus; milk antibody titers were below the protective threshold; formalin inactivation may have destroyed antigens needed to elicit protective antibodies; or the immunizing cell culture-propagated bovine rotavirus and virulent challenge rotavirus may have represented different strains of serotypes. In a subsequent study of this vaccine under farm conditions in a single herd, vaccination resulted in reduced rotavirus shedding and diarrhea in calves.[92] However, the rotavirus present on the farm was the same serotype as the vaccine virus,[92] so heterotypic protection could not be evaluated.

We were able to amplify and extend the initial vaccination studies of Snodgrass et al.[90] A summary of our results describing optimal methods for vaccination of pregnant cows with bovine rotavirus to enhance antibody titers in colostrum is in Table 3. Our major findings were as follows. All normal (control) cows sampled were seropositive for bovine rotavirus antibodies. IgG_1 antibodies were prevalent in mammary secretions, followed in prevalence by IgG_2, IgA, and IgM.[5,78,86,91] Following vaccination, in cows which showed significantly enhanced antibody titers in colostrum and milk, the greatest increase in antibody titers was associated with IgG_1, followed by IgG_2, IgA, and IgM.[5,78,86,91] Maximal enhancement of colostral antibody titers (23- to 148-fold) occurred in cows (Gp 1) vaccinated IM (7 d prior to drying off) and IMm (7 d after drying off) with modified live (GMT = 360,205) or binary ethylenimine (BEI) inactivated bovine rotavirus (Gp 8, GMT = 55,956) in incomplete Freund's adjuvant (IFA). Administering inactivated bovine rotavirus in IFA, IM prior to and IMm with the dry treatment (Gp 9), resulted in ~30-fold lower colostral antibody titers (GMT = 11,738) compared with the prior vaccination scheme. The IM vaccination of cows with either live attenuated (Gp 2) or BEI inactivated bovine rotavirus (GP 6) in IFA resulted in similar colostral GMT (~24,000) which were ~15-fold lower than the maximal titers. The following vaccines were ineffective in significantly enhancing bovine rotavirus antibody titers in colostrum (GMT = 2,828 to 3,775): (1) the commercial vaccine without adjuvant (as recommended by the manufacturer, Gp 3); (2) B-propiolactone-inactivated bovine rotavirus (Gp 5); and (3) BEI-inactivated bovine rotavirus with A10H adjuvant (Gp 7). Colostral bovine rotavirus antibody titers in all cows were associated primarily with IgG_1. Similar results, reviewed recently,[86] were noted by others using formalin-inactivated rotavirus with oil adjuvant. In addition, commercial coronavirus vaccines also failed to elicit enhanced coronavirus antibody titers in bovine mammary secretions.[5,32]

Passive protection studies were conducted in newborn unsuckled calves fed pooled colostrum supplements from the vaccinated cows[5,8] (results summarized in Table 3). Complete protection against rotavirus diarrhea and infection during the experimental feeding period was achieved by feeding 1% supplemental pooled colostrum from cows vaccinated IM + IMm using

TABLE 3
Evaluation of Rotavirus Vaccines in Dairy Cows for Increasing Antibody Titers in Colostrum and Providing Passive Protection in Calves

Cow group no.	Rotavirus vaccine				Significantly increased antibody GMT in colostrum[b]	Passive protection in calves[c] against	
	Virus form[a]	(PFU/ml)	Route[a]	Adjuvant[a]	(increase over control)	Infection	Diarrhea
	Modified Live						
1	OARDC	(10^8)	IM + IMm	IFA	+ (148×)	100%	100%
2	OARDC	(10^7)	IM (2×)	IFA	+ (12×)	NT	NT
3	Calf Guard	(10^4)	IM (2×)	None	—	0	0
4	Calf Guard	(10^4)	IM (2×)	IFA	+ (6×)	NT	NT
	Inactivated						
5	OARDC BPL	(10^7)	IM (2×)	IFA	—	NT	NT
6	OARDC BEI	(10^7)	IM (2×)	IFA	+ (10×)	100%	100%
7	OARDC BEI	(10^7)	IM (2×)	A1OH	—	NT	NT
8	OARDC BEI	(10^7)	IM + IMm	IFA	+ (23×)	NT	NT
9	OARDC BEI	(10^7)	IM + IMm	IFA	+ (5×)	NT	NT
10	None			None	—	0	0

[a] Five cows in each group were vaccinated intramuscularly (IM) or IM and intramammarily (IMm) at 3 to 9 weeks prepartum.[5,78,91] Groups 1 and 8 were vaccinated IM at 9 weeks prepartum and IMm at 7 weeks prepartum and group 9, IM at 9 weeks and IMm with the dry treatment; OARDC = Ohio Agricultural Research and Development Center; Calf Guard = Rota-coronavirus vaccine, Norden Laboratories, U.S.; BPL = 5% B-propiolactone; BEI = 10% binary ethylenimine; IFA = incomplete Freund's adjuvant.

[b] GMT = virus neutralization geometric mean titer.

[c] Newborn unsuckled calves were fed 1% supplemental colostrum, 2 to 3 ×/d for 5 d and challenged with virulent rotavirus; NT = not tested.

live bovine rotavirus (Gp 1) or cows vaccinated IM using BEI-inactivated bovine rotavirus (Gp 6). Moreover feeding 0.1% colostrum from Gp 1 cows resulted in partial protection, characterized by delayed onset and shortened duration of diarrhea and virus shedding.[5,8] No passive protection was evident following feeding of 1% colostrum from unvaccinated or commercial vaccine-inoculated cows (Gps 3 and 10). These results demonstrate that high titers of IgG_1 rotavirus antibodies in colostrum correlate with passive protection against challenge with homotypic (NCDV) bovine rotavirus. They support the concept that in ruminants IgG_1 may supplement the role of SIgA in lactogenic immunity.[5,59] Other data which emphasize similarities between SIgA in other species and IgG_1 in ruminants include: (1) resistance of IgG_1 to certain proteolytic enzymes;[93] (2) the predominance of IgG_1 in milk and its specificity against enteric viruses;[5,90,91] and (3) the increase in IgG_1 but not IgA milk antibodies after intestinal administration of antigens in pregnant cows.[5,59,94]

The development of antibodies in calves fed various levels of passive antibodies was also investigated.[5,29,95] These studies revealed that colostrum-deprived calves or calves fed control colostrum developed rapid (5 to 7 d post-exposure) but transient fecal IgM and IgA rotavirus antibodies after challenge, and appearance of peak IgM and IgA titers correlated with termination of diarrhea and rotavirus shedding. Most calves fed "immune' colostrum did not develop detectable IgM and IgA rotavirus antibodies in feces, but transient titers of each were evident in serum, suggesting development of subclinical infections in such calves in the face of high levels of passive IgG_1 antibodies.[5,29]

Further studies are needed to examine the effect of other important variables on passive immunity to enteric viruses before optimal vaccination or control strategies for these infections can be implemented. These include: (1) examining passive immunity to heterotypic and repeated enteric viral challenge both experimentally and by conducting and evaluating field challenge studies; (2) analyzing the effect of passive immunity (comparing colostral and milk antibodies) on development of active immunity; and (3) defining the relative effectiveness of colostral versus milk antibodies in providing passive immunity.

V. FUTURE VACCINE APPROACHES

Recombinant DNA (rDNA) techniques have made possible the development of a new generation of viral vaccines. Subunit, synthetic peptide, generally modified live viruses, and recombinant vector vaccines have been used to successfully immunize animals against viral diseases. Although each of these new vaccines offers certain advantages over conventional attenuated and inactivated vaccines, they are not always effective in eliciting protective immunity to a particular pathogen. Consideration must be given to viral epitope conformation in developing subunit and synthetic peptide vaccines. In addition, adjuvants are usually necessary when using these latter two types of vaccines. Live rDNA vaccines are not always practical because they have the potential to be genetically unstable, which could result in the ability to cause disease. The immune status of the host must also be considered when using live vaccines, since a compromised immune system can allow an otherwise safe vaccine to cause disease. However, rDNA live vaccines provide several advantages over subunit and synthetic peptide vaccines. They stimulate both humoral and cellular immunity and usually provide a longer lasting immunity than inactivated or subunit vaccines. More importantly, rDNA live viral vaccines may stimulate a more effective mucosal immunity, which is extremely important for controlling enteric pathogens. Oral administration of soluble protein antigens has historically resulted in poor mucosal immune responses or even tolerance induction.[14,15] Thus, the oral administration of subunit or synthetic peptide enteric viral vaccines will likely evoke poor musocal immunity, unless new mucosal adjuvants and delivery systems are also designed for use with these vaccines.

A. LIVE VACCINES PRODUCED USING GENETIC ENGINEERING

Two basic categories of live vaccines have been produced using rDNA technology. Viral pathogens have been attenuated by deletion of specific portions of the viral genome. In some cases, new genetic material has also been added to the genomes of these viruses to make them easier to identify using nucleic acid probes. The other broad category of live genetically engineered vaccines includes viruses and bacteria which have been used as vectors to express genes coding for protective antigens of other pathogens. Examples of the vectors which potentially could be used to prepare viral vaccines are numerous. They include viruses and bacteria, most notably, poxviruses, herpesviruses, papillomaviruses, adenoviruse, *Escherichia coli,* and *Salmonella* species. Each of these agents has the potential to become vectors because they have the ability to accommodate additional genetic material in their large genomes and have one or more nonessential regions which could serve as sites for insertion of foreign genes. Although these vectors are in various stages of development, the two which appear to hold the most promise for immunizing against enteric pathogens are the adenovirus and Salmonella vectors.

B. ADENOVIRUS VECTORS

Initially, adenovirus vectors required wild type adenovirus helpers which could supply functions lost as a result of gene insertion.[96] The development of helper-free adenovirus vectors[97,98] has made the use of these vectors much more practical. These helper-free viral vectors replicate to high copy numbers in host cells and appear to be relatively stable. The nonessential E3 region of the genome has been used to insert foreign genes in these viruses. This region of adenovirus type 2 encodes a 19 kilodalton (kDa) glycoprotein which blocks the expression of specific major histocompatibility antigens on the host cell surface.[9,100] Disruption of the E3 region by insertion of a foreign gene reduces the ability of the adenovirus vector to replicate *in vivo* compared to viruse with a functional E3 region.[101] Thus, these recombinant adenoviruses are expected to be les virulent than wild-type viruses.

Saito and co-workers[102] reported on the first gene to be transcribed by its own promoter in the adenovirus type 5 genome. The gene used in their studies encoded the hepatitus B virus surface antigen (HBsAg). Although transcription of the HBsAg was observed, translation of the mRNA for HBsAg was relatively low. The complex post-transcriptional controls of adenoviruses have been implicated in the lack of HBsAg expression. Using the E3 promoter instead of the hepatitis B viral promoter, Morin and co-workers[101] have obtained efficient translation and transcription of the HBsAg gene.

Adenoviruses have a relatively strict host range. Development of attenuated virus vectors for specific hosts may be required. Many attenuated adenoviruses already exist and could potentially be developed into vectors. Attenuated adenoviruses already exist and could potentially be developed into vectors. Attenuated adenovirus vectors for use in humans have been developed.[103] These type 4 adenoviruses produced an enteric infection in human volunteers. Furthermore, they were relatively safe and lacked any detectable oncogenicity. Thus, this adenovirus strain may represent a potential vector for use as a live rDNA vaccine containing and expressing gene inserts for various human enteric viruses.

C. Salmonella vectors

Strains of *Salmonella typhimurium* have been developed which are attenuated but still immunogenic.[104] *Salmonella typhimurium* SR-11 mutants invade and proliferate in the cells of the gut-associated lymphoid tissue (GALT) but have a reduced ability to invade mesenteric lymph nodes and the spleen.[105] Invasion of the FALT generally elicits a cellular and humoral immune response.[106] The humoral immunity which develops includes both systemic and local secretory antibodies.[105] Attenuated and immunogenic *Salmonella* have the potential to target antigens directly to lymphoid tissues of the gut and therefore are an attractive organism for development of a live vaccine vector.

The development of a *Salmonella* vector system requires that the foreign gene be carried in a plasmic in the bacterium. Constitutive expression of the foreign gene is dependent on the stability of the plasmid in the organism. Nakayama et al.[107] indicate that plasmids used as vectors generally contain an antibiotic resistance gene. The use of antibiotics *in vivo* to maintain plasmid stability, however, is not practical and not permitted by the Food and Drug Administration for live vaccines. This limitation has been overcome by the development of a lethal mutation in the *asd* gene of *Salmonella typhimurium* which is complemented by a plasmic carrying the normal gene.[107]

The *asd* gene codes for the enzyme aspartate beta-semialdehyde dehydrogenase which is required for the synthesis of diaminopimelic acid (DAP) an essential component of peptidoglycan in Gram-negative bacterial cell walls. A mutation in the *asd* gene (*asd*–) is lethal for *Salmonella*. This lethal mutation can be complemented by a plasmid carrying a normal *asd* gene (*asd*+). Since mammalian cells do not contain DAP, *asd*– *Salmonella* must retain the *asd*+ plasmid to survive *in vivo*.

Nakayama and co-workers[107] reported on the development of a plasmid which contains a functional *asd* gene from *S. mutans*. This plasmid was used to complement *asd*- *E. coli* and *S. typhimurium*. It contains the promoter trc[108] which was used to express the cell envelope protein of *S. sobrinus* in a vaccine strain of *S. typhimurium*.

Other *Salmonella* strains have also been used as vectors to express various antigens. One *S. typhimurium* strain has a stable deletion in the *aro*A gene which destroys the synthesis of enterochelin and folic acid. This mutation is not lethal, but does significantly lower virulence in these organisms.[104] *Salmonella* containing the *aro*A mutation have been used as vectors for the expression of beta-galactosidase[109] and the B subunit of *E. coli* heat-labile enterotoxin.[110]

Adenovirus and *Salmonella* vectors have a tremendous potential for being developed into safe and effective vaccines for enteric pathogens. The pathogenesis of each enteric pathogen is different and, therefore, the use of these vectors may not be applicable in every case. Adenovirus and *Salmonella* vectors will have to be compared carefully with other vector systems, recombinant DNA vaccines, and conventional vaccines, to develop the most efficacious and safe enteric viral vaccines for future use in man and animals.

REFERENCES

1. **Melnick, J. L.,** Enteroviruses: polioviruses, coxsackieviruses, echoviruses and newer enteroviruses, in *Virology,* Fields, B. N., Ed., Raven Press, New York, 1985, 739.
2. **Ogra, P.L., Karzon, D. T., Righthand, F., and McGillivray, M.,** Immunoglobulin response in serum and secretions after immunization with live and inactivated polio vaccine and natural infection, *N. Engl. J. Med.,* 279, 831, 1968.
3. **Ghendon, Y. U. Z. and Sankoyeva, L. I.,** Comparison of the resistance of the intestinal tract to polio virus (Sabin's strain) in persons after naturally and experimentally acquired immunity, *Acta Virol.,* 5, 265, 1961.
4. **Bohl, E.H.,** Enteric viral infections as related to diarrhea in swine, in *Proc. 3rd Int. Symp. on Neonatal Diarrhea,* Acres, S. D., Forman, A. J., and Fast H., Eds., VIDO, University of Saskatchewan, Saskatoon, Canada, 1981, 1.
5. **Saif, L. J. and Smith, K. L.,** Enteric viral infections and passive immunity, *J. Dairy Sci.,* 68, 206, 1985.
6. **Ogra, P. l., Coppola, P. R., MacGillivray, M. H., and Dzierba,** Mechanisms of mucosal immunity to viral infections in IgA-immunoglobulin deficiency syndrome, *Proc. Soc. Exp. Biol. Med.,* 145, 811, 1974.
7. **Snodgrass, D. R. and Wells, P. W.,** Passive immunity in rotaviral infections, *JAVMA,* 173, 565, 1978.
8. **Saif, L. J., Redman, D. R., Smith, K. L., and Theil, K. W.,** Passive immunity to bovine rotavirus in newborn calves fed colostrum supplements from immunized or non-immunized cows, *Infect. Immun.,* 41, 1118, 1983.
9. **Riepenhoff-Talty, M., Dhcnakul, T., Kowalski, E., Sterman, D., and Ogra, P. L.,** Rotavirus infection in mice: pathogenesis and immunity, *Adv. Exp. Med. Biol.,* 216B, 1015, 1987.
10. **Offitt, P. A., Shaw R. D., and Greenberg, H. B.,** Passive protection against rotavirus-induced diarrhea by monoclonal antibodies to surface proteins VP3 and VP7, *J. Virol.,* 58, 700, 1986.
11. **Eiden, J., Lederman, H. M., Vonderfecht, S., and Yolken, R.,** T-cell-deficient mice display normal recovery from experimental rotavirus infection, *J. Virol.,* 57, 706, 1986.
12. **Saulsbury, F. T., Winkelstein, J. A., and Yolken, R. H.,** Chronic rotavirus infection in immunodeficiency, *J. Pediatr.,* 97, 61, 1980.
13. **Offitt, P. A. and Dudzik, K. I.,** Rotavirus-specific cytotoxic T lymphocytes cross-react with target cells infected with different rotavirus serotypes, *J. Virol.,* 62, 127, 1988.
14. **Andre, C., Bazin, H., and Hermans, J. F.,** Influence of repeated administration of antigen by the oral route on specific antibody-producing cells in the mouse spleen, *Digestion,* 9, 166, 1973.
15. **Rubin, D., Weiner, H. L., Fields, B. N., and Green, M. I.,** Immunologic tolerance after oral administration of reovirus: requirement for two viral gene products for tolerance induction, *J. Immunol.,* 127, 1697, 1981.
16. **Hooper, B. E. and Haelterman, E. O.,** Concepts of pathogensis and passive immunity in transmissible gastroenteritis in swine, *J. Am. Vet. Med. Assoc.,* 149, 1580, 1966.
17. **Woode, G. N., Jones, J., and Bridger, J.,** Levels of colostral antibody against neonatal calf diarrhea virus,*Vet. Rec.,* 97, 148, 1975.
18. **Besser, T. E., Gay, C. C., MacGuire, T. C., and Evermann, J. F.,** Passive immunity to bovine rotavirus infection associated with transfer of serum antibody into the intestinal lumen, *J. Virol,* 62, 2238, 1988.
19. **Saif, L. J., Saif, Y. M., and Theil, K. W.,** Enteric viruses in diarrheic turkey poults, *Avian Dis.,* 29, 798, 1985.
20. **Reynolds, D. L., Saif, Y. M., and Theil, K. W.,** Enteric viral infections of turkey poults: incidence of infection, *Avian Dis.,* 31, 272, 1986.
21. **Flewett, T. H.,** Clinical features of rotavirus infections, in *Virus Infections of the Gastrointestinal Tract,* Tyrrell, D. A. J. and Kapikian, A. Z., Eds., Marcel Dekker, New York, 1982. 125.
22. **Tzipori, S., Ed.,** *Infectious Diarrhoea in the Young: Strategies for Control in Humans and Animals,* Elsevier Science Publ., Amsterdam, The Netherlands, 1985.
23. **Middleton, P. J.,** Role of viruses in pediatric gastrointestinal disease and epidemiologic factors, in *Virus Infections of the Gastrointestinal Tract,* Tyrrell, D. A. J. and Kapikian, A. Z., Eds., Marcel Dekker, New York, 1982, 211.

24. **Woode, G. N., Kelso, N. E., Simpson, T. F., Gaul, S. K., Evans, L. E., and Babiuk, L.,** Antigenic relationships among some bovine rotaviruses: serum neutralization and cross-protection in gnotobiotic calves, *J. Clin. Microbiol.,* 18, 358, 1983.
25. **Bohl, E. H., Theil, K. W., and Saif, L. J.,** Isolation and serotyping of porcine rotaviruses and antigenic comparison with other rotaviruses, *J. Clin. Microbiol.,* 19, 105, 1984.
26. **Bishop, R. F., Barnes, G. L., Cipriani, E., and Lund, J. S.,** Clinical immunity after neonatal rotavirus infection, A prospective longitudinal study in young children, *N. Engl. J. Med.,* 309, 72, 1983.
27. **Saif, L. J. and Bohl, E. H.,** Transmissible gastroenteritis, in *Diseases of Swine,* 6th ed., Leman, A. D., Ed., Iowa State University Press, Ames, 1986, 225.
28. **Mebus, C. A., White, R. G., Bass, E. P., and Twiehaus, M. J.,** Immunity to neonatal calf diarrhea virus, *JAVMA,* 163, 880, 1973.
29. **Van Zaane, D., Ijzerman, D., and DeLeeuw, P. W.,** Intestinal antibody response after vaccination and infection with rotavirus of calves fed colostrum with or without rotavirus antibody, *Vet. Immunol. Immunopathol.,* 11, 45, 1986.
30. **Acres, S. D. and Radostits, O. M.,** The efficacy of a modified live reo-like virus vaccine and an *E. coli* bacteria for prevention of acute neonatal diarrhea of beef calves, *Can. Vet. J.,* 17, 197, 1976.
31. **DeLeeuw, P. W., Ellens, D. J., Talmon, F. P., et al.,** Rotavirus infections in calves: efficacy of oral vaccination in endemically infected herds, *Res. Vet. Sci.,* 29, 142, 1980.
32. **Myers, L. L. and Snodgrass, D. R.,** Colostral and milk antibody titers in cows vaccinated with a modified live rotavirus-coronavirus vaccine, *JAVMA,* 181, 486, 1982.
33. **Hancock, D. D., Saif, L. J., Bohl, E. H., and Redman, D. R.,** Evaluation of response to field use of commercial oral rotavirus-coronavirus vaccine in 10 Ohio dairy herds, Abstr. 242, Conf. of Res. Workers in Animal Diseases, Chicago, IL, November 9 to 10, 1981.
34. **DeLeeuw, P. W. and Tiessinik, J. W. A.,** Laboratory experiments on oral vaccination of calves against rotavirus or coronavirus-induced diarrhea, *Zentralbl. Vet. Med. B,* 32, 55, 1985.
35. **Hoblet, K. H., Saif, L. J., Kohler, E. M., Theil, K. W., Bech-Nielsen, S., and Stitzlein, G. A.,** Efficacy of an orally administered modified live porcine origin rotavirus vaccine against postweaning diarrhea in pigs, *Am. J. Vet. Res.,* 47, 1697, 1986.
36. **Furuuchi, S., Shimizu, Y., and Kumagai, T.,** Vaccination of newborn pigs with an attenuated strain of TGE virus, *Am. J. Vet. Res.,* 37, 1401, 1976.
37. **Greenberg, H., Offitt, P., Tran, C. T., Kapikian, A. Z., Robinson, W., Shaw, R., Greta, R., and Bellamy, R.,** Vaccine strategies for prevention of rotavirus diarrhea, in *Infectious Diarrhea in the Young,* Tzipori, S., Ed., Elsevier Science Publ., Amsterdam, The Netherlands, 1985, 447.
38. **Kapikian, A. Z., Flores, J., Hoshino, Y., Glass, R. I., Midthun, K., Gorziglia, M., and Chanock, R. M.,** Rotavirus: the major etiologic agent of severe infantile diarrhea may be controllable by a "Jennerian" approach to vaccination, *J. Infect. Dis.,* 153, 185, 1986.
39. **Vesikari, T., Isolauri, E., Delem, A., d'Hondt, E., Andre, F. E., Beards, G. M., and Flewett, T. H.,** Clinical efficacy of the RIT 4237 live attenuated bovine rotavirus vaccine in infants vaccinated before a rotavirus epidemic, *J. Pediatr.,* 107:189, 1985.
40. **DeMol, P., Zissis, G., Butzler, J. P., Mutwewingabo, A., and Andre, R. E.,** Failure of live, attenuated oral rotavirus vaccine, *Lancet,* 2, 108, 1986.
41. **Hanlon, P., Marsh, V., Shenton, F., Jobe, O., Hayes, R., Whittle, H. C., Hanlon, L., Byass, P., Hasson-King, M., Sillah, H., M'Boge, B. H., and Greenwood, B. M.,** Trial of an attenuated bovine rotavirus vaccine (RIT 4237) in Gambian infants, *Lancet,* 1, 1342, 1987.
42. **Clark, F. H., Borian, F. E., Bell, L. M., Modesto, K., Gowvea, V., and Plotkin, S. A.,** Protective effect of WC3 vaccine against rotavirus diarrhea in infants during a predominantly serotype 1 rotavirus season, *J. Infect. Dis.,* 158, 570, 1988.
43. **Rennels, M. B., Losonsky, G. A., Levine, M. M., and Kapikian, A. Z.,** Preliminary evaluation of the efficacy of rhesus rotavirus vaccine strain MMU 18006 in young children, *Pediatr. Infect. Dis.,* 5, 587, 1986.
44. **Flores, J., Gonzales, M., Perez,** *et al.,* Protection against severe rotavirus diarrhea by rhesus rotavirus vaccine in Venezuelan infants, *Lancet,* 1, 882, 1987.
45. **Brambell, F. W. R.,** The Transmission of Passive Immunity from Mother to Young, North-Holland Research Monographs, *Frontiers of Biology,* Vol. 18, North-Holland, Amsterdam, 1970.
46. **Ogra, S. S., Weintraub, D., and Ogra, P. L.,** Immunologic aspects of human colostrum and mik. III. Fate and absorption of cellular and soluble components in the gastrointestinal tract of the newborn, *J. Immunol.,* 119, 245, 1977.
47. **Porter, P.,** Immunoglobulins in bovine mammary secretions, Quantitative changes in early lactation and absorption by the neonatal calf, *Immunology,* 23, 225, 1972.
48. **Porter, P. and Allen, W. D.,** Classes of immunoglobulins related to immunity in the pig: a review, *J. Am. Vet. Med. Assoc.,* 160, 511, 1972.
49. **Bourne, F. J., Newby, T. J., Evans, P., and Morgan, K.,** The immune requirements of the newborn pig and calf, *Ann. Rech. Vet.,* 9, 239, 1978.

53. **Goldblum, R. M., Ahlstedt, S., Carlsson, B., Hanson, L. A., Jodal, U., Lidinjanson, G., and Sohl-Akerlund, A.**, Antibody-forming cells in human colostrum after oral immunization, *Nature (London)*, 257, 797, 1975.
54. **Bohl, E. H., Gupta, R. P. K., Olquin, F. M. W., and Saif, L. J.**, Antibody responses in serum, colostrum and milk of swine after infection or vaccination with transmissible gastroenteritis virus, *Infect. Immun.*, 6, 289, 1972.
55. **Saif, L. J., Bohl, E. H., and Gupta, R. K. P.**, Isolation of porcine immunoglobulins and determination of the immunoglobulin classes of transmissible gastroenteritis viral antibodies, *Infect. Immun.*, 6, 289, 1972.
56. **Weisz-Carrington, P., Roux, M. E., McWilliams, M., Phillips-Quagliata J. M., and Lamm, M. E.**, Hormonal induction of the secretory immune system in the mammary gland, *Proc. Natl. Acad. Sci.*, 75, 2928, 1978.
57. **Montgomery, P. C., Cohn, J., and Lally, E. T.**, The induction and characteristics of secretory IgA antibodies, in *The Immunoglobulin A System*, J. Mestecky and A. R. Lawton, Eds., Plenum Press, New York, 1974, 453.
58. **Saif, L. J. and Bohl, E. H.**, Role of secretory IgA in passive immunity of swine to enteric viral infections, in *Immunology of Breast Milk*, Ogra, P. L. and Dayton, D., Eds., Raven Press, New York, 1979, 237.
59. **Saif, L. J. and Bohl, E. H.**, Passive immunity against enteric viral infections, in *Proc. 3rd Int. Symp. on Neonatal Diarrhea*, Acres, S. D., Forman, A. J., and Fast H., Eds., VIDO, University of Saskatchewan, Saskatoon, Canada, 1981, 83.
60. **Parmley, M. J., Beer, A. E., and Billingham, R. F.**, *In vitro* studies on the T lymphocyte population of human milk, *J. Exp. Med.*, 144, 358, 1976.
61. **Kohler, E. M.**, Protection of pigs against neonatal enteric colibacillosis with colostrum and milk from orally vaccinated sows, *Am. J. Vet. Res.*, 35, 331, 1971.
62. **Evans, P. A., Newby, T. J., Stokes, C. R., Patel, D., and Bourne, F. J.**, Antibody response of the lactating sow to oral immunization with *E. coli*, *Scand. J. Immunol.*, 11, 419, 1980.
63. **Halsey, J. F., Johnson, B. H., and Cebra, J. J.**, Transport of immunoglobulins from serum into colostrum, *J. Exp. Med.*, 151, 767, 1980.
64. **Jackson, G. D. F., Lemaitre-Coelho, I., Vaerman, J. P., Bozin, H., and Beckers, A.**, Rapid disappearance from serum of intravenously injected rat myeloma IgA and its resecretion into bile, *Eur. J. Immunol.*, 8, 123, 1978.
65. **Husband, A. J. and Watson, D. L.**, Immunity in the intestine, *Vet. Bull.*, 48, 911, 1978.
66. **Lascelles, A. K. and McDowell, G. H.**, Localized humoral immunity with particular reference to ruminants, *Transplant Rev.*, 19, 170, 1974.
67. **Mata, L. J., Kronmal, R. A., Garcia, B., Butler, W., Urrutia, J. J., and Murillo, S.**, Breast-feeding, weaning and the diarrhoeal syndrome in a Guatemalan Indian Village, in *Acute Diarrhoea in Childhood*, Ciba Foundation Symp. 42, Elsevier Sci. Publ., North Holland, Amsterdam, The Netherlands, 1976, 311.
68. **Hanson, L. A. and Winberg, J.**, Breast milk and defense against infection in the newborn, *Arch. Dis. Child.*, 47, 845, 1972.
69. **Bullen, C. L. and Willis, A. T.**, Resistance of the breast fed infant to gastroenteritis, *Br. Med. J.*, 3, 338, 1971.
70. **Berger, R., Hadziselimorrc, F., Just, M., and Reigel, P.**, Effect of feeding human milk on nosocomial rotavirus infections in an infants ward, *Dev. Biol. Stand.*, 53, 219, 1983.
71. **Lachman, P. J.**, Does complement play a role in mucosal immunity, *Monogr. Allergy*, 24, 121, 1988.
72. **Nagy, L. K., Bhogal, B. S., and McKenzie, T.**, The effect of colostrum or past colibacillosis on the adhesion of *E. coli* to the small intestine of the pig, *Res. Vet. Sci.*, 21, 303, 1976.
73. **Nguyen, T. D., Bottreua, E., Bernard, S., Lantier, I., and Aynaud, J. M.**, Neutralizing secretory IgA and IgG do not inhibit attachment of transmissible gastroenteritis virus, *J. Gen. Virol.*, 67, 939, 1986.
74. **McNabb, P. C.**, Host defense mechanisms at mucosal surfaces, *Annu. Rev. Microbiol.*, 35, 477, 1981.
75. **Lowell, G. H., Smith, L. F., Griffiss, J. M., Brandt, B. L., and McDermott, R. P.**, ADCC-Comparison of convalescent and post-immunization IgG, IgM and IgA, *J. Clin. Invest.*, 66, 260, 1980.
76. **Shen, L. and Fanger, M. W.**, SIgA's synergize with IgG in promoting ADCC by human PMN monocytes and lymphocytes, *Cell. Immunol.*, 59, 75, 1981.
77. **Voets, M. T., Pensaert, M., and Rondhius, P. R.**, Vaccination of pregnant sows against transmissible gastroenteritis with two attenuated virus strains and different inoculation routes, *Vet. Quart.*, 2, 211, 1980.
78. **Saif, L. J.**, Passive immunity to coronavirus and rotavirus infections in swine and cattle: enhancement by maternal vaccination, in *Infectious Diarrhea in the Young*, Tzipori, S., Ed., Elsevier Science Publ., Amsterdam, The Netherlands, 1985.
79. **Woods, R. D.**, Efficacy of vaccination of sows with serologically-related coronaviruses for control of transmissible gastroenteritis virus in nursing pigs, *Am. J. Vet. Res.*, 45, 1726, 1984.
80. **Underdown, B. J. and Dorrington, K. J.**, Studies on the structural and conformational basis for the relative resistance of serum and secretory immunoglobulin A to proteolysis, *J. Immunol.*, 112, 949, 1974.
81. **Nagura, H., Nakane, D. K., and Brown, W. R.**, Breast milk IgA binds to jejunal ephithelial in suckling rates, *J. Immunol.*, 120, 1330, 1978.

82. **Saif, L. J. and Bohl, E. H.,** Passive immunity to TGE virus: Intramammary viral inoculation of sows, *Ann. N. Y. Acad. Sci.,* 409, 708, 1982.
83. **Snodgrass, D. R., Ojeh, C. K., Campbell, I., and Herring, A. J.,** Bovine rotavirus serotypes and their significance for immunization, *J. Clin. Microbiol.,* 20, 342, 1984.
84. **Kapikian, A. Z., Wyatt, R. G., Levine, M. M., Black, R. E., Greenberg, H. B., Flores, J., Kalica, A. R., Hoshino, Y., and Chanock, R. M.,** Studies in volunteers with human rotavirus, *Dev. Biol. Stand.,* 53, 209, 1983.
85. **Svennerholm, A. M., Holmgren, J., Hanson, L. A., Lindblad, B. S., Quereshi, F., and Rahimtoola, R. J.,** Boosting of secretory IgA antibody responses in man by parenteral cholera vaccination, *Scand. J. Immunol.,* 6, 1345, 1977.
86. **Saif, L. J. and Smith, K. L.,** A review of rotavirus immunization of cows and passive protection in calves, keynote address, *Proc. 4th Int. Symp. on Neonatal Diarrhea,* Acres, S. D., Ed., VIDO, Saskatoon, Saskatchewan, Canada, 1984, 394.
87. **Pierce, A. E. and Feinstein, A.,** Biophysical and immunological studies on bovine immunoglobulins with evidence for selective transport within the mammary gland from maternal plasma to colostrum, *Immunology,* 8, 106, 1965.
88. **Mackenzie, D. D. S. and Lascelles, A. K.,** The transfer of ^{131}I-labelled immunoglobulins and serum albumin from blood into milk of lactating ewes, *Aust. J. Exp. Biol. Med. Sci.,* 46, 285, 1968.
89. **Newby, T. J. and Bourne, F. J.,** The nature of the local immune system of the bovine mammary gland, *J. Immunol.,* 118, 461, 1977.
90. **Snodgrass, D. R. Fahey, K. J., Wells, P. W., Campbell, I., and Whitelaw, A.,** Passive immunity in calf rotavirus infections. Maternal vaccination increases and prolongs immunoglobulins G_1 antibody secretion in milk, *Infect. Immun.,* 28, 344, 1980.
91. **Saif, L. J., Smith, K. L., Landmeier, B. J., Bohl, E. H., and Theil, K. W.,** Immune response of pregnant cows to bovine rotavirus immunization, *Am. J. Vet. Res.,* 45, 49, 1984.
92. **McNulty, M. S. and Logan, E. F.,** Effect of vaccination of the dam on rotavirus infection in young calves, *Vet. Rec.,* 120, 250, 1987.
93. **Brock, J. H., Fanny, R., Arzabe, F., Ortega, F., and Pineiro, A.,** The effect of limited proteotysis by trypsin and chymotrypsin on bovine colostral IgG, *Immunology,* 32, 215, 1977.
94. **Chang, C. C., Winter, A. J., and Norcross, N. L.,** Immune response in the bovine mammary gland after intestinal, local, and systemic immunization, *Infect. Immun.,* 31, 650, 1981.
95. **Saif, L. J., Weilnau, P., Miller, K., and Stitzlein, L.,** Isotypes of intestinal and systemic antibodies in colostrum-fed and colostrum-deprived calves challenged with rotavirus, *Adv. Exp. Med. Biol.,* 216B, 1815, 1987.
96. **Thummel, C., Tjian, R., and Grodzicker, T.,** Construction of adenovirus expression vectors by site-directed *in vivo* recombination, *J. Mol. Appl. Genet.,* 1, 435, 1981.
97. **Berkner, K. L. and Sharp, A.,** Preparation of adenovirus recombinants using plasmids of viral DNA, in *Eukaryotic Viral Vectors,* Gluzman, Y., Ed., Cold Spring Harbor Laboratory, Cold Spring Harbor, NY, 1982, 193.
98. **Gluzman, Y., Reichl, H., and Solnick, D.,** Helper-free adenovirus type-5 vectors, in *Eukaryotic Viral Vectors,* Gluzman, Y., Ed., Cold Spring Harbor Laboratory, Cold Spring Harbor, NY, 1982, 187.
99. **Anderson, M., Paabo, S., Nilsson, T., and Peterson, P. A.,** Impaired intracellular transport of class 1 MHC antigens as a possible means for adenoviruses to evade immune surveillance, *Cell,* 43, 215, 1985.
100. **Burgert, H. and Kvist, S.,** An adenovirus type 2 glycoprotein blocks cell surface expression of human histocompatibility class 1 antigens, *Cell,* 41, 987, 1985.
101. **Morin, J. E., Lubeck, M. D., Barton, J. E., Conley, A. J., Davis, A. R., and Hung, P. P.,** Recombinant adenovirus induces antibody to hepatitis B virus surface antigen in hamsters, *Proc. Natl. Acad. Sci., U.S.A.,* 84, 4626, 1987.
102. **Saito, I., Oya, Y., Yamamoto, D., Yuasa, T., and Shimojo, H.,** Construction of nondefective adenovirus type 5 bearing a 2.8 kilobase hepatitis B virus DNA near the right end of its genome, *J. Virol.,* 54, 711, 1985.
103. **Chanock, R. M., Murphy, B. R., Collins, P. L., Coelingh, K. V. W., Olmsted, R. A., Snyder, M. H., Spriggs, M. K., Prince, G. A., Moss, B., Flores, J., Gorzigla, M., and Kapikian, A. Z.,** Live viral vaccines for respiratory and enteric tract diseases, *Vaccine,* 6, 129, 1988.
104. **Curtiss, R. and Kelly, S. M.,** *Salmonella typhimurium* deletion mutants lacking adenylate cyclase and cyclic amp receptor protein are avirulent and immunogenic, *Infect. Immun.,* 55, 3035, 1987.
105. **Carter, P. B. and Collins, F. M.,** The route of enteric infection in normal mice, *J. Exp. Med.,* 139, 1189, 1974.
106. **McCaughan, G. and Basten, A.,** Immune system of the gastrointestinal tract, *Int. Rev. Physiol.,* 28, 131, 1983.
107. **Nakayama, K., Kelly, S. M., and Curtiss, R.,** Construction of an asd+ expression-cloning vector: stable maintenance and high level expression of cloned genes in a *Salmonella* vaccine strain, *Biotechnology,* 6, 693, 1988.

108. **Amann, E. and Brosius, J.,** ATG vectors for regulated high-level expression of cloned genes in *Escherichia coli*, *Gene*, 40, 183, 1985.

109. **Brown, A., Hormaeche, C. E., Demarco de Hormaeche, R., Winther, M., Dougan, G., Maskell, D. J., and Stocker, B. A. D.,** An attenuated *aro*A *Salmonella typhimurium* vaccine elicits humoral and cellular immunity to cloned beta-galactosidase in mice, *J. Infect. Dis.*, 155, 86, 1987.

110. **Clements, J. D., Lyon, F. L., Lowe, K. L., Farrand, A. L., and ElMorshidy, S.,** Oral immunization of mice with attenuated *Salmonella enteritidis* containing a recombinant plasmid which codes for production of the B subunit of heat-labile *Escherichia coli* enterotoxin, *Infect. Immun.*, 53, 685, 1986.

Index

INDEX

A

F

G

H

I

J

K

L

M

N

O

R

S